Springer Series in Information

Edit

Springer
Berlin
Heidelberg
New York
Barcelona
Hong Kong
London
Milan
Paris
Singapore
Tokyo

Physics and Astronomy | ONLINE LIBRARY

http://www.springer.de/phys/

Springer Series in Information Sciences

Editors: Thomas S. Huang   Teuvo Kohonen   Manfred R. Schroeder

Professor Teuvo Kohonen

Helsinki University of Technology Neural Networks Research Centre
P.O. Box 5400
02015 HUT, Espoo, FINLAND

Series Editors:

Professor Thomas S. Huang

Department of Electrical Engineering and Coordinated Science Laboratory,
University of Illinois, Urbana, IL 61801, USA

Professor Teuvo Kohonen

Helsinki University of Technology Neural Networks Research Centre
P.O. Box 5400
02015 HUT, Espoo, FINLAND

Professor Dr. Manfred R. Schroeder

Drittes Physikalisches Institut, Universität Göttingen, Bürgerstrasse 42-44,
37073 Göttingen, GERMANY

ISSN 0720-678X
ISBN 3-540-67921-9 3rd Edition Springer-Verlag Berlin Heidelberg New York

ISBN 3-540-62017-6 2nd Edition Springer-Verlag Berlin Heidelberg New York

Library of Congress Cataloging-in-Publication Data applied for.

Die Deutsche Bibliothek - CIP-Einheitsaufnahme

Kohonen, Teuvo:
Self-organizing maps / Teuvo Kohonen.– 3. ed. – Berlin; Heidelberg; New York; Barcelona;
Hong Kong; London; Milan; Paris; Singapore; Tokyo: Springer, 2001
  (Springer series in information sciences; 30)
  ISBN 3-540-67921-9

Springer-Verlag Berlin Heidelberg New York
a member of BertelsmannSpringer Science+Business Media GmbH

Typesetting: Data conversion by K. Mattes, Heidelberg
Cover design: *design & production* GmbH, Heidelberg
Printed on acid-free paper   SPIN: 10890059   56/3111 – 5 4 3 2 1

Teuvo Kohonen

Self-Organizing Maps

Third Edition
With 129 Figures and 22 Tables

 Springer

Preface to the Third Edition

Since the second edition of this book came out in early 1997, the number of scientific papers published on the Self-Organizing Map (SOM) has increased from about 1500 to some 4000. Also, two special workshops dedicated to the SOM have been organized, not to mention numerous SOM sessions in neural-network conferences. In view of this growing interest it was felt desirable to make extensive revisions to this book. They are of the following nature.

Statistical pattern analysis has now been approached more carefully than earlier. A more detailed discussion of the eigenvectors and eigenvalues of symmetric matrices, which are the type usually encountered in statistics, has been included in Sect. 1.1.3; also, new probabilistic concepts, such as factor analysis, have been discussed in Sect. 1.3.1. A survey of projection methods (Sect. 1.3.2) has been added, in order to relate the SOM to classical paradigms.

Vector Quantization is now discussed in one main section, and derivation of the point density of the codebook vectors using the calculus of variations has been added, in order to familiarize the reader with this otherwise complicated statistical analysis.

It was also felt that the discussion of the neural-modeling philosophy should include a broader perspective of the main issues. A historical review in Sect. 2.2, and the general philosophy in Sects. 2.3, 2.5 and 2.14 are now expected to especially help newcomers to orient themselves better amongst the profusion of contemporary neural models.

The basic SOM theory in Chap. 3 has now first been approached by a general qualitative introduction in Sect. 3.1. Other completely new concepts discussed in Chap. 3 are the point density of the model vectors (Sect. 3.12) and the interpretation of the SOM mapping (Sect. 3.16).

Only modest revisions have been made to Chap. 4.

Among the new variants in Chap. 5, the SOM of symbol strings (and other nonvectorial items) has been discussed in Sect. 5.7, and a generalization of the SOM in the direction of evolutionary learning has been made in Sect. 5.9.

To Chap. 6, the batch-computation scheme of the LVQ1 has been added.

In Chap. 7, a major revision deals with a new version of WEBSOM, the SOM of large document files, by which it has been possible to implement one of the most extensive ANN applications ever, namely the SOM of seven million patent abstracts. The amount of text thereby mapped is 20 times that of the Encyclopaedia Britannica!

The most visible and most requested addition to the third edition is the new Chap. 8 on software tools which we hope will be useful for practitioners.

It was not possible, however, to extend the survey of new SOM applications much beyond that already published in the second edition. A new hardware implementation has been discussed at the end of Chap. 9, but the main change made to the literature survey in Chap. 10 is its reorganization: the taxonomy and indexing of its contents is now more logical than in the second edition.

In the Preface to the first edition I gave advice for the first reading of this book. Since the numbering of the sections has now changed, the new recommended reading sequence for a short introductory course on the SOM is the following: 2.2, 2.9, 2.10, 2.12, 3.1–3, 3.4.1, 3.5.1, 3.6, 3.7, 3.13, 3.14, 3.15, 6.1, 6.2 (skipping the derivation), 6.4–9, 7.1 and 7.2.

Again I got much help in the revision from my secretary, Mrs. Leila Koivisto; I am very much obliged to her. Some parts of the word processing were done by Mr. Mikko Katajamaa. Naturally I have benefited from the good scientific work done by all my younger colleagues in our university.

Dr. Samuel Kaski has kindly allowed me to use in Sect. 1.3.2 material compiled by him.

After writing all three editions, I want to once again express my gratitude to the Academy of Finland for allowing me to carry out this activity under their auspices.

Espoo, Finland   *Teuvo Kohonen*
October, 2000

Preface to the Second Edition

The second, revised edition of this book came out sooner than originally planned. The main reason was that new important results had just been obtained.

The ASSOM (Adaptive-Subspace SOM) is a new architecture in which invariant-feature detectors emerge in an unsupervised learning process. Its basic principle was already introduced in the first edition, but the motivation and theoretical discussion in the second edition is more thorough and consequent. New material has been added to Sect. 5.9 and this section has been rewritten totally. Correspondingly, Sect. 1.4 that deals with adaptive-subspace classifiers in general and constitutes the prerequisite for the ASSOM principle has also been extended and rewritten totally.

Another new SOM development is the WEBSOM, a two-layer architecture intended for the organization of very large collections of full-text documents such as those found in the Internet and World Wide Web. This architecture was published after the first edition came out. The idea and results seemed to be so important that the new Sect. 7.8 has now been added to the second edition.

Another addition that contains new results is Sect. 3.15, which describes acceleration of computing of very large SOMs.

It was also felt that Chapter 7, which deals with SOM applications, had to be extended.

To recapitulate, Chaps. 1, 3, 5, 7, and 9 contain extensive additions and revisions, whereas Chaps. 2, 4, 6, 8, and 10 are identical with those of the first edition, let alone a few minor corrections to their text.

In the editing of this revision I got much help from Mr. Marko Malmberg.

Espoo, Finland   *Teuvo Kohonen*
September, 1996

Preface to the First Edition

The book we have at hand is the fourth monograph I wrote for Springer-Verlag. The previous one named "Self-Organization and Associative Memory" (Springer Series in Information Sciences, Volume 8) came out in 1984. Since then the self-organizing neural-network algorithms called SOM and LVQ have become very popular, as can be seen from the many works reviewed in Chap. 9. The new results obtained in the past ten years or so have warranted a new monograph. Over these years I have also answered lots of questions; they have influenced the contents of the present book.

I hope it would be of some interest and help to the readers if I now first very briefly describe the various phases that led to my present SOM research, and the reasons underlying each new step.

I became interested in neural networks around 1960, but could not interrupt my graduate studies in physics. After I was appointed Professor of Electronics in 1965, it still took some years to organize teaching at the university. In 1968–69 I was on leave at the University of Washington, and D. Gabor had just published his convolution-correlation model of autoassociative memory. I noticed immediately that there was something not quite right about it: the capacity was very poor and the inherent noise and crosstalk were intolerable. In 1970 I therefore suggested the autoassociative correlation matrix memory model, at the same time as J.A. Anderson and K. Nakano.

The initial experiences of the application of correlation matrix memories to practical pattern recognition (images and speech) were somewhat discouraging. Then, around 1972–73, I tried to invert the problem: If we have a set of pairs of input-output patterns, what might be the *optimal* transfer matrix operator in relation to smallest residual errors? The mathematical solution that ensued was the optimal associative mapping, which involved the Moore-Penrose pseudoinverse of the input observation matrix as a factor. As an associative memory, this mapping has a capacity three times that of the networks discussed by J. Hopfield. The *recognition accuracy* for natural data, however, was essentially not improved, even when we used nonlinear (polynomial) preprocessing! Obviously there was still something wrong with our thinking: associative memory and pattern recognition could not be the same thing!

During 1976–77 I had a new idea. In the theory of associative memory I had worked, among other problems, with the so-called Novelty Filter, which is an adaptive orthogonal projection operator. I was trying to conceive a neuron that would represent a whole *class* of patterns, and I was playing with the idea of a neuron, or rather a small set of interacting neurons describable as a Novelty Filter. If that would work, then an arbitrary linear combination of the stored reference patterns would automatically belong to the same class (or manifold). It turned out that the so-called linear-subspace formalism known in pattern recognition theory was mathematically equivalent to my idea. Then I went one step further: since according to our own experiments the basic subspace method was still too inaccurate for classification, how about trying some *supervised* "training" of the subspaces, or their basis vectors? I soon invented *the first supervised competitive-learning algorithm*, the Learning Subspace Method (LSM), and it worked almost three times more accurately than the previous ones! Its handicap was a slower speed, but we developed a fast co-processor board to cope with "neural network" computations to make the LSM work in real time. We based our first speech-recognition hardware system on that idea, and this algorithm was in use for several years. Half a dozen Ph.D. theses were done on that system in our laboratory.

Our research on the Self-Organizing Map (SOM) did not begin until in early 1981, although I had already jotted down the basic idea into my notebook in 1976. I just wanted an algorithm that would effectively map similar patterns (pattern vectors close to each other in the input signal space) onto contiguous locations in the output space. Ch. v.d. Malsburg had obtained his pioneering results in 1973, but I wanted to generalize and at the same time ultimately simplify his system description. We made numerous similarity (clustering) diagrams by my simplified but at the same time very robust SOM algorithm, including the map of phonemes. When we tentatively tried the SOM for speech recognition in 1983, we at first got no improvement at all over that already achieved by LSM. Then, in 1984, I again thought about supervised learning, and the Supervised SOM described in Sect. 5.8 solved the problem. We had developed an algorithm that was best so far.

During 1985–87 our laboratory had a cooperative project on speech recognition with Asahi Chemical Co., Ltd., the biggest chemical company in Japan. During the first phase I introduced two new algorithms (based on research that I had started a couple of years earlier): the Learning Vector Quantization (LVQ), which is a supervised version of SOM particularly suitable for statistical pattern recognition, and the Dynamically Expanding Context (DEC), both of which will be described in this book. For many years thereafter they formed the basis of our speech recognition systems.

Over the years we worked on numerous other practical applications of the SOM, and these projects will be found among the references.

The present monograph book contains a brand-new, so far unpublished result, the *Adaptive-Subspace SOM (ASSOM)*, which combines the old Learning Subspace Method and the Self-Organizing Map. It does also something more than most artificial neural network (ANN) algorithms do: it detects *invariant* features, and to achieve, e.g., translational invariance to elementary patterns, the "winner take all" function had to be modified fundamentally. The sophisticated solutions described in Sects. 5.9 and 5.10 could not have been invented at once; they had to be acquired during a long course of development from many small steps. The ideas of "representative winner" and "competitive episode learning" had not been conceived earlier; with these ideas the generally known wavelet and Gabor-filter preprocessing can now be made to emerge automatically.

This book contains an extensive mathematical introduction as well as a Glossary of 555 relevant terms or acronyms. As there exists a wealth of literature, at least 1500 papers written on SOM, it was also felt necessary to include a survey of as many works as possible to lead the readers to the newest results and save them from painstaking hours in the library.

I have to say quite frankly, however, that the SOM and LVQ algorithms, although being potentially very effective, are not always applied in the correct way, and therefore the results, especially benchmarkings reported even in respectable journals and by respectable scientists, are not always correct. I felt it necessary to point out *how* many details have to be taken into account, before the problem is approached in the proper way. Let me also emphasize the following facts: 1) The SOM has not been meant for statistical pattern recognition; it is a clustering, visualization, and abstraction method. Anybody wishing to implement decision and classification processes should use LVQ in stead of SOM. 2) Preprocessing should not be overlooked. The ANN algorithms are no sausage machines where raw material (data) is input at one end and results come out at the other. Every problem needs a careful selection of feature variables, which so far is mostly done by hand. We are just at the dawn of automated feature extraction, using ANN models such as ASSOM. 3) In benchmarkings and practical problems one should compare the speed of computation, too, not only ultimate accuracies. A relative difference in accuracy of a few per cent can hardly be noticed in practice, whereas tiny speed differences during the actual operation are very visible.

People who are reading about the SOM for the first time may feel slightly uneasy about its theoretical discussion: so sophisticated, and yet only leading to partial results. Therefore, let me quote three well-known French mathematicians, Professors M. Cottrell, J.-C. Fort, and G. Pagès: "Despite the large use and the different implementations in multi-dimensional settings, the Kohonen algorithm is surprisingly resistant to a complete mathematical study." Perhaps the SOM algorithm belongs to the class of "ill posed" problems, but so are many important problems in mathematics. In *practice* people have applied many methods long before any mathematical theory existed for them and even if none may exist at all. Think about walking; theoretically we know that we could not walk at all unless there existed gravity and friction, by virtue of which we can kick the globe and the other creatures on it in the opposite direction. People and animals, however, have always walked without knowing this theory.

This book is supposed to be readable without any tutor, and it may therefore serve as a handbook for people wanting to apply these results in practice. I have tried to anticipate many problems and difficulties that readers may encounter, and then to help them clear the hurdles.

Can this book be used as a university textbook? Yes, many efforts have been made to this end so that it could be useful. It should serve especially well as collateral reading for neural-network courses.

If only a short introductory course on the SOM is given, the teaching material might consist of the following sections: 2.6–8, 2.12, 3.1, 3.2, 3.3.1, 3.4.1, 3.5, 3.9, 3.10, 3.11, 6.1, 6.2 (thereby skipping the derivation of Eq. (6.6)), 6.4–7, 7.1, and 7.2. This is also the recommendable sequence for the first reading of this book.

To a lecturer who is planning a whole special course on the SOM, it might be mentioned that if the audience already has some prior knowledge of linear algebra, vector spaces, matrices, or systems theory, Chapter 1 may be skipped, and one can start with Chapter 2, proceeding till the end of the book (Chap. 8). However, if the audience does not have a sufficient mathematical background, Chapter 1 should be read meticulously, in order to avoid trivial errors in the application of the algorithms. Matrix calculus has plenty of pitfalls!

Acknowledgements. This book would have never been completed without the generous help of many people around me. First of all I would like to mention Mrs. Leila Koivisto who did a very big job, magnificiently typing out my pieces of text and then had the stamina to produce the numerous revisions and the final layout. The nice appearance of this book is very much her achievement.

Dr. Jari Kangas should be mentioned for being responsible for the extensive library of literature references on SOM and LVQ. His efforts made it possible to include the References section in this book.

I would like to thank the following people for their help in making many simulations, part of which have also been published as research papers or theses. Mr. Samuel Kaski simulated the physiological SOM model. Mr. Harri Lappalainen helped me in the Adaptive-Subspace SOM (ASSOM) research.

Mr. Ville Pulkki collected works on hardware implementations of SOM, and I have used some of his illustrations. I have also used the nice pseudocolor image of cloud classification that was obtained in the research project led by Dr. Ari Visa. Some other results have been utilized, too, with due credits given in the corresponding context.

The following people have helped me in word processing: Taneli Harju, Jussi Hynninen, and Mikko Kurimo.

I am also very much obliged to Dr. Bernard Soffer for his valuable comments on this text.

My research was made possible by the generous support given to our projects by the Academy of Finland and Helsinki University of Technology.

Espoo, Finland   *Teuvo Kohonen*
December, 1994

Contents

1. Mathematical Preliminaries

In this book we shall be dealing with neural networks, artificial and biological. For their qualitative and quantitative description we must set up analytical (mathematical) system models. Many cognitive functions cannot be interpreted at all until one understands the nature of collective interactions between elements in a complex system.

When dealing with spatially and temporally related samples of signal values that constitute *patterns*, we need a mathematical framework for the description of their quantitative interrelations. This framework is often provided by the generalized *vector formalism*. The operations in vector spaces, on the other hand, can conveniently be manipulated by *matrix algebra*. These concepts are first introduced in Sect. 1.1.

For the comparison of patterns, be they vectorial or nonvectorial, one has to introduce a measure that describes either the *distance between* or the *similarity of* their representations. Various standard measures are reviewed in Sect. 1.2.

Exploratory data analysis is a discipline whose aim is the formation of simplified, usually visual, overviews of data sets. Such overviews are often provided by various *projection methods* discussed in Sect. 1.3.2.

Identification and classification of natural observations may be based on the statistical properties of the signal patterns. The classical formalisms of probabilistic decision and detection theories that take into account statistical knowledge about the occurrences of the patterns are therefore supposed to serve as a good starting point for more advanced theories. Statistical decision-making and estimation methods are discussed in Sects. 1.3.3 through 1.3.4. A newer formalism that underlies one of the very recent neural-network algorithms in this area (Sects. 5.11 and 5.12) is the *subspace classification method* discussed in Sect. 1.4.

This book concentrates on *Self Organizing Maps (SOMs)*, which have a bearing on *Vector Quantization (VQ)* introduced in Sect. 1.5.

Neural-network theories may also deal with higher-level *symbolic rules*, and the "learning grammar" introduced in Sect. 1.6 implements a fundamental and very effective mapping for structured sequences of symbols.

This chapter thus mainly contains various formalisms that will be referred to in network theories later on. It is difficult to recommend any single textbook for collateral reading of the mathematical part. Maybe the books of *Albert* [1.1] and *Lewis and Odell* [1.2] could serve for that purpose.

1.1 Mathematical Concepts and Notations

1.1.1 Vector Space Concepts

Representation Vectors. Especially in the physical theory of information processes, spatially or temporally adjacent signal values are thought to form *patterns*, which can be regarded as ordered sets of real numbers. In pursuance of the methods developed in the theories of pattern analysis and pattern recognition, such sets are usually described by *representation vectors*, which are generalizations of the vectors of plane and space geometry. If there are n independently defined and somehow related real numerical values $\xi_1, \xi_2, \ldots, \xi_n$, they can be regarded as coordinates in a space that has n dimensions. This space is denoted $\Re^n$ and it is the set of all possible n-tuples of real numbers, each of which is from the interval $(-\infty, +\infty)$. (Note that a *scalar* is defined in the space $\Re^1 = \Re$.) Scalars are written by lower-case Greek letters and vectors in lower-case italics, respectively, unless otherwise stated. Matrices (Sect. 1.1.2) and scalar-valued functionals are usually written in capital italics. A vector x is a point in $\Re^n$ (expressed $x \in \Re^n$), the coordinates of which are $(\xi_1, \xi_2, \ldots, \xi_n)$. To visualize a vector (although this may be difficult in a multidimensional space) it may be imagined as a directed line from the origin to the point in question. *In automatic data processing, however, a vector is regarded just as an array of numbers.*

Consider a representation vector that is put to stand for optical patterns. These patterns are usually composed of picture elements (*"pixels"*) like a mosaic, and every pixel attains a numerical value. The indexing of the elements can now be made in an arbitrary fashion. In Fig. 1.1, three different examples of indexing are shown.

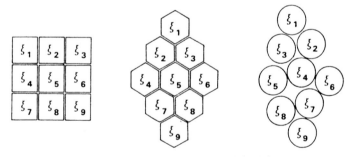

Fig. 1.1. Examples of pattern "vectors"

In all of them, the formal representation vector attains the form $(\xi_1, \xi_2, \ldots, \xi_9)$. It should be noticed that this is a *linear array of numbers* in spite of the original images being two-dimensional.

Linear Vector Spaces. For analytical purposes, it will be necessary to define the concept of a vector space. A *linear vector space* V (over the reals) is generally defined as a set of elements, the vectors, in which the operations of vector addition $(+)$ and (scalar) multiplication $(\cdot)$ have been defined, and the following facts hold true: if $x, y, z, \in V$ are vectors and $\alpha, \beta \in \Re$ are scalars, then

A1) $x + y = y + x \in V$ (commutativity)

A2) $\alpha \cdot (x + y) = \alpha \cdot x + \alpha \cdot y \in V$ $\left.\right\}$

A3) $(\alpha + \beta) \cdot x = \alpha \cdot x + \beta \cdot x$ (distributivity)

A4) $(x + y) + z = x + (y + z)$ $\left.\right\}$

A5) $(\alpha \cdot \beta) \cdot x = \alpha \cdot (\beta \cdot x)$ (associativity)

A6) There exists a zero vector 0 such that for every $x \in V$, $x + 0 = x$.

A7) For scalars 0 and 1, $0 \cdot x = 0$ and $1 \cdot x = x$.

An example of V is $\Re^n$. The sum of two vectors is then defined as a vector with elements (coordinates, components) that are obtained by summing up the respective elements of the addends. The scalar multiplication is an operation in which all elements of a vector are multiplied by this scalar. (For simplicity of notation, the dot may be dropped.)

Inner Product and Dot Product. The concept of *inner product* refers to a two-argument, scalar-valued function (x, y), which has been introduced to facilitate an analytic description of certain geometric operations. One very important case of inner product of vectors $x = (\xi_1, \ldots, \xi_n)$ and $y = (\eta_i, \ldots, \eta_n)$ is their *scalar product* defined as

$$(x, y) = \xi_1 \eta_1 + \xi_2 \eta_2 + \ldots + \xi_n \eta_n , \tag{1.1}$$

and unless otherwise stated, the inner product is assumed to have this functional form. The scalar product is also called *"dot product"* and denoted $x \cdot y$. It should be pointed out, however, that there are infinitely many choices for inner products, for instance, variants of (1.1) where the elements of the vectors are weighted differently, or enter the function (x, y) with different powers.

In general, the inner product of two elements x and y in a set, by convention, must have the following properties. Assume that the addition of the elements, and multiplication of an element by a scalar have been defined. If the inner product function is in general denoted by (x, y), there must hold

B1) $(x, y) = (y, x)$

B2) $(\alpha x, y) = \alpha(x, y)$

B3) $(x_1 + x_2, y) = (x_1, y) + (x_2, y)$

B4) $(x, x) \geq 0$, where equality holds only if x is the zero element.

Metric. Observable vectors must usually be represented in a space that has a *metric*. The latter is a property of any set of elements characterized by another function called *distance* between all pairs of elements, and denoted $d(x, y)$. For the choice of the distance, the following conditions must hold:

C1) $d(x, y) \geq 0$, where equality holds if and only if $x = y$.
C2) $d(x, y) = d(y, x)$
C3) $d(x, y) \leq d(x, z) + d(z, y)$.

An example of distance is the *Euclidean distance $d_E(x, y)$* in a rectangular coordinate system, almost exclusively used in this book: for the vectors $x = (\xi_1, \ldots, \xi_n)$ and $y = (\eta_1, \ldots, \eta_n)$,

$$d_E(x, y) = \sqrt{(\xi_1 - \eta_1)^2 + (\xi_2 - \eta_2)^2 + \ldots + (\xi_n - \eta_n)^2} . \qquad (1.2)$$

Another example of distance is the *Hamming distance* that in the simplest case is defined for binary vectors; a binary vector has elements that are either 0 or 1. The Hamming distance indicates in how many positions (elements) the two vectors are different. Clearly the rules C1 to C3 are valid for Hamming distances, too.

Norm. The magnitude of a vector can be defined in different ways. The name *norm* is used for it, and in general, the norm, in a set of elements for which scalar multiplication, addition, and the zero element have been defined, is a function $||x||$ of an element x for which the following rules must be valid:

D1) $||x|| \geq 0$, and the equality holds if and only if $x = 0$.
D2) $||\alpha x|| = |\alpha| \, ||x||$ where $|\alpha|$ is the absolute value of α.
D3) $||x_1 + x_2|| \leq ||x_1|| + ||x_2||$.

The *Euclidean norm* can be defined by the scalar product

$$||x||_E = \sqrt{(x, x)} = \sqrt{\xi_1^2 + \xi_2^2 + \ldots + \xi_n^2} . \qquad (1.3)$$

Henceforth we shall always mean the Euclidean norm when we write $|| \cdot ||$. Notice that the Euclidean distance $d_E(x, y)$ is equivalent to the Euclidean norm $||x - y||$. A space in which Euclidean distance and norm have been defined is called *Euclidean space*. Vectors of the Euclidean space are called *Euclidean vectors*.

Angles and Orthogonality. Generalization of the usual concept of *angle* for higher-dimensional spaces is straightforward. The angle θ between two Euclidean vectors x and y is defined by

$$\cos \theta = \frac{(x, y)}{||x|| \, ||y||} . \qquad (1.4)$$

Accordingly, the two vectors are said to be *orthogonal* and denoted $x \perp y$ when their inner product vanishes.

Linear Manifolds. The vectors $x_1, x_2, \ldots, x_k$ are said to be *linearly independent* if their weighted sum, or the linear combination

$$\alpha_1 x_1 + \alpha_2 x_2 + \ldots + \alpha_k x_k \tag{1.5}$$

cannot become zero unless $\alpha_1 = \alpha_2 = \ldots = \alpha_k = 0$. Accordingly, if the sum expression can be made zero for some choice of the α coefficients, all of which are not zero, the vectors are *linearly dependent*. Some of the vectors can then be expressed as linear combinations of the others. Examples of linear dependence can be visualized in the three-dimensional space $\Re^3$: three or more vectors are linearly dependent if they lie in a plane that passes through the origin, because then each vector can be expressed as a weighted sum of the others (as shown by the dashed-line constructions in Fig. 1.2).

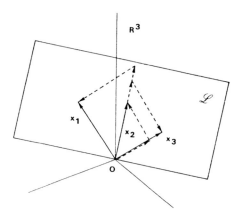

Fig. 1.2. Exemplification of linear dependence

Consider all possible linear combinations of the vectors $x_1, x_2, \ldots, x_k$ where k is at most equal to the dimensionality n; they are obtained when the α coefficients take on all real values from $(-\infty, +\infty)$. The set of all linear combinations is called a *linear subspace* in $\Re^n$ and denoted by $\mathcal{L}$. Examples of linear subspaces are planes and straight lines in $\Re^3$ that pass through the origin. In higher-dimensional spaces, very important linear subspaces are *hyperplanes* that are defined as linear combinations of $n-1$ linearly independent vectors, and which divide $\Re^n$ into two half spaces. All linear subspaces, including $\Re^n$, are also named *linear manifolds*, and the set of vectors that define the manifold is said to *span* it. It is to be noted that the manifold spanned by the above vectors is k-dimensional if the vectors are linearly independent. In this case the vectors $x_1, x_2, \ldots, x_k$ are called the *basis vectors of $\mathcal{L}$*.

Orthogonal Spaces. A vector is said to be *orthogonal to subspace* $\mathcal{L}$ if it is orthogonal to every vector in it (abbreviated $x \perp \mathcal{L}$). Two subspaces $\mathcal{L}_1$ and $\mathcal{L}_2$ are said to be orthogonal to each other ($\mathcal{L}_1 \perp \mathcal{L}_2$) if every vector of $\mathcal{L}_1$ is orthogonal to every vector of $\mathcal{L}_2$. An example of orthogonal subspaces in $\Re^3$ are a coordinate axis and the plane spanned by the two other axes of a rectangular coordinate system.

Orthogonal Projections. Below it will be pointed out that if $\mathcal{L}$ is a subspace of $\Re^n$, then an arbitary vector $x \in \Re^n$ can be uniquely decomposed into the sum of two vectors of which one, $\hat{x}$, belongs to $\mathcal{L}$ and the other, $\tilde{x}$, is orthogonal to it.

Assume tentatively that there exist two decompositions

$$x = \hat{y} + \tilde{y} = \hat{z} + \tilde{z} , \tag{1.6}$$

where $\hat{y}$ and $\hat{z}$ belong to $\mathcal{L}$ and $\tilde{y} \perp \mathcal{L}, \tilde{z} \perp \mathcal{L}$. Now $\tilde{z} - \tilde{y} \perp \mathcal{L}$, but since $\tilde{z} - \tilde{y} = \hat{y} - \hat{z}$, then $\tilde{z} - \tilde{y} \in \mathcal{L}$. Consequently, $\tilde{z} - \tilde{y}$ has thus been shown to be orthogonal to itself, or $(\tilde{z} - \tilde{y}, \tilde{z} - \tilde{y}) = 0$, which cannot hold true in general unless $\tilde{z} = \tilde{y}$. This proves that the decomposition is unique.

The proof of the existence of the decomposition will be postponed until the concept of an *orthogonal basis* has been introduced. Let us tentatively assume that the decomposition exists,

$$x = \hat{x} + \tilde{x}, \quad \text{where } \hat{x} \in \mathcal{L} \text{ and } \tilde{x} \perp \mathcal{L} . \tag{1.7}$$

Hereupon $\hat{x}$ will be called the *orthogonal projection* of x on $\mathcal{L}$; it will also be useful to introduce the space $\mathcal{L}^\perp$, which is named the *orthogonal complement* of $\mathcal{L}$; it is the set of all vectors in $\Re^n$ that are orthogonal to $\mathcal{L}$. Then $\tilde{x}$ is called the orthogonal projection of x on $\mathcal{L}^\perp$. The orthogonal projections are very fundamental to the theories of subspace classifiers and the adaptive-subspace SOM discussed in this book.

Theorem 1.1.1. *Of all decompositions of the form $x = x' + x''$ where $x' \in \mathcal{L}$, the one into orthogonal projections has the property that $||x''||$ is minimum.*

To prove this theorem, use is made of the definition $||x'||^2 = (x', x')$ and the facts that $\hat{x} - x' \in \mathcal{L}, x - \hat{x} = \tilde{x} \perp \mathcal{L}$, whereby $(\hat{x} - x', x - \hat{x}) = 0$. The following expansion then yields

$$||x - x'||^2 = (x - \hat{x} + \hat{x} - x', x - \hat{x} + \hat{x} - x') = ||x - \hat{x}||^2 + ||\hat{x} - x'||^2 . \tag{1.8}$$

Because the squared norms are always positive or zero, there can be written

$$||x - x'||^2 \geq ||x - \hat{x}||^2 , \tag{1.9}$$

from which it is directly seen that $x'' = x - x'$ is minimum for $x' = \hat{x}$ whereby $x'' = \tilde{x}$. ∎

Orthogonal projections in a three-dimensional space are visualized in Fig. 1.3.

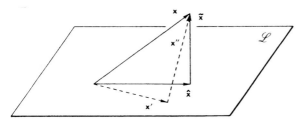

Fig. 1.3. Orthogonal projections in $\Re^3$

The Gram-Schmidt Orthogonalization Process. For the computation of the orthogonal projections, and to prove that an arbitrary vector can be decomposed uniquely as discussed above, a classical method named *Gram-Schmidt orthogonalization process* can be used. Its original purpose is to construct an *orthogonal vector basis* for any linear space $\Re^k$, i.e., to find a set of basis vectors that are mutually orthogonal and span $\Re^k$.

To start with, consider the nonzero vectors $x_1, x_2, \ldots, x_p$, $p \geq k$ that span the space $\Re^k$. For a vector basis, one direction can be chosen freely, and the first new basis vector is customarily selected as $h_1 = x_1$. Unless x_2 has the same direction as x_1, it is easily found that the vector

$$h_2 = x_2 - \frac{(x_2, h_1)}{||h_1||^2} h_1 \tag{1.10}$$

is orthogonal to $h_1 = x_1$ (but not yet normalized); the inner product (h_2, h_1) is zero. Therefore, h_2 can be chosen for the new basis vector. If, on the other hand, x_2 had the same direction as x_1, it would be represented by x_1, and can thus be ignored. Consider now the sequence of vectors $\{h_i\}$ where each new member is constructed according to the recursive rule

$$h_i = x_i - \sum_{j=1}^{i-1} \frac{(x_i, h_j)}{||h_j||^2} h_j \, , \tag{1.11}$$

where the sum over j shall include terms with nonzero h_j only. In this way a set of vectors h_i is obtained. To prove that all the h_i are mutually orthogonal, the method of complete induction can be used. Assume that this rule is correct up to $i - 1$, i.e., the $h_1, h_2, \ldots, h_{i-1}$ are mutually orthogonal, which means that for all $q < i$, and for nonzero $h_j, h_q, (h_j, h_q) = ||h_j|| \cdot ||h_q|| \cdot \delta_{jq}$ where δ_{jq} is the Kronecker delta ($= 1$ for $j = q, = 0$ for $j \neq q$). By a substitution there follows

$$(h_i, h_q) = \begin{cases} (x_i, h_q) - (x_i, h_q) = 0 \text{ if } h_q \neq 0 \, , \\ (x_i, h_q) = 0 \text{ if } h_q = 0 \, . \end{cases} \tag{1.12}$$

Therefore the rule is correct up to i, and thus generally.

From the way in which the h-vectors were constructed, there directly follows that the $h_1, \ldots, h_i$ span exactly the same space as the $x_1, \ldots, x_i$ do. When the process is carried out for the vectors $x_1, x_2, \ldots, x_p$ among which there are known to exist k linearly independent vectors, the space $\Re^k$ will be spanned exactly.

If a further vector $x \in \Re^n$ that may or may not belong to space $\Re^k$ shall be decomposed into its orthogonal projections $\hat{x} \in \Re^k$ and $\tilde{x} \perp \Re^k$, the Gram-Schmidt process is simply continued one step further, whereby $\tilde{x} = h_{p+1}$, $\hat{x} = x - h_{p+1}$.

Since the above process is always possible, and it was earlier shown that the decomposition is unique, the following *Decomposition Theorem* can now be expressed.

Theorem 1.1.2. *An arbitrary vector $x \in \Re^n$ can be decomposed uniquely into two vectors of which one is in a subspace $\mathcal{L}$ of $\Re^n$ and the other is orthogonal to it.*

1.1.2 Matrix Notations

Matrices, Their Sums, and Matrix Products. The concept of a *matrix* may be generally known. It might be necessary to point out that formally matrices are ordered sets of numbers or scalar variables indexed by a pair of indices. If index k runs through $(1, 2, \ldots, p)$, l through $(1, 2, \ldots, q)$, and m through $(1, 2, \ldots, r)$, the following ordered sets of elements α_{kl} and β_{lm} can be named matrices. The sum of two matrices (of the same dimensionality) is a matrix in which every element is a sum of the respective elements of the addends. The *matrix product* of sets $A = (\alpha_{kl})$ and $B = (\beta_{lm})$ is defined by

$$AB = C \;, \tag{1.13}$$

where C is a further ordered set (γ_{km}) with elements defined by

$$\gamma_{km} = \sum_{l=1}^{q} \alpha_{kl}\beta_{lm} \;. \tag{1.14}$$

Definition of the matrix product has been introduced for the description of *linear transformations*, eventually cascaded transformations describable by *operator products*. A matrix is regarded as a *linear operator*.

A matrix product is defined only for such pairs of indices in which an element like l above is common. Multiplication of a matrix by a scalar is defined as multiplication of all elements of the matrix by it.

Array Representation for Matrices. For better visualization, matrices can be represented as rectangular arrays of numbers. Below, only matrices consisting of real numbers are considered. The dimensionality of a matrix is expressed by the product of the numbers of rows and columns; for instance an $m \times n$ matrix has m (horizontal) rows and n (vertical) columns. Matrices are denoted by capital italic letters, and when they are written explicitly, brackets are used around the array.

The *transpose* of any matrix is obtained by rewriting all columns as rows, whereby the notation X^T is used for the transpose of X. For instance,

$$X = \begin{bmatrix} 1 & 5 \\ 2 & 1 \\ 3 & 3 \end{bmatrix}, \qquad X^{\mathrm{T}} = \begin{bmatrix} 1 & 2 & 3 \\ 5 & 1 & 3 \end{bmatrix} . \tag{1.15}$$

Row and Column Vectors. Since one-row and one-column matrices are linear arrays of numbers, they can be understood as vectors, called *row* and *column vectors*, respectively. Such vectors are usually denoted by lower-case italic letters.

Matrix-Vector Products. Consider the matrix $A = (\alpha_{kl})$, and four vectors $b = (\beta_k)$, $c = (\gamma_l)$, $d = (\delta_l)$, and $e = (\varepsilon_k)$, where k runs through $(1, 2, \ldots, p)$ and l through $(1, 2, \ldots, q)$, respectively. Here b and c are row vectors, and d and e column vectors, respectively. The following matrix-vector products are then defined:

$$c = bA, \quad \text{where} \quad \gamma_l = \sum_{k=1}^{p} \beta_k \alpha_{kl} ,$$

$$e = Ad, \quad \text{where} \quad \varepsilon_k = \sum_{l=1}^{q} \alpha_{kl} \delta_l . \tag{1.16}$$

Notice again the compatibility of indices.

In a matrix-vector product, the row vector always stands on the left, and the column vector on the right. For reasons that may become more apparent below, representations of patterns are normally understood as column vectors. For better clarity, column vectors are normally denoted by simple lower-case letters like x, whereas row vectors are written in the transpose notation as x^{T}. Within the text, for typographic reasons, column vectors are usually written in the form $[\xi_1, \xi_2, \ldots, \xi_n]^{\mathrm{T}}$ where commas may be used for clarity, and accordingly, a row vector then reads $[\xi_1, \xi_2, \ldots, \xi_n]$.

Linear Transformations. The main reason for the introduction of matrices is to convert successive linear transformations into operator products. The transformation of vector x into vector y is generally denoted by function $y = T(x)$. In order to call T a *linear transformation*, it is a necessary and sufficient condition that for $x_1, x_2 \in \Re^n$,

$$T(\alpha x_1 + \beta x_2) = \alpha T(x_1) + \beta T(x_2) . \tag{1.17}$$

The general linear transformation of vector $x = [\xi_1, \xi_2, \ldots, \xi_n]^{\mathrm{T}}$ into vector $y = [\eta_1, \eta_2, \ldots, \eta_m]^{\mathrm{T}}$ can be expressed in the element form as

$$\eta_i = \sum_{j=1}^{n} \alpha_{ij} \xi_j , i = 1, 2, \ldots, m . \tag{1.18}$$

Equation (1.18) can also be expressed symbolically as a matrix-vector product. If A is the ordered set of parameters (α_{ij}) that defines the transformation, there follows from the above definitions that

$$y = Ax . \tag{1.19}$$

The matrix-vector-product notation now makes it possible to describe successive transformations by transitive operators: if $y = Ax$ and $z = By$, then $z = BAx$.

Symmetric, Diagonal, and Unit Matrices. A matrix is called *symmetric* if it is identical with its transpose, i.e., it is symmetric with respect to the main diagonal whereupon it also must be square. A matrix is called *diagonal* if it has zeroes elsewhere except on the main diagonal. If all diagonal elements of a diagonal matrix are unities, the matrix is *unit matrix*, denoted by I.

Partitioning of Matrices. Sometimes it will be necessary to compose a rectangular matrix from rows or columns that have a special meaning: for instance, in a so-called *observation matrix* X, representation vectors $x_1, x_2, \ldots, x_n$ may appear as columns and the matrix can then be written as $X = [x_1, x_2, \ldots, x_n]$.

Matrices can also be partitioned into rectangular submatrices. In the transposition of a partitioned matrix, the submatrices change their positions like the scalars in a normal matrix, and in addition they are transposed:

$$\begin{bmatrix} A & B \\ C & D \end{bmatrix}^{\mathrm{T}} = \begin{bmatrix} A^{\mathrm{T}} & C^{\mathrm{T}} \\ B^{\mathrm{T}} & D^{\mathrm{T}} \end{bmatrix} . \tag{1.20}$$

In the product of partitioned matrices, submatrices are operated according to the same rules as elements in matrix products. Submatrices must then have dimensionalities for which the matrix products are defined. For instance,

$$\begin{bmatrix} A & B \\ C & D \end{bmatrix} \begin{bmatrix} E \\ F \end{bmatrix} = \begin{bmatrix} AE + BF \\ CE + DF \end{bmatrix} . \tag{1.21}$$

Comment. It is possible to regard a real $m \times n$ matrix as an element of the real space $\Re^{m \times n}$.

Some Formulas for Matrix Operations. In general, matrix products are not commutative, i.e., one can not change the order of the matrices in the product, but they are associative and distributive according to the following formulas. These can be proven when written in component form:

E1) $IA = AI = A$
E2) $(AB)C = A(BC)$
E3) $A(B + C) = AB + AC$
E4) $(A^{\mathrm{T}})^{\mathrm{T}} = A$
E5) $(A + B)^{\mathrm{T}} = A^{\mathrm{T}} + B^{\mathrm{T}}$
E6) $(AB)^{\mathrm{T}} = B^{\mathrm{T}} A^{\mathrm{T}}$.

It should be noticed that in a product of two matrices the former must always have as many columns as the latter one has rows, otherwise the summation over all indices is not defined.

1.1.3 Eigenvectors and Eigenvalues of Matrices

Consider a vector equation of the type

$$Ax = \lambda x ,\tag{1.22}$$

where A is a square matrix of dimensionality $n \times n$ and λ is a scalar; any solutions for x are called *eigenvectors* of A. Notice that the "direction" of an eigenvector corresponding to the relative magnitudes of its elements is not changed in the multiplication by A. Notice also that (1.22) is a homogeneous equation: correspondingly, we may solve it only for the ratio of the components of x.

Methods for the construction of eigenvectors for a general matrix can be found in numerous textbooks of matrix algebra, but they must be skipped here. Instead, we shall discuss below solutions in the case that A is symmetric; this result will be needed in this book.

It is far easier to find the values of λ associated with the eigenvectors. They are called *eigenvalues* of A. Let the determinant formed of the elements of a square matrix M be denoted $|M|$. If the equation $(A - \lambda I)x = 0$ must have solutions other than the trivial one $x = 0$, the familiar condition known from systems of linear equations is that

$$|A - \lambda I| = 0 .\tag{1.23}$$

Clearly the determinant can be written in powers of λ, and (1.23) is of the form

$$\lambda^n + \gamma_1 \lambda^{n-1} + \ldots + \gamma_n = 0 ,\tag{1.24}$$

where the $\gamma_i, i = 1 \ldots n$ are parameters that depend on the elements of matrix A. This polynomial equation is also called the *characteristic equation* of A, and has n roots $\lambda_1, \lambda_2, \ldots, \lambda_n$ some of which may be equal. These roots are the *eigenvalues* of A.

The *spectral radius* of matrix A is defined as $\rho(A) = \max_i\{\lambda_i(A)\}$, where the $\lambda_i(A)$ are the eigenvalues of A.

Next we shall show an important property of real symmetric matrices.

Theorem 1.1.3. *Two eigenvectors of a real symmetric matrix belonging to different eigenvalues are orthogonal.*

To prove the theorem, let u_k and u_l be two eigenvectors of the real symmetric matrix C, belonging to the eigenvalues λ_k and λ_l, respectively. Then we have the equations

$$Cu_k = \lambda_k u_k ,\tag{1.25}$$
$$Cu_l = \lambda_l u_l .\tag{1.26}$$

Taking into account that C is symmetric and transposing both sides of (1.25) we get

$$u_k^T C = \lambda_k u_k^T \ . \tag{1.27}$$

Multiplication of the sides of (1.27) on the right by u_l gives

$$u_k^T C u_l = \lambda_k u_k^T u_l \tag{1.28}$$

and multiplication of (1.26) by u_k^T on the left gives

$$u_k^T C u_l = \lambda_l u_k^T u_l \ . \tag{1.29}$$

Hence, subtracting, we obtain

$$(\lambda_k - \lambda_l) u_k^T u_l = 0 \ , \tag{1.30}$$

which shows that, if $\lambda_k \neq \lambda_l$, then $u_k^T u_l = 0$ and thus the two eigenvectors u_k and u_l are orthogonal. ∎

Expression of a Symmetric Matrix in Terms of Eigenvectors. Let us assume that all eigenvectors of matrix C are orthonormal (orthogonal and normalized) with different eigenvalues, so that the u_k constitute a complete basis of $\Re^n$. Construct the matrix U that has the eigenvectors as it columns:

$$U = [u_1, \ldots, u_n] \ , \tag{1.31}$$

and similarly construct a diagonal $(n \times n)$ matrix with the λ_k as its diagonal elements:

$$D = \begin{bmatrix} \lambda_1 & & 0 \\ & \ddots & \\ 0 & & \lambda_n \end{bmatrix} . \tag{1.32}$$

Since for all k, $Cu_k = \lambda_k u_k$, one can easily show by forming the products of partitioned matrices that

$$CU = UD \tag{1.33}$$

(notice that the diagonal matrix commutes with the square U). Because the u_k are orthonormal, it is also easy to show that $UU^T = I$. Multiplying the sides of (1.33) on the right by U^T we get

$$CUU^T = C = UDU^T = [u_1, \ldots, u_n] D [u_1, \ldots, n_n]^T \ . \tag{1.34}$$

But when the last matrix product is written explicitly in terms of the partitioned matrices, one gets

$$C = \sum_{k=1}^{n} \lambda_k u_k u_k^T \ . \tag{1.35}$$

Eigenvectors of a Real Symmetric Matrix. Several iterative methods for the approximation of the eigenvectors of a real symmetric matrix have been devised: cf., e.g., the book of *Oja* [1.3]. A neural-network type solution has been given in [1.4].

A very lucid and simple way to approximate the eigenvectors that works reasonably fast even for relatively high-dimensional vectors is the following. Multiply the expression of C in (1.35) by itself:

$$C^2 = \sum_{k=1}^{n} \sum_{l=1}^{n} \lambda_k \lambda_l u_k u_k^T u_l u_l^T ; \tag{1.36}$$

but $u_k^T u_l = \delta_{kl}$ (Kronecker delta), so

$$C^2 = \sum_{k=1}^{n} \lambda_k^2 u_k u_k^T . \tag{1.37}$$

Now the relative magnitudes of the terms of C^2 have changed from those in C: if $\lambda_1 > \lambda_2 > \ldots > \lambda_n$, then the term with $k = 1$ starts to dominate. In continued squaring $(C^2)^2 \ldots$ one gets

$$\lim_{p \to \infty} C^{2p} = \lambda_1^{2p} u_1 u_1^T ; \tag{1.38}$$

this expression may be scaled properly, because in the first place we are only interested in the relative values of the components of u_1. The latter are obtained multiplying an arbitrary vector $a \in \Re^n$ by both sides of (1.38), whereby

$$u_1 \approx \text{const.} \; C^{2p} a . \tag{1.39}$$

Here the only restriction is that a must not be orthogonal to u_1. Notice that the constant factor is inversely proportional to a dot product where a is a member, so the arbitrary choice of a is compensated for. Notice also that C^{2p} is approximately singular. After that u_1 is obtained by normalizing $C^{2p} a$. Once u_1 is obtained, λ_1 follows from (1.38).

In order to obtain u_2 one may first compute

$$C' = C - \lambda_1 u_1 u_1^T , \tag{1.40}$$

from which the eigenvector u_1 has been eliminated, and the highest eigenvalue of C' is now λ_2. The same squaring procedure is repeated for C' to obtain u_2, and so on for the rest of the eigenvectors.

1.1.4 Further Properties of Matrices

The Range and the Null Space of a Matrix. The *range* of a matrix A is the set of vectors Ax for all values of x. This set is a linear manifold denoted $\mathcal{R}(A)$, and it is the *subspace spanned by the columns of A*. This can be shown by writing $A = [a_1, a_2, \ldots, a_k]$ and $x = [\xi_1, \xi_2, \ldots, \xi_k]^T$ and noticing by (1.16) that $Ax = \xi_1 a_1 + \xi_2 a_2 + \ldots + \xi_k a_k$, whereby Ax is found to be the general linear combination of the columns of A.

The *null space* of A is the set of all vectors x for which $Ax = 0$. It has at least one element, namely, the zero vector. The null space is a linear manifold, too, denoted $\mathcal{N}(A)$.

The Rank of a Matrix. The *rank* of matrix A, abbreviated $r(A)$, is the dimensionality of the linear manifold $\mathcal{R}(A)$. Especially an $m \times n$ matrix A is said to be *of full rank* if $r(A) = \min(m, n)$. Without proof it is stated that for any matrix A,

$$r(A) = r(A^{\mathrm{T}}) = r(A^{\mathrm{T}} A) = r(AA^{\mathrm{T}}) \,. \tag{1.41}$$

In particular, $r(A) = r(A^{\mathrm{T}} A)$ implies that if the columns of A are linearly independent, then $A^{\mathrm{T}} A$ is of full rank. In a similar way, $r(A^{\mathrm{T}}) = r(AA^{\mathrm{T}})$ implies that if the rows of A are linearly independent, then AA^{T} is of full rank.

Singular Matrices. If A is a square matrix and its null space consists of the zero vector only, then A is called *nonsingular*. A nonsingular matrix has an *inverse* A^{-1}, defined by $AA^{-1} = A^{-1}A = I$. Otherwise a square matrix A is *singular*. If $Ax = b$ is a vector equation which, when written for each element of b separately, can be regarded as a set of linear equations in the elements ξ_i of vector x, then the singularity of A means that in general a unique solution for this system of equations does not exist.

The *determinant* formed of the elements of a singular matrix is zero. All square matrices of the form ab^{T} are singular, which can easily be verified when writing out the determinant explicitly.

When dealing with matrix equations, extreme care should be taken when multiplying matrix expressions by matrices that may become singular. There is a possibility to end up with false solutions, in a similar way as when multiplying the sides of scalar equations by zeroes.

Idempotent Matrices. The matrix A is called *idempotent* if $A^2 = A$; from the iterative application of this condition it follows that for any positive integer n there holds $A^n = A$. Notice that $(I - A)^2 = I - A$ whereby $I - A$ is idempotent, too. The identity matrix I is idempotent trivially.

Positive Definite and Positive Semidefinite Matrices. If A is a square matrix and for all nonzero $x \in \Re^n$ there holds that the scalar expression $x^{\mathrm{T}} Ax$ is positive, then by definition A is *positive definite*. If $x^{\mathrm{T}} Ax$ is positive or zero, then A is called *positive semidefinite*. The expression $x^{\mathrm{T}} Ax$ is named a *quadratic form* in x.

Without proof it is stated [1.1] that *any* of the following conditions can be applied for the definition of a positive semidefinite (also named nonnegative definite) matrix, where the definition is restricted to symmetric matrices only.

F1) $A = HH^{\mathrm{T}}$ for some matrix H.
F2) There exists a symmetric matrix R such that $R^2 = A$,
      whereupon $R = A^{1/2}$ is the square root of A.
F3) The eigenvalues of A are positive or zero.
F4) $x^{\mathrm{T}} Ax \geq 0$, as already stated.

If A is further nonsingular, it is positive definite.

Positive semidefinite matrices occur, for instance, in linear transformations. Consider the transformation $y = Mx$, which yields a vector; the inner product $(y, y) = y^{\mathrm{T}} y$ must be nonnegative, whereby $x^{\mathrm{T}} M^{\mathrm{T}} M x$ is nonnegative, and $M^{\mathrm{T}} M$ is a positive semidefinite matrix for any M.

Elementary Matrices. There exist matrices that have been used extensively in the solution of systems of linear equations and also occur in the projection operations. They are named *elementary matrices*, and in general, they are of the form $(I - uv^{\mathrm{T}})$ where u and v are column vectors of the same dimensionality.

Matrix Norms. The norm of a matrix can be defined in different ways, and clearly it must satisfy the general requirements imposed on the norm in any set. The *Euclidean matrix norm* is by definition the square root of the sum of squares of its elements.

The *trace* of a square matrix S, denoted $\mathrm{tr}(S)$, is defined as the sum of all diagonal elements. The Euclidean norm of any matrix A is then easily found, by writing it out explicitly by elements, as

$$\|A\|_{\mathrm{E}} = \sqrt{\mathrm{tr}(A^{\mathrm{T}} A)} \ .$$

Another definition of matrix norm that is different from the Euclidean norm can be derived from any definition of vector norm denoted by $\| \cdot \|$. Such a matrix norm is said to be *consistent* with the vector norm, and the definition reads

$$\|A\| = \max_{\|x\|=1} \|Ax\| \ . \tag{1.42}$$

Notice that the Euclidean matrix norm is not consistent with the Euclidean vector norm.

Hadamard Products. There exists another type of matrix product that has a simpler multiplication rule and applications in some nonlinear problems. The *Hadamard product* $C = (\gamma_{ij})$ of matrices $A = (\alpha_{ij})$ and $B = (\beta_{ij})$ is defined as

$$C = A \otimes B, \quad \text{where } \gamma_{ij} = \alpha_{ij} \beta_{ij} \ . \tag{1.43}$$

In other words, the respective matrix elements are multiplied mutually.

1.1.5 On Matrix Differential Calculus

Algebraic matrix equations are more problematic than scalar equations. In a similar way it may be expected that matrix differential equations behave differently from the scalar differential equations. This is due to several reasons: matrix products in general do not commute, matrices may become singular, and first of all, a matrix differential equation is a system of coupled equations of the matrix elements whereupon stability conditions are more complicated. Extreme care should therefore be taken when dealing with matrix differential equations.

Derivatives of Matrices. If the matrix elements are functions of a scalar variable, for instance time, then the derivative of a matrix is obtained by taking the derivatives of the elements. For instance, for the matrix A,

$$A = \begin{bmatrix} a_{11} & a_{12} \\ a_{21} & a_{22} \end{bmatrix}, \quad dA/dt = \begin{bmatrix} da_{11}/dt & da_{12}/dt \\ da_{21}/dt & da_{22}/dt \end{bmatrix}. \tag{1.44}$$

Partial derivatives of a matrix are obtained by taking the partial derivatives of the elements.

In the differentiation of products of matrices and other matrix functions, the noncommutativity must be kept in mind. For instance,

$$d(AB)/dt = (dA/dt)B + A(dB/dt). \tag{1.45}$$

This rule is important in the derivatives of the powers of matrices: e.g., if A is a square matrix,

$$dA^3/dt = d(A \cdot A \cdot A)/dt = (dA/dt)A^2 + A(dA/dt)A + A^2(dA/dt), \tag{1.46}$$

and in general the above form cannot be simplified because the terms are not combinable.

The formulas of derivatives of general integral powers are found if the following fact is considered: if the inverse matrix A^{-1} exists, then there must hold

$$d(AA^{-1})/dt = (dA/dt)A^{-1} + A(dA^{-1}/dt) = 0 \text{ (the zero matrix)};$$
$$dA^{-1}/dt = -A^{-1}(dA/dt)A^{-1}. \tag{1.47}$$

In general it is obtained:

$$dA^n/dt = \sum_{i=0}^{n-1} A^i(dA/dt)A^{n-i-1}, \text{ when } n \geq 1,$$
$$dA^{-n}/dt = \sum_{i=1}^{n} -A^{-i}(dA/dt)A^{i-n-1}, \text{ when } n \geq 1, |A| \neq 0. \tag{1.48}$$

Gradient. The *gradient* of a scalar is a vector. In matrix calculus, the gradient operator is a column vector of differential operators of the form

$$\nabla_x = [\partial/\partial\xi_1, \partial/\partial\xi_2, \ldots, \partial/\partial\xi_n]^{\mathrm{T}}, \tag{1.49}$$

and differentiation of a scalar α is formally equivalent to a matrix product of vector ∇_x and α:

$$\nabla_x\alpha = [\partial\alpha/\partial\xi_1, \partial\alpha/\partial\xi_2, \ldots, \partial\alpha/\partial\xi_n]^{\mathrm{T}}. \tag{1.50}$$

Other notations for the gradient, also used in this book, are $\nabla_x\alpha = grad_x\alpha = \partial\alpha/\partial x$ (notice that here x is a vector).

If a scalar-valued function is a function of vector x, then the differentiation rules are most easily found when writing by elements, e.g.,

$$\nabla_x(x^{\mathrm{T}}x) = [\partial/\partial\xi_1, \partial/\partial\xi_2, \ldots, \partial/\partial\xi_n]^{\mathrm{T}}(\xi_1^2 + \xi_2^2 + \ldots + \xi_n^2) = 2x \ . \quad (1.51)$$

Since ∇_x is a vector, it is applicable to all row vectors of arbitrary dimensionality, whereby a matrix results. For instance, $\nabla_x x^{\mathrm{T}} = I$. In some cases it is applicable to products of vectors or vectors and matrices if the expression has the same dimensionality as that of a scalar or row vector. The following examples can be proven when writing by elements: if a and b are functions of x, and p and q are constants,

$$\nabla_x[a^{\mathrm{T}}(x)b(x)] = [\nabla_x a^{\mathrm{T}}(x)]b(x) + [\nabla_x b^{\mathrm{T}}(x)]a(x) \ , \quad (1.52)$$

$$\nabla_x(p^{\mathrm{T}}x) = p \ , \quad (1.53)$$

$$\nabla_x(x^{\mathrm{T}}q) = q \ . \quad (1.54)$$

Consider a quadratic form $Q = a^{\mathrm{T}}(x)\psi a(x)$ where ψ is symmetric. Then

$$\nabla_x Q = 2[\nabla_x a^{\mathrm{T}}(x)]\psi a(x) \ , \quad (1.55)$$

which can be proven by writing $\psi = \psi^{1/2}\psi^{1/2}$, where $\psi^{1/2}$ is symmetric.

1.2 Distance Measures for Patterns

1.2.1 Measures of Similarity and Distance in Vector Spaces

In a way *distance* and *similarity* are reciprocal concepts. It is a matter of terminology if we call distance *dissimilarity*. Below we shall give concrete examples of both.

Correlation. Comparison of signals or patterns is often based on their *correlation*, which is a trivial measure of similarity. Assume two ordered sets, or sequences of real-valued samples $x = (\xi_1, \xi_2, \ldots, \xi_n)$ and $y = (\eta_1, \eta_2, \ldots, \eta_n)$. Their *unnormalized correlation* is

$$C = \sum_{i=1}^{n} \xi_i \eta_i \ . \quad (1.56)$$

If x and y are understood as Euclidean (real) vectors, then C is their *scalar* or *dot product*.

In case one of the sequences may be *shifted* with respect to the other by an arbitrary amount, the comparison may better be based on a translationally invariant measure, the *maximum correlation* over a specified interval:

$$C_m = \max_k \sum_{i=1}^{n} \xi_i \eta_{i-k}, \quad k = -n, -n+1, \ldots, +n \ . \quad (1.57)$$

It will be necessary to emphasize that the correlation methods are most suitable for the comparison of signals contaminated by *Gaussian noise*; since the distributions of natural patterns may often not be Gaussian, other criteria of comparison, some of which are discussed below, must be considered, too.

Direction Cosines. If the relevant information in patterns or signals is contained only in the *relative magnitudes* of their components, then similarity can often be better measured in terms of *direction cosines* defined in the following way. If $x \in \Re^n$ and $y \in \Re^n$ are regarded as Euclidean vectors, then

$$\cos \theta = \frac{(x, y)}{||x|| \, ||y||} \tag{1.58}$$

is by definition the direction cosine of their mutual angle, with (x, y) the scalar product of x and y, and $||x||$ the Euclidean norm of x. Notice that if the norms of vectors are standardized to unity, then (1.56) complies with (1.58), or $\cos \theta = C$. Expression (1.58) is concordant with the traditional definition of *correlation coefficient* in statistics, provided that we understand the vectors x and y as sequences of stochastic numbers $\{\xi_i\}$ and $\{\eta_j\}$, respectively.

The value $\cos \theta = 1$ is often defined to represent the *best match* between x and y; vector y is then equal to x multiplied by a scalar, $y = \alpha x \; (\alpha \in \Re)$. On the other hand, if $\cos \theta = 0$ or $(x, y) = 0$, vectors x and y are said to be *orthogonal*.

Euclidean Distance. Another measure of similarity, actually that of *dissimilarity*, closely related to the previous ones, is based on the *Euclidean distance* of x and y defined as

$$\rho_E(x, y) = ||x - y|| = \sqrt{\sum_{i=1}^{n} (\xi_i - \eta_i)^2} \; . \tag{1.59}$$

Measures of Similarity in the Minkowski Metric. The *Minkowski metric* is a generalization of (1.59). This measure has been used, e.g., in experimental psychology. The distance in it is defined as

$$\rho_M(x, y) = \left(\sum_{i=1}^{n} |\xi_i - \eta_i|^\lambda \right)^{1/\lambda} , \quad \lambda \in \Re \; . \tag{1.60}$$

The so-called *"city-block distance"* is obtained with $\lambda = 1$.

Tanimoto Similarity Measure. Some experiments have shown [1.5–9] that determination of similarity between x and y in terms of the measure introduced by *Tanimoto* [1.10] sometimes yields good results; it may be defined as

$$S_T(x, y) = \frac{(x, y)}{||x||^2 + ||y||^2 - (x, y)} \; . \tag{1.61}$$

The origin of this measure is in the comparison of sets. Assume that A and B are two unordered sets of distinct (nonnumerical) elements, e.g., identifiers or descriptors in documents, or discrete features in patterns. The similarity of

A and B may be defined as the ratio of the number of their common elements to the number of all different elements; if $n(X)$ is the number of elements in set X, then the similarity is

$$S_T(A, B) = \frac{n(A \cap B)}{n(A \cup B)} = \frac{n(A \cap B)}{n(A) + n(B) - n(A \cap B)} \, . \tag{1.62}$$

Notice that if x and y above were binary vectors, with components $\in \{0, 1\}$ the value of which corresponds to the exclusion or inclusion of a particular element, respectively, then $(x, y), \|x\|$, and $\|y\|$ would be directly comparable with $n(A \cap B), n(A)$, and $n(B)$, correspondingly. Obviously (1.61) is an extension of (1.62) to the components of real-valued vectors.

The Tanimoto measure has been used with success in the evaluation of relevance between documents [1.8]; the descriptors can thereby be provided with individual weights. If α_{ik} is the weight assigned to the kth descriptor of the ith document, then the similarity of two documents denoted by x_i and x_j is obtained by defining

$$(x_i, x_j) = \sum_k a_{ik} a_{jk} = \alpha_{ij} \, , \tag{1.63}$$

and

$$S_T(x_i, x_j) = \frac{\alpha_{ij}}{\alpha_{ii} + \alpha_{jj} - \alpha_{ij}} \, .$$

Weighted Measures for Similarity. We shall not introduce probabilistic concepts until in Sect. 1.3; however, let it be mentioned already at this point that if ψ is a certain weight matrix, the inner product may be defined as

$$(x, y)_\psi = (x, \psi y) \, , \tag{1.64}$$

and the *distance* may be defined as

$$\rho_\psi(x, y) = \|x - y\|_\psi = \sqrt{(x - y)^T \psi (x - y)} \, . \tag{1.65}$$

In the theory of probability, the weighting matrix ψ is the *inverse of the covariance matrix* of x or y (cf. Sect. 1.3.1) and T denotes the transpose. This measure is then named the *Mahalanobis distance*.

Since ψ can then be shown to be symmetric and positive semidefinite, it can be expressed as $\psi = (\psi^{1/2})^T \psi^{1/2}$, and then x and y can be thought to be preprocessed by the transformations $x' = \psi^{1/2} x, y' = \psi^{1/2} y$. After that, comparison can be based on Euclidean measures (scalar product or distance) of x' and y'.

Unfortunately there are some drawbacks with this method: (i) In order to evaluate the covariance matrix for patterns of high dimensionality n, an immense number of samples ($\gg n^2$) may have to be collected. (ii) Computation of matrix-vector products is much heavier than formation of scalar products.

Comparison by Operations of Continuous-Valued Logic. The basic operations of multiple-valued logic were first introduced by *Lukasiewicz* [1.11] and *Post* [1.12], and later extensively utilized in the theory of fuzzy sets by *Zadeh* [1.13], as well as others. Here we shall adopt only a few concepts, believed to be amenable to fast and simple computation in comparison operations.

The application of continuous-valued logic to comparison operations is here based on the following reasoning. The "amount of information" carried by a scalar signal is assumed proportional to its difference from a certain "neutral" reference level. For instance, consider a continuous scale in which ξ_i and η_i belong to the interval $0 \ldots +1$. The signal value $1/2$ is assumed indeterminate, and the representation of information is regarded the more reliable or determinate the nearer it is to either 0 or +1. The *degree of matching* of scalars ξ_i and η_i is expressed as a generalization of *logical equivalence*. The usual definition of logical equivalence of Boolean variables a and b (denoted $a \equiv b$) is

$$(a \equiv b) = (\bar{a} \wedge \bar{b}) \vee (a \wedge b) , \tag{1.66}$$

where $\bar{a}$ is the *logical negation* of a, $\wedge$ means the *logical product*, and $\vee$ the *logical sum*, respectively. In continuous-valued logic, $(\wedge)$ is replaced by minimum selection (min), $(\vee)$ by maximum selection (max), and negation by complementation with respect to the scale, respectively. In this way, (1.66) is replaced by "equivalence" $e(\xi, \eta)$,

$$e(\xi, \eta) = \max\{\min(\xi, \eta), \min[(1 - \xi), (1 - \eta)]\} . \tag{1.67}$$

The next problem is how to combine the results $e(\xi_i, \eta_i)$. One possibility is to generalize the logical product using the operation $\min_i$. However, the effect of mismatching at a single element would thereby be fatal. Another possibility is the linear sum of $e(\xi_i, \eta_i)$ as in (1.59). A compromise would be to define the similarity of x and y with the aid of some function that is symmetric with respect to its arguments, e.g.,

$$S_M(x, y) = \varphi^{-1} \left\{ \sum_{i=1}^{n} \varphi[e(\xi_i, \eta_i)] \right\} , \tag{1.68}$$

where φ is some monotonic function, and φ^{-1} is its inverse. For instance,

$$S_M(x, y) = \left(\sum_{i=1}^{n} [e(\xi_i, \eta_i)]^p \right)^{1/p} , \tag{1.69}$$

with p some real value is one possibility. Notice that with $p = 1$ the linear sum is obtained, and with $p \to -\infty$, $S_M(x, y)$ will approach $\min_i[e(\xi_i, \eta_i)]$.

This method has two particular advantages when compared with, say, the correlation method: (i) Matching or mismatching of low signal values is taken into account. (ii) The operations max and min are computationally, by digital or analog means, much simpler than the formation of products needed in correlation methods. For this reason, too, the value $p = 1$ in (1.69) might be considered.

It has turned out in many applications that there is no big difference in the comparison results based on different similarity measures, and computational simplicity should therefore be taken into account next.

1.2.2 Measures of Similarity and Distance Between Symbol Strings

Words in a written text may be understood as patterns, too. Their elements are symbols from some alphabet, but such patterns can hardly be regarded as vectors in vector spaces. Other similar examples are *codes* of messages, which have been studied for a long time in information theory. We shall not discuss how these strings are formed; on the other hand, it may be understandable that they can be contaminated by errors that may have been generated in the conversion process itself, or in subsequent handling and transmission. In the statistical comparison of strings, the first task is to define some reasonable distance measure between them.

Hamming Distance. The most trivial measure of similarity, or in fact dissimilarity between coded representations is the *Hamming distance*. Originally this measure was defined for binary codes [1.14], but it is readily applicable to comparison of any *ordered sets* that consist of discrete-valued elements.

Consider two ordered sets x and y consisting of discrete, nonnumerical symbols such as the logical 0 and 1, or letters from the English alphabet. Their comparison for dissimilarity may be based on the *number of different symbols in them*. This number is known as the *Hamming distance* ρ_H that can be defined for sequences of equal length only: e.g.,

$$x = (1, 0, 1, 1, 1, 0)$$
$$y = (1, 1, 0, 1, 0, 1) \qquad \rho_H(x, y) = 4$$

and

$$u = (p, a, t, t, e, r, n)$$
$$v = (w, e, s, t, e, r, n) \qquad \rho_H(u, v) = 3 \ .$$

For binary patterns $x = (\xi_1, \ldots, \xi_n)$ and $y = (\eta_1, \ldots, \eta_n)$, assuming ξ_i and η_i as Boolean variables, the Hamming distance may be expressed formally as an arithmetical-logical operation:

$$\rho_H(x, y) = \text{bitcount} \ \{(\bar{\xi}_i \wedge \eta_i) \vee (\xi_i \wedge \bar{\eta}_i) | i = 1, \ldots, n\} \qquad (1.70)$$

where $\wedge$ is the logical product and $\vee$ is the logic sum (denoted OR); $\bar{\xi}$ is the logic negation of ξ. The function bitcount $\{S\}$ determines the number of elements in the set S that attain the logical value 1; the Boolean expression occurring in the set variables is the EXCLUSIVE OR (EXOR) function of ξ_i and η_i.

Distance Between Unordered Sets. The strings of symbols to be compared may have different lengths, whereupon the Hamming distance cannot be defined. Nonetheless it is possible to regard the strings as *unordered* sets of symbols. Consider two unordered sets A and B that consist of discrete elements such as symbols. Denote the number of elements in set S by $n(S)$. The following distance measure has been found to yield a simple and effective comparison measure between unordered sets:

$$\rho(A, B) = \max\{n(A), n(B)\} - n(A \cap B) . \tag{1.71}$$

This measure bears a certain degree of similarity to the Tanimoto similarity measure discussed above.

Distance Between Histograms of Symbols. An even better distance measure for strings of different lengths is based on their histograms over the occurring symbols. The histograms can be regarded as Euclidean vectors that have as many components as there are symbols in the alphabet. Such histogram vectors can be compared according to their Euclidean distance, dot product, or another vectorial similarity measure.

Levenshtein or Edit Distance. The statistically most accurate distance between strings of symbols is the *Levenshtein distance (LD)*, which for strings A and B is defined [1.15] as

$$LD(A, B) = \min\{a(i) + b(i) + c(i)\} . \tag{1.72}$$

Here B is obtained from A by $a(i)$ replacements, $b(i)$ insertions, and $c(i)$ deletions of a symbol. There exists an indefinite number of combinations $\{a(i), b(i), c(i)\}$ to do this, and the minimum is sought, e.g., by a *dynamic-programming* method as shown below.

Since the various types of error (change of a symbol to another, insertion of a new symbol, and deletion of a symbol) occur with different probabilities and depend on the occurring symbols, a statistical decision based on distance is more reliable if, in the distance measure, the editing operations are provided with different *weights*. This then leads to the *weighted Levenshtein distance (WLD)*:

$$WLD(A, B) = \min\{pa(i) + qb(i) + rc(i)\} \tag{1.73}$$

where the scalar coefficients p, q and r may be obtained from the so-called *confusion matrix* of the alphabet, or the inverse probability for a particular type of error to occur.

Notice that for the unweighted Levenshtein distance one can take

$$p(A(i), B(j)) = \begin{array}{ll} 0 & \text{if } A(i) = B(j) , \\ 1 & \text{if } A(i) \neq B(j) , \end{array}$$
$$q(B(j)) = 1 ,$$
$$r(A(i)) = 1 ,$$

where A(i) is the ith symbol of string A, and B(j) is the jth symbol of string B. Table 1.1 shows the dynamic-programming algorithm for the computation of WLD.

Table 1.1. Algorithm for the computation of *WLD*

---

begin
$D(0,0)$: $= 0$;
for i: $= 1$ step 1 until length (A) do $D(i,0)$: $= D(i-1,0) + r(A(i))$;
for j: $= 1$ step 1 until length B do $D(0,j)$: $= D(0,j-1)+q(B(j))$;
for i: $= 1$ step 1 until length A do
    for j: $= 1$ step 1 it until length (B) do begin
        $m1$: $= D(i-1,j-1) + p(A(i), B(j))$;
        $m2$: $= D(i,j-1)+q(B(j))$;
        $m3$: $= D(i-1,j) + r(A)(i))$;
        $D(i,j) = \min(m1, m2, m3)$;
end
WLD: $= D(\text{length}(A)), \text{length}(B))$;
end

Maximum Posterior Probability Distance. Closely related to *WLD* is the *maximum posterior probability distance (MPR)* that refers to the most probable sequence of events in which A is changed into B [1.16]. It is defined as

$$MPR(A, B) = \text{prob}\{B|A\} .$$

If the errors can be assumed to occur independently, a dynamic-programming method resembling that of *WLD* (Table 1.2) is applicable to the computation of *MPR*, too.

Table 1.2. Algorithm for the computation of *MPR*

---

begin
$D(0,0)$: $= 1.0$;
for i: $= 1$ step 1 until length (A) do $D(i,0)$: $= D(i-1,0) * r(A(i))$;
for j: $= 1$ step 1 until length B do $D(0,j)$: $= D(0,j-1) * q(B(j))$;
for i: $= 1$ step 1 until length A do
    for j : $= 1$ step 1 it until length (B) do begin
        $m1$: $= D(i-1,j-1) * p(A(i), B(j))$;
        $m2$: $= D(i,j-1) * q(B(j))$;
        $m3$: $= D(i-1,j) * r(A)(i))$;
        $D(i,j) = m1 +m2 + m3$
end
MPR: $= D(\text{length}(A)), \text{length}(B))$;
end

Here p, q and r are now the probabilities for a particular type of error (replacement, insertion, and deletion, respectively) to occur.

The Basic Hash-Coding Method. In the early 1950s, when the first IBM compilers were made, it was found that an item could be located in a normal addressed computer memory almost directly on the basis of its contents, whereupon the number of accesses to the memory is not significantly greater than one on the average. There was no need to order contents in any way, as the case is with most of the other searching methods. This idea, named *hash-coding* or *hash-addressing*, has been used extensively in compiling techniques (for the management of symbol tables) and in data base management. For a textbook account of these methods, cf. [1.17]. We shall see below that hash coding is an operation that can be used effectively for the shortcut comparison of given symbol strings with reference strings stored in memory.

The basic idea in hash coding is to determine the address of the stored item as some simple arithmetic function of its contents. If the record is identifiable by a single descriptor or phrase, called the *keyword*, the literal form of the keyword may be regarded as an integer in some number base. For instance, if the letters A, B, ..., Z were put to correspond to digits in the base of 26, the keyword, e.g., DOG, would correspond to the number $a = \text{D} \cdot 26^2 + \text{O} \cdot 26 + \text{G} = 2398$, where D $= 3$, O $= 14$, and G $= 6$.

The first problem encountered is that the address space of all legitimate keywords may be very large and occupied sparsely. There exist certain "perfect" or "optimal" functions that directly map the keywords in a more compressed way, but they are computationally rather heavy. As will be seen, this problem can be solved by other, simpler methods. The first step is to map the above arithmetic expression onto the space of allocated addresses using a *hashing function*, which generates a pseudorandom number $h = h(a)$, named *hash address*. The *division algorithm* is one of the simplest randomizers and also very efficient. Assume that the allocated addresses range from b to $n + b - 1$; for the hashing function one may then take

$$h = (a \bmod n) + b \, . \tag{1.74}$$

In order to create good pseudorandom numbers, n ought to be a prime. The most important property of function h is that it is supposed to distribute the hash addresses (often named "home addresses") as uniformly as possible over the allocated memory area, the *hash table*.

The next problem is that since the hash addresses are computed as residues, two different keywords may have equal hash addresses, and this occasion is named a *collision* or *conflict*. One step toward the solution is to partition the memory into groups of contiguous locations, called *buckets*, and to address each bucket, i.e., its first location by the hash address. A fixed number of spare slots is available for colliding items in each bucket. This organization is frequently used when the hash tables are held in backup memories, e.g., disk units, where a whole sector can be reserved for the bucket. (Data transfers are usually made by sectors.) Even the buckets, however, may overflow, and some *reserve locations* for the colliding items must be assigned. They must be derivable from the home address in a simple way. One frequently used method is to allocate reserve locations in a separate memory area, the addresses of which are defined using *address pointers* stored as the last items in buckets. The reserve locations can form a single linear list, because their contents are directly found from the place indicated by the pointer.

Another method for handling collisions is to use empty locations of the original hash table as reserve locations for the home addresses. No buckets need thereby be used; every location in the hash table corresponds to one possible hash address. The simplest choice is to store every colliding item in the first vacant location following the home address in numerical order (Fig. 1.4). As long as the hash table is no more than, say, half full, the reserve location is found rather quickly, in a few successive probings.

A new problem thereby arising is how the right item could be *identified* during reading. Because different items may have the same calculated hash address, it is obviously necessary to store a copy of the keyword, or some shorter unique identifier of it, in each location. When searching for a keyword, the home address and its reserve locations must be scanned sequentially until the external search argument agrees with the stored keyword or its identifier. If the item does not exist in the table, no matches are found in the identifier field.

Many sophisticated improvements to the basic hash coding method have been developed. For instance, one may use shortcut methods to check whether the item is in the table or not. The reserve locations may be chained to the home address with pointers for faster reference. It is also important to note that only the keyword is encoded in the hash table; the associated records, especially if they are longer can be kept in a separate memory area, which then may be utilized more efficiently. It is only necessary to set pointers to the associated records in the hash table. This arrangement is called *hash index table*.

Another important problem concerns multiple keywords in the hash coding method. Let it suffice to refer to [1.17] for its handling.

Using hash coding, it is then possible to locate all those strings in memory that *match exactly* with the given (key) string.

Redundant Hash Addressing (RHA). It seems necessary to point out that contrary to a common belief, searching on the basis of *incomplete* or *erroneous* keywords, thereby taking into account *similarity measures* as explained in Sect. 1.2.2, is also readily amenable to software methods that use hash coding [1.17, 18] . This solution, which is little known, can cope efficiently with the hardware methods, and being readily accessible to a common user of computers, ought to draw considerable attention. It may also be applied for to implement an effective encoding of neural networks [1.19].

The central idea in this method, named *redundant hash addressing (RHA)*, is to extract *multiple features* from the same piece of text (or other structured

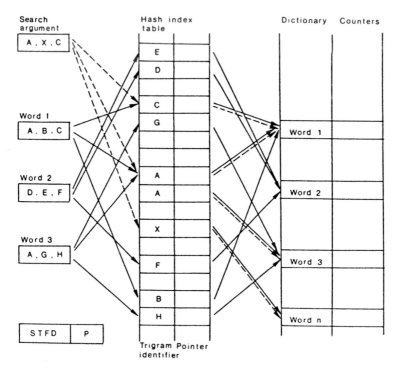

Fig. 1.4. Illustration of the RHA method. (A, B, ..., X): features. Notice that there is a collision of feature A, and the next empty address in the hash index table is used as reserve location

description), e.g, from the keyword by which the associated text is encoded. The original record or document may reside in a separate memory area, named the *dictionary*, and addressed by pointers stored in the hash index table. Unlike in the basic hash coding method, *several hash addresses* are now derived from the *same* keyword, whereas all of the corresponding home addresses contain an identical pointer to the stored record (Fig. 1.4). The handling of collisions and other organizational solutions of the hash table can be done the same way as in the basic method.

This procedure normally tolerates all the usual types of string errors. Let groups of contiguous characters (called *n-grams*) be used as local features of the text. There is no significant difference between bigrams ($n = 2$) and trigrams ($n = 3$) used as the features. For each *n*-gram, *a separate hash address is now computed*. In addition, the *within-word coordinate* of the *n*-gram must be stored in the hash table because, for matching, the relative displacement of the *n*-gram in the search argument and the stored word may be shifted by a few character positions, due to insertion and deletion errors, and one has to check whether this shift is within allowed limits.

If the search argument is erroneous but contains at least two correct n-grams within the specified displacement, it is highly probable that the majority of the pointers (of matching features) found from the hash table point to the correct item, whereas the erroneous n-grams give rise to random stray pointers to wrong items. Obviously some kind of voting procedure is necessary. The simplest method is to collect all pointers found during the searching process into a list and to find their majority by posterior counting, whereupon the word lengths can also be taken into account in computing the best match. It has been shown by extensive tests that a rather useful measure for practical searching of "nearest" strings is given by the *feature distance (FD)*, which is related to the expression (1.71) and needs very little computation [1.17, 18]. The original string is first represented by all its n-grams, i.e., consecutive groups of n symbols (which overlap). For instance, the bigrams of the word 'table' are 'ta', 'ab', 'bl', and 'le'. If the ends of symbol strings are included in the definition and indicated by the blank symbol '_', the bigrams '_t' and 'e_' may be added. Assume now that the matching of a "local feature" in the words A and B means that two similar n-grams are found in symbol positions with a relative displacement of at most d positions. Further denote the number of such elementary matches by n_d, and let n_A, n_B be the numbers of n-grams in A and B, respectively. The feature distance, relating to the allowed displacement d, is then defined as

$$FD_d(A, B) = \max(n_A, n_B) - n_d . \tag{1.75}$$

For example, let $A = $ '_table_' and $B = $ '_tale_'. The matching bigrams, for $d = 1$ are '_t', 'ta', 'le', and 'e_'. Then $FD_1(A, B) = \max(6, 5) - 4 = 2$.

If the keywords or phrases are long and if the number of stored items and, thus, collisions is fairly large, it is quicker to carry out voting by an "auction" method during the search. Each word in the dictionary is then provided with a *counter field* (Fig. 1.4) that accumulates the number of pointers to this item found so far during the searching operation. In addition, two auxiliary variables, *the smallest tentative feature distance (STFD)* and the pointer of the corresponding location *(P)*, respectively, are updated continuously; if the tentative FD, computed from the count number (plus one) at the location corresponding to the present pointer, and taking into account the word length according to (1.75) is smaller than the current value in $STFD$, the new FD is copied into $STFD$ and the corresponding pointer into P. When all the n-grams of the keyword have been operated, the pointer of the best-matching item is left in P.

Notice that because the errors are always generated in a stochastic process, the correct item cannot be identified with hundred per cent certainty; there is always a possibility for the keyword being changed into another legitimate one by the errors.

We have used the *RHA* method with much success in lexical access associated with automatic speech recognition, whereupon the phonetic transcriptions constructed by the equipment contain a certain number of stochastic errors. Another extensively studied application has been encoding sentences into a semantic structure where the nodes (words) are located by *RHA*.

It may be necessary to remark that the *RHA* method can also be used for prescreening a number of the best candidate items, whereupon some more accurate string similarity method, e.g., the *LD*, or some probabilistic analysis, can be applied to the final comparison of the best candidates.

1.2.3 Averages Over Nonvectorial Variables

It is clear that the arithmetic operations do not apply to nonvectorial items such as strings of symbols; notice that one of the problems in the comparison of the strings is their variable length. However, it is always possible to define the "middlemost" member or members in any set, if some distance function is definable for all pairs of its members, and at least some kind of *average* of the members is then computable [1.20].

Assume a fundamental set $\mathcal{S}$ of some items $x(i)$ and let $d[x(i), x(j)]$ be some distance measure between $x(i), x(j) \in \mathcal{S}$. The *set median m* over $\mathcal{S}$ shall minimize the expression

$$\mathcal{D} = \sum_{x(i) \in \mathcal{S}} d[x(i), m] \,. \tag{1.76}$$

The reason for calling m the median is that it is relatively easy to show that the usual (set) median of real numbers is defined by (1.76), if $d[x(i), x(j)] = |x(i) - x(j)|$. If, namely, $d[x(i), m] = |x(i) - m|$, if m is tentatively regarded as a free variable, and if the gradient of $\mathcal{D}$ with respect to m is taken, the latter consists of the sum of the signs of $[x(i) - m]$, which attains a value closest to zero if the number of the $x(i)$ on both sides of m is as closely the same as possible.

Above, we assumed that m belongs to the fundamental set $\mathcal{S}$. However, it may often be possible to find a hypothetical item m such that $\mathcal{D}$ attains its absolute minimum value. In contrast to the set median, we shall then use the term *generalized median* to denote the value of m that gives the absolutely minimum value for $\mathcal{D}$.

Let us now restrict ourselves to *strings of discrete symbols*.

Let us recall that the three basic types of error that may occur in strings are: (1) replacement, (2) insertion, (3) deletion of a symbol. (Interchange of two consecutive symbols can be reduced to two of the above basic operations in many ways.) An insertion or deletion error changes the relative position of all symbols to the right of it, whereupon, e.g., the Hamming distance is not applicable. We discussed in Sect. 1.2.2 two categories of distance measures that in an easy way take into account the "warping" of strings: (1) *Dynamic programming*, by which the Levenshtein distance or the weighted Levenshtein distance, i.e. the minimum (weighted) number of editing operations (replacements, insertions, and deletions of symbols) needed to change one string into another is computed; (2) Comparison of strings by their *local features*, e.g., substrings of N consecutive symbols (N-grams), whereupon the respective local features are said to match only if their relative position in the two strings differs in no more than, say, p positions; the distance of the strings is then a function of their lengths and the matching score.

The set median is found easily, by computing all the mutual distances between the given strings, and searching for the string that has the minimum sum of distances from the other elements. The generalized median is then found by systematically varying each of the symbol positions of the set median, making artificial string errors of all the basic types (replacement, insertion, and deletion of a symbol), over the whole alphabet, and accepting the variation if the sum of distances from the other elements is decreased. The computing time is usually quite modest; in the following examples, the generalized median was found in one or a few cycles of variation.

Examples of Medians and Generalized Medians of Garbled Strings. Table 1.3 gives two typical examples of sets of erroneous strings produced by a random-number-controlled choice of the errors. The probabilities for each type of error to occur were thereby equal. The set medians, mean and the generalized medians have been computed using the Levenshtein distance and the bigram feature distance method, respectively.

One should not draw too far-reaching conclusions from this couple of examples, but it can be stated quite generally that the most accurate distance measure for symbol strings is the weighted Levenshtein distance; in this example the unweighted LD was used, since the probabilities for different errors were assumed identical.

1.3 Statistical Pattern Analysis

1.3.1 Basic Probabilistic Concepts

Let us now proceed from deterministic to *stochastic* vector space concepts: most natural occurrences are random, with some probability values assigned to them. The observations, often called *samples*, are *distributed* over some domain of values that is called *support* or *manifold*. The samples may attain either discrete or continuously distributed values. In the latter case the samples are said to have a *density function*.

Probability Density Functions. In the mathematical theory of probability, the density function is usually assumed to have some analytical or at least otherwise theoretically definable form, and it is then called the *probability density function (pdf)*. One has to realize that the pdf and the observed distribution are not identical concepts, and must not be confused. In general, the values of the probabilistic functions can usually only be *approximated* by means of sets of samples.

Table 1.3. Set medians and generalized medians of garbled strings.
LD: Levenshtein distance; *FD*: feature distance

Correct string: **MEAN**
Garbled versions (50 per cent errors):

1. MAN	6. EN
2. QPAPK	7. MEHTAN
3. TMEAN	8. MEAN
4. MFBJN	9. ZUAN
5. EOMAN	10. MEAN

Set median *(LD)*:	**MEAN**
Generalized median *(LD)*:	**MEAN**
Set median *(FD)*:	**MEAN**
Generalized median *(FD)*:	**MEAN**

Correct string: **HELSINKI**
Garbled versions (50 per cent errors):

1. HLSQPKPK	6. HOELSVVKIG
2. THELSIFBJI	7. HELSSINI
3. EOMLSNI	8. DHELSIRIWKJII
4. HEHTLSINKI	9. QHSELINI
5. ZULSINKI	10. EVSDNFCKVM

Set median *(LD)*:	HELSSINI
Generalized median *(LD)*:	**HELSINKI**
Set median *(FD)*:	HELSSINI
Generalized median *(FD)*:	HELSSINI

Assume first a process in which only a finite number of distinct events $X_1, X_2, \ldots, X_s$ may occur. The relative frequencies of their occurrences are denoted by $P(X_k), k = 1, 2, \ldots, s$, where $P(\cdot)$ may be identified with the usual *probability* value. On the other hand, if x is a continuous-valued stochastic variable, then the differential probability for x falling in an interval dV_x in the x space and divided by dV_x is called the probability density function and denoted $p(x)$. If the domain of x, being a subset of the Euclidean space $\Re^n$, is divided into volume differentials dV_x, then, obviously, $p(x)dV_x$ is the usual probability for the occurrence of a value x within this differential volume.

For brevity, everywhere in the sequel, dV_x is now written dx, although this is actually an incorrect usage (coming from theoretical physics); note that dx would mean a vectorial differential, whereas it is intended to mean a volume differential!

In the following, probabilities for discrete events are denoted by the capital letter $P(\cdot)$, and the probability for X on the condition that Y occurs is $P(X|Y)$. The probability densities are denoted by lower-case letters p, and the probability density of variable x on the condition that Y occurs is $p(x|Y)$.

It should be noted that quite consistently in this convention $P(Y|x)$ is the probability for Y as a function of x, i.e., on the condition that the continuous variable x attains a certain value.

Expectation Value, Correlation Matrix and Covariance Matrix. The statistical average, or the *expectation value* of a scalar or vectorial variable x is denoted $\bar{x}$ and defined as

$$\bar{x} = \int xp(x)dx \stackrel{\text{def.}}{=} \text{E}\{x\} . \tag{1.77}$$

The concepts "correlation" and "correlation coefficient" in Sect. 1.2.1 were still used in a limited, deterministic sense. They were just computed from sequences of values, without any stochastic properties attributed to the variables. As a matter of fact, we do not need those concepts any longer in this book.

The second statistical function we define is the *correlation matrix* C_{xx}, which for stochastic vector $x \in \Re^n$ reads

$$C_{xx} = \int xx^\text{T}p(x)dx . \tag{1.78}$$

If a sufficiently large set of samples of x, denoted $\{x(t) \mid t = 1, 2, \ldots, N\}$ is available, the correlation matrix can be approximated by

$$C_{xx} \approx \frac{1}{N} \sum_{t=1}^{N} x(t)x^\text{T}(t) . \tag{1.79}$$

The probabilistic discussions are often simplified, without loss of generality, if the considerations are carried out in a coordinate system where the expectation value of each stochastic variable is zero. Instead of the correlation matrix one is then using the *covariance matrix* , usually denoted by Ψ and defined as

$$\Psi = \text{E}\{(x - \bar{x})(x - \bar{x})^\text{T}\} = \int (x - \bar{x})(x - \bar{x})^\text{T}p(x)dx . \tag{1.80}$$

Computation of the eigenvectors and eigenvalues of the correlation matrix or the covariance matrix are needed in many statistical problems. This task is simpler than for a general square matrix, because C_{xx} and Ψ are symmetric, and the methods introduced in Sect. 1.1.3 can then be used.

Principal Components. It was shown in Sect. 1.1.3 that every symmetric matrix can be expressed in terms of its orthonormal eigenvectors $u_k \in \Re^n$ and eigenvalues λ_k. Since C_{xx} is symmetric, we can write

$$C_{xx} = \sum_{k=1}^{n} \lambda_k u_k u_k^\text{T} , \tag{1.81}$$

whereupon the eigenvectors can be computed as in Sect. 1.1.3. The u_k come in handy for the approximation of an arbitrary vector $x' \in \Re^n$:

$$x' = \sum_{k=1}^{p} (u_k^{\mathrm{T}} x') u_k + \varepsilon \ , \tag{1.82}$$

where $p \leq n$ and ε is a residual that is minimized in the sense of least squares. The coefficients $u_k^{\mathrm{T}} x'$ are called the *principal components* of x', and by using an increasing number of the largest terms in (1.82), x' will be approximated by an increasing accuracy; if $p = n$, ε will become zero. This approximation will further be related to the *orthogonal projections* to be discussed in Sect. 1.3.2. Notice that (1.82) holds for *any* orthonormal basis $\{u_k\}$ of $\Re^n$.

The idea of using principal components for data analysis stems from the 1930s [1.21]. Their importance in data compression was realized in the 1950s [1.22]. *Principal-component analysis (PCA)* is even today widely used in data compression, and several neural-network approaches for it have been suggested [1.3, 23–25].

Factor Analysis. The purpose of PCA is to reduce the dimensionality of the observation vectors by trying to span the data in those dimensions where it has most variance: the eigenvector corresponding to the first principal component is oriented in the direction of the highest variance, the second eigenvector is oriented in the direction that is orthogonal to that of the first one and where the residual variance is highest, etc. (For a thorough discussion of the variance aspects, see, e.g., [1.3, 22].) Accordingly, in analogy with point mechanics, these directions are said to be those of the *principal axes*. Notice that the eigenvectors have the same dimensionality as $x \in \Re^n$, whereas their number, and that of the principal components is $m \leq n$. It is then possible to collect all the principal components into the vector

$$f = [(u_1^{\mathrm{T}} x), (u_2^{\mathrm{T}} x), \dots, (u_m^{\mathrm{T}} x)]^{\mathrm{T}} \in \Re^m \tag{1.83}$$

that is called the vector of the principal components and often also the *vector of the principal factors*. In image analysis and statistical pattern recognition it is also called the *feature vector*, although only very elementary features, viz. reduced-dimensionality components of x are thereby meant.

The name *factor* comes from an old analytical discipline that was developed in behavioral sciences, such as psychology and sociology, in order to explain the dependence between observed scalar variables in terms of their correlations. It was thought that the correlated fluctuations in the variables, say, $\xi_i, i = 1, \dots, n$ could be explained by corresponding fluctuations in a much smaller set of latent scalar variables, say, $\phi_j, j = 1, \dots, m, m < n$ that, however, are not directly observable. The ϕ_j are called *factors*. Of course, in order to be able to explain the observations, it was hoped that some factors would coincide with some of the real, meaningful observed variables, which could then be used as *determining variables* or *predictors*. As this is usually not fully possible, the next-to-best objective is to find those variables that have the highest correlation with the selected factors; the "weight" of a factor on an original observed variable is then called *factor loading*.

The basic assumption made in *factor analysis* is that in the first approximation the observables and the factors are linearly dependent:

$$\forall i, \quad \xi_i = \sum_{j=1}^{m} \alpha_{ij}\phi_j + \varepsilon_i , \tag{1.84}$$

where the α_{ij} are scalar coefficients and ε_i is a statistically independent residual error, in the simplest theoretical approach at least. In more refined analyses, the ε_i may be estimated in iterations, but if no a priori information about them is available, the ε_i may be assumed to be uncorrelated with the ϕ_j and mutually.

As the factor analysis has been used extensively in humanistic sciences and there has been much pressure to aim at meaningful explanations, the analytical procedures were later modified in many ways. Notwithstanding the basic philosophy and understanding of the method have remained rather vague. Therefore, it may be most proper to relate the factor analysis to the PCA, which was the first approach, indeed [1.21].

Let us put (1.84) into the form of a matrix-vector product and denote the estimate of x by $\hat{x}$:

$$\hat{x} = Af , \tag{1.85}$$

where A is the matrix formed of the α_{ij} and f the vector of the ϕ_j, respectively. *Now we have to determine mn unknown elements of A and m unknown elements of f simultaneously!* This is not possible without some extra strategy. If we only want to explain the variances, the most fundamental approach is to resort to the PCA. Assume tentatively that f can be identified with the vector of the principal factors, and recall that

$$\hat{x} = \sum_{k=1}^{m} (u_k^{\mathrm{T}} x) u_k \tag{1.86}$$

was the approximation of x used in the PCA. Now (1.86) can be put into the form

$$\hat{x} = [u_1, u_2, \ldots, u_m] \begin{bmatrix} (u_1^{\mathrm{T}} x) \\ (u_2^{\mathrm{T}} x) \\ \vdots \\ (u_m^{\mathrm{T}} x) \end{bmatrix} , \tag{1.87}$$

where the first member of the product is the matrix with the u_k as its columns. *This expression has already the form of (1.85),* while constituting a solution to the variance problem. Therefore, it is an *acceptable* approach to the factor analysis.

It is possible, however, that other optimal solutions exist as well: as a matter of fact, it can be shown that the same minimum-variance conditions are met although the vector f is *rotated orthogonally* in $\Re^m$. Thus we need not be interested in the rank order of the principal factors, like in the PCA. One can also choose a particular rotation such that the axes satisfy certain criteria of "simple structure," but we shall not discuss such or other refinements in this book.

In order to describe the significance of an observable ξ_i, i.e., the correlation between the factors and the variable, one can compute its *factor loadings*: in the PCA, the loading of factor ϕ_j on variable ξ_i is simply α_{ij}. The so-called *communality* of all factors on ξ_i is defined as

$$\text{communality of the factors on } \xi_i : \sum_{j=1}^{m} \alpha_{ij}^2 \ . \tag{1.88}$$

1.3.2 Projection Methods

The objective of *exploratory data analysis* is to produce simplified descriptions and summaries of large data sets. Clustering (which we shall discuss in Sect. 1.3.4) is one standard method; another alternative is to *project* high-dimensional data sets as *points* on a low-dimensional, usually 2D display.

The purpose of the projection methods is to reduce the dimensionality of the data vectors. The projections represent the input data items in a lower-dimensional space in such a way that the clusters and the metric relations of the data items are preserved as faithfully as possible. Projections can also be used to *visualize* the data sets if a sufficiently small dimensionality, such as one, two, or three, for the output display is chosen.

Linear Projection. Before proceeding further, we have to recall the concept of the *linear subspace* from Sect. 1.1.1. In the three-dimensional Euclidean space such a subspace can easily be illustrated as being a plane (two-dimensional) or a line (one-dimensional). Each vector in an m-dimensional linear subspace $(m \leq n)$ is a *linear combination* of m independently selected *basis vectors*.

The principal-component analysis displays high-dimensional data items as a linear projection consisting of points on a much smaller-dimensional subspace such that the variance of the original data is preserved as well as possible. Linear projection simply means that each component of the projected vector is a linear combination of the components of the original data item; the projection is formed multiplying each component by a certain scalar coefficient and summing up the results. This is formally a matrix-vector product, and the coefficients form the *projection matrix*. A demonstration of PCA is presented in Fig. 1.5.

Multidimensional Scaling. If the data set is high-dimensional and its distribution is highly unsymmetric or otherwise structured, it may be difficult to visualize the structures of its distribution using linear projections. Several nonlinear methods have been introduced for reproducing high- dimensional structures of data distributions on a low-dimensional display. The most common idea is to find a mapping such that the *distances between the image points* of the data items would be as similar as possible to the distances of the corresponding data items in the original metric space. The various methods only differ in how the different distances are weighted in optimization. No PCA theory is thereby needed.

Multidimensional scaling (MDS) signifies a class of methods used widely in behavioral, econometric, and social sciences. There exists a multitude of variants of MDS with slightly different goals and optimization algorithms [1.27–33]. The methods have been later generalized for analyzing nonmetric data, for which only the order of the distances between the data items is important. A matrix consisting of the pairwise dissimilarities of the entities is assumed to be available.

Sammon's Mapping. A widely used basic nonlinear projection that belongs to the so-called metric MDS methods is *Sammon's mapping* [1.34] that tries to match the pairwise distances of the lower-dimensional representations of the data items with their original distances. The special feature in Sammon's mapping is that the errors are divided by the distances in the original data space, which emphasizes small distances. The cost function in Sammon's mapping is (omitting a constant normalizing factor)

$$E_S = \sum_{K \neq l} \frac{[d(k,l) - d'(k,l)]^2}{d(k,l)} . \tag{1.89}$$

A practical algorithm for the computation of Sammon's mapping can be derived in the following way.

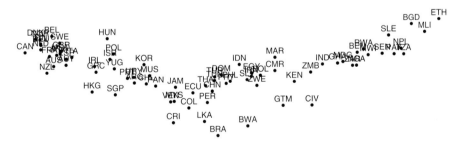

Fig. 1.5. The data set used in this illustration consisted of statistical indicators of 77 countries picked up from the World Development Report published by the World Bank [1.26]. Each component of the 39-dimensional data vectors describes a different aspect of the welfare and poverty of one country. Missing data values were neglected when computing the principal components, and zeroed when forming the projections. The three-letter codes for the countries, abbreviations of their names, may be self-explanatory. The data set was projected linearly as points onto the two-dimensional linear subspace obtained with PCA [1.27]

Consider a finite set of vectorial or even nonvectorial samples $\{x_i\}$. Let $d_{ij} = d(x_i, x_j)$ be any defined distance between x_i and x_j. The *distance matrix* is defined by its elements d_{ij}. Let $r_i \in \Re^2$ be the location or coordinate vector of the *image* of x_i on a display plane. The principle is to place the $r_i, i = 1, 2, \ldots, k$, onto the plane in such a way that all their mutual Euclidean distances $\|r_i - r_j\|$ have as closely as possible the corresponding values d_{ij}. Clearly this can only be made approximately.

To approximate the d_{ij} by the $\|r_i - r_j\|$ one may make iterative corrections. For instance, the elements $d_{ij}, i \neq j$ may be drawn from the distance matrix at random, and the *corrections* may be defined (heuristically) as

$$
\begin{aligned}
\Delta r_i &= \lambda \cdot \frac{(d_{ij} - \|r_i - r_j\|)}{\|r_i - r_j\|} \cdot (r_i - r_j) , \\
\Delta r_j &= -\Delta r_i .
\end{aligned}
\tag{1.90}
$$

Here $0 < \lambda < 1$, and λ can be made to decrease monotonically to zero during the process. An appreciable number of corrections, at least on the order of 10^4 to 10^5 times the number of x_i samples, is needed for acceptable statistical accuracy.

Although (1.90) already works reasonably well, it may be more advisable to use the following mathematically strict derivation suggested in [1.34].

Consider the objective or error function E_s:

$$
E_s = \frac{1}{\sum_i \sum_{j>i} d_{ij}} \sum_i \sum_{j>i} \frac{(d_{ij} - \|r_i - r_j\|)^2}{d_{ij}} .
\tag{1.91}
$$

The basic idea is to adjust the r_i and r_j vectors on the $\Re^2$ plane so as to minimize the objective function E_s; this leads to a configuration of the r_i points such that the clustering properties of the x_i data are visually discernible from this configuration.

To minimize E_s by a steepest-descent iterative process, it is advisable to write the optimizing equations in component form. (Our notations are slightly different from [1.34].) Denote now

$$
\begin{aligned}
c &= \sum_i \sum_{j>i} d_{ij} ; \\
d'_{pj} &= \|r_p - r_j\| ; \\
r_p &= [y_{p1}, y_{p2}]^{\mathrm{T}} .
\end{aligned}
\tag{1.92}
$$

Then, denoting the iteration index by m, we have

$$
y_{pq}(m+1) = y_{pq}(m) - \alpha \cdot \frac{\dfrac{\partial E_s(m)}{\partial y_{pq}(m)}}{\left| \dfrac{\partial^2 E_s(m)}{\partial y_{pq}^2(m)} \right|}
\tag{1.93}
$$

where again

$$\frac{\partial E_s}{\partial y_{pq}} = -\frac{2}{c}\sum_j \sum_{p \neq j} \left(\frac{d_{pj} - d'_{pj}}{d_{pj}d'_{pj}} \right) (y_{pq} - y_{jq}) ,$$

$$\frac{\partial^2 E_s}{\partial y^2_{pq}} = -\frac{2}{c}\sum_j \sum_{p \neq j} \frac{1}{d_{pj}d'_{pj}} \tag{1.94}$$

$$\cdot \left[(d_{pj} - d'_{pj}) - \frac{(y_{pq} - y_{jq})^2}{d'_{pj}} \left(1 + \frac{d_{pj} - d'_{pj}}{d'_{pj}} \right) \right] .$$

The iteration gain parameter α may attain the value 0.3 to 0.4, and it is called the "magic factor", obviously because this value, found completely experimentally, guarantees a fairly good convergence.

Sammon's mapping is particularly useful for preliminary analysis in all statistical pattern recognition, because it can be made to roughly visualize class distributions, especially the degree of their overlap. It is always recommended for a preliminary test of the data to be used for Self-Organizing Maps.

A demonstration of Sammon's mapping is presented in Fig. 1.6. Clearly it spreads the data in a more illustrative way than PCA.

Nonmetric Multidimensional Scaling. Preservation of the original metric distances in the projection may not always be the best goal, however. The *nonmetric MDS* [1.35, 36] introduces a monotonically increasing function f of the original metric distances that tries to preserve the *rank order* of the mapped data in the projection, while allowing deviations in the exact reproduction of the distances. The cost function then reads

Fig. 1.6. Sammon's mapping of the same data set that was projected using the PCA in Fig. 1.5 [1.27]

$$E_N = \frac{1}{\sum_{k \neq l}[d'(k,l)]^2} \sum_{k \neq l}[f(d(k,l)) - d'(k,l)]^2 , \qquad (1.95)$$

where $d(k,l)$ is the distance between the original data items indexed by k and l, respectively, and $d'(k,l)$ is the distance in the two-dimensional space between the corresponding image points. The function $f(\cdot)$ is optimized adaptively, during computation, for each mapping [1.30]. A demonstration of nonmetric MDS, applied to a dimensionality-reduction task, is given in Fig 1.7.

Principal Curves and Curvilinear Component Analysis. The PCA can be generalized in the nonlinear domains as forming nonlinear manifolds. While in the PCA a good projection of a data set onto a linear manifold, a "hyperplane," was constructed, the goal in constructing a principal curve or a principal surface is to project the set onto a nonlinear manifold, a curved "hypersurface." The *principal curves* [1.38] are smooth curves that are defined by the property that each point of the curve is the average of all data points that project to it, i.e., for which that point is the closest point on the curve. The principal curves are generalizations of principal components extracted using PCA in the sense that a linear principal curve is a principal component.

The conception of continuous principal curves may aid in understanding how principal components could be generalized. The algorithm of Hastie and Stuetzle [1.38] proposed for finding discretized principal curves resembles the batch version of the Self-Organizing Map (SOM) algorithm that will be discussed in Sect. 3.6 of this book, although the details are different.

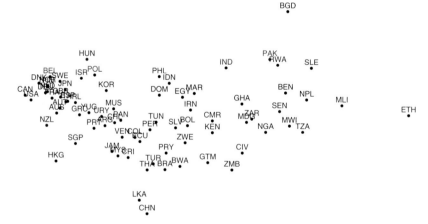

Fig. 1.7. A nonlinear projection constructed using nonmetric MDS [1.27] The data set is the same as in Fig. 1.5 and Fig. 1.6 Missing data values were treated by the following simple method which has been demonstrated to produce good results at least in the pattern recognition context [1.37]. When computing the distance between a pair of data items, only the (squared) differences between component values that are available are computed. The rest of the differences are then set to the average of the computed differences

To be useful in practical computations, however, the curves must always be discretized. It has turned out [1.39, 40] that discretized principal curves are then essentially equivalent to the SOMs, introduced long before Hastie and Stuetzle introduced the principal curves. It thus seems that the conception of principal curves is only useful in providing one possible viewpoint to the properties of the SOM algorithm.

There also exists another method related to the principal curves, namely, the *curvilinear component analysis* [1.41, 42]. The data items are first clustered and then the clusters are mapped with a fast algorithm in which the relative emphasis on the large and small distances, respectively, can be regulated.

The Self-Organizing Map (SOM) discussed in this book combines the clustering and projection operations. The most essential differences with respect to the other methods discussed above are: 1. The SOM represents an open set of multivariate items by a *finite set* of model items m_i, like in the K-means clustering, 2. The SOM represents the items in an orderly fashion on a two-dimensional regular grid, and the model items are then associated with the nodes of the grid, 3. It is not necessary to recompute the whole mapping for every new sample, because, if the statistics can be assumed as stationary, the new sample can be directly mapped onto the closest old model item.

1.3.3 Supervised Classification

Comparison Methods. The most trivial of all classification or pattern recognition methods is to compare the unknown pattern x with all known reference patterns x_k on the basis of some criterion for the degree of similarity, in order to decide to which class the pattern belongs. The opposite of similarity between two patterns x and x_k is their mutual distance $d(x, x_k)$ in some metric. For instance, the Euclidean metric is often quite useful, but any of the distance measures in Sect. 1.2.1 can be applied. Since natural patterns usually have a high dimensionality, it is obvious that plenty of computations are needed, unless a small representative set of reference patterns from each class can be used for comparison. In the *Nearest Neighbor (NN) method*, $d(x, x_k)$ is computed for all reference patterns and pattern x is classified according to the smallest value.

The K-Nearest-Neighbor (KNN) Method. The distributions of the reference patterns of different classes usually overlap, whereupon the simple *NN* method does not define the class borders reliably. The traditional method is to consider K nearest reference vectors to the sample to be classified, and to identify the sample according to the *majority* of classes encountered among the K nearest neighbors. It can be shown that the class borders thereby defined are very closely the same as the statistically optimal borders derived below according to the Bayesian theory of probability.

Classification of Patterns by the Angles. If the structures in patterns are defined by the *relative* intensities of the pattern elements, one might use the *"directions"* of pattern vectors in comparing their similarity. Because the patterned representations may have different norms, it is often reasonable to define for the degree of similarity of the key with stored patterns the *cosine of the angle* between these vectors, called the *direction cosine* (1.58): in n-dimensional space, the angle between vectors x and x_k is defined as

$$\cos\theta_k = \frac{(x, x_k)}{||x||\,||x_k||} . \tag{1.96}$$

Taking A Priori Information Into Account. The most trivial classification or pattern recognition method was to compare the unknown item with all the reference data, and to determine the best match according to some similarity or distance measure. The accuracy of classification is usually improved significantly, however, if all available a priori knowledge is utilized. In many cases there are good reasons to assume on the basis of physical or other properties of observations that their distribution has some familiar theoretical form; it is then possible to derive mathematical expressions for the discriminating surfaces that are optimal from the statistical point of view.

Pattern classification is a process that is related to decision making, detection theory, etc., and may be discussed in these settings. One of the motives for the following discussion is to point out that certain forms of separating surfaces, for instance, the linear and quadratic ones, may have a statistical justification.

Discriminant Function. For each class S_i one can define the so-called *discriminant function* $\delta_i(x)$, which is a continuous, scalar-valued function of the pattern vector x. If $x \in S_i$, then for all $j \neq i$ there shall hold $\delta_i(x) \geq \delta_j(x)$. The tie situation, with a continuous probability density function of x, is assumed to have the zero probability. There are infinitely many possible choices for the discriminant function: linear, quadratic, etc. In the following we shall define them by statistical arguments. The class border, or the separating surface between two classes i and j is defined in terms of the $\delta_i(x)$ as

$$\delta_i(x) - \delta_j(x) = 0. \tag{1.97}$$

Then

$$\delta_c(x) = \max_i\{\delta_i(x)\} \tag{1.98}$$

defines the class S_c of x.

Statistical Definition of the Discriminant Functions. Let the a priori probability for any observed pattern vector belonging to class S_k be denoted $P(S_k)$. Let the conditional density function of those pattern vectors that belong to class S_k be $p(x|S_k)$. In some rare cases, the mathematical form of $p(x|S_k)$ can be assumed. More often, however, it must be estimated on the basis of exemplary samples selected in preliminary studies.

The theories of decision processes are centrally based on the concept of a *loss function* or *cost function*, the value of which depends on factors set up heuristically, or after extensive experience, for instance, on the importance of detection of a particular occurrence, or on a miss to detect it. Let $C(S_i, S_k)$ be the unit cost of one decision that indicates the punishment of classification of a pattern to S_i when it was actually from S_k. If all patterns were considered equally important and the cost of a correct classification were denoted by 0, then all false classifications may be assumed to cause a unit cost. (In classification, only relative costs are significant.) Therefore, one simple case of unit cost function might be $C(S_i, S_k) = 1 - \delta_{ik}$. The *average conditional loss*, or the cost of classifying patterns into class S_k their actually being statistically distributed into classes $S_i, i = 1, 2, \ldots, m$ is then

$$L(x, S_k) = \sum_{i=1}^{m} P(S_i|x)C(S_k, S_i) , \tag{1.99}$$

where $P(S_i|x)$ is the probability of pattern with value x belonging to class S_i. This expression is related to $p(x|S_i)$, and may be thought to result from the fact that the density functions $p(x|S_i)$ may overlap in $\Re^n$, whereupon there is a definite probability for x belonging to any class. There is a fundamental identity in the theory of probability that states that if $P(X,Y)$ is the joint probability for the events X and Y to occur, then

$$P(X,Y) = P(X|Y)P(Y) = P(Y|X)P(X) . \tag{1.100}$$

This is the foundation of the so-called *Bayesian philosophy of probability* that deals with probabilities dependent on new observations. When applied to the present case, the above formula yields

$$L(x, S_k) = \sum_{i=1}^{m} p(x|S_i)P(S_i)C(S_k, S_i) . \tag{1.101}$$

It will now be seen that the negative of the loss function satisfies the requirements set on the discriminant functions: since the cost of classifying x into S_i any other than S_k is higher, $-L(x, S_k) = \max_i\{-L(x, S_i)\}$. On the other hand, when considering candidates for discriminant functions, any factors or additive terms in this expression that are common to all cases can be dropped. It may now be a good strategy to select the discriminant functions as

$$\delta_k(x) = \sum_{i=1}^{m} p(x|S_i)P(S_i)C(S_k, S_i) . \tag{1.102}$$

One basic case of discriminant functions is obtained by taking $C(S_k, S_i) = 1 - \delta_{ki}$, i.e., having the cost of correct classification zero and that of misclassification unity, respectively, whereby

$$\delta_k(x) = \sum_{i=1}^{m} p(x|S_i)P(S_i) + p(x|S_k)P(S_k) \ . \tag{1.103}$$

Since now all possible classes occur among the $S_i, i = 1, 2, \ldots, m$, then the identity

$$\sum_{i=1}^{m} p(x|S_i)P(S_i) = p(x) \tag{1.104}$$

holds, and since $p(x)$ is independent of k, this term can be dropped. For a new discriminant function it is then possible to redefine

$$\delta_k(x) = p(x|S_k)P(S_k) \ . \tag{1.105}$$

It should be realized that according to the definition of discriminant functions, the form of separating surface is not changed if any monotonically increasing function of $p(x|S_k)P(S_k)$ is selected for *every* $\delta_k(x)$.

An Example of Parametric Classification. Assume that the pattern vectors of every class have *normal distributions* described by multivariate (normalized) Gaussian probability density functions, but each class is characterized by different parameters:

$$p(x|S_k) = \frac{1}{\sqrt{|\Psi_k|}} \exp\left[-\frac{1}{2}(x - m_k)^{\mathrm{T}}\Psi_k^{-1}(x - m_k)\right] \ , \tag{1.106}$$

where m_k is the mean of x in class k, Ψ_k is the covariance matrix of x in class k, and $|\Psi_k|$ is its determinant. For the discriminant function, $2\log[p(x|S_k)P(S_k)]$ can now be selected:

$$\delta_k(x) = 2\log P(S_k) - \log|\Psi_k^{-1}| - (x - m_k)^{\mathrm{T}}\Psi_k^{-1}(x - m_k) \ , \tag{1.107}$$

which is a *quadratic function* of x. Consequently, the separating surfaces are *second-degree surfaces*, which are defined by a finite set of parameters. This method is therefore called *parametric classification*.

It may happen with a restricted set of samples that Ψ_k^{-1} does not exist. The so-called *Moore-Penrose pseudoinverse matrix* Ψ_k^{+} [1.1] may then be used instead of Ψ_k^{-1}.

For a special case, discriminant functions that have equal covariance matrices $\Psi_k = \Psi$ may be taken. This may be the case, e.g., with patterns derived from univariate time series. All terms independent of k are now dropped. Moreover, the symmetry of covariance matrices is utilized for simplification, whereby

$$\delta_k(x) = m_k^{\mathrm{T}}\Psi^{-1}x - \frac{1}{2}m_k^{\mathrm{T}}\Psi^{-1}m_k + \log P(S_k) \ . \tag{1.108}$$

Surprisingly this is *linear* in x, in spite of the density functions being described by nonlinear (Gaussian) forms.

The Robbins-Monro Stochastic Approximation. In many optimization problems including learning processes, the solution is defined as the set of parameter values that minimizes a given *performance index* or *objective function*, usually expressed as an integral. There are theoretical cases when such an optimum can be determined in closed form, whereas practical problems usually tend to be more difficult. As early as 1951 a practical method of approximate optimization, called the *stochastic approximation*, was devised by Robbins and Monro [1.43]. In a general sense this problem belongs to the discipline of mathematical statistics, and discussion of stochastic processes with mathematical rigor in an interdisciplinary book like this is very difficult. Therefore the central ideas of stochastic approximation are only presented and applied in this context in an exemplary case.

One of the earliest adaptive neural models was the *Adaline* of Widrow [1.44]. It is a nonlinear device, equipped with a thresholding stage at its output, but before that is operates linearly. It receives a set of real-valued scalar signals $\{\xi_j\}$, which together form the input signal vector $x = [\xi_1, \xi_2, \ldots, \xi_n]^T \in \Re^n$, and yields a scalar response η to them. Vector x is a stochastic variable. A linear dependence is assumed: $\eta = m^T x + \varepsilon$, where $m = [\mu_1, \mu_2, \ldots, \mu_n]^T \in \Re^n$ is the *parameter or weight vector*, and ε is a stochastic error with zero expectation value. (Unlike in the original Adaline, we do not consider any bias in the x values here. Addition of the bias to the input vector x is trivial.) The scalar-valued objective function J is defined as the average expected value of the quadratic error,

$$J = E\{(\eta - m^T x)^2\} , \tag{1.109}$$

where $E\{\cdot\}$ denotes the mathematical expectation over an infinite set of samples of x. If $p(x)$ denotes the *probability density function* of x, (1.109) can be written

$$J = \int (\eta - m^T x)^2 p(x) dx , \tag{1.110}$$

where dx is a differential volume element in the n-dimensional hyperspace of signal values over which the integral is taken. The problem is to find the value m^* for m that minimizes J; the condition for local extremality is

$$\nabla_m J = \left[\frac{\partial J}{\partial \mu_1}, \frac{\partial J}{\partial \mu_2}, \ldots, \frac{\partial J}{\partial \mu_n} \right]^T = 0 . \tag{1.111}$$

As the probability density function $p(x)$ may not be known in analytical form (for instance, when the distribution of samples of x is only experimentally known), we must approximate J using available sample values of x. Let us denote the *sample function* of J by $J_1(t)$:

$$J_1(t) = [\eta(t) - m^T(t) x(t)]^2 . \tag{1.112}$$

Here $t = 0, 1, 2, \ldots$ is the sample index, and $m(t)$ is an approximation (more or less good) of m at time t.

The idea of Robbins and Monro was to determine $m(t)$ at each step from a recursion, where the gradient of J is approximated by the gradient of $J_1(t)$ with respect to $m(t)$. It was shown that starting with arbitrary initial values $m(0)$, the sequence $\{m(t)\}$ converges to the neighborhood of the optimal vector m^*. This recursion is defined as a series of gradient steps

$$m^{\mathrm{T}}(t+1) = m^{\mathrm{T}}(t) - G_t \nabla_{m(t)} J_1(t) , \tag{1.113}$$

where G_t is a *gain matrix*. It was shown that a necessary condition for converge is G_t being positive definite. The matrix norm of G_t has some bounds, too. One important choice in practice is $G_t = \alpha(t)I$, where I is the identity matrix (this means steepest descent in the m space for small $\alpha(t)$), and $\alpha(t)$ is a scalar.

Equation (1.113) now reads

$$m^{\mathrm{T}}(t+1) = m^{\mathrm{T}}(t) + \alpha(t)[\eta(t) - m^{\mathrm{T}}(t)x(t)]x^{\mathrm{T}}(t) . \tag{1.114}$$

Robbins and Monro proved that this sequence converges with probability one toward the locally optimal value $m^{*\mathrm{T}}$ for which J is minimum, if and only if the following conditions are valid:

$$\sum_{t=0}^{\infty} \alpha(t) = \infty , \ \sum_{t=0}^{\infty} \alpha^2(t) < \infty . \tag{1.115}$$

For instance, $\alpha(t) = \mathrm{const.}/t$ satisfies these conditions. The following choice [1.45] has often been found to yield an even faster convergence:

$$\alpha(t) = \left(\sum_{p=0}^{t} ||x(p)||^2 \right)^{-1} . \tag{1.116}$$

It has to be mentioned that in this simple linear case the convergence limit is *exactly* $m^{*\mathrm{T}}$, whereas the same cannot be said generally of nonlinear processes. The convergence conditions (1.115) are generally valid, however, but there may exist several local minima of J, any of which is a possible convergence limit.

1.3.4 Unsupervised Classification

Assume that we do not know the classes *a priori*. If the samples nevertheless fall in a finite set of categories, say, according to their similarity relations, the problem is called *unsupervised classification*, and it can be formalized mathematically as follows. The *clustering methods* discussed in this section are able to discover such relationships between the items.

It may not be necessary to aim at a complete survey of clustering methods: let it suffice to mention some of the most central references [1.46–56].

Simple Clustering. The clustering problem may be set up, e.g., in the following way. Assume that $A = \{a_i; i = 1, 2, \ldots, n\}$ is a finite set of representations of items. This set has to be partitioned into disjoint subsets $A_j, j = 1, 2, \ldots, k$, such that with respect to this division, some functional that describes the "distance" between items attains an extremum value. This functional might describe the *quality of groupings* in the sense that the mutual distances between all a_i belonging to the same A_j are as small as possible while the distances between different A_j are as large as possible.

In the simplest case, for the distance between two vectorial items, the (weighted) Euclidean distance measure may be chosen; more general Minkowski metrics are frequently used, too. The functional describing the grouping may contain, e.g., sums of some powers (like squares) of the distances. Determination of the subsets A_j is a global optimization problem, whereupon a set of simultaneous algebraic equations describing the optimality criterion must be solved by direct or iterative methods.

Hierarchical Clustering (Taxonomy Approach). A significant reduction in computational load is achieved if the subsets A_j are determined in several sequential steps; then, however, the optimality criterion can be applied only at each individual step whereby the final grouping usually remains suboptimal. This kind of *hierarchical clustering*, on the other hand, is able to find generic relationships that exist between the resulting clusters; in classification, encoding, and retrieval of empirical data this advantage is often of great importance. There is one major application area in which hierarchical clustering is a standard method, namely, *numerical taxonomy*.

The main types of hierarchical clustering are the *splitting* (divisive, partitioning) methods, and the *merging* (agglomerative, coalescence) methods.

Splitting Methods. These are straightforward methods that are related to the simple approach. The set A is first partitioned into disjoint subsets A_1 and A_2, whereby some *interclass distance* $d = d(A_1, A_2)$ is maximized. There are many choices for d. One of the simplest is the one using Euclidean metrics in the representation space,

$$d = \sum_{j=1}^{k} n_j \|\bar{x}_j - \bar{x}\|^2 , \tag{1.117}$$

where n_j is the number of representations in A_j, $\bar{x}_j$ is the mean of A_j, and $\bar{x}$ is the mean of A, respectively; usually $k = 2$. The partitioning is continued for A_1 and A_2 to produce the subsets A_{11}, A_{12}, A_{21}, and A_{22}, and so on. A stopping rule (e.g., if d falls below a given limit) may be applied for to automatically prevent a further division of a subset. This method then directly produces a *binary tree structure*.

Instead of applying distance measures of the *representation space* directly, one might first construct a tree structure, the so-called *minimal spanning tree* (that links all representation "points" by the shortest path; cf. the next paragraph). This structure, by means of its topology, then defines the mutual distance of any two points in it. Using an arc-cutting algorithm [1.56] the tree can then be partitioned optimally.

Merging Methods. These are more common than the splitting methods. One starts with single-element subsets $A_i^{(0)} = \{a_i\}$. At every successive step (s), k most similar subsets $A_{j1}^{(s-1)}, \ldots, A_{jk}^{(s-1)}$ are merged into a larger subset $A_j^{(s)}$. Usually $k = 2$. It is to be noted that the following cases may occur: the closest subsets consist of (a) single elements, (b) one single element and another set with > 1 members, (c) two sets with > 1 members. In the merging method, the distance between two disjoint subsets A_i and A_j can conveniently be defined, e.g., as the minimum of $d(a_k, a_l)$, where $a_k \in A_i$ and $a_l \in A_j$ (so-called *single linkage*). In fact, this measure also directly defines the so-called *minimal spanning tree*, if a link (arc) between the nearest a_k and a_l is set at the same time when they are merged into a new subset.

Figure 1.8 illustrates the division of a set of points into two subsets, and their representation by the minimal spanning tree, respectively.

(a)   (b)

Fig. 1.8. Examples of clustering. **(a)** Division into subsets. **(b)** Minimal spanning tree

1.4 The Subspace Methods of Classification

1.4.1 The Basic Subspace Method

Subspaces As Classes. A less well known approach in statistical pattern recognition techniques is to model the data in pattern classes directly as analytically definable *manifolds*. For instance, in practice, many important transformation groups are automatically taken into account if the classes are put to correspond to *linear subspaces* [1.3, 57] . Thereby the class-affiliation of an unknown input pattern is expressed in terms of the *distance of its representation from such a subspace*. The subspace itself is defined by its *basis vectors*, for which one can select, for instance, representative sample vectors or some statistically defined variables as discussed below. The input pattern is classified according to its similarity with the *general linear combination of the basis vectors*. This makes a big difference with respect to direct comparison of unknown vectors with weight vectors, usual in most neural-network approaches.

Examples of this are encountered in the classification of various *spectra*. Consider a source of acoustic signals; in most cases the signals are produced by various modes of mechanical vibration. The resonant frequencies are fixed, but the intensities of the modes depend, among other things, on the exciting conditions that vary more or less stochastically. A template comparison of spectra based on some previous similarity measure (e.g., Hamming or Euclidean distance, direction cosines, or correlation) may therefore not be expected to produce particularly good results. On the other hand, a general linear combination of these modes of vibration might be regarded as a class. If we would thus describe each class by at least as many linearly independent prototype vectors (spectra) as there are vibratory modes, then it is expected that if a linear combination of the prototypes on the unknown vector is fitted to each class separately, and the residual with respect to a particular class is smaller than for the other classes, then the vector can be regarded to belong to that class. The *distance* of a vector from a class may be defined by the smallest residual, whereas no single prototype need be similar. This analysis can be formulated very simply in terms of vector space concepts as shown below.

The subspace method, however, is not restricted to spectral patterns: for instance, it has been used successfully for the classification of optic patterns on the basis of their statistical features. Examples are classification of textures [1.58, 59] and handwritten characters [1.60].

In pattern classification, the subspace methods were applied for the first time in the 1960s [1.57].

Let a subspace $\mathcal{L}$ of the real space $\Re^n$ be defined as the subset of vectors $x \in \Re^n$ expressible as the general linear combination of the basis vectors $(b_1, \ldots, b_K, K < n)$:

$$\mathcal{L} = \mathcal{L}(b_1, b_2, \ldots, b_K) = \left\{ x \middle| x = \sum_{k=1}^{K} \alpha_k b_k \right\} , \tag{1.118}$$

where the $\alpha_1 \ldots \alpha_K$ are arbitrary real scalars from the domain $(-\infty, +\infty)$.

The linear subspace is defined uniquely by its basis vectors, but the converse is not true: there exist infinitely many combinations of basis vectors that define the same $\mathcal{L}$. To make the computational subspace operations most economical, the basis vectors can be selected as *orthonormal*, i.e., mutually orthogonal and each of them having the norm of unity. In artificial models, such a new basis can be constructed for arbitrary basis vectors by, e.g., the *Gram-Schmidt method* discussed in Sect. 1.1.1.

Projection. The basic subspace operation is the *projection* discussed in Sect. 1.1.1. The closest point in $\mathcal{L} \subset \Re^n$ to any vector $x \in \Re^n$ is called the projection of x on $\mathcal{L}$ and denoted $\hat{x} \in \mathcal{L}$. Among all the decompositions of the type

$$x = y + z , \tag{1.119}$$

where $y \in \mathcal{L}$, the one with $y = \hat{x}$ was earlier found to be in a special position, because z is then orthogonal to all vectors of $\mathcal{L}$. This value of z is denoted $\tilde{x}$, and it is also said that $\tilde{x} \perp \mathcal{L}$. Since $||\tilde{x}||$ is the minimum of all $||z||$, it is defined to be the *distance* of x from $\mathcal{L}$. If we have constructed an orthonormal basis $(u_1, u_2, \ldots, u_K)$ for $\mathcal{L}$, where often K is much smaller than n, then as can be seen from the Gram-Schmidt process of Sect. 1.1.1 with $||u_k|| = 1$,

$$\hat{x} = \sum_{k=1}^{K} (u_k^{\mathrm{T}} x) u_k \quad \text{and} \quad \tilde{x} = x - \hat{x} \ . \tag{1.120}$$

If parallel computers or neural networks are used to implement matrix operations, it is often convenient to define the *orthogonal projection operator*

$$P = \sum_{k=1}^{K} u_k u_k^{\mathrm{T}} \ , \tag{1.121}$$

whereby

$$\hat{x} = Px \quad \text{and} \quad \tilde{x} = (I - P)x \ ; \tag{1.122}$$

here I means the identity matrix.

Classification. Consider now N classes of patterns denoted $S_1, \ldots, S_N$. Each class is represented by its own subspace $\mathcal{L}^{(i)}$ and basis vectors $b_1^{(i)}, \ldots,$ $b_{K(i)}^{(i)}$. Like in many basic statistical pattern recognition methods, x is classified according to its distances from the various classes, and, as said earlier, it is expedient to define the distances as the $||\tilde{x}^{(i)}||$, evaluated by decomposing the input vector x with respect to *each class* as

$$x = \hat{x}^{(i)} + \tilde{x}^{(i)} \ . \tag{1.123}$$

The class S_c to which x is assigned can be defined by

$$c = \arg\min_i \{||\tilde{x}^{(i)}||\} \quad \text{or}$$

$$c = \arg\max_i \{||\hat{x}^{(i)}||\} \ . \tag{1.124}$$

Notice now that in the subspace method one may define the discriminant function as

$$\delta_i(x) = ||\hat{x}^{(i)}||^2 = ||x||^2 - ||\tilde{x}^{(i)}||^2 \ . \tag{1.125}$$

It can be deduced from (1.97), (1.121), (1.122), and (1.125) that if we use projection operators (matrices) $P^{(i)}$ to represent the various $\mathcal{L}^{(i)}$, the boundaries between the classes S_i and S_j are also defined by the equations

$$x^{\mathrm{T}}(P^{(i)} - P^{(j)})x = 0 \ . \tag{1.126}$$

These are quadratic equations in x, and the boundaries are second-degree cones (cf. [1.3]).

Obviously the subspaces can be designed to minimize the classification errors, and this will be discussed in this section.

There have existed several earlier statistical and geometric methods for constructing optimized subspaces, whereupon the dimensions of the latter in general vary (e.g., [1.61, 62]).

Network Architecture of the Subspace Classifier. It is possible to think of a neural-network implementation of the subspace classifier consisting of two layers, as in Fig. 1.9, followed by some kind of *winner-take-all (WTA)* circuit that selects the maximum of its inputs. The input "neurons" are linear, and their weight vectors are identified with the basis vectors $b_h^{(i)}$, where i is the index of one of the modules or units separated by the dashed lines and h is the index on an input neuron within that unit. The transfer function of each output neuron has some quadratic form (Q). In practical computations we tend to keep the basis vectors normalized and mutually as orthogonal as possible (although this is not necessary in principle), whereby the output "neuron" only has to form the sum of squares of its inputs. In this case it is also possible to think that the output neurons form linear sums of terms, the values of which have already been squared in the output stages of the first layer. In mathematical terms the output of unit i is always $||\hat{x}^{(i)}||^2$, the squared projection of input x on subspace $\mathcal{L}^{(i)}$. The maximum of the $||\hat{x}^{(i)}||^2$ can then be selected by the WTA circuit.

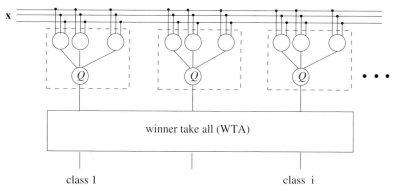

Fig. 1.9. Neural network consisting of linear-subspace modules. Q: quadratic neuron

1.4.2 Adaptation of a Model Subspace to Input Subspace

Assume that the dimensionality of a subspace $\mathcal{L}$, i.e. the number of its linearly independent basis vectors is fixed a priori. Nonetheless $\mathcal{L}$ may be made to adapt to stochastic input data x in order to minimize the average expected projection of x on $\mathcal{L}$, denoted by

$$E = \int \|\tilde{x}^2\| p(x) dx .$$

(1.127)

In other words, $\mathcal{L}$ may be *fitted* to $p(x)$ in the sense of minimum E, which comprises one kind of *regression*.

One attempt to minimize E is a stepwise corrective process, in which, for each new input sample x, its error $\|\tilde{x}\|$ is reduced. Consider that $b_k \in \Re^n$ is one of the basis vectors of $\mathcal{L}$, and form the new value b'_k as the matrix-vector product

$$b'_k = (I + \lambda x x^T) b_k ,$$

(1.128)

where I is the identity matrix, and λ is a free scalar ($\lambda > 0$). The multiplication (1.128) is performed for every basis vector b_k of $\mathcal{L}$. Then we can show that the projection of x on x on $\mathcal{L}$ increases, and the residual $\tilde{x} = x - \hat{x}$ decreases, respectively.

As a matter of fact, we shall prove an even stronger result: if the vectors x that form a stochastic sequence are all confined to a linear subspace $\mathcal{X}$, then, in continual application of (1.128) to all the basis vectors of an arbitrary subspace $\mathcal{L}$ that has at most the dimensionality of $\mathcal{X}$, the subspace $\mathcal{L}$ will converge to a subspace $\mathcal{L}^* \subseteq \mathcal{X}$. We shall need this result in Sect. 5.9.

Let the input vectors $x \in \Re^n$ be derived from some subspace $\mathcal{X}$ in a stochastic process specified below. Without loss of generality we can assume that the x vectors are normalized. To lay the foundations for a theorem we first derive two lemmata.

Lemma 1. *If any vector $b \in \Re^n$, $b \notin \mathcal{X}$ is multiplied by $P = I + \lambda x x^T$, $x \in \mathcal{X} \subset \Re^n$, $\lambda > 0$, then $\|\hat{b}\|^2 / \|b\|^2$ increases, unless $x \perp b$.*

Proof. Denote by $b^{(r)}$ the vector obtained by multiplying b with P,

$$b^{(r)} = b + \lambda (x^T b) x ,$$

(1.129)

and denote the projection of any vector a onto $\mathcal{X}$ by $\hat{a}$. Then a can be decomposed as $a = \hat{a} + \tilde{a}$, where $\tilde{a} \perp \hat{a}$.

Using the above decomposition on b, and noting that $\lambda (x^T b) x \in \mathcal{X}$, we get the following decomposition for $b^{(r)}$:

$$\hat{b}^{(r)} = \hat{b} + \lambda (x^T b) x , \text{ and}$$

(1.130)

$$\tilde{b}^{(r)} = \tilde{b} .$$

(1.131)

Without loss of generality it may be assumed that $\|b\| = 1$. Using the identity $x^T \hat{b} = x^T b$ and the assumption $\|x\| = 1$ we get, omitting some details,

$$\|\hat{b}^{(r)}\|^2 = \|\hat{b}\|^2 [1 + (2\lambda + \lambda^2) \cos^2 \phi] = \|\hat{b}\|^2 + (2\lambda + \lambda^2) \cos^2 \psi ,$$

(1.132)

where ϕ is the angle between the vectors x and $\hat{b}$, while ψ is the angle between x and b, respectively. Thus, $\|\hat{b}\|$ increases whereas $\|\tilde{b}\|$ remains constant when b is multiplied with P. Obviously, the ratio $\|\hat{b}\|^2 / \|b\|^2$ then increases, unless $x \perp b$, Q.E.D.

Lemma 2. *Assume that $\mathcal{X} \subseteq \Re^n$ is a subspace with dimensionality of at least k, the $b_1 \ldots b_k$ are normalized vectors that span the subspace $\mathcal{B}$, and there holds that $b_1 \ldots b_{k-1} \in \mathcal{X}$ are orthogonal, whereas $b_k \notin \mathcal{X}$. Assume further that all the $b_1 \ldots b_{k-1}$ are approximately orthogonal with b_k, so that we can set $b_j^T b_k = \lambda_j$, $j = 1 \ldots k-1$, where every λ_j is at most on the same order of magnitude as the learning-rate factor λ below. If b_k is multiplied by $I + \lambda x x^T$, where $0 < \varepsilon \le \lambda \ll 1$ and $x \in \mathcal{X}$, and b_k is subsequently orthonormalized with respect to all the b_j, $j = 1 \ldots k-1$, then the squared norm of the projection of b_k on $\mathcal{X}$, denoted $\|\hat{b}_k\|^2$, increases, let alone terms of the order of λ^2 and higher, and unless $x \perp b_k$.*

Proof. Denote by $b_k^{(r)}$ the value of b_k after rotation by the operator $P = I + \lambda x x^T$, and by b_k' its value after both the rotation and orthogonalization with respect to the b_i, $i = 1 \ldots k-1$. The orthogonalization can be done, e.g., using the Gram-Schmidt process:

$$b_k' = b_k^{(r)} - \delta b_k , \tag{1.133}$$

where the correction is $\delta b_k = \sum_{i=1}^{k-1}((b_k^{(r)})^T b_i)b_i$. According to the assumption on approximate orthogonality, $\delta b_k = O(\lambda)$.

Since $\delta b \in \mathcal{X}$, the component of b_k orthogonal to $\mathcal{X}$ is not transformed at all,

$$\tilde{b}_k' = \tilde{b}_k^{(r)} = \tilde{b}_k , \tag{1.134}$$

where the last equality follows from (1.131). The (squared) magnitude of the component of b_k' in $\mathcal{X}$ can be obtained from (1.133) using the fact that $\delta b_k \perp b_k'$.

$$
\begin{aligned}
\|\hat{b}_k'\|^2 &= \|\hat{b}_k^{(r)}\|^2 - \|\delta b_k\|^2 \\
&= \|\hat{b}_k^{(r)}\|^2 + O(\lambda^2) \\
&= \|\hat{b}_k\|^2 + 2\lambda \cos^2 \psi + O(\lambda^2) ,
\end{aligned}
\tag{1.135}
$$

where ψ is the angle between vectors x and b_k. The last equality follows from (1.132).

Since $\hat{b}_k$ increases while $\tilde{b}_k$ remains constant during the rotation and orthogonalization operations, after normalization of $\|b_k\|$ the projection on $\mathcal{X}$ has increased, Q.E.D.

We are now ready to express the *Adaptive-Subspace (AS) Theorem.* Let us assume that the following conditions hold:

(i) The successive values of the input vector x are obtained in a stochastic stationary process from a support $\mathcal{D} \subseteq \mathcal{X}$, where the dimensionality of $\mathcal{D}$ is at least k, and all values of the probability density function $p(x)$ are nonzero.

(ii) The initial values of the $b_1 \ldots b_k$ are not orthogonal to all $x \in \mathcal{X} \subseteq \mathcal{D}$.

Theorem 1.4.1. Let $b_1 \ldots b_k \in \Re^n$ be orthonormal vectors that span the linear subspace $\mathcal{B} \subset \Re^n$, and let $\mathcal{X} \subset \Re^n$ be another linear subspace with the dimensionality of at least that of $\mathcal{B}$. Let $x = x(t_p) \in \mathcal{X}$ satisfy Conditions (i) and (ii), where t_p is the sampling index of x. If the $b_1 \ldots b_k$ are multiplied by $I + \lambda x(t_p) x^{\mathrm{T}}(t_p)$, where $0 < \varepsilon \leq \lambda \ll 1$, and mutually orthonormalized in the same order, and this cycle is repeated an unlimited number of times, then the sequence of subspaces $\{\mathcal{B}(t_p)\}$, with the $\mathcal{B}(t_p)$ always spanned by the new values of $b_1 \ldots b_k$, will converge into subspace $\mathcal{B}^* \subseteq \mathcal{X}$ with an accuracy that can be made arbitrarily good by selecting λ sufficiently small.

Proof. It is a trivial task to prove first that b_1 converges to $\mathcal{X}$, even when λ is not small. According to Lemma 1, the relative squared projection of b_1 on $\mathcal{X}$, denoted by $r = \|\hat{b}_1\|^2 / \|b_1\|^2 \leq 1$ increases monotonically for successive rotations. If r were to converge to some value $r^* < 1$ we would have a contradiction, because after any finite number of steps, according to Condition (i), there will occur with a positive probability a value of x such that relating to (1.132), $\hat{b}_1$ increases by an amount larger than a fixed small positive quantity ε.

Since we can thus show that b_1, independent of $b_2 \ldots b_k$, converges to $\mathcal{X}$ with probability one, and even monotonically, we only need to show what the $b_2 \ldots b_k$ are doing *after that*.

Let now $b_1 \in \mathcal{X}$, whereby its corrected value is $b_1' \in \mathcal{X}$, and let $b_1^{\mathrm{T}} b_2 = \lambda_1 \ll 1$. According to Lemma 2, the squared norms $\|\hat{b}_2\|^2$ (neglecting effects of the order of λ^2 and the case $x \perp b_2$) are monotonically increasing. Then b_2 will converge to $\mathcal{X}$, too, by similar additional arguments as with b_1.

Let us next consider the case that $b_1, b_2 \in \mathcal{X}$, whereby their corrected values are $b_1', b_2' \in \mathcal{X}$. Let $b_1^{\mathrm{T}} b_3 = \lambda_1 \ll 1$, $b_2^{\mathrm{T}} b_3 = \lambda_2 \ll 1$. In applying Lemma 2 to b_3, we may notice that b_1' and b_2' span exactly the same subspace as their orthonormalized counterparts, and orthonormalization of b_1' and b_2' is anyway carried out before orthonormalization of the corrected value of b_3. Therefore the corrections to $\|\hat{b}_3\|^2$ behave in the same way as under the conditions stated in Lemma 2, and by similar additional arguments as with $\|\hat{b}_1\|^2$ and $\|\hat{b}_2\|^2$, the squared norm $\|\hat{b}_3\|^2$ increases monotonically, and thus b_3 converges to $\mathcal{X}$.

This same deduction is repeated for $b_4 \ldots b_k$, always considering the last b vector in relation to the previous ones. When all the $b_1 \ldots b_k$ have converged to $\mathcal{X}$, the subspace $\mathcal{B}^*$ spanned by any values of $b_1 \ldots b_k$ obtainable in learning after that will always be such that $\mathcal{B}^* \subseteq \mathcal{X}$, Q.E.D.

Comment 1: Even after convergence of $\mathcal{B}$ to $\mathcal{X}$, its basis vectors will vary within $\mathcal{X}$ in an unrestricted way.

Comment 2: Taking into account how Lemma 2 was derived, the condition that b_1 etc. have converged to $\mathcal{X}$ can be relaxed to mean that the distances of b_1 etc. from $\mathcal{X}$ have reached values that are at most of the order of $\lambda^* \ll \lambda$. As a matter of fact, the approximation b_1 etc. $\in \mathcal{X}$ only affects equation

(1.134), whereupon the component of b_k orthogonal to $\mathcal{X}$ could still change a little and thus change the relative projection of b_k on $\mathcal{X}$. Since we can make b_1 etc. *arbitrarily close* to $\mathcal{X}$, these adverse changes can be made *arbitrarily small*, whereby convergence of the $b_2 \ldots b_k$ to $\mathcal{X}$ holds true with a probability that can be made arbitrarily close to unity.

1.4.3 The Learning Subspace Method (LSM)

In the simple adaptative subspace methods of classification, a least-square regression of each class subspace onto the distribution of respective class samples is determined. This may be justifiable if the class probability density functions are symmetric with the same variance. In most practical applications this principle does not quarantee good classification accuracy because if the density functions are very dissimilar or unsymmetric, the decision surfaces between the classes may be severely offset from their optimal values.

This author has shown [1.63] that the basic subspace method, which is computationally very light, can easily be modified to approximate Bayesian decision surfaces, at least in the first approximation. This modification is named the *Learning Subspace Method* (LSM), and it has been used with considerable success in our earlier speech recognition systems [1.64, 65] . For additional aspects of LSM, see [1.3, 66], .

The LSM is a *supervised classification method*; it is in fact a *parametric* pattern recognition method (Sect. 1.3.3) since the forms of the discrimination functions are derived from the basic subspace method whereas their parameters are modified in a training procedure, using training vectors with known class-affiliation. The aim of training is to make the classification of at least all training vectors correct. Nonetheless, the modifications applied in the LSM have a deeper-going statistical meaning than just heuristically defined correction; the central idea is to replace the set of true prototypes of each class by another set of effective basis vectors *that tunes the decision surfaces for best classification results*.

The basic idea in LSM is that the classification accuracy can be improved by *rotating the* $\mathcal{L}^{(i)}$ *in a decision-directed or supervised way*. This rotation is a so-called *competitive-learning process*. The Self-Organizing Map and Learning Vector Quantization discussed in this book are competitive-learning processes, too; the LSM may be regarded as a historical precursor to them.

Let the closest subspace to x be $\mathcal{L}^{(c)}$ (called the *winner*), and let the next-to-closest subspace (*runner up*) be $\mathcal{L}^{(r)}$. If classification was wrong but $\mathcal{L}^{(r)}$ would have been the correct subspace, then the projection of x on $\mathcal{L}^{(c)}$ should be *decreased*, while the projection of x on $\mathcal{L}^{(r)}$ should be *increased*, respectively. This criterion is logically very similar to that used in the LVQ2 algorithm to be discussed in Sect. 6.4. The projection of x on a particular $\mathcal{L}^{(i)}$ can be changed by rotating $\mathcal{L}^{(i)}$, *or its basis vectors*. We shall determine the "optimal" directions and amounts of rotation. Before that, we have to discuss how the subspace dimensionalities can be selected.

Dimensionalities of the Subspaces. The discriminant functions defined by (1.125) are *quadratic*; accordingly, the decision surfaces in the general case are quadratic, too. Only in the case that two classes have the same dimensionality, i.e., if they are spanned by an equal number of linearly independent prototypes, the decision surface between these classes is a hyperplane. It will be clear that the form of the decision surface should comply with the statistical properties of the neighbouring classes; accordingly, the dimensionalities of the class subspaces should be determined first.

A straightforward learning procedure for the determination of dimensionalities is *decision-controlled choice of the prototypes*; for this, as well as the other learning methods, we have first to define a *training set of vectors* with known class-affiliation. The initial set of prototypes may be started with one prototype per class, which may be selected from the training set at random. After that, new prototypes are accepted only if a training vector is classified incorrectly. The training vector itself is thereby added to the set of basis vectors of the correct subspace. In orded to avoid numerical instability in practical calculations, the new prototype is not accepted if it is very close being linearly dependent on the previous prototypes. The value $\|\tilde{x}_i\|/\|x\|$ can be used as an indicator; for the acceptance of a prototype, this figure has to exceed, say, five to ten per cent. In order to prevent the dimensionality from growing too high, an upper limit, say, a number that is still significantly smaller than the vector dimensionality, can also be set to the number of prototypes per class.

When a new prototype is added to the correct class, from the subspace into which the training vector is wrongly classified one of its prototypes is at the same time deleted, namely that one which makes the smallest angle with the training vector. This deletion is not performed unless a certain minimum number, e.g., one or two prototypes, is always left in the class.

After each addition and deletion, a new orthonormal basis for the subclass should be computed.

It has also been suggested [1.57, 67] that the initial subspaces $\mathcal{L}^{(i)}$ might be spanned optimally by the largest eigenvectors of the covariance matrix of the prototypes, and the dimensionality of $\mathcal{L}^{(i)}$ could then be determined optimally by choosing a subset of eigenvectors with the largest eigenvalues, the latter exceeding a heuristically determined limit.

We have found out that the final results depend only weakly on the absolute dimensionalities of the subspaces; on the other hand, the relative dimensionalities are more decisive, which is understandable as they strongly affect the forms of the decision surfaces [1.3].

A further suggestion has been to exclude intersections of the $\mathcal{L}^{(i)}$ [1.68]. There are indications for that at least in the LSM algorithm, the latter modification does not bring about significant improvements, whereas the eigenvalue method may in some cases be even better than the decision-controlled choice of prototypes.

Decision-Controlled Rotation of the Subspaces. If the basic subspace
method produces errors, the recognition results can be improved in a straightforward way by modifying the subspaces to change the relative lengths of
projections on them. This is done simplest by *rotating* the subspaces. The
rotating procedure may be derived in the following way.

First we determine the relative directions of rotation of the basis vectors
$b_h^{(i)}$ of subspace $\mathcal{L}^{(i)}$ for which the changes in the $\hat{x}^{(i)}$ or $\tilde{x}^{(i)}$ are most rapid.
Obviously they are the same as (or opposite to the directions) for which $b^{(i)}$
can be completely *orthogonalized* with respect to x. Consider the elementary
matrix

$$P_1 = \left(I - \frac{xx^{\mathrm{T}}}{x^{\mathrm{T}}x}\right) ; \tag{1.136}$$

if $b \in \Re^n$ is an arbitrary vector, $P_1 b$ will be orthogonal to x, which is obvious,
because $x^{\mathrm{T}} P_1 b = 0$. Therefore P_1 may be called the elementary orthogonal
projection matrix. If we multiply all the basis vectors of any subspace $\mathcal{L}^{(i)}$
by P_1, they become orthogonal to x, and thus the projection of x on $\mathcal{L}^{(i)}$ is
zero.

Let us henceforth assume that the fastest direction of rotation is indeed
the same as or opposite to the direction of complete orthogonalization. The
amount of rotation can be moderated and its sign can be controlled by a
scalar factor α: Let us write

$$P_2 = \left(I + \alpha\frac{xx^{\mathrm{T}}}{x^{\mathrm{T}}x}\right) . \tag{1.137}$$

Here P_2 is the graded projection matrix that *reduces* the projection $\hat{x}^{(i)}$ if
$-2 < \alpha < 0$ and *increases* it in the same direction if $\alpha > 0$. Often α is a
function of time. We may then call α in (1.137) the *learning-rate factor*.

Comment. The word "rotation" is here used only in the sense that the
direction of a vector is changed. In general its norm is thereby changed,
too. In subspace methods, the projections are independent of the lengths of
the basis vectors, but for improved numerical accuracy, a new (equivalent)
orthonormal basis for each subspace should be computed intermittently.

However, correction of subspace $\mathcal{L}^{(i)}$ with respect to only one vector does
not guarantee good performance for all vectors. Therefore, corrections must
be made iteratively over the whole set of training vectors, and the degree of
rotation must be moderated, e.g., by selecting suitable values for α which are
usually $\ll 1$.

The principle aimed at in training should be such that the projection
on the wrong subspace is decreased, and the one on the correct subspace
increased, respectively. As with any learning method, the adaptation gain
α also ought to be some suitable function of the training step to warrant a
good compromise between final accuracy and speed of convergence. A possible
strategy is to make α at each training step *just about sufficient* to correct for

the misclassification of the last training vector. Such a rule is called *absolute correction rule*.

Let the length of the relative projection of x on subspace $\mathcal{L}^{(i)}$ be defined as

$$\beta_i = \|\hat{x}_i\|/\|x\|. \tag{1.138}$$

Assume a rotation operator which, when applied to all the basis vectors of a given subspace, changes the relative projection of x from β_i to β_i'. We shall show below that the following relation holds:

$$\alpha = 1 - \frac{\beta_i'}{\beta_i}\sqrt{\frac{1 - \beta_i^2}{1 - \beta_i'^2}}. \tag{1.139}$$

Let us call that subspace to which the training vector belongs the "own" subspace, and the subspace which the training vector is closest to the "rival" subspace, respectively. Then, in order to determine the value of α that is just sufficient for corrective action, assume that the classification of a training vector was wrong in the sense that its relative projection on the own subspace, denoted by β_o, was smaller than its relative projection β_r on the rival subspace. If a rotating procedure is now applied by which the lengths of the new projections become

$$\beta_o' = \lambda\beta_o \quad \text{and} \quad \beta_r' = \beta_r/\lambda, \tag{1.140}$$

then a correct classification $(\beta_o' > \beta_r')$ is obtained if

$$\lambda = \sqrt{\beta_r/\beta_o} + \Delta, \tag{1.141}$$

where Δ is a small positive constant. Notice that β_r and β_o define λ which, by (1.140 and 1.141), again determines α. A proper choice of Δ in our experiments was 0.005 to 0.02. The correcting procedure has to be iterated over the training set.

Now we shall revert to the derivation of (1.139). It will be useful to show first that the orthogonal projection of x on the subspace $\mathcal{L}'$, obtained from $\mathcal{L}$ by rotating can be denoted by $\hat{x}'$; this projection can be computed as a projection on a certain line [1.69].

Theorem 1.4.2. *The orthogonal projection of x on $\mathcal{L}'$ is identical to the orthogonal projection of x on a line spanned by the vector*

$$z = \hat{x} + \alpha\beta^2 x,$$

provided that $\hat{x}$ was nonzero and x does not belong to $\mathcal{L}$.

Proof. Let us tentatively denote the latter projection by $\hat{x}_z$ whereby $x = \hat{x}_z + \tilde{x}_z$. The other orthogonal component $\tilde{x}_z$ will first be shown to be orthogonal to $\mathcal{L}'$. To this end it has to be orthogonal to all the new basis vectors

$$a_i' = \left(I + \alpha\frac{xx^{\mathrm{T}}}{\|x\|^2}\right)a_i.$$

The following formulas are fundamental for orthogonal projections:

$$
\begin{aligned}
\tilde{x}_z &= \left(I - \frac{zz^{\mathrm{T}}}{z^{\mathrm{T}}z}\right)x, \\
\hat{x}^{\mathrm{T}}x &= \hat{x}^{\mathrm{T}}\hat{x} = \beta^2 x^{\mathrm{T}}x, \\
\hat{x}^{\mathrm{T}}a_i &= x^{\mathrm{T}}a_i.
\end{aligned}
\tag{1.142}
$$

Denote tentatively

$$
z = \hat{x} + \xi x.
\tag{1.143}
$$

By substitution one readily obtains

$$
\tilde{x}_z^{\mathrm{T}}\left(I + \alpha\frac{xx^{\mathrm{T}}}{\|x\|^2}\right)a_i = \frac{(\xi - \alpha\beta^2)(\beta^2 - 1)}{\beta^2 + 2\xi\beta^2 + \xi^2}x^{\mathrm{T}}a_i.
\tag{1.144}
$$

If $\xi = \alpha\beta^2$ as assumed, then $\forall i, \tilde{x}_z^{\mathrm{T}}a_i' = 0$, or $\tilde{x}_z \in \mathcal{L}'$.

Next it is shown that z belongs to $\mathcal{L}'$:

$$
z = \left(\hat{x} + \alpha\frac{x^{\mathrm{T}}\hat{x}}{x^{\mathrm{T}}x}x\right) = \left(I + \alpha\frac{xx^{\mathrm{T}}}{\|x\|^2}\right)\hat{x}.
\tag{1.145}
$$

Now, because $\hat{x}$ is a linear combination of the a_i (or belongs to $\mathcal{L}$) and it is rotated by the same operator as the a_i to get z, then it can be seen that z is a linear combination of the a_i', or it belongs to $\mathcal{L}'$.

Since $\hat{x}' \in \mathcal{L}'$, then also $\tilde{x}_z \perp \hat{x}'$ holds true, and because $z \in \mathcal{L}'$, then also $\tilde{x}' \perp \hat{x}_z$ holds true. It is to be recalled that $x = \hat{x}_z + \tilde{x}_z = \hat{x}' + \tilde{x}'$ (where $\tilde{x}'$ is the orthogonal distance vector to $\mathcal{L}'$). Now

$$
\begin{aligned}
(\hat{x}_z - \hat{x}')^{\mathrm{T}}(\hat{x}_z - \hat{x}') &= (x - \tilde{x}_z - x + \tilde{x}')^{\mathrm{T}}(\hat{x}_z - \hat{x}') \\
&= \tilde{x}'^{\mathrm{T}}\hat{x}_z - \tilde{x}'^{\mathrm{T}}\hat{x}' - \tilde{x}_z^{\mathrm{T}}\hat{x}_z + \tilde{x}_z^{\mathrm{T}}\hat{x}' = 0,
\end{aligned}
\tag{1.146}
$$

which cannot hold unless $\hat{x}_z = \hat{x}'$.   Q.E.D.

Now

$$
\hat{x}' = \frac{x^{\mathrm{T}}z}{z^{\mathrm{T}}z}z,
\tag{1.147}
$$

and by substitution

$$
\beta'^2 = \frac{\hat{x}'^{\mathrm{T}}\hat{x}'}{x^{\mathrm{T}}x} = \frac{\beta^2(1+\alpha)^2}{1 + 2\alpha\beta^2 + \alpha^2\beta^2}.
\tag{1.148}
$$

Solution for α yields

$$
\alpha = \pm\frac{\beta'}{\beta}\sqrt{\frac{1 - \beta^2}{1 - \beta'^2}} - 1,
\tag{1.149}
$$

where the positive sign for the square root has to be selected.   ∎

Another strategy that has produced slightly improved results in the classification results from the following rather natural conditions. Let $\hat{x}_o$ and $\hat{x}_r$ be the projections of x on the own and rival subspaces before correction, and after that they shall be denoted $\hat{x}'_o$ and $\hat{x}'_r$, respectively. A proper choice for α obviously involves at least the following considerations:

i) $\|\hat{x}'_o\| - \|\hat{x}'_r\| > \|\hat{x}_o\| - \|\hat{x}_r\|$.

ii) The corrections must form a monotonically converging series.

We have applied the following computationally simple rule for the determination of α at each step; its derivation has been presented elsewhere [1.65]. Let λ be a parameter that controls the amount of correction such that

$$\begin{aligned} \|\hat{x}'_o\| &= \|\hat{x}_o\| + \lambda/2, \\ \|\hat{x}'_r\| &= \|\hat{x}_r\| - \lambda/2; \end{aligned} \qquad (1.150)$$

$$\lambda = \begin{cases} \|\hat{x}_r\| - \|\hat{x}_o\| + \Delta\|x\|, & \text{if } \|\hat{x}_o\| \le \|\hat{x}_r\|, \\ \|\hat{x}_r\| - \|\hat{x}_o\| + \sqrt{(\|\hat{x}_r\| - \|\hat{x}_o\|)^2 + \Delta^2\|x\|^2}, & \text{if } \|\hat{x}_o\| > \|\hat{x}_r\|, \end{cases}$$

with Δ a small numerical constant (e.g., $\Delta = 0.002$). The values of α_o and α_r, referring to the own and rival subspaces, respectively, follow from the projections:

$$\alpha_o = \frac{\|\hat{x}'_o\|}{\|\hat{x}_o\|} \sqrt{\frac{\|x\|^2 - \|\hat{x}_o\|^2}{\|x\|^2 - \|\hat{x}'_o\|^2}} - 1, \qquad (1.151)$$

and a similar expression is obtained for α_r, with subscript r replaced for o. (Notice that $\alpha_o > 0, \alpha_r < 0$.)

In our experiments, this procedure has always provided fast learning and large separation between subspaces. It has also ensured at least moderate classification accuracy for "rare" classes.

A particular caution, however, is due. Although the subspace methods have been very effective in particular applications, and they are relatively easy to apply and program, for best results (and fastest convergence) the α-sequence ought to be designed even more carefully than devised above. It is a very interesting academic question and one of practical importance, too, to find out these sequences. Although for reasonably good classification accuracy the subspace methods easily lend themselves to practical application, for ultimate results some expertise in these methods is needed.

Finally we would like to comment on a very important property of the subspace algorithms that is influential from the computational point of view. If all input and prototype vectors are normalized to unit length, then the projection operations produce new vectors *all components of which are in the same range* $[-1, +1]$. Accordingly, fixed-point (or integer) arithmetic operations are then feasible, and due to this property these algorithms are readily programmable for special microprocessors, thereby achieving a very high speed of computation.

1.5 Vector Quantization

1.5.1 Definitions

Vector quantization (VQ) is a classical signal-approximation method (for reviews, see [1.70–74]) that usually forms a quantized approximation to the distribution of the input data vectors $x \in \Re^n$, using a finite number of so-called *codebook vectors* $m_i \in \Re^n$, $i = 1, 2, \ldots, k$. Once the "codebook" is chosen, the approximation of x means finding the codebook vector m_c closest to x (in the input space), usually in the Euclidean metric:

$$||x - m_c|| = \min_i \{||x - m_i||\} \ , \ \ \text{or}$$
$$c = \arg \min_i \{||x - m_i|| \ . \tag{1.152}$$

One kind of optimal selection of the m_i minimizes the average expected square of the *quantization error*, often also called the *distortion measure*, which is defined as

$$E = \int ||x - m_c||^2 p(x) dx \ , \tag{1.153}$$

where the integral is taken over the complete metric x space, dx is a shorthand notation for the n-dimensional volume differential of the integration space, and $p(x)$ is the probability density function of x.

Notice that the subscript c is a function of x and all the m_i, so this time the gradient of E with respect to any of the $m_i, i = 1, 2, \ldots, k$, cannot be formed so easily. As a matter of fact, if m_i is varied, the index c may make a discontinuous jump from one discrete value to another, because for the m_i closest to x, a new closest m_i may then become swapped abruptly.

It has been shown in [1.70] that when the m_i values minimize E in (1.153), also the average quantization error in each Voronoi set is the same.

Voronoi Tessellation. A concept that is useful for the illustration of the vector quantization methods in pattern recognition and for neural networks in general is called *Voronoi tessellation* [1.76]. Figure 1.10 exemplifies a two-dimensional space where a finite number of codebook or *reference vectors* is shown as points, corresponding to their coordinates. This space is partitioned into regions, bordered by lines (in general, hyperplanes) such that each partition contains a reference vector that is the "nearest neighbor" to any vector within the same partition. These lines, or the "midplanes" of the neighboring reference vectors, together constitute the Voronoi tessellation.

Voronoi Set. All x vectors that have a particular reference vector as their closest neighbor, i.e, all x vectors in the corresponding partition of the Voronoi tessellation, are said to constitute the *Voronoi set*.

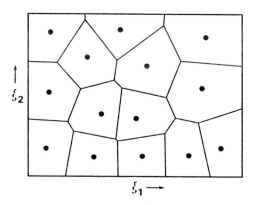

Fig. 1.10. Voronoi tessellation partitions here a two-dimensional (ξ_1, ξ_2) "pattern space" into regions around reference vectors, shown as points in this coordinate system. All vectors (ξ_1, ξ_2) in the same partition have the same reference vector as their nearest neighbor

1.5.2 Derivation of the VQ Algorithm

As closed-form solutions for the determination of the m_i are not available, at least with general $p(x)$, one must resort to iterative approximation schemes. The derivation introduced by this author and presented next [1.75] is believed to be the first mathematically fully rigorous one.

Assume that $p(x)$ is continuous. The problem caused by the discontinuity of $c = c(x; m_1, \ldots, m_N)$ can be circumvented if use is made of the following identity. Let $\{a_i\}$ be a set of positive real scalar numbers; then it holds generally that

$$\min_i \{a_i\} \equiv \lim_{r \to -\infty} \left[\sum_i a_i^r \right]^{\frac{1}{r}} . \tag{1.154}$$

A further result we need concerns the functional form

$$f(x, r) = (1 + |x|^r)^{\frac{1}{r}} . \tag{1.155}$$

Excluding all values of x at which f or $\lim_{r \to -\infty} f$ are not differentiable, i.e., $x \in \{-1, 0, 1\}$, there holds (even generally):

$$\lim_{r \to -\infty} \frac{\partial f}{\partial x} = \frac{\partial}{\partial x} \left(\lim_{r \to -\infty} f \right) . \tag{1.156}$$

In forming the gradient $\nabla_{m_j} E$, use is now made of the fact that a function of the form $(\sum_i \|x - m_i\|^r)^{\frac{2}{r}}$ is continuous, single-valued, well-defined, and continuously differentiable in its arguments, except when x is exactly equal to some m_i. Further we have to exclude the cases when one of the $\|x - m_i\|^r$ is exactly equal to the sum of the other terms (cf. (1.155) and (1.156)). With stochastic x and continuous $p(x)$, all these singular cases have zero probability in the process discussed anyway. Under these conditions one can easily become convinced that in the following the gradient and limit operations can be exchanged, whereby

$$||x - m_c||^2 = \left[\min_i \{||x - m_i||\}\right]^2 = \lim_{r \to -\infty} \left(\sum_i ||x - m_i||^r\right)^{\frac{2}{r}} , \quad (1.157)$$

and

$$\nabla_{m_j} E = \int \lim_{r \to -\infty} \nabla_{m_j} \left(\sum_i ||x - m_i||^r\right)^{\frac{2}{r}} p(x) dx . \quad (1.158)$$

Denote

$$\sum_i ||x - m_i||^r = A , \quad (1.159)$$

whereby

$$\nabla_{m_j} A^{\frac{2}{r}} = \frac{2}{r} A^{(\frac{2}{r})-1} \cdot \nabla_{m_j} (||x - m_j||^r)$$
$$= \frac{2}{r} A^{(\frac{2}{r})-1} \cdot \nabla_{m_j} (||x - m_j||^2)^{\frac{r}{2}} . \quad \text{(Note: no sum over } j.) \quad (1.160)$$

After simple operations and rearrangement, we obtain

$$\nabla_{m_j} A^{\frac{2}{r}} = -2 \cdot \left(A^{\frac{2}{r}}\right) \cdot \frac{(||x - m_j||^2)^{(r/2)-1}}{A} \cdot (x - m_j) . \quad (1.161)$$

Due to (1.157),

$$\lim_{r \to -\infty} A^{\frac{2}{r}} = ||x - m_c||^2 . \quad (1.162)$$

Denote

$$B = \frac{(||x - m_j||^2)^{(r/2)-1}}{A} = \frac{||x - m_j||^r}{\sum_i ||x - m_i||^r} \cdot ||x - m_j||^{-2}$$

$$= \left(\sum_i \frac{||x - m_i||^r}{||x - m_j||^r}\right)^{-1} \cdot ||x - m_j||^{-2} . \quad (1.163)$$

Notice that when $r \to -\infty$, the term $||x - m_i||^r / ||x - m_j||^r$, for any given m_j, is maximum when $m_i = m_c$, and starts to predominate progressively over the other terms; therefore

$$\lim_{r \to -\infty} B = \lim_{r \to -\infty} \left(\frac{||x - m_j||}{||x - m_c||}\right)^r \cdot ||x - m_j||^{-2} = \delta_{cj} ||x - m_j||^{-2} , \quad (1.164)$$

where δ_{cj} is the Kronecker delta ($= 1$ for $c = j$, 0 otherwise). Combining the partial results we obtain

$$\lim_{r \to -\infty} \nabla_{m_j} A^{\frac{2}{r}} = -2 \cdot ||x - m_c||^2 \cdot \delta_{cj} \cdot ||x - mj||^{-2} \cdot (x - m_j)$$
$$= -2 \cdot \delta_{cj} \cdot (x - m_j) , \quad (1.165)$$

and

$$\nabla_{m_j} E = \int \lim_{r \to -\infty} \nabla_{m_j} A^{\frac{2}{r}} \cdot p(x)dx = -2 \int \delta_{cj}(x - m_j)p(x)dx . \quad (1.166)$$

The sample function of this gradient at time t is

$$\nabla_{m_j} E|_t = -2 \cdot \delta_{cj}[x(t) - m_j(t)] , \quad (1.167)$$

and the steepest descent in the "E landscape" occurs in the direction of $-\nabla_{m_j} E|_t$; after change of indices, and definition of the step size by the factor $\alpha(t)$ that includes the constant -2, we get

$$m_i(t + 1) = m_i(t) + \alpha(t) \cdot \delta_{ci}[x(t) - m_i(t)] . \quad (1.168)$$

The K-Means (Linde-Buzo-Gray) Algorithm. For a general $p(x)$, optimal placement of the m_i in the input space is usually not possible in closed form, but some iterative solutions converge very fast. In practice, a convenient computing scheme for the LVQ is batch computation, usually named the K-means algorithm. The *Linde-Buzo-Gray (LBG) algorithm* [1.71] is essentially the same, and it can be shown to hold for many different metrics. Its steps are defined below.

1. For the initial codebook vectors take, for instance, the first K training samples.
2. For each map unit i, collect a list of all those training samples whose nearest codebook vector is m_i.
3. Take for each new codebook vector the mean over the respective list.
4. Repeat from 2 a few times.

1.5.3 Point Density in VQ

Very thorough treatments of VQ can be found in [1.70, 74]. Consider that the distortion measure were defined using the rth power of the quantization error,

$$E = \int \|x - m_c\|^r p(x)dx , \quad (1.169)$$

where r is some real-valued exponent, then under rather general conditions one can determine the *point density* $q(x)$ of the m_i as in the following expression:

$$q(x) = \text{const.} \left[p(x)^{\frac{n}{n+r}} \right] . \quad (1.170)$$

This result is valid only in the continuum limit, i.e., when the number of m_i approaches infinity. Another condition for obtaining this result is that the configuration of the m_i is reasonably regular, as the case usually is in VQ when $p(x)$ is smooth.

Usually the derivation of (1.170) is a very cumbersome task that involves many results of the error theory. For the purpose of this book and also from a general interest, this author has constructed a much shorter derivation based on the classical *calculus of variations*.

If $p(x)$ is smooth and the placement of the m_i in the signal space is reasonably regular, as the VQ solutions usually are, one may try to approximate the Voronoi sets, which are polytopes (general polyhedrons) in the n-dimensional space, by n-dimensional hyperspheres centered at the m_i. This, of course, is a rough approximation, but it was in fact used already in the classical VQ papers [1.70, 74], and no better treatments exist for the time being.

Denoting the radius of the hypersphere by R, its hypervolume has the expression kR^n, where k is a numerical factor. We have to assume that $p(x)$ is approximately constant over the polytope. The *elementary integral of the distortion* $\|x - m_i\|^r = \rho^r$ *over the hypersphere* is

$$D = nk \int_0^R p(x) \cdot \rho^r \cdot \rho^{n-1} d\rho = \frac{nk}{n+r} \cdot p(x) \cdot R^{n+r} \; ; \qquad (1.171)$$

notice that if $v(\rho)$ is the volume of the n-dimensional hypersphere with radius ρ, then $dv(\rho)/d\rho = nk\rho^{n-1}$ is the "hypersurface area" of the hypersphere. Over a polytope of equal size, of course, the $(r + n - 1)$th moment of $p(x)$ would be slightly different.

Now, however, we also encounter the problem that the hyperspheres do not fill up the signal space exactly. Following Gersho's argument [1.70] we shall anyway sum up the elementary integrals of distortion over the signal space. Thereupon we end up with an approximation of the distortion measure that differs from the true one by a numerical factor. A similar approximation, although with a slightly different error, will then be made in the restrictive condition (1.174). Following Gersho we argue that a reasonable approximation of optimization may anyway be obtained in this way.

Notice that according to our earlier conventions the point density $q(x)$ is defined as $1/kR^n$. What we aim at first is the approximate "distortion density" that we denote by $I[x, q(x)]$, where $q(x)$ is the point density of the m_i at the value x:

$$I[x, q(x)] = \frac{D}{kR^n} = \frac{n}{n+r} \cdot p(x) \cdot R^r = \frac{np(x)}{n+r}[kq(x)]^{-\frac{r}{n}} \; . \qquad (1.172)$$

Using the concept of "distortion density" referred to x, we approximate, in the continuum limit, the total distortion measure by the integral of the "distortion density" over the complete signal space:

$$\int I[x, q(x)]dx = \int \frac{np(x)}{n+r}[kq(x)]^{-\frac{r}{n}} dx \; . \qquad (1.173)$$

This integral is minimized under the restrictive condition that the sum of all quantization vectors shall always equal N; in the continuum limit the condition reads

$$\int q(x)dx = N \; . \qquad (1.174)$$

In the classical *calculus of variations* one often has to optimize a functional which in the one-dimensional case with one independent variable x and one dependent variable $y = y(x)$ reads

$$\int_a^b I(x, y, y_x)dx \; ; \tag{1.175}$$

here $y_x = dy/dx$, and a and b are fixed integration limits. If a restrictive condition

$$\int_a^b I_1(x, y, y_x)dx = \text{const.} \tag{1.176}$$

has to hold, the generally known Euler variational equation reads, using the Lagrange multiplier λ and denoting $K = I - \lambda I_1$,

$$\frac{\partial K}{\partial y} - \frac{d}{dx}\frac{\partial K}{\partial y_x} = 0 \; . \tag{1.177}$$

In the present case x is vectorial, denoted by x, $y = q(x)$, and I and I_1 do not depend on $\partial q/\partial x$. In order to introduce fixed, finite integration limits one may assume that $p(x) = 0$ outside some finite support. Now we can write

$$
\begin{aligned}
I &= \frac{nk^{-\frac{r}{n}}}{n+r} \cdot p(x) \cdot [q(x)]^{-\frac{r}{n}} \; , \tag{1.178}\\
I_1 &= q(x) \; ,\\
K &= I - \lambda I_1 \; ,
\end{aligned}
$$

and obtain

$$\frac{\partial K}{\partial q(x)} = -\frac{rk^{-\frac{r}{n}}}{n+1} \cdot p(x) \cdot [q(x)]^{-\frac{n+r}{n}} - \lambda = 0 \; . \tag{1.179}$$

At every location x there then holds

$$q(x) = C \cdot [p(x)]^{\frac{n}{n+r}} \; , \tag{1.180}$$

where the constant C can be solved by substitution of $q(x)$ into (1.174). Clearly (1.180) is identical with (1.170).

1.6 Dynamically Expanding Context

The algorithm described in this section is actually an artificial-intelligence method. It was introduced by this author in 1986 [1.77] for the postprocessing stage of the "neural phonetic typewriter." The information-theoretic *minimum description length principle (MDL)* introduced by *Rissanen* in 1986 [1.78] contains some related ideas. This algorithm finds a great number of context-dependent production rules from sequences of discrete data, such as phonetic transcriptions obtained from practical speech recognition devices, and uses them for the correction of erroneous strings.

1.6.1 Setting Up the Problem

We shall demonstrate in this section that a set of complex grammatical transformation or production rules, automatically found from natural data, can
be utilized to correct errors from symbol strings.

Assume that for training purposes we have available both erroneous transcriptions and corresponding ideal transcriptions that are correct by definition; for instance, the latter may be prepared manually by experts. Whatever
the source data is, resurrection of its correct form from erroneuous versions
can be carried out by a transformation that is defined in terms of a set of
specific *context-dependent grammatical mappings*.

Let S be a symbol string that represents the erroneous source data, and T
be its corrected transformation after application of the mapping. It is assumed
that the mappings are *local*, i.e., one can identify corresponding segments in
S and T obtained from each other by a set of *production rules*. These rules
have to be deduced automatically from a representative set of training pairs
(S, T), where T now stands for the ideal form.

For instance, if we intend to pronounce the (Finnish) diphtong /au/, in a
speech recognition device it is usually realized as the transcription /aou/. One
of the corrective production rules would then be /aou/ → /au/. It is more
common, however, that the errors depend on the context of the neighboring phonemic units, whereupon the corrective productions must be defined
using similar context. Here, however, we encounter the basic problem: *what
is the minimum sufficient amout of context needed to define all the usual
productions uniquely?*

Notational Convention. Throughout this section we shall use the notation
x(A)y → (B) for the context-dependent mapping where A is a segment of
the source string S, B is the corresponding segment in the transformed string
T, and x(.)y is the *context* in string S where A occurs. In other words, A is
replaced by B on the condition x(.)y.

Comment. In the theory of formal languages, a context-sensitive production would be written xAy → xBy. We shall not write the context on the
right-hand side since the results are later simply *concatenated*. The notation
convention made above must be understood as that of a simple mapping.

If the context were selected too wide, one would have to record a great
number of examples of the type x(A)y → (B) before these rules have a coverage that is sufficient for practical purposes; on the other hand, if the context
x(.)y were too narrow, there might remain conflicts in the set of rules, i.e., one
would observe cases of the type x(A)y→(B) as well as x(A)y→(B'), where
B≠B'.

The philosophy advanced in this section is to determine *just a sufficient
amount of context for each individual segment A such that all conflicts in the
set of training examples will be resolved*; this is then expected to constitute

the optimal compromise between accuracy and generality, or selectivity and coverage of the rules.

The practical implementation of this principle is based on the concept that we call *dynamically expanding context*. The central idea underlying it is that for any segment A we first define a series of stepwise expanding *frames*, each frame, e.g., consisting of an increasing number of symbol positions on either or both sides of A. The nth frame in this series is then said to correspond to the *nth level of context*. The due level is determined for each production automatically, as explained below. The reader is adviced to take a quick look at the algorithm in Sect. 1.6.5 where such frames are needed.

Consider, for instance, the string "eisinki" (erroneous form of "helsinki"). It may first be divided into segments of contiguous vowels and consonants, as discussed in Sect. 1.6.2; alternatively, each symbol itself may constitute a segment. Consider the segment (s). Possible definitions of frames and contextual levels relating to "eisinki" around (s) are exemplified in Table 1.4.

Table 1.4. Two examples of contextual levels

Level	Context I	Context II
0 (Context-free)	–	–
1	i(.)	(.)i
2	i(.)i	i(.)i
3	ei(.)i	i(.)in
4	ei(.)in	ei(.)in

Contexts of this type can be defined up to some highest level, after which the remaining conflicts may be handled by an estimation procedure explained in Sect. 1.6.6.

The central idea in the present method is that we always first try to find a production referring to the lowest contextual level, i.e., (A)→(B). Only if contradictory cases are encountered in the given examples, e.g., there would exist cases of the types (A)→(B) as well as (A)→(B') in the source data, where B' ≠B, we have to resort to contexts of successively higher levels, in order to resolve for the conflicts, like x(A)y→(B) and x'(A)y'→(B'); here x(.)y and x'(.)y' relate to different strings.

1.6.2 Automatic Determination of Context-Independent Productions

To exemplify this method, we have used phoneme strings obtained from an automatical speech recognition device for the primary symbol strings [1.69]. They must first be divided into segments denoted by the symbol A above. The simplest segments consist of the phonemes themselves. Another choice, used in the following examples, is to segment the phoneme strings into *groups of contiguous vowels and consonants*, respectively, corresponding to the symbol A. In the processing of natural text, there has been no marked difference in the performance for these choices.

Before one is able to define the *context-dependent* production rules, one has first to find the *context-independent* productions, i.e., to determine which segments in T (the correct string) match best with the given segments in S (source string). We have used the Levenshtein distance method introduced in Sect. 1.2.2 (cf. also [1.77]) that globally optimizes the alignment of respective portions in S and T. The algorithm computes the so-called distance lattice D(i,j) and finds an optimal path through it from node (i,j) = (0,0) to node (i,j) = (length(S), length(T)) that defines the *WLD*.

The alignment of S and T now follows from the Levenshtein algorithm. For a rather arbitrary, noninteger choice of the weight parameters p, q, and r, a *unique* optimal path through the distance lattice is usually obtained, whereby this path directly defines the alignment of S and T that is optimal with respect to the minimum number of editing operations. Assume that S="ekhnilinen" and T="teknillinen"; if we fill hyphens in S to indicate the deleted symbols, and similarly use hyphens in T at places where S has an inserted symbol, the optimal alignment would look like

$$
\begin{array}{rccccccccccc}
S=\text{``} & - & e & k & h & n & i & l & - & i & n & e & n & \text{''} \\
T=\text{``} & t & e & k & - & n & i & l & l & i & n & e & n & \text{''}
\end{array}
$$

The hyphens in S, corresponding to deletion errors, are first assigned into vowel or consonant segments. For instance, the leading hyphen may be assigned to the next segment and in all other cases, the hyphen is assigned to the previous segment. The productions would then be /e/→/te/, /khn/→/kn/, /i/→/i/, /l/→/ll/, /n/→/n/, /e/→/e/. (A detailed discussion of this kind of problems would make the present section rather lenghty, and must be abandoned here.)

1.6.3 Conflict Bit

We do not know *a priori* how wide a context is necessary to resolve all conflicts; it must be determined experimentally, after all data have been collected. In fact, it may even be necessary to present the data a few times reiteratively. In order to allow free determination of context, we have first to construct a series of tentative productions, starting at the lowest contextual level, and proceeding higher upon need. The tentative productions x(A)y→(B), with increasing amount of context x(.)y, are stored in memory in the form of ordered sets (xAy, B, conflict bit), where xAy is simply the concatenation of x, A and y, and a *conflict bit* is further associated with each set. This bit is 0 initially, and as long as the rule is valid. If it becomes necessary to invalidate the rule, the conflict bit is set equal to 1. The rule must not be deleted completely from memory, because it is needed in the hierarchical search where the final contextual level is determined automatically.

1.6.4 Construction of Memory
for the Context-Dependent Productions

For this method, an effective search memory is needed where a great number of sets of the type (xAy, B, conflict bit) is made accessible on the basis of the search argument xAy. It may be implemented by software (list structures) or hardware (content-addressable memories).

When the pairs of strings (S, T) in the training data are scanned and the context-independent productions (A)→(B) formed as in Sect. 1.6.2, they are first stored in the form (A, B, 0) where 0 denotes the conflict bit value 0. Before that, however, a search with A as the search argument must be made in order to check (a) whether a similar production was already stored, (b) whether a conflict would arise from the new production. In the former case no storing is necessary; in the latter case, the old production stored in memory must first be *invalidated* by setting its conflict bit equal to 1. After that, the next-higher level of context around argument A is defined, and the new production is stored in the memory in the form (xAy, B, 0), where x(.)y is the next-higher context. Naturally, also this step must be preceded by a check of the conflict bit. The value 1 of the conflict bit indicates that checking must be continued at the still-next-higher level of context, and so on. Notice, however, that during one presentation of the training data, only the last member of the conflicting productions can be handled in this way; the original string relating to the former member has not been stored, and the new context for it cannot be defined. However, if we apply the training data afterwards reiteratively, the first conflicting member that was stored in the memory in its original form will be restored during the next cycle of iteration in the form (x'Ay', B' ,0) where x'(.)y' is its new context relating to the original data. The total number of iterations needed for the construction of the memory is thus equal to half of the number of contextual levels plus one; in the present case it means five cycles.

Example. Assume, for instance, that for the string "eisinki", conflicts relating to segment (s) up to the third level have occured. There would then exist in the memory productions of all the types ("s", B, 1), ("is", B', 1), ("isi", B", 1), ("eisi", B"', 1), ("eisin", B"", 0) and the value of 0 of the conflict bit indicates that only the last production is valid.

1.6.5 The Algorithm for the Correction of New Strings

The following steps define the algorithm by which T is obtained from S by transformations applied to the successive segments of S:

0. The erroneous string S to be handled is first divided into segments X_i, $i = 1, 2, \ldots, n$, in the natural order of their appearance, in accordance with the segmentation used in the construction of memory.

1. A search for X_1 from the memory is first made. If the search in unsuccessful, the production is simply replaced by the identity transformation $Y_1=X_1$; in other words, the best strategy is to keep a novel segment as such. If the search, on the other hand, is successful, a set $(X_1, B_1,$ conflict bit) is found from the memory. If the conflict bit is then 0, one should take for the corrected segment $Y_1=B_1$.

2. If the search for X_1 is successful but the conflict bit is 1, one has to try the new context one level up. Denote the new search argument by xX_1y. If the new search is now successful and the conflict bit is 0, the new B value is assigned to Y_1. On the other hand, if the conflict bit is still 1, the still-next-higher level of context x'(.)y' must be tried and the search repeated, and so on, until the conflict bit becomes 0, in which case the B value found at that level is accepted for Y_1.

3. In the case that during some step of this procedure the search started successfully at step 1 but later became unsuccessful, one has encountered a strange context (strange piece of string). In this case one has to resort to the *estimation procedure* as explained more closely in Sect. 1.6.6. The same is due if the highest contextual level is exceeded, and the conflicts are not yet resolved.

4. This procedure is repeated for segments X_2, X_3, etc. starting at step 1, after which the corrected string T is the concatenation $Y_1 Y_2 \ldots Y_n$.

1.6.6 Estimation Procedure for Unsuccessful Searches

It may be obvious that the production rules can only be applied to such segments in the source strings S that are familiar on the basis of given examples. There exist, however, some possibilities for an "intelligent guess" on the basis of strange segments, too.

The following simple procedure has been found to increase the correction accuracy by a few per cent (referring to the percentage of errors corrected). For each segment X_i to be processed we list down all the tentative productions B starting from context level zero up to that level where a strange segment was found. For the most probable Y_i one may then take that value that is obtained by majority voting over the above set of tentative productions. It has been found necessary to include X_i in the set of candidates, too.

1.6.7 Practical Experiments

The algorithm has been tested by extensive experiments using natural data, Finnish and Japanese speech. Our speech recognition system was used to produce the erroneous phoneme strings. Its natural transcription accuracy was between 65 and 90 percent, depending on the vocabulary used and the number of phonemic classes defined. It has to be noted that this method was not restricted to isolated words, but connected speech can be used for

Table 1.5. String correction experiments

	Transcription accuracy before correction, per cent	Transcription accuracy after correction, per cent	Percentage of errors corrected
Most frequent Finnish words	68.0	90.5	70
Japanese names	83.1	94.3	67
Business letters (in Finnish)	66.0	85.1	56

training and testing as well. The performance tests were always carried out using statistically independent test data, spoken on a different day than when data was used to construct the memory.

The memory was constructed completely automatically on the basis of this natural material, whereby typically 10000 to 20000 production rules, examples of which are shown in Table 1.6, were established. All experiments were performed on a standard personal computer without special hardware, using less than 500 kilobytes of mainframe memory; programming was made in high-level languages, except for the search memory that was programmed in assembly language. The transformation time in our experiments was less than 100 milliseconds per word.

Table 1.6. Examples of context-dependent productions, automatically obtained from speech data

a(hk)i	→	(ht)	ai(jk)o	→	(t)
a(hkk)o	→	(ht)	ai(k)aa	→	(t)
a(hko)o	→	(ht)	ai(k)ae	→	(k)
a(hl)o	→	(hd)	ai(k)ah	→	(k)
a(h)o	→	(hd)	ai(k)ai	→	(k)
ai(j)e	→	(h)	ai(k)ia	→	(kk)

Comment. It should be noted that the rather low figures for speech recognition accuracy given in Table 1.5 were obtained by our old speech recognition system from 1985–1987. The present transcription accuracy (as of 1995) for an average speaker and business letters would be over 80 per cent before postprocessing, and about 95 per cent after postprocessing, relating to the correctness of individual letters in the Finnish orthography.

2. Neural Modeling

Two different motives are discernible in neural modeling. The original one is an attempt to describe biophysical phenomena that take place in real biological neurons, whereupon it may be expected that some primitives or basic elements of information processing by the brain could be isolated and identified. Another one is a direct attempt to develop new devices based on heuristically conceived, although biologically inspired simple components such as threshold-logic units or formal neurons. The circuits thereby designed are usually called *artificial neural networks (ANNs)*.

One thing ought to be made clear: Are the contemporary ANN models meant for the description of the brain, or are they practical inventions for new components and technologies? It seems safe to say that the models of the 1940s and 1950s, although still primitive, were definitely meant to be of the former type, whereas most of the present ANNs seem to have been elaborated for new generations of information-processing devices. These two motives, however, may have been around all the time, at least subconsciously.

2.1 Models, Paradigms, and Methods

We are modeling our experiences all the time. Our thinking is based on mental images and ideas, which are projections of some internal representations from the brain into the exterior world. In that process our nervous system carries out the modeling. Our languages, spoken and written, make simplified models of various occurrences. In the history of mankind, mathematical modeling was first used in counting, then in geometry relating to land use and astronomy, and finally in all exact and even less exact sciences.

Modern neurophysiology, describes each neural cell as a complex dynamical system controlled by neural signals, slow electric fields, and numerous chemical transmitter and messenger molecules.

However, if one tries to take all known neurophysiological and neurochemical facts into account in a "model," for the description of even a single neural cell one would need more than all the computing resources that are available to a theorist nowadays. This would also put a stop to theoretical brain research.

For a prospective scientist who intends to select "neural networks" as a research topic it would therefore be advisable to first distinguish between three kinds of approach, any of which is possible. We call them *models*, *paradigms*, and *methods*.

A *model*, especially an analytical one, usually consists of a finite set of variables and their quantitative interactions that are supposed to describe, e.g., states and signals in a real system, often assumed to behave according to known, simplified laws of nature. A *paradigm* is a traditional or exemplary simplifying approach to complex problems, and it usually in a general way directs the selection of variables and processes to be studied, setting up of models (or in experimental research, other hypotheses), and interpretation of research data. One might say that a paradigm is a model or case example of a general theory. A *method* may ensue from models or paradigms, or be developed with no regard to them, and there is no other justification for a method than that it works effectively in applications.

The *statistical models* are usually not meant to describe the physical structures of the systems, but the statistical processes that are at work in producing the observable phenomena.

It is not always clear which ones of the above aspects are meant in the contemporary theoretical "neural networks" research. The researchers seem to be split into two parties: I call these *scholars* and *inventors*. The scholars have learned from classical science that a theory or model must always be verifiable by experiments; but in neural-network research this principle is bound to lead them to add more and more details, such as structural assumptions, nonlinearities, dynamic effects, and of course, plenty of parameters to the models to make the latter comply with experiments. The inventors, on the other hand, want to utilise their materials and tools most effectively, so if they use silicon for artificial neural circuits, they may not care, for instance, about metabolism and active triggering that are essential to the operation of biological neural cells, or synchronized oscillations that are known to take place in and between some brain areas. In the classical scientific terminology their "models" are then not "right." We should say, however, that the scholars and the inventors follow different paradigms, even if their objectives happened to be the same (to explain or artificially implement some parts of human information processing).

Different objectives then also ought to be distinguished: Are we trying to understand the brain more deeply, or to develop new technologies? Both of these aspects are important, of course.

2.2 A History of Some Main Ideas in Neural Modeling

Some theoretical views of thinking that might be regarded as modeling were already held by the Greek philosophers, first of all *Aristotle* (384–322 B.C.) [2.1]. Of the later philosophers who had mechanistic views of the nervous system one may mention empiricist philosophers of the 16th century and *Descartes* (1596–1650).

In a modern systematic quest, some mathematical biologists, e.g. *Nicholas Rashevsky* in Chicago in the 1930s, were trying to describe systems of neural cells in a similar way to that of *many-particle interactions* as described in physics. One of the main objectives was to demonstrate the spreading of activity and the propagation of waves in the cell mass. This physical view probably originally ensued from the microscopic studies made by *Camillo Golgi* about neural cells of various forms, and those by *Ramón y Cajal* who found that these cells, the neurons, are connected densely through the neural fibers, the axons. Both Golgi and Cajal won the Nobel prize in medicine in 1906. It has also been known for a long time that the neurons are active components capable of triggering electrical impulses.

Another research paradigm, which started in 1943 with the famous article of *Mc Culloch and Pitts* [2.2] was to regard the neurons as *threshold-logic switches*, i.e., elements that form weighted sums of signal values and elicit an active response when the sum exceeds a threshold value. It was speculated that arbitrary computing circuits could then be built of such elements, in principle at least. About this early research tradition one has to mention at least the adaptive network of *Farley and Clark* in 1954 [2.3], and the "Perceptron" of *Rosenblatt* [2.4], the "Adaline" of *Widrow and Hoff* [2.5], the time-delay networks of *Caianiello* [2.6], and the "learning matrices" of *Steinbuch* [2.7] around 1960. The inherent difficulty in trying to design structured, eventually multilayered networks of such elements was their switching threshold, which made the phenomena discontinuous; no calculus, e.g., to optimize the pattern classification accuracy, seemed to apply. It was not until the steep threshold was replaced by a smooth "sigmoid function" when one could use the gradient-descent techniques for the minimization of the average expected error. *Paul Werbos* [2.8] developed in 1975 the famous chain differentiation method, called error back-propagation, for layered feedforward networks. Today these networks are called "multilayer Perceptrons (MLPs)".

One of the oldest misinterpretations of the physiological observations is indeed the "all-or-none" principle, i.e., that the neurons are triggering standard impulses. Even if one does not consider individual neural impulses, one has often regarded the silent state of the neuron as "0" and its triggering at full frequency as "1," respectively. However, one can also find slow potentials, graded responses, graded triggering frequencies, and volleys of spikes that represent usual modes of operation. The analogy with statistical mechanics is neither completely realistic for another reason. Even if we take into consideration that the neuron states are continuous-valued, in most ANN paradigms the networks or interactions are only thought to consist of discrete nodes and arcs, with some processing properties assigned to both. Two remarks can immediately be made: 1. There exist plenty of different neuron types and dozens of chemical transmitters, with a different effect and at least very different time constants for each, 2. There exists a plethora of chemical agents, messengers and modulators that process information and control the level of activity and rate of learning in different subsystems of the brain in a diffuse, often nonspecific way. Quite honestly, one should admit that the brain is a mixture of a vast number of different nonlinear dynamical systems.

The networks built of the threshold-logic switches were also called "associative nets". It is true that elementary associative-memory effects can be implemented by them as well as the statistical-mechanics models. However, the neural-network researchers did not know that very effective *content-addressable memories (CAMs)* had been developed in computer technology, starting in the mid-1950s [1.6]. The purpose of the CAMs, however, was very practical and simple: to find from a database those stored, standard-format records, each of which contains a subset of component parts specified in the query. Such a CAM can be implemented by traditional digital components, albeit at higher cost than the addressed memories: for every stored bit, about half a dozen logic gates at minimum are needed. The CAMs have become important parts in virtual memory systems of contemporary computers, as the cache memories that are used to speed up the overall operations. For large databases, however, the CAM is a much too expensive solution, and can be replaced easily by programming tricks [1.6]. The "neural" associative memories would be even more costly.

Ever since the mid-1950s and even nowadays, the use of ANNs has also been suggested for performing *pattern recognition*, i.e., identification or categorization of observation vectors. Nonetheless it was realized already in the early 1960s that at least pattern recognition is *not* a content-addressing problem, because the natural input patterns, such as those that occur in visual scenes and speech vary in wide limits, for many different and obvious reasons. This was verified in the first practical tasks for which pattern recognition was used: analysis of medical images and multispectral images taken from satellites. Neither are the newer statistical-mechanics models particularly good at pattern recognition for the same reason. It has turned out that quite different approaches, statistical and structural, are then necessary. In statistical pattern recognition one first extracts a sufficient set of characteristic *features* from the primary input patterns, and then applies statistical decision theory for the identification and classification of the latter. In the structural methods one usually constructs some kind of syntactic representation based on the relations between the features. At present, the ANNs that are used for pattern recognition are provided with some kind of heuristically or mathematically designed preprocessing.

The third category of ANNs has its deepest roots in digital signal processing. For the effective encoding of the continuous-valued input data (signal) space one had to develop various *vector quantization (VQ)* methods (Sect. 1.5): the signal space was divided into cells, each one represented by a quantized model called the codebook vector, and the distance of an input vector from the closest codebook vector in the input space is the quantization error [2.9] [2.10]. In an optimal design of the "codebook" the average expected quantization error over all occurring signals is minimized. Vector quantization performs an elementary unsupervised clustering and classification of the input items, and is one of the cornerstones of digital telecommunications.

There were mathematical psychologists and biologists who in the mid-1970s independently arrived at similar results starting from modeling theories. The selection of the closest codebook vector called the "winner" in these modeling theories was not based on direct metric comparison, but on a collective interaction mechanism, which is found in certain brain-like laterally interconnected networks. The whole adaptive process is called *competitive learning*. One has to mention at least the following works in which this method was developed: [2.11–23] A concrete neurophysiological function explained by the competitive-learning theory is the emergence of the so-called *feature-specific cells*, i.e., neurons or neuron systems that respond to certain input signal patterns selectively. In the "learning subspace method" [1.3, 63, 64], which is a supervised competitive-learning method, the input pattern is not compared with a set of model patterns but a set of manifolds of model patterns; cf. also Sect. 1.4.3.

Above I have actually described the three main categories or paradigms that dominate artificial neural-network research. We may call them 1. The state transfer models, 2. The signal transfer models, 3. Competitive learning; we shall revert to them in Sect. 2.10

As a matter of fact, the state transfer models are special cases of nonlinear feedback circuits, and the signal transfer models are very similar to the expressions in mathematical approximation theory, which discusses nonlinear, multistage functional expansions. As mentioned above, competitive learning is related to vector quantization. In all the traditional fields there exist plenty of mathematical results that could be transferred to neural-networks research.

2.3 Issues on Artificial Intelligence

During the early phases of neural modeling up to about 1960, there did not yet exist any distinction between the objectives of artificial-intelligence (AI) research and modeling of neural networks. It was perhaps hoped that artificial intelligence would ensue from networks of neural models, when they were made extensive enough. However, since the capacities of the computers of that time were not sufficient for any large-scale simulations, and it was not yet known theoretically, how basic ANN circuits should be scaled up, ANN research entered a state of quiescence in the 1960s and early 1970s. In the mid-1960s the interest of most computer scientists was directed towards heuristic programming of the big third-generation computers, and it was more straightforward to implement logic and decision tables than learning theories using digital computers. Rule-based artificial-intelligence research dominated the scene since then up to the 1980s.

Nonetheless the final objective of AI research, as well that of the ANNs has always been to create autonomous intelligence. One can even discern two main lines of philosophies that have been called *strong AI* and *weak AI*, respectively. In the strong AI line of thought a computer is believed to be made to think, understand, and feel like a human being. According to weak AI, a computer is only regarded as a good tool in the development and testing of theories about the mind and cognitive processes.

I would like to call logic and rule-based AI *hard information processing*, and artificial neural networks as well as statistical learning theories *soft information processing*, respectively. This terminology is concordant with the term "soft computing," which includes ANNs, fuzzy logic, genetic algorithms, and some further formalisms: other names are "real-world computing" and "natural computing." At any rate, it has already been shown that many high-level cognitive functions can be implemented by intensive computation, whereas there remain fewer and fewer people who believe in hard AI.

One of the bottlenecks in rule-based AI is the combinatorial problem. There exist natural tasks such as image understanding, where it is simply impossible to create complete rule systems for any more complex natural scene. However, there also exists another simpler and more familiar task that could be used for the testing of AI methods, and that is crossword-puzzles solving. The way in which humans proceed is to refer to a vast knowledge data base associatively, but the associative recall cannot be based on any logic or simple similarity comparison; the matching conditions may relate to many different levels of abstraction. It is obvious that such associative structures exist in the human brain, but one does not understand them in detail; at least there are no counterparts for them in logic reasoning.

Also the models of neural networks must be developed further in the hierarchical direction before they can cope with this problem and many others.

2.4 On the Complexity of Biological Nervous Systems

It should be realized that the biological nervous systems are defined along the *cell membranes*, which often constitute active transmission media, and the operation of which is conditioned by chemical processes taking place within and between the cells. Such systems ought to be described by a vast number of very complicated nonlinear spatiotemporal partial differential equations. We have to take into account the dimensionality of the problem: the number of neural cells in the human brain is on the order of 10^{11}, and the length of neural wiring, the *axons*, of a single person would stretch several dozen times the circumference of the earth, or about two times to the moon and back. This description, however, is still very modest. The wiring is made in a very specific way. The main signal connections between the cells, the *synapses*, are like small biological organs, and there are on the order of 10^{15} in number and dozens of different types of them in the human nervous systems. The cell membranes are coated with molecular agglomerations forming electro-chemical pumps for light ions, and the cells and their parts communicate through many kinds of chemical *messenger* molecules, some of which are diffused into the medium and into the circulation. Part of the interactions is mediated through electric fields by so-called *electrotonic processing*. To describe the global meteorological system around the earth would be an extremely trivial task compared with description of the complete behavior of the brain.

Of course, by the same token as general physiology *compartmentalizes* the biological systems of animals and man into discrete functions for to study their interactions, one may try to compartmentalize the brain into macroscopic parts (not necessarily the same as anatomical parts), each of which seems to have a definite *function*. Thus one may talk of certain nuclei (formations of the brain) controlling the "arousal", other formations being in central position in controlling "emotions" or "feelings", part of the cortex (the so-called supplementary motor cortex) activating the "intentions", etc., and on a very general level one may be able to set up abstract system models for them that interact according to general control-theoretic principles.

It is true that experimental brain research has produced masses of new data in quite recent years, and certainly our present understanding of the brain has progressed far beyond that of, say, the 1960s; but another plain truth is that unlike in theoretical physics, no clear, unique, and generally accepted paradigm has emerged in the brain theory. One of the obvious reasons for this is that whereas the experimental and theoretical physicists have been educated by the same schools, the experimental and theoretical brain scientists represent quite different cultures and do not understand each others' language, objectives, and conclusions. Another, more concrete reason is that the targets of research, that is, all the existing nervous systems with their different phylogenetic and ontogenetic phases of development, are not describable by traditional mathematical models. For instance, in quantum electrodynamics, theoretical predictions and measurements can be made to agree to an accuracy of six to seven decimals even when very simple mathematical models are used, and in organic quantum chemistry, quantitative computerized models can be set up for molecules consisting of thousands of atoms; these models are able to predict what role a single electron in the structure has. In a system-theoretic approach to the neural functions it first seems that the biological neurons operate like simple frequency-modulated *pulse oscillators*, and the classical Hodgkin-Huxley equations [2.24] and dynamic models based on them describe the electrochemical phenomena at the cell membranes rather accurately. However, the *geometric forms* of the cells are already so complicated that even if the cell membrane is described by a great number of finite elements and represented by one thousand nonlinear spatiotemporal differential equations, this "model" does not yet describe any *information-processing unit*: the interaction with neighboring cells through electric fields and chemical reactions has thereby not yet been taken into account, and as a matter of fact there does not exist any general agreement about what analytical law for the neural control, i.e., what combined quantitative effect of various synaptic inputs onto the cell activity is valid. It may be justified to state that *in the classical sense of exact sciences there do not yet exist any quantitatively accurate information processing models for complete neural cells and nervous systems.* The most reliable, experimentally verified models in biophysics and biological cybernetics are still only *models of elementary phenomena.*

2.5 What the Brain Circuits Are Not

It is much easier to state with certainty what the biological neural systems cannot be, than to discover any details of their information processing functions. Nonetheless these statements can already help in formulating questions.

In this section I shall first present six assertions and then give the most convincing arguments in a concise form.

(i) The biological nervous systems do not apply principles of logic or digital circuits.
Arguments: If the alleged computing principle were asynchronous, the duration of the neural signals (impulses) should be variable, eventually indefinite, which is not true. If the principle were synchronous, one would need a global clock, to which the frequency of the pulses must be synchronized. Neither of these principles are possible. The neurons cannot be threshold-logic circuits, because there are thousands of variable inputs at every neuron, and the "threshold" is time-variable. The accuracy and stability of such circuits is simply not sufficient to define any Boolean functions. Moreover, the *collective* processes which are centrally important in neural computing are not implementable by simple logic circuits. Accordingly, the brain must be an *analog* computer, although a complex and adaptive one.

No digital arithmetic circuits have been found from the brain. One should realize that we are learning even the simplest arithmetic functions such as the addition and multiplication tables visually or acoustically, by rote learning.

(ii) Neither the neurons, nor the synapses are bistable memory elements. Corollary: Spin analogies are not relevant to neural networks.
Arguments: All physiological facts speak in favour of neurons acting as analog integrators, and the efficacies of the synapses change gradually. At least they do not flip back-and-forth.

(iii) No machine instructions or parallel or temporal control codes occur in neural computing. Corollary: There does not exist "brain codes."

Arguments: The most important reasons were mentioned in (i). The format of such codes in neural transmission cannot be maintained for any significant periods of time. It is not sufficient to encode the signals, but they should also be decoded; no physiological evidence for such "codec" protocol exists. Coded representations and "codec" circuits cannot be maintained during growth processes.

(iv) The brain circuits are not automata, and do not implement recursive computation.

Arguments: According to (i), the circuits are not stable enough to allow *recursive* definition of functions as in digital computers. In the latter, to implement the programmed procedures, recursions may consist of billions of steps. In neural circuits, the number of recursions can hardly be more than half a dozen. Further, unlike in digital computers, all internal states are not equally accessible, which would make the automatic operation unsatisfactory.

(v) The brain networks do not intrinsically contain circuits for problem-solving, abstract optimization, or decision making.

Arguments: High-level tasks belong to civilized behaviour for which external instruments are needed, and the strategies have been found by trial and error during a long time. The formalisms for problem solving must be learnt. Decision making needs analog estimation operations, and the final resolution, a kind of state of consciousness, is a global "epiphenomenon." It would also be very difficult to program the abstract optimization problems into neural circuits, because they would need *programming of the network structure*, as in classical analog computers.

(vi) Even on the highest level, the nature of information processing is different in the brain and in the digital computers.

Arguments: In order to emulate each other, at least on some level of abstraction the internal states of two computing systems must be equally accessible. Such an equivalence does not exist between the brain and the programming systems, as argued above. Artificial computers can neither collect and interpret *all human experiences* on which the assessment of values is based.

2.6 Relation Between Biological and Artificial Neural Networks

Let us first try to list down some of those characteristics that seem to be common to both the biological neural networks and most of the advanced ANNs:

1. Analog representation and processing of information (which allows any degree of asynchronism of the processes in massively parallel computation without interlocking).
2. Ability to average conditionally over sets of data (statistical operations).
3. Fault tolerance, as well as graceful degradation and recovery from errors; this ability has a great *survival value*.
4. Adaptation to changing environment, and emergence of "intelligent" information processing functions by self-organization, in response to data.

One should notice carefully that in order to implement these properties, very different technical solutions can be used (as they are also very divergent in nature); therefore the ANNs need not be faithful models of any biological circuits, but can utilize, e.g., specific properties of silicon. Also, while the analog representation seems to imply certain architectural solutions, it may be *emulated* digitally, especially if some kind of circuit technology is particularly profitable for it.

It has frequently been claimed, especially by biologically inclined scientists, that the ANN models are not effective until they imitate nature as accurately as possible. It is true that nature has had time and resources to "invent" many optimized mechanisms that we are supposed to be aware of. On the other hand, in technology it is possible to utilize many solutions that are impossible for nature: for instance, the freely rotating wheel does not exist in biology (because there cannot be any blood vessels through the bearing, and the wheel would remain without metabolism). Thus, nature may have to use complicated solutions for some simple auxiliary functions such as energy supply and stabilization, which may be solved by much more straightforward means, say, in electronics, *especially if servicing need not be automated*. Further it does not seem reasonable to copy details from nature to the models, unless one fully understands what they are meant for.

The simplified ANN models may only describe a piece of a tissue, say, not more than a couple millimeters in diameter. The network structures are thereby usually assumed regular, e.g., layered. In the brain networks, however, a high degree of specificity (different types of neurons and network structures) already exists in the microscopic scale. Often the interactions between neurons "skip" over neighboring cells. These facts have given rise to unjustified claims that the simplified, especially uniform network models are not "accurate." This is a fundamental misunderstanding. Such models have *not* been meant to describe the *complete anatomy*, but a *functional structure*, where the elements picked up from the tissue and activated for the functional operation only assume their role *in a particular situation*. Such a selection or sampling of the cells can be implemented by many kinds of known control, biasing, and presynaptic inputs at the neurons [2.25].

My personal view is that especially many supervised-learning models, although they look like neural networks, may not describe low-level neural anatomy or physiology at all: they should rather be regarded as *behavioral* models, or general models of learning, where the nodes and links represent abstract processors and communication channels, respectively.

2.7 What Functions of the Brain Are Usually Modeled?

The purpose of any biological nervous system is *centralized control* of vital functions such as sensorimotor functions, rhythms, blood circulation, respiration, metabolism, emergency functions, etc. For more developed forms of behavior, higher animals (including man) possess consciousness and the ability of planning various actions, referring to memory and imagination. In neurology and psychiatry it is further customary to distinguish two particular control systems in the brain, namely, that of *cognitive functions* and that of *emotional reactions*, respectively.

   Unless the modeling directly aims at simulation of psychological or physiological effects, the scope of artificial neural networks is usually more restricted. Some researchers tend to think that the ultimate goal of the ANN research is to develop *autonomous robots*; accordingly, the main functions to be implemented thereby are:

− sensory functions (or more generally, artificial perception)
− motor functions (manipulation, locomotion)
− decision making (reasoning, estimation, problem solving)
− verbal behavior (speech understanding and production, reading and writing, question-answering, annunciation).

2.8 When Do We Have to Use Neural Computing?

Several computational formalisms, traditional as well as newer, have been developed to cope with real-world situations: probabilistic reasoning, fuzzy logic, fuzzy set theory, fuzzy reasoning, artificial intelligence, genetic algorithms, and neural networks. As all of them have been claimed to be associated with thinking, it is not easy to specify what would put the artificial neural networks in a special position. Neural networks, however, should be considered at least in the following situations:

Lots of Noisy and Ill-Defined Data. Natural data are not always describable by low-order (first and second order) statistical parameters; their distributions are non-Gaussian; and their statistics are nonstationary. The functional relations between natural data elements are often nonlinear. Under these conditions adaptive neural-network computing methods are more effective and economic than the traditional ones.

Signal Dynamics. The more advanced neural-network models take into account signal dynamics, whereas, say, in artificial intelligence techniques and fuzzy logic, dynamic conditions are mainly dealt with on the level of symbols and abstract attributes.

Collective Effects. Although statistical dependencies between signal and control variables may be considered in any of the above-mentioned formalisms, only neural-network models rely on *redundancy of representations in space and time*. In other words, they normally disregard individual signal or pattern variables and concentrate on *collective properties* of sets of variables, e.g., on their correlations, conditional averages, and eigenvalues. Neural networks are often suitable for nonlinear estimation and control tasks in which the classical probabilistic methods fail.

Massively Parallel Computation. When signal and data elements are dynamic variables that change asynchronously, their analysis is best done by analog massively parallel computation, most naturally implemented by neural-network computing architectures. (Naturally analog computation can be *emulated* by digital computations.)

Adaptation. In order to deal with nonstationary data, the signal transfer and transformation properties of the networks should be made *adaptive*. In the simplest case the transfer parameters must depend on the transmitted signals for their optimal scaling and stabilization. The performance of the systems should be improved with use. More generally, the accuracy and selectivity of the complete system should be optimized with respect to some global performance index. Other desirable adaptive effects are optimal allocation of computing resources, and the associative memory function that corrects and complements noisy and incomplete input data patterns toward their ideal or standard forms and recalls relevant items. All these properties are inherent in neural networks.

Emergence of Intelligent Information Processing Functions. Only neural networks are able to create new information processing functions, such as specific feature detectors and ordered internal representations for structured signals, in response to frequently occurring signal patterns. Also, only neural networks can create higher abstractions (symbolisms) from raw data completely automatically. Intelligence in neural networks ensues from abstractions, not from heuristic rules or manual logic programming.

2.9 Transformation, Relaxation, and Decoder

To facilitate understanding of the very different philosophies advanced in the context of neural-network literature, it may be helpful to first clarify the three elementary functions that frequently occur in neural-network architectures, namely, *signal transformation*, *relaxation* of network activity, and *decoding* or *detection* of signal patterns.

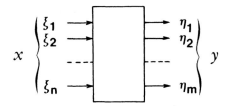

Fig. 2.1. Signal transformation

Signal Transformation. One of the basic operations in a signal network is transformation of the signal values (Fig. 2.1). Consider a set of real-valued (nonnegative) input signals $\{\xi_i\}$, which in biological neural networks are usually encoded or represented by impulse frequencies. These signals together form an n-dimensional real signal vector $x = [\xi_1, \xi_2, \ldots, \xi_n]^T \in \Re^n$ (for notations, cf. Chap. 1). The output signals from the network form another set $\{\eta_j\}$, where the encoding is similar as for the input. The output signals are regarded as an m-dimensional signal vector $y = [\eta_1, \eta_2, \ldots, \eta_m]^T \in \Re^m$. In the stationary state of signals, every output is a function of all the inputs, i.e.,

$$\eta_j = \eta_j(x) \,. \tag{2.1}$$

In general the η_j functions are nonlinear, whereas in the basic signal-transformation networks there is no sequential memory: even if the signals are changing in time, and there are signal delays in the network, every output $\eta_j(x) = \eta_j(x, t)$ is a single-valued function of the present and earlier input values $x(t'), t' \leq t$. In other words, although the signals are transmitted via different delays through the network, the output signals in signal-transformation networks are not assumed to *latch* to alternative states like in the well-known flip-flop circuits of computer engineering, or the multistable relaxation circuits discussed next.

Relaxation. In *feedback systems* part of the output signals or eventually all of them are connected back to the input ports of the network, and typically in ANNs the transfer relations are very nonlinear. In the simplest case the delay is assumed identical for all feedback signals (Fig. 2.2). Without much loss of generality in description, time-continuous signals are often approximated by constant values between the sampling instants, and in the *discrete-time representation* the system equation for the network reads

$$y(t+1) = f[x(t), y(t)] \,, \quad t = 0, 1, 2, \ldots \,. \tag{2.2}$$

The vector-valued input-output transfer function f is usually very non-linear, and its typical form is such that each component η_j of y is bounded at low and high limits, while between these values f may reflect almost linear properties. Systems with the above properties are often *multistable*: for the same input x the steady-state output y may have *many stable solutions*,

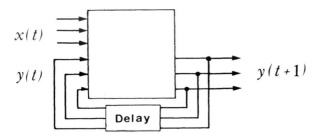

Fig. 2.2. Feedback system

which are then called *attractors* of this network. In a more general case, attractor networks are described by continuous-time differential equations of the type

$$dy/dt = f_1(x, y) \,, \tag{2.3}$$

where f_1 is such that its components tend to zero at low and high values of the respective components of y. In this book we shall not analyze the convergence properties of attractor networks except in some special cases: their discussions can be found in numerous textbooks of nonlinear systems theory, catastrophe theory, etc.

In the context of neural networks, the attractor networks are usually thought to operate in the following way. One first defines the initial conditions $y = y(0)$ representing some input information, whereafter the state of the system makes a sequence of transitions, or is *relaxed*, eventually ending up in an attractor. The latter then represents the outcome of computation, often some basic statistical feature. Attractor networks are used as associative memories and for finding solutions to simple optimization problems. Many competitive-learning networks such as the physiological SOM model discussed in Chap. 4 of this book belong partly to attractor networks.

Decoder. *Decoders* may be familiar from computer technology. An addressed memory location is activated for reading or writing by a special logic circuit, called the *address decoder*, which with a certain combination of binary input values (address) selects one of the locations for reading or writing (Fig. 2.3).

The so-called *feature-sensitive cells* in neural networks operate in a somewhat similar way, except that combinations of *real-valued signals* are used in place of the binary ones. Accordingly, for an input vector $x \in \Re^n$, only one of the m outputs of the network is supposed to give an "active" response, whereas the rest of the outputs remains passive. Many kinds of "decoders" or feature-sensitive cells will be discussed in this book, where they are used for feature detection, pattern recognition, etc. One has to realize that in such a circuit one does not consider amplitude relations of input-output transformations in the first place, like in feedforward networks. The exact output *value*

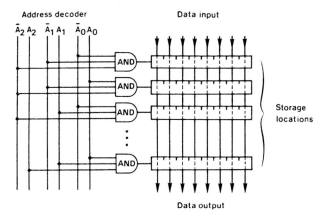

Fig. 2.3. Location-addressed memory provided with address decoder. A_0, A_1, A_2: address bits. $\overline{A}_0$, $\overline{A}_1$, $\overline{A}_2$: their logical negations, respectively. AND: logical product

is of little importance: it is the *location* of the response in the network that interprets the input.

2.10 Categories of ANNs

Traditionally, the following three categories of models are regarded as "pure" ANNs: *signal transfer networks*, *state transfer networks*, and *competitive-learning*.

Signal-Transfer Networks. In the *signal-transfer networks*, the output signal values depend uniquely on input signals; these circuits are thus designed for signal transformations. The mapping is parametric: it is defined by fixed "basis functions" that depend, e.g., on the available component technology. The basis functions can be fitted to data by algebraic computation or gradient-step optimization. Typical representatives are layered feedforward networks such as the *multilayer Perceptron* [2.26], the *Madaline* [1.44], and the feedforward network in which learning is defined by an *error-back-propagation algorithm* [2.8, 27]. The *radial-basis-function networks* [2.28] can be counted as signal-transfer neural networks, too.

Signal-transfer networks are used for the identification and classification of input patterns, for control problems, for coordinate transformations, and for the evaluation of input data.

State-Transfer Networks. In the *state-transfer networks*, which are based on relaxation effects, the feedbacks and nonlinearities are so strong that the activity state very quickly converges to one of its stable values, an *attractor*. Input information sets the initial activity state, and the final state represents the result of computation. Typical representatives are the *Hopfield network*

[2.29] and the *Boltzmann machine* [2.30]. The *bidirectional associative memory (BAM)* [2.31] also belongs to this category.

The main applications of state-transfer networks are in various associative memory functions and optimization problems. Although they have also been used for pattern recognition, their accuracy remains much below that of other ANNs.

Competitive Learning. In the *competitive-learning networks* the cells, in the simplest structures at least, receive identical input information, on which they compete. By means of lateral interactions, positive and negative, one of the cells becomes the "winner" with full activity, and by negative feedback it then suppresses the activity of all the other cells. This kind of action will be described later in Sect. 4.2.1 in more detail. For different inputs the "winners" alternate. If only the active cell is learning the input, in the long run each cell becomes sensitized to a different domain of vectorial input signal values, and acts as a decoder of that domain [2.11–14].

In the context of neural modeling, some researchers have pursued ideas similar to vector quantization discussed in Sect. 1.5 to explain the formation and operation of *feature-sensitive neural cells* [2.15, 16, 22, 23]. These modeling approaches have laid foundations to the theory of one of the most important categories of neural functions.

Many parts of the brain tissue, for instance the cerebral cortex, are organized spatially in such a way that there is discernible a clear geometric *local order* of the neural functions along with some feature coordinate values in that particular brain area. Of early theoretical works that aimed at the explanation of such an organisation one may mention [2.32–34].

In an attempt to implement a learning principle that would work reliably in practice, effectively creating *globally ordered maps* of various sensory features onto a layered neural network, this author formalized the self-organizing process in 1981 and 1982 into an algorithmic form that is now being called the *Self-Organizing (Feature) Map (SOM)* [2.35–37]. The SOM algorithm is the main topic of this book. In the pure form, the SOM defines an "elastic net" of points (parameter, reference, or codebook vectors) that are fitted to the input signal space to approximate its density function *in an ordered fashion*. The main applications of the SOM are thus in the *visualization of complex data in a two-dimensional display*, and *creation of abstractions* like in many clustering techniques.

Another development of vector quantization methods is the supervised-learning algorithm called *Learning Vector Quantization (LVQ)* [2.38–41]. In it, each class of vectorial input samples is represented by its own set of codebook vectors. The only purpose of LVQ is to describe class borders by the nearest-neighbour rule. The main applications of LVQ are thus in *statistical pattern recognition* or *classification*.

Other. It is difficult to specify what characteristics in general are distinctive to artificial neural networks. It seems, for instance, that lots of new algorithms of mathematical statistics are reported in journals and conferences as ANN methods. One may mention, for instance, the principal-component analyzer (PCA) networks [1.23–25] and independent-component analysis (ICA) [2.42, 43].

2.11 A Simple Nonlinear Dynamic Model of the Neuron

Up to this time, most ANN models, especially the signal-transfering feed-forward networks have assumed neurons similar to those introduced over 30 years ago, namely, multiple-input, single-output static elements that form a weighted sum of input signal values ξ_i called *input activation I_i*, and then amplify I_i in a nonlinear output circuit into η_i (Fig. 2.4).

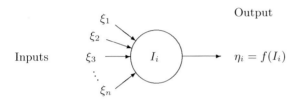

Fig. 2.4. Static nonlinear model for a neuron. Here $f(\cdot)$ is some monotonic function that has low and high saturation limits

Such components are easy to fabricate, but it would be almost as easy to add an *integrator* like in the classical analog computers (so-called differential analyzers) to the output circuit to make this component also solve *dynamical* problems.

At the other extremum of biophysical theories, there also exist neuron models that try to describe what electro-chemical triggering phenomena take place at the active cell membranes of biological neurons: the classical *Hodgkin-Huxley equations* [2.24] are most often cited in this context. Although the observable rhythms of input and output pulses become described rather faithfully in that way, the level of analysis is computationally too heavy for the description of signal transformations in complicated neural networks, not to talk about collective adaptive effects and computational architectures.

Over the years, biophysicists and biomathematicians have tried to reach a compromise between accuracy and simplicity in writing the neuron equations. One may mention, e.g., the *FitzHugh equation* [2.44] that was meant to simplify the Hodgkin-Huxley equations and facilitate a simple description of the axons as active transmission lines. Adaptive signal processing in systems of neurons, in which the transmission-line effects are no longer taken into account, but where the effect of synapses depolarizing the membrane is described rather realistically, can be described by various kinds of *Grossberg membrane equations* [2.15, 16, 20].

What this author suggested in 1988 [2.40] is no new biophysical theory in the same sense as the above works, but an ultimately simplified, practical *(phenomenological)* equation, the solution of which still exhibits the typical nonlinear dynamic behavior for many kinds of neuron rather faithfully, while being computationally light, from the point of view of both simulation and electronic implementation. The steady-state solution of this equation also defines a kind of "sigmoid" nonlinearity $f(\cdot)$ in a very straightforward way, as will be seen below. Contrary to many other physically oriented models that identify neural activities with membrane potentials, in this simple model *the nonnegative signal values directly describe impulse frequencies*, or more accurately, *inverse values of intervals* between successive neural impulses.

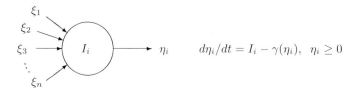

Fig. 2.5. Nonlinear dynamic model for a neuron

Consider Fig. 2.5. Here the ξ_j and η_i are nonnegative scalar variables, and the input activation I_i is some function of the ξ_j and a set of internal parameters. If I_i is linearized, the input coupling strengths μ_{ij}, frequently also called *synaptic weights* then enter the model. The neuron now acts like a "leaky integrator," and the leakage effect in this model is *nonlinear*. In a similar way as in many other simple models, the input activation I_i may be approximated by

$$I_i = \sum_{j=1}^{n} \mu_{ij}\xi_j \ . \tag{2.4}$$

Other laws for input activation that are nonlinear functions of the ξ_j may also be considered. The system equation for a "nonlinear leaky integrator" of Fig. 2.5 can be written

$$d\eta_i/dt = I_i - \gamma_i(\eta_i) \quad \text{with } \eta_i \geq 0 \ , \tag{2.5}$$

where $\gamma_i(\eta_i)$ is the leakage term, a nonlinear function of output activity. At least for large values of η_i, in order to guarantee good stability especially in feedback networks, this function must be *convex*, i.e., its second derivative with respect to the argument η_i must be positive. Such a leakage term describes the resultant of all the different losses and dead-time effects in the neuron rather realistically, as a progressive function of activity. One further remark must be made: (2.5) only holds if η_i is positive or zero. Because there is a lower bound for η_i that is assumed to be zero, $d\eta_i/dt$ must also be zero if η_i is zero and the right-hand side is negative. However, if η_i is zero and the right-hand side is positive, $d\eta_i/dt$ must be positive. To be formally more accurate, (2.5) should thus be written in a compact form:

$$d\eta_i/dt = \phi[\eta_i, I_i - \gamma_i(\eta_i)] \,, \quad \text{where } \phi[a,b] = \max\{H(a)b, b\} \,, \tag{2.6}$$

and H is the Heaviside function: $H(a) = 0$ when $a < 0$, and $H(u) = 1$ when $a \geq 0$.

An interesting result is obtained if the input signals are steady for a sufficient time, whereupon the system converges to the stationary state, $d\eta_i/dt = 0$. Then, in the domain in which γ_i is defined, we can solve for η_i:

$$\eta_i = \gamma_i^{-1}(I_i) \,, \tag{2.7}$$

where γ_i^{-1} is the *inverse function* of γ_i. By a suitable choice of the function γ_i, (2.7) can be made to represent various "sigmoid" forms of the neural transfer function in the stationary state (Fig. 2.6). Conversely, if the form of the "sigmoid" could be measured, then (2.7) would define the $\gamma_i(\cdot)$ to be used in this simple model. An element that is described by (2.7) could then be used to implement *static* neural networks in any of the categories introduced in Sect. 2.10. The dynamic formulation of the model, however, makes these circuits even more general than the static ones.

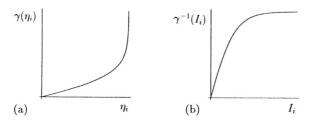

Fig. 2.6. Illustration of how the loss function determines the saturable nonlinearity in the steady state. **(a)** Plot of $\gamma(\eta_i)$ vs. η_i. **(b)** Plot of $\eta_i = \gamma^{-1}(I_i)$

2.12 Three Phases of Development of Neural Models

Another attempt to classify generations of models in neural-network theory is made in the following. We call these phases of modeling *memoryless* models, *adaptive* models, and *plasticity-control* models, respectively.

Memoryless Models. In the first modeling phase, starting with the classical *McCulloch-Pitts* network [2.2] the transfer properties of the network were assumed fixed. If feedback connections were added, such as in some laterally interconnected networks already devised by *Rosenblatt* [2.26], and also in some contemporary state-transfer models, the main effect thereby materialized was *relaxation* of activity distributions. Consider a neural medium in which signal activity A is a function of location; A is thereby mathematically similar to a scalar field. Let I be the external input "field" acting on the same locations. The dynamic state equation may then be written generally as

$$dA/dt = f(I, A) , \tag{2.8}$$

where f is a general function of I and A, and of location.

Adaptive Models. *Adaptation* and *memory* are properties that ensue from *parametric changes* in the network. In a general way, denoting the set of system parameters by a variable M, one may write

$$\begin{aligned} dA/dt &= f(I, A, M) , \\ dM/dt &= g(I, A, M) , \end{aligned} \tag{2.9}$$

where f and g are general functions of I, A, and M.

These equations describe the adaptive signal-transfer circuits. The variable M may be a function of location and represent an adaptive *bias*, but more generally it is regarded to be a function of *two* locations, whereby it describes the adaptive signal connection between these locations. A general property of adaptive circuits is that the time constants of the M equation are much larger than those of the A equation. The first attempts to model emergence of feature- sensitive cells and elementary forms of self-organizing mappings were based on these equations.

Plasticity-Control Models. It was felt by this author around 1980 that a model with adaptive connectivity parameters, with a more or less fixed value assigned to each, was not satisfactory enough to capture all aspects of self-organization. In neurophysiology, the learning rate of an interconnect, such as a synaptic connection, is called *plasticity*, and it is known that plasticity is a function of age but may also change temporarily; it is apparently controlled chemically. A recent theory of this author [2.45] advances the idea that the plasticity should be described and controlled by a third group of state variables called P below. There are no restrictions to separation of the dynamics of M and P. Relating to this third phase in modeling, we write the system equations as

$$\begin{aligned} dA/dt &= f(I, A, M) , \\ dM/dt &= g(I, A, M, P) , \\ dP/dt &= h(I, A, M, P) , \end{aligned} \tag{2.10}$$

where f, g, and h are still general functions. *Notice that P, unlike M, does not take part in the control of activity A.* This idea has been represented in the SOM equations of this author ever since 1981, and it results in the most effective forms of self-organization. One of the main purposes of the present book is to advance this kind of thinking and modeling.

The assumption of controllable plasticity solves one particular problem that concerns memory time constants. It has been difficult to understand how the memory traces can be formed promptly but stay for indefinite periods, if memory is a simple integral of signals. Now it could be natural to assume that the memory traces are formed only when the plasticity control P is on: this could be determined by special afferent inputs (I) corresponding to, say, the attentive state. When the conrol of P is off, the memory state can stay constant indefinitely.

2.13 Learning Laws

Before stipulating how the signals should modify the adaptive "synaptic" input weights μ_{ij} or other parameters of the neurons in learning, it will be necessary to make clear *what* the neural networks are supposed to do. For instance, one of such functions is associative memory, and another is detection of elementary patterns or features, but also many other functions occur in the neural realms.

2.13.1 Hebb's Law

Consider first the simplest classical learning law for formal neurons, the latter being defined by Fig. 2.4. If the networks made of such neurons are supposed to reflect simple *memory effects*, especially those of *associative* or *content-addressable memory*, a traditional model law that describes changes in the interconnects is based on *Hebb's hypothesis* [2.46]:

"When an axon of cell A is near enough to excite a cell B and re-peatedly or persistently takes part in firing it, some growth process or metabolic change takes place in one or both cells, such that A's efficiency, as one of the cells firing B, is increased."

Analytically this means that the weight μ_{ij} is changed according to

$$d\mu_{ij}/dt = \alpha\eta_i\xi_j \,, \tag{2.11}$$

where ξ_j is the presynaptic "activity", η_i is the postsynaptic "activity", and α is a scalar parameter named *learning-rate factor*. This law, generally called *Hebb's law*, has given rise to some elementary associative memory models, named *correlation matrix memories*, for instance in [2.47–49]. Many learning equations in ANN theory are called Hebbian, although the form of (2.11) may thereby be radically modified.

There are, however, certain severe weaknesses in (2.11), such as: since ξ_j and η_i describe frequencies and are thus nonnegative, the μ_{ij} can only change monotonically. For $\alpha > 0$, for instance, the μ_{ij} would grow without limits. Naturally some saturation limits might be set. In biology and the more developed ANNs, however, the changes of the μ_{ij} should be *reversible*. Apparently some *forgetting* or other normalizing effects should be added to (2.11).

In this context the associative memory function is omitted, and even if it were not, its modeling might benefit from the modifications made below. We will be more concerned about *feature-sensitive cells*, which have central roles in the classification functions both at the input layers of the neural networks, as well as inside them. It may suffice to consider here two basic modifications of Hebb's law only, the *Riccati-type* learning law, and the *principal-component-analyzer (PCA) type* law, respectively.

2.13.2 The Riccati-Type Learning Law

The first revision we make to Hebb's law is to question whether the modifications in μ_{ij} should be exactly proportional to the output activity of the *same* cell. There is some recent evidence that synaptic plasticity depends on the activity in *nearby* neurons (of numerous works, cf., e.g., [2.50, 51]), and maybe the output activity of neuron i itself (at which μ_{ij} is defined) could thereby even be zero. The latter kind of learning would be called *presynaptic*, and it has been measured long ago in primary animals such as the Aplysia [2.52]. Starting from a general point of view we now introduce a *scalar-valued plasticity-control function* P that may depend on many factors: activities, diffuse chemical control, etc. This function shall have a time-dependent *sampling effect* on the *learning* of the signals ξ_j. On the other hand, it seems safe to assume, in the first approximation at least, that the μ_{ij} are affected proportionally to ξ_j. The first term in the learning equation can then be written as $P\xi_j$, where P is a general (scalar) functional that describes the effect of activity in the surroundings of neuron i.

The second revision is inclusion of some kind of "active forgetting" term that guarantees that the μ_{ij} remain finite, and preferably the whole weight vector $m_i = [\mu_{i1}, \mu_{i2}, \ldots, \mu_{in}]^T \in \Re^n$ *or major parts of it* should become *normalized* to constant length. To this end we introduce a scalar-valued *forgetting-rate functional* Q, which is some function of synaptic activities at the membrane of cell i. However, we should beware of using a learning equation in which, with *passive* inputs, the μ_{ij} would *converge to zero*. Therefore the plasticity control P must affect the *total learning rate*, and we can describe this kind of "active learning and forgetting" by

$$d\mu_{ij}/dt = P(\xi_j - Q\mu_{ij}) . \tag{2.12}$$

In this expression, P may be thought to describe *extracellular* effects and Q *intracellular* effects, respectively, as will be specified more closely below.

Now P controls the overall *speed of learning* and can assume arbitrary scalar values in modeling: it mainly defines the *learning-rate time constant* $1/P$. Somewhat less freedom exists for the choice of Q. We shall resort to an idea that active synapses probably disturb other synapses of the same cell, at least in the long run, and somehow share the finite amount of local molecular and energetic resources available within the cell or some of its branches (although their amounts need not be conserved exactly). The *disturbing effect* may approximately be assumed proportional to *signal coupling* or activation, which at synapse j is described by $\mu_{ij}\xi_j$. Naturally we should take into account differences of individual synapses, their relative location at the cell's membrane, etc. Somewhere, however, we have to draw a line in modeling, and it seems proper to assume approximately that the "active forgetting" effect at synapse j is proportional to $\sum_{r=1}^{n} \mu_{ir}\xi_r$, on the average at least, where the sum extends over the whole cell, or a major part of it such as an apical dendrite (on of the most distant dendrites in, say, a principal neuron), including synapse j itself. Now (2.12) can be written explicitly as

$$d\mu_{ij}/dt = P \left(\xi_j - \mu_{ij} \sum_{r=1}^{n} \mu_{ir}\xi_r \right) . \tag{2.13}$$

This equation resembles the *Riccati-type equation* that this author introduced and solved in 1984 [2.53]: in vector form, (2.13) can be written

$$dm_i/dt = \alpha x - \beta m_i m_i^{\mathrm{T}} x , \tag{2.14}$$

where $\alpha = P$ and $\beta = PQ$. For a general functional form $x = x(t)$ (2.14) cannot be integrated in closed form; however, if x is assumed to have certain well-defined *statistical properties*, the so-called "most probable," or "averaged" trajectories of $m_i = m_i(t)$ can be solved. (For a general setting of such problems, cf. [2.54].)

It has been shown by this author in [2.53] that under rather general conditions, with stochastic $x = x(t)$, the solution $m_i = m_i(t)$ converges to a vectorial value, the *norm* of which only depends on α and β and not at all on the initial values $m_i(0)$, exact values of $x(t)$ during the process, or exact expression of P. This result is stated in terms of the following theorem.

Theorem 2.13.1. *Let* $x = x(t) \in \Re^n$ *be a stochastic vector with stationary statistical properties, let* $\bar{x}$ *be the mean of* x, *and let* $m \in \Re^n$ *be another stochastic vector. Let* $\alpha, \beta > 0$ *be two scalar parameters. Consider the solutions* $m = m(t)$ *of the Riccati differential equation*

$$\frac{dm}{dt} = \alpha x - \beta m(m^T x) . \tag{2.15}$$

For arbitrary nonzero x *and* $m^T x > 0$, *the Euclidean norm* $\|m\|$ *always approaches the asymptotic value* $\sqrt{\alpha/\beta}$, *and for stochastic* x *the most probable trajectories* $m = m(t)$ *tend to*

$$m^* = \bar{x}\sqrt{\alpha}/\|\bar{x}\|\sqrt{\beta} . \tag{2.16}$$

Proof. Before proceeding with the statistical problem, we shall derive an important auxiliary result. It is possible to analyze the convergence of $\|m\|$, the Euclidean norm of m, for rather general values of $x = x(t)$ multiplying both sides of (2.16) by $2m^T$, thereby obtaining, with $\dot{m} = dm/dt$,

$$2m^T \dot{m} = \frac{d}{dt}(||m||^2) = 2m^T x(\alpha - \beta ||m||^2) . \tag{2.17}$$

Notice that this equation is still expressed in the deterministic form. If we now had $m^T x > 0$, it would be immediately clear that $||m||^2$ would converge to $\alpha/\beta \overset{\text{def}}{=} ||m^*||^2$. In other words, $||m||$, or the length of vector m, would tend to the value $||m^*||$ for any $x = x(t)$ such that $m^T x > 0$. Mathematically, of course, the expression $m^T x$ could be ≤ 0, too. In neural models nonnegativity of $m^T x$ may mean that we are mainly working with *excitatory* synapses, for which at least most of the components of m are positive. For a mixture of excitatory and inhibitory synapses, it is possible to restrict the operation to the case $m^T x \geq 0$ by a suitable choice of $||x||$ and the initial conditions of m.

Statistical problems are in general very cumbersome mathematically; even after reasonable simplifications, the discussion may be disguised by complicated proofs, unless we rest content with the understanding of the basic operation. The first reasonable simplification made here is to consider only such input vectors $x(t)$, the statistical properties of which are constant in time, i.e., stationary stochastic processes, the subsequent values of which are statistically independent. In practice, this means that the "averaged" trajectories $m = m(t)$ are obtained by taking the expectation value of the sides of (2.15) conditional on m,

$$E\{\dot{m}|m\} = E\{\alpha x - \beta m^T x m|m\} . \tag{2.18}$$

Because x and m were assumed independent, we can denote

$$E\{x|m\} = \bar{x} \quad (\text{mean of } x) . \tag{2.19}$$

Since now $\bar{x}$ is a constant vector in time, we can analyze how the direction of m will change with respect to that of $\bar{x}$. Let us denote the angle between m and $\bar{x}$ by θ, whereupon one readily obtains

$$E\{d(\cos\theta)/dt|m\} \overset{\text{def.}}{=} E\left\{\frac{d}{dt}\left(\frac{\bar{x}^T m}{||\bar{x}|| \cdot ||m||}\right)\Big|m\right\}$$
$$= E\left\{\frac{d(\bar{x}^T m)/dt}{||\bar{x}|| \cdot ||m||} - \frac{(\bar{x}^T m)d(||m||)/dt}{||\bar{x}|| \cdot ||m||^2}\Big|m\right\} . \tag{2.20}$$

The first term of the last line of (2.20) can be calculated multiplying both sides of (2.15) by $\bar{x}^T(||\bar{x}|| \cdot ||m||)$, and the second term is obtained from (2.17) taking into account that $d(||m||^2)/dt = 2||m|| \cdot d(||m||)/dt$. Then

$$E\{d(\cos\theta)/dt|m\} = \frac{\alpha||\bar{x}||^2 - \beta(\bar{x}^T m)^2}{||\bar{x}|| \cdot ||m||} - \frac{(\bar{x}^T m)^2(\alpha - \beta ||m||^2)}{||\bar{x}|| \cdot ||m||^3} . \tag{2.21}$$

After simplification this equation becomes

$$E\left\{\frac{d\theta}{dt}\Big|m\right\} = -\frac{\alpha||\bar{x}||}{||m||}\sin\theta ; \tag{2.22}$$

for nonzero $||\bar{x}||$, the "averaged" direction of m can be seen to tend monotonically to that of $\bar{x}$.

Since by (2.17) the magnitude of m approaches the value $||m^*|| = \sqrt{\alpha/\beta}$, the asymptotic state must then be

$$m(\infty) = m^* = \frac{\sqrt{\alpha}}{\sqrt{\beta} \cdot ||\bar{x}||} \cdot \bar{x} ,\tag{2.23}$$

a vector that has the direction of $\bar{x}$, and length normalized to $\sqrt{\alpha/\beta}$.

This value is also obtained by determining the so-called *fixed point of m*, denoted m^*, for which $\mathrm{E}\{\dot{m}|m\} = 0$. Notice that mm^T is a matrix; then

$$\alpha\bar{x} - (\beta m^* m^{*T})\bar{x} = 0 .\tag{2.24}$$

A trial of the form $m^* = p\bar{x}$ with p a scalar constant will yield

$$a\bar{x} = \beta p^2 \bar{x}(\bar{x}^T\bar{x}) ,$$
$$m^* = \frac{\sqrt{\alpha}}{\sqrt{\beta}||\bar{x}||}\bar{x} .\tag{2.25}$$

2.13.3 The PCA-Type Learning Law

Another candidate for the learning law, introduced by *E. Oja* [2.55], will be discussed next. It is otherwise similar as (2.14), except that its right-hand side is multiplied by the expression

$$\eta_i = \sum_{j=1}^{n} \mu_{ij}\xi_j .\tag{2.26}$$

It has to be noted that this law was introduced in the context of *linear adaptive models*, whereupon η_i above would represent the true output, and for the first term we would then get a genuine Hebbian term, proportional to $\eta_i\xi_j$. The second term gives rise to normalization similar to that in Sect. 2.13.2. We shall only consider this model in mathematical form, because the same kind of physiological justification as with the Riccati-type learning (in which no linear transfer relation $\eta_i = \sum_{j=1}^{n} \mu_{ij}\xi_j$ need be assumed) is no longer valid.

Theorem 2.13.2. *Let $x = x(t) \in \Re^n$ be a stochastic vector with stationary statistical properties, let $\bar{x}$ be the mean of x, let C_{xx} be the covariance matrix of x, and let $m \in \Re^n$ be another stochastic vector. Let $\alpha, \beta > 0$ be two scalar parameters. Consider the solutions $m = m(t)$ of the differential equation*

$$\frac{dm}{dt} = \alpha\eta x - \beta\eta^2 m = \alpha x x^T m - \beta(m^T x x^T m)m .\tag{2.27}$$

For arbitrary nonzero x and $\eta > 0$, the norm $||m||$ always approaches the asymptotic value $\sqrt{\alpha/\beta}$, and for stochastic x the most probable trajectories $m = m(t)$ tend to the direction coinciding with that of the eigenvector of C_{xx} with the largest eigenvalue.

The original proof was presented in [2.55]. An alternative proof can be derived in a way analogous with Theorem 2.13.2, as pointed out in [2.53], pp. 96–98.

It may be interesting to note that the Riccati-type law learns *first-order statistics*, while the PCA-type law *second-order statistics*, respectively.

2.14 Some Really Hard Problems

The following long-standing problems have resisted the attempts to solve them satisfactorily. They have earlier been recognized in the fields of psychology, cybernetics, artificial intelligence, and pattern recognition. I shall concentrate on the following issues: 1. Invariances in perception. 2. Generalization. 3. Abstraction. 4. Hierarchical information processing. 5. Representation of dynamic events. 6. Integration of perceptions and actions. 7. Consciousness.

The traditional artificial neural networks as such perform only template matching between their input patterns and weight vector patterns, and have a very limited *invariance*, if any, to translation, rotation, size changes, or deformation. One trivial attempt to achieve the invariance with respect to *one* transformation group at a time is to pretransform the input pattern: for instance, in some technical problems translational invariance might suffice, and the Fourier transform of the input pattern has the wanted property. However, it has turned out very difficult to combine several transformations of different kinds in succession. A natural continuation is then to use *locally* defined transforms, such as those based on the *Gabor functions* [2.56]: the input pattern is thought to be decomposed into a great number of different local features, each one of which separately can be detected by a special "wavelet filter" invariantly to some transformation. We shall later see in Sect. 5.11 how more general wavelet filters for different kinds of transformations ensue automatically.

The neural network then carries out the classification on the basis of the mixture of such detected features. This is one of the prevailing strategies in contemporary image analysis. However, even this does not yet comply with, say, biological vision. Imagine that you were looking at letters or other objects of varying size. Your field of attention will adjust and focus to the object: it is as if first some dynamically adjusted "channels" were built up, and after that, the perceptive systems start using these channels. In my laboratory a small unpublished experiment has been carried out, where the test subject was looking at a computer screen where letters of different size appeared, and the time to recognize them was measured. This time was independent of the size of the letters if it remained constant, but increased proportionally to size difference, if the sizes of subsequent letters varied irregularly.

Many people identify *generalization* with clustering, i.e., grouping a set of representations together on the basis of some metric. The most trivial generalization of two or more metric representations is their average. However, this cannot be a sufficiently "general" view. First of all, generalization has to do something with the invariances: all members belonging to the same transformation group of an object are generalizations of this object. However, even such a similarity between the representations is not "general" enough. The concept of *functional similarity*, for instance, means that two items are similar if they have the same utility value, i.e., "fitness" to a subject. Two pairs of shoes, such as sandals and boots, may look completely different and have very few parts in common, and still they have almost the same utility value. Now it seems that the *fitness function* used, e.g., in evolutionary learning can serve as a basis of generalization, and graphs for such generalized similarity have been constructed automatically [2.57]. We shall revert to this in Sect. 5.9.2.

The problem of automatic *abstraction* can be taken quite pragmatically, segmenting and classifying an image (such as an aerial or satellite picture) at a given topical level, e.g. concentrating on land use, or one can have a more ambitious goal of trying to interpret the meaning of an observation in terms of increasingly more invariant or general concepts. Part of the problem is that these concepts, that is, their essential contents and characteristic properties ought to be found automatically.

The brain is without doubt *hierarchically* organized. For instance, in the sensorimotor control the flow of information from input to output follows the order: sensory organs → sensory nerves → central nervous system pathways with nuclei → sensory cortex → associative cortices → motor cortex → motor pathways with motor nuclei → motor nerves → muscles. However, in vision alone there exist many parallel paths of a different nature at each stage and feedback between the stages. Massive feedback exists, for instance, between the cortex and the thalamus. It seems that information processing is first "compartmentalized" into specialized functions, and there is some communication system between the compartments. The above example referred to one integrated function, which is to a great extent known, but there probably exist hierarchies also on the cortical level where the cognitive and emotional information processes take place. It would be overwhelming to build up these structures into models, or to expect that they would emerge in artificial systems, such as robots. Perhaps the most that one can do artificially at the moment is to try to implement some kind of active sensory processing: for instance, active computer vision would mean adaptation of the visual organs to the environment, controlling the attention and gaze to places with the highest information content, segmentation and identification of objects, and readjustment of the visual organs for improved perception at a higher level of abstraction.

The subjects, such as humans, animals and robots are usually moving in a natural environment, from which they obtain very variable and mutually occluding observations. The question is how one can encode the *dynamic* observations and produce representations of the objects somewhere on the basis of such observations. When I say "somewhere," it can only mean a place, and the representation must be locally confined. That is, at least some parameters in the representations of the objects and motion *must* be static, although, by means of feedback such as that in finite automata one can *reproduce* dynamic events, or recollections of the sensory occurrences. It is impossible to believe that memories of dynamic events could be stored in dynamic form, at least in such an unstable and heterogeneous medium as the brain mass.

From physiological observations especially on lower animals it has transpired that the sensory systems cannot be isolated from the motor functions: both of them must be embedded in some kind of *integrated* functional architecture. This does not support the idea, however, that such a nervous system would have a simple feedforward (MLP) structure, because the behavior of even the lowest species is already so complex and dynamic that some kind of planning functions are needed for that. Understanding how a complex sensorimotor behavior ensues from a small network of ganglia, which is the only "brain" that these animals have, remains an intriguing specific problem for further studies.

One of the big issues that has recently drawn attention is the nature of *consciousness*. Some thirty years ago it was a general notion among the philosophers that consciousness means the ability to make verbal statements of the observations or of the subject itself. However, the verbal statements produced by computers, even if they would describe the observation, are originally based on the consciousness of the programmers and not of the machine. On the other hand, although the ability of humans to make verbal statements is evidence of their consciousness, it cannot be used as a definition, because it is experimentally verified that animals, at least the higher ones, are also conscious. Neither can one say that the activity of a part of the brain means its being conscious: there are parts in the nervous system that are often inevitably active without any consciousness being associated with them. Maybe changes in activity are consequences of consciousness and not vice versa. Consciousness, as well as the emotional states do not have a clear physiological definition, and without such a definition it is impossible to construct network models for them.

One has to realize, however, that there are many concepts that are usually regarded as mental but which have a concrete explanation, without having to refer to consiousness. Take, for instance, the mental images: they differ from actual perceptions in that there are no sensory stimuli present. The mental images, however, are similar to the *virtual images* when one is looking at the so-called Fresnel hologram, illuminated by a spotlight. A virtual image of an object can be seen behind or in front of the hologram. Nonetheless this object is no longer there: only the optical traces on the film, which are not three-dimensional, are preserved. The illusion of the object is constructed by the viewer in his sensory system and the brain on the basis of the wavefronts entering the eye, but the "memories" exist on the hologram only. This author has experimented on models of neural networks in which similar "virtual images" have been generated [2.58], [2.53].

2.15 Brain Maps

It has been known for a long time that the various areas of the brain, especially of the cerebral cortex, are organized according to different sensory modalities: there are areas performing specialized tasks, e.g., speech control and analysis of sensory signals (visual, auditory, somatosensory, etc.) (Fig. 2.7). Between the primary sensory areas that comprise only ten per cent of the total cortical area, there are less well known *associative areas* onto which signals of different modalities convergence. The planning of actions takes place in the frontal lobe. More recent experimental research has revealed a *fine-structure* within many areas: for instance, the visual, somatosensory, etc. response signals are obtained in the same topographical order on the cortex in which they were received at the sensory organs. These structures are called *maps*. See, e.g., the somatotopic map shown in Fig. 2.8.

Before proceeding further, it will be necessary to emphasize that one can discern three different kinds of "maps" in the brain: 1. Feature-specific cells, the spatial location of which does not seem to correlate with any feature value. Examples of such cells are those which respond to human faces. 2. Anatomical projection of some receptor surface onto, say, the cortex. Examples are the areas in the visual and somatosensory cortices. 3. Ordered maps of some abstract feature, for which no receptor surface exist. An example is the color map in the visual area V4.

Without doubt the main structures of the brain networks are determined genetically. However, there also exists experimental evidence for sensory projections being affected by experience. For instance, after ablation of sensory

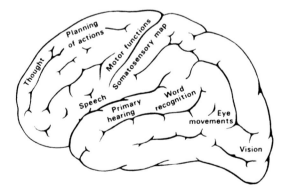

Fig. 2.7. Brain areas

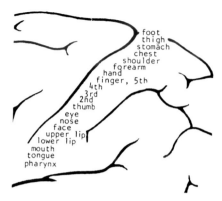

Fig. 2.8. The somatotopic map

organs or brain tissues, or sensory deprivation at young age, some projections are not developed at all, and the corresponding territory of the brain is occupied by remaining projections [2.59–61]. Recruitement of cells to different tasks depending on experience is well known. These effects should be explained by neural plasticity and they exemplify simple self-organization that is mainly controlled by sensory information.

Models for feature-specific cells are produced by the competitive-learning theories mentioned in Sect. 2.10. A pure topographical or anatomical order of the neural connections might also be explained rather easily: the axons of the neural cells, when growing toward their destinations, may be kept apart by the cell structures, and the destination is found by following the control action of chemical markers (e.g., [2.62–64]). Nonetheless, there is a somewhat confusing aspect that these connections are not always one-to-one, because there also exist in the signal-transmission paths processing stations (nuclei) in which signals are mixed. The above growth explanation breaks down completely in some maps where the primary signal patterns are ordered in a more abstract way. For instance, in the auditory cortex there exists the *tonotopic map* in which the spatial order of cell responses corresponds to the pitch or acoustic frequency of tones perceived (Fig. 2.9); although some researchers claim that this order corresponds to the location of resonances on the basilar membrane on the inner ear, the neural connections are no longer direct due to many nuclei in the auditory pathways. Some more confusion is caused by the fact that the neural signals corresponding to the lowest tones are not encoded by the position of the resonance; nonetheless the map of acoustic frequencies on the auditory cortex is perfectly ordered and almost logarithmic with respect to frequency, i.e., to the average statistical occurrence of the different tones.

Some evidence also exists for more abstract maps being formed elsewhere in the brain systems, even according to sensory experiences. Some maps of the geographic environment have been measured in the *hippocampus*, which

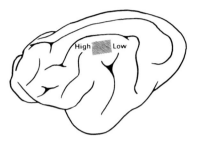

Fig. 2.9. The tonotopic map (of cat)

is a two-piece formation in the middle of the *cerebrum*; when, for instance, a rat has learned its location in a maze, then certain cells on the hippocampal cortex respond only when the animal is in a particular corner [2.65]. It is believed that many other kinds of maps exist in the cortex, in the thalamus, in the hippocampus, and other parts of the brain system. Examples of other feature maps include the directional-hearing map in the owl midbrain [2.66] and the target-range map in the auditory cortex [2.67]. These maps are usually rather small (2–3 mm) in diameter. As no receptive surface exists for such abstract features, the spatial order of representations must be produced by some self-organizing process, which occurs mainly postnatally.

It is amazing that although masses of experimental data and observations convincingly demonstrate the existence of a meaningful *spatial order and organization* of the brain functions, and this order seems to be ubiquitous in the nervous systems, the majority of works on artificial neural networks does not take it into account in any way. This order is useful for many different reasons: 1. By bringing mutually relevant functions close to each other spatially, the wiring can be minimized. 2. If the responses are spatially segregated (although the underlying network may be distributed), there will be minimal "crosstalk" between the functions, and the brain architecture can be made more logical and robust. 3. It seems that for effective representation and processing of knowledge one anyway needs some kind of metric "conceptual space" [2.68] to facilitate the emergence of natural concepts. If a logic concept were defined only in terms of its attributes, as made in the classical philosophy, one would run into the "property inheritance" problem [2.69], because a concept should contain all the attributes of its superordinates; but where could they be stored? It would be more natural that a concept is represented in terms of its relations to the most relevant concepts that are located in the neighborhood in the ordered "representation space."

The first work in which ordered representations, viz. ordered orientation-specific neural cells were produced in simulations was due to v.d. Malsburg in 1973 [2.32]. Later, topographically ordered anatomical projections between neuronal layers were analyzed, e.g., by Amari [2.34] and many others.

The possibility that the representation of knowledge in a particular category of things in general might assume the form of a feature map that is geometrically organized over the corresponding piece of the brain originally motivated the series of theoretical investigations reported in this book. The results thereby obtained have gradually led this author into believing that one and the same general functional principle might be responsible for self-organization of widely different representations of information. Furthermore, some results indicate that the very same functional principle that operates on a uniform, singly-connected, one-level medium is also able to represent *hierarchically related data*, by assigning different subareas of a homogeneous medium to different abstract levels of information. The representations over such maps then resemble *tree structures* that are obtainable by conventional taxonomic and clustering methods (Sect. 1.3.4). This result is theoretically very fundamental, since it reveals a new aspect of hierarchical representations; it is not necessary, and it may not even be always possible to arrange the processing units to form subsequent levels, as usually thought; although such levels may still exist, a more fundamental hierarchical organization may follow from *structured occupation and utilization of a uniform memory territory*.

The self-organizing process can be realized in any set of elements, illustrated schematically in Fig. 2.10, where only a few basic operational conditions are assumed. For simplicity, let the elements (for example, single neurons or groups of closely cooperating neurons) form a regular planar array and let each element represent a set of numerical values M_i, called a *model*. These values can correspond to some parameters of the neuronal system, and in contemporary computational neuroscience it is customary to relate these to synaptic efficacies. It is also assumed that each model is modified by the messages the element receives.

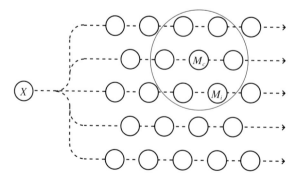

Fig. 2.10. A self-organizing model set. An input message X is broadcast to a set of models M_i, of which M_c best matches X. All models that lie in the vicinity of M_c (larger circle) improve their matching with X. Note that M_c differs from one message to another

Let there exist some mechanism by which an ingoing message X, a set of parallel signal values, can be compared with all models M_i. In brain theory it is customary to speak of "competition" between the elements, when they are stimulated by common input, and the element, whose parameters are fittest to this input is activated most. This element is called the "winner" if it succeeds in suppressing the activity in the neighboring neurons by, e.g., lateral inhibition. The "winner" model is denoted by M_c. Another requirement for self-organization is that the models shall be modified only in the local vicinity of the winner(s) and that all the modified models shall then resemble the prevailing message better than before.

When the models in the neighborhood of the winner simultaneously start to resemble the prevailing message X better, they also tend to become more similar mutually, i.e., the differences between all the models in the neighborhood of M_c are smoothed. Different messages at different times affect separate parts of the set of models, and thus the models M_i, after many learning steps, start to acquire values that relate to each other smoothly over the whole array, in the same way as the original messages X in the "signal space" do; in other words, maps related topologically to the sensory occurrences start to emerge as will be proven mathematically in this book. These three subprocesses – broadcasting of the input, selection of the winner, and adaptation of the models in the spatial neighborhood of the winner – seem to be sufficient, in the general case, to define a self-organization process that then results in the emergence of the topographically organized "maps."

We shall approach the basic self-organizing processes as adaptive phenomena that take place in simple physical systems. Somewhat similar, although simpler phenomena occur in the so-called retinotectal mappings [2.33], which are often obviously implemented by chemical labeling during the growth of the tissue. The present discussion, on the other hand, is related to an idealized hypothetical neural structure that is affected by sensory signals, and its purpose is to show what kind of structural and functional properties of systems are sufficient for the implementation of self-organization. It is thus the *process* in which we are primarily interested, not its particular implementation. If enough knowledge about accurate chemical phenomena would be available, the same principles might be demonstrable by a purely chemical model, too. However, it seems that the neural structures have ample details and mechanisms that are particularly favorable for the formation of feature maps with very high resolution and at an arbitrary level of abstraction.

In some investigations [2.36, 45] it has turned out that certain layered networks that consist of interconnected adaptive units have an ability of changing their responses in such a way that the location of the cell in the network where the response is obtained becomes specific to a certain characteristic feature in the set of input signals. This specification occurs *in the same topological order that describes the metric (similarity) relations of the input signal patterns*. As pointed out above, the cells or units do not *move* anywhere; it is the set of their internal *parameters* that defines this specificity and is made to change. Since such networks are usually mainly planar (two-dimensional) arrays, this result also means that there exist mappings that are able *to preserve the topological relations while performing a dimensionality reduction of the representation space.*

From a theoretical point of view, our main task is thus to find abstract self-organizing processes in which maps resembling the brain maps are formed, whereas it is of less interest whether the maps are formed by evolution, postnatal growth, or learning. Most self-organizing algorithms discussed in this book are mathematical processes for which the biological implementation need not be specified. Nonetheless the theoretical learning conditions seem to have counterparts in certain biological effects that we know; the particular physiological modeling attempt made in Ch. 4 must be understood as one case example, which does not necessarily exclude other explanations.

Perhaps the most intriguing theoretical finding is that a regular network structure, such as a two-dimensional lattice of neurons, can create ordered maps of any (metric) high-dimensional signal space, often in such a way that the main dimensions of the map thereby formed correspond to the most prominent features of the input signals. Examples are: maps of

- acoustical frequencies [2.37]
- phonemes of speech [2.37]
- color (hue and saturation) [2.70]
- elementary optical features [2.37]
- textures [2.71]
- geographic location [2.37]

etc.

In another important category of maps, the locations describe some *contextual features* that occur in *symbol strings.* For instance, if the symbols represent words of a natural language, such maps can identify the words as belonging to categories like nouns, verbs, adverbs, and adjectives [2.73–75] etc. There is some experimental evidence for such category maps existing in the brain, too [2.76–78].

In addition to serving as models for brain maps, the SOMs can be used as mathematical tools and they have practical applications ranging from speech recognition, image analysis, and industrial process control [2.72, 79] to the automatic ordering of document libraries, the visualization of financial records and the categorization of the electric signals in the brain; Ref. [2.80] lists almost 4000 works based on the SOM algorithm. The computed SOMs are very similar to many brain maps and they also behave dynamically in the same way, for example, their magnification is adjusted in proportion to the occurrences of the stimuli.

3. The Basic SOM

We shall now demonstrate a very important phenomenon that apparently has a close relationship to the brain maps discussed in Sect. 2.15 and occurs in certain spatially interacting neural networks. While being categorizable as a special kind of *adaptation*, this phenomenon is also related to *regression*. In regression, some simple mathematical function is usually fitted to the distribution of sample values of input data. The "nonparametric regression" considered in this chapter, however, involves fitting a number of *discrete, ordered reference vectors*, similar to the codebook vectors discussed in Sect. 1.5, to the distribution of vectorial input samples. In order to approximate continuous functions, the reference vectors are here made to define the nodes of a kind of hypothetical "elastic network," whereby the topological order characteristic of this mapping, and a certain degree of regularity of the neighboring reference vectors ensue from their *local interactions*, reflecting a kind of "elasticity." One possibility to implement such an "elasticity" would be to define the local interactions between the nodes *in the signal space* [3.1–7], whereas more realistic spatial interactions, from a neural modeling point of view, are definable between the neurons *along the neural network*. The latter approach is mainly made in this text.

Formation of a topologically ordered mapping from the signal space onto the neural network by this "regression" is already one interesting and important result, but from a practical point of view an even more intriguing result is that the various neurons develop into specific *decoders* or *detectors* of their respective signal domains in the input space. These decoders are formed onto the network *in a meaningful order*, as if some *feature coordinate system* were defined over the network.

The process in which such mappings are formed is defined by the *Self-Organizing Map (SOM)* algorithm that is the main subject of this book. The "feature maps" thereby realized can often effectively be used for the preprocessing of patterns for their recognition, or, if the neural network is a regular two-dimensional array, *to project and visualize high-dimensional signal spaces on such a two-dimensional display*. As a theoretical scheme, on the other hand, the adaptive SOM processes, in a general way, may explain the organizations found in various brain structures.

3.1 A Qualitative Introduction to the SOM

The Self-Organizing Map (SOM) is a new, effective software tool for the visualization of high-dimensional data. In its basic form it produces a *similarity graph of input data*. It converts the nonlinear statistical relationships between high-dimensional data into simple geometric relationships of their image points on a low-dimensional display, usually a regular two-dimensional grid of nodes. As the SOM thereby compresses information while preserving the most important topological and/or metric relationships of the primary data elements on the display, it may also be thought to produce some kind of *abstractions*. These two aspects, visualization and abstraction, can be utilized in a number of ways in complex tasks such as process analysis, machine perception, control, and communication.

In its present form the SOM was conceived by the author in 1982. At present, about 4000 research papers have been published on it; a list of them is available in the Internet [2.80]. Many tutorial texts and surveys on the SOM have appeared (Chap. 10). Several software packages that contain all the central procedures, a number of monitoring, diagnostic, and display programs, and exemplary data are freely available in the Internet, too (Chap. 8).

The SOM may be described formally as a nonlinear, ordered, smooth mapping of high-dimensional input data manifolds onto the elements of a regular, low-dimensional array. This mapping is implemented in the following way, which resembles the classical vector quantization (Sect. 1.5). Assume first for simplicity that the set of input variables $\{\xi_j\}$ is definable as a real vector $x = [\xi_1, \xi_2, \ldots, \xi_n]^T \in \Re^n$. With each element in the SOM array we associate a parametric real vector $m_i = [\mu_{i1}, \mu_{i2}, \ldots, \mu_{in}]^T \in \Re^n$ that we call a *model*. Assuming a general distance measure between x and m_i denoted $d(x, m_i)$, the *image* of an input vector x on the SOM array is defined as the array element m_c that matches best with x, i.e., that has the index

$$c = \arg\min_i \{d(x, m_i)\} \, . \tag{3.1}$$

Differing from the traditional vector quantization, our task is to define the m_i in such a way that the mapping is *ordered* and descriptive of the distribution of x. Before proceeding further, it must be emphasized that the "models" m_i need not be vectorial variables. It will suffice if the distance measure $d(x, m_i)$ is defined over all occurring x items and a sufficiently large set of models m_i.

Consider Fig. 3.1 where a two-dimensional ordered array of nodes, each one having a general model m_i associated with it, is shown. The initial values of the m_i may be selected as random, preferably from the domain of the input samples. Then consider a list of input samples $x(t)$, where t is an integer-valued index. Let us recall that in this scheme, the $x(t)$ and m_i may be vectors, strings of symbols, or even more general items. Compare each $x(t)$ with all the m_i and copy each $x(t)$ into a sublist associated with that

Inputs

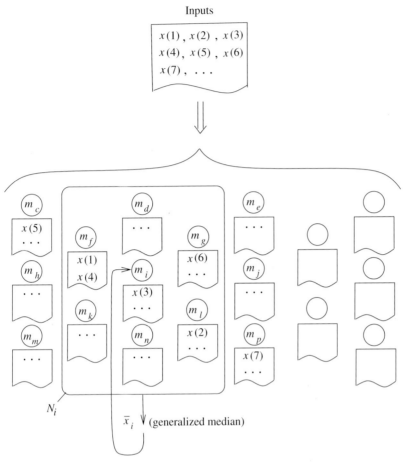

Fig. 3.1. Illustration of the batch process in which the input samples are distributed into sublists under the best-matching models, and then the new models are determined as (generalized) medians of the sublists over the neighborhoods N_i

node, the model vector of which is most similar to $x(t)$ relating to the general distance measure. When all the $x(t)$ have been distributed into the respective sublists in this way, consider the *neighborhood set* N_i around model m_i. Here N_i consists of all nodes up to a certain radius in the grid from node i. In the union of all sublists in N_i, the next task is to find the "middlemost" sample $\bar{x}_i$, defined as that sample that has the smallest sum of distances from all the samples $x(t), t \in N_i$. This sample $\bar{x}_i$ is now called the *generalized median* in the union of the sublists. (cf. Sect. 1.2.3) If $\bar{x}_i$ is restricted to being one of the samples $x(t)$, we shall indeed call it the *generalized set median*; on the other hand, since the $x(t)$ may not cover the whole input domain, it may be possible to find another item $\bar{x}'_i$ that has an even smaller sum of distances from the $x(t), t \in N_i$. For clarity we shall then call $\bar{x}'_i$ the *generalized median*.

Notice too that for the Euclidean vectors the generalized median is equal to their *arithmetic mean* if we look for an arbitrary Euclidean vector that has the smallest *sum of squares* of the Euclidean distances from all the samples $x(t)$ in the union of the sublists.

The next phase in the process is to form $\bar{x}_i$ or $\bar{x}'_i$ for each node in the above manner, always considering the neighborhood set N_i around each node i, and to replace each old value of m_i by $\bar{x}_i$ or $\bar{x}'_i$, respectively, in a simultaneous operation.

The above procedure shall now be iterated: in other words, the original $x(t)$ are again distributed into the sublists (which now change, because the m_i have changed), and the new $\bar{x}_i$ or $\bar{x}'_i$ are computed and made to replace the m_i, and so on. This is a kind of *regression process*.

There is an important question that we are not able to answer completely at the moment: even if we keep the $x(t)$ the same all the time, does this process converge? Do the m_i finally coincide with the $\bar{x}_i$? If the $x(t)$ and m_i are are Euclidean vectors, and a slightly modified (locally smoothed) distance measure is used in place of $d(x, m_i)$, the convergence has been proved by *Cheng* [3.8]. On the other hand, the original formulation of the SOM process that we shall discuss at length below is closely related to the above description.

Either the above process or the original SOM algorithm will then likely produce asymptotically converged values for the models m_i, the collection of which will approximate the distribution of the input samples $x(t)$, *even in an ordered fashion*. Let us look at Fig. 3.2, which represents the self-organized map of short-time acoustic spectra, viz. those taken from the (Finnish) speech at 20 ms intervals. The round symbols stand for the nodes of the SOM, and the curves inside them are *models* of spectra: low frequencies are on the left, high frequencies on the right, and the ordinates of the curve correspond to the voice intensity at the various spectral channels. One can immediately discern the similarity of the spatially adjacent models: the neighboring models look more similar than those farther apart. The collection of the models is also supposed to approximate the distribution of the input spectra.

At the first glance one might also think that the global order in the "map" reflects some kind of *harmonicity*: it seems as if every model were the average of the neighboring models, like in the theory of the harmonic functions. On a closer look, however, one can find several properties of the SOM that differ from the harmonicity. First of all, *there are no fixed boundary values for the models at the edges*: the values are determined freely in the regression process, when the neighborhood set N_i of the edge nodes is made to contain nodes from the inside of the grid. The nature of regression is slightly different at the edges and in the inside of the grid, resulting in certain border effects, Another deviation from the harmonicity, as later demonstrated by numerous examples, is that there are areas in the map where the models are very similar, but then there are again places where a bigger "jump" between the neighboring

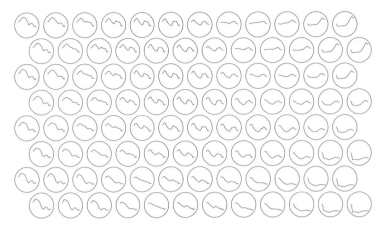

Fig. 3.2. In this exemplary application, each processing element in the hexagonal grid holds a model of a short-time spectrum of natural speech (Finnish). Notice that neighboring models are mutually similar

models is discernible. If the collection of the models has to approximate the distribution of the inputs, such uneven areas must exist in the map.

On the other hand, if one considers the process depicted in Fig. 3.1, one can easily realize how the definition of "harmonicity" must be modified in order to describe the input data: *the collection of models is ordered by definition, if each model is equal to the average of input data mapped to its neighborhood.*

3.2 The Original Incremental SOM Algorithm

We shall now proceed to the theory of the Self-Organizing Maps using the original SOM algorithm as the starting point. This algorithm can be seen to define a special recursive regression process, in which only a subset of models is processed at every step.

Consider Fig. 3.3. The SOM here defines a mapping from the input data space $\Re^n$ onto a two-dimensional array of nodes. With every node i, a parametric *model vector*, also called *reference vector* $m_i = [\mu_{i1}, \mu_{i2}, \ldots, \mu_{in}]^\mathrm{T} \in \Re^n$ is associated. Before recursive processing, the m_i must be *initialized*. In the preliminary examples we select random numbers for the components of the m_i to demonstrate that starting from an arbitary initial state, in the long run the m_i will attain two-dimensionally ordered values. This is the basic effect of self-organization. Later in Sect. 3.7 we shall point out that if the initial values of the m_i are selected as regular, the process can be made to converge much faster.

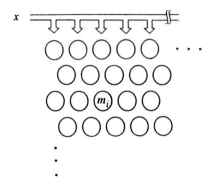

Fig. 3.3. The array of nodes (neurons) in a two-dimensional SOM array. The underlying mechanisms for parallel comparison of x and the m_i are not shown

The lattice type of the array can be defined to be rectangular, hexagonal, or even irregular; hexagonal is effective for visual display. In the simplest case, an input vector $x = [\xi_1, \xi_2, \ldots, \xi_n]^{\mathrm{T}} \in \Re^n$ is connected to all neurons in parallel via variable scalar weights μ_{ij}, which are in general different for different neurons. In an abstract scheme it may be imagined that the input x, by means of some parallel computing mechanisms, is compared with all the m_i, and the *location of best match* in some metric is defined as the location of the "response." We shall later in Chap. 4 demonstrate what kind of physical, eventually physiological parallel computing mechanism is able to do this. By computer programming, of course, location of the best match is a trivial task. The exact *magnitude* of the response need not be determined: the input is simply mapped onto this *location*, like in a set of *decoders*.

Let $x \in \Re^n$ be a stochastic data vector. One might then say that the SOM is a "nonlinear projection" of the probability density function $p(x)$ of the high-dimensional input data vector x onto the two-dimensional display. Vector x may be compared with all the m_i in any metric; in many practical applications, the smallest of the *Euclidean distances* $||x - m_i||$ can be made to define the *best-matching node*, signified by the subscript c:

$$c = \operatorname*{argmin}_{i} \{||x - m_i||\} \,, \quad \text{which means the same as}$$

$$||x - m_c|| = \min_{i} \{||x - m_i||\} \,. \tag{3.2}$$

In Sect. 3.3 and Chap. 4 we shall consider another, more "biological" matching criterion, based on the *dot product* of x and m_i.

During *learning*, or the process in which the "nonlinear projection" is formed, those nodes that are *topographically close* in the array up to a certain *geometric distance* will activate each other to learn something from the same input x. This will result in a local *relaxation* or *smoothing effect* on the weight vectors of neurons in this neighborhood, which in continued learning leads to *global ordering*. Consider the eventual convergence limits of the following learning process, whereupon the initial values of the $m_i(0)$ can be arbitrary, e.g., random:

$$m_i(t+1) = m_i(t) + h_{ci}(t)[x(t) - m_i(t)] , \qquad (3.3)$$

where $t = 0, 1, 2, \ldots$ is an integer, the discrete-time coordinate. In the relaxation process, the function $h_{ci}(t)$ has a very central role: it acts as the so-called *neighborhood function*, a smoothing kernel defined over the lattice points. For convergence it is necessary that $h_{ci}(t) \to 0$ when $t \to \infty$. Usually $h_{ci}(t) = h(||r_c - r_i||, t)$, where $r_c \in \Re^2$ and $r_i \in \Re^2$ are the location vectors of nodes c and i, respectively, in the array. With increasing $||r_c - r_i||$, $h_{ci} \to 0$. The average width and form of h_{ci} define the "stiffness" of the "elastic surface" to be fitted to the data points.

In the literature, two simple choices for $h_{ci}(t)$ occur frequently. The simpler of them refers to a *neighborhood set* of array points around node c (Fig. 3.4). Let their index set be denoted N_c (notice that we can define $N_c = N_c(t)$ as a function of time), whereby $h_{ci}(t) = \alpha(t)$ if $i \in N_c$ and $h_{ci}(t) = 0$ if $i \notin N_c$. The value of $\alpha(t)$ is then identified with a *learning-rate factor* $(0 < \alpha(t) < 1)$. Both $\alpha(t)$ and the radius of $N_c(t)$ are usually decreasing monotonically in time (during the ordering process).

Another widely applied, smoother neighborhood kernel can be written in terms of the Gaussian function,

$$h_{ci}(t) = \alpha(t) \cdot \exp\left(-\frac{||r_c - r_i||^2}{2\sigma^2(t)}\right) , \qquad (3.4)$$

where $\alpha(t)$ is another scalar-valued "learning-rate factor," and the parameter $\sigma(t)$ defines the width of the kernel; the latter corresponds to the radius of $N_c(t)$ above. Both $\alpha(t)$ and $\sigma(t)$ are some monotonically decreasing functions of time.

The algorithm chosen here for preliminary simulations is only representative of many alternative forms. If the SOM network is not very large (say, a few hundred nodes at most), selection of process parameters is not very crucial. We can also use the simple neighborhood-set definition of $h_{ci}(t)$.

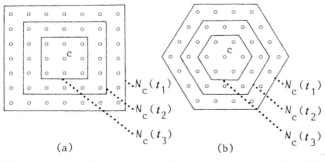

(a)   (b)

Fig. 3.4. a, b. Two examples of topological neighborhood $(t_1 < t_2 < t_3)$

Special caution, however, is required in the choice of the size of $N_c = N_c(t)$. If the neighborhood is too small to start with, the map will not be ordered globally. Instead various kinds of mosaic-like parcellations of the map are seen, between which the ordering direction changes discontinuously. This phenomenon can be avoided by starting with a fairly wide $N_c = N_c(0)$ and letting it *shrink* with time. The initial radius of N_c can even be more than half the diameter of the network! During the first 1000 steps or so, when the proper ordering takes place, and $\alpha = \alpha(t)$ is fairly large, the radius of N_c can shrink linearly to, say, one unit; during the fine-adjustment phase, N_c can still contain the nearest neighbors of cell c.

If the initial values have been selected at random, for approximately the first 1000 steps, $\alpha(t)$ should have reasonably high values (close to unity), thereafter decreasing monotonically. An accurate time function is not important: $\alpha = \alpha(t)$ can be linear, exponential, or inversely proportional to t. For instance, $\alpha(t) = 0.9(1 - t/1000)$ may be a reasonable choice. The *ordering* of the m_i occurs during this initial period, while the remaining steps are only needed for the fine adjustment of the map. After the ordering phase, $\alpha = \alpha(t)$ should attain small values (e.g., of the order of or less than .02) over a long period. Neither is it crucial whether the law for $\alpha(t)$ decreases linearly or exponentially during the final phase.

With very large maps, however, it may be important to minimize the total learning time. Then, selection of an optimal $\alpha(t)$ law may be crucial; cf. Sect. 3.8, where we shall consider "optimal" choices, essentially inversely proportional to t. Effective choices for these functions and their parameters have so far only been determined experimentally.

Since learning is a stochastic process, the final statistical accuracy of the mapping depends on the number of steps in the final convergence phase, which must be reasonably long; there is no way to circumvent this requirement. A "rule of thumb" is that, for good statistical accuracy, the number of steps must be at least 500 times the number of network units. On the other hand, the number of components in x has no effect on the number of iteration steps, and if hardware neural computers are used, a very high dimensionality of input is allowed. Typically we have used up to 100000 steps in our simulations, but for "fast learning," e.g., in speech recognition, 10000 steps and even less may sometimes be enough. Note that the algorithm is computationally extremely light. If only relatively few samples are available, they must be recycled for the desired number of steps.

Comment. In the preliminary examples described first our motive is to demonstrate that, starting from a random initial state, the mapping will be ordered in a finite number of learning steps. However, in practical applications one can start from an initial state that is already ordered and roughly complies with the input density function. If this is done (cf. Sect. 3.7), the learning process converges rapidly even if the neighborhood function were very narrow (of the order of its final form) and $\alpha(t)$ would start with low values, say, .2 or .1.

Examples of Ordering. It may be quite surprising that when starting with random $m_i(0)$, the reference vectors will attain *ordered* values in the long run, even in high-dimensional spaces. This ordering is first illustrated by means of two-dimensional input data $x = [\xi_1, \xi_2]^{\mathrm{T}} \in \Re^2$ that have some arbitrarily structured distribution. For simplicity, if x is a stochastic vector, its probability density function $p(x)$ is in this example assumed uniform within the framed areas in Fig. 3.5 and zero outside them. The topological relations between the neurons in a square array can be visualized by auxiliary lines that are drawn between the neighboring reference or codebook vectors (points in the signal space). The reference vectors in these graphs now correspond to the crossings and end points of this network of auxiliary lines, whereby the relative topological order becomes immediately visible.

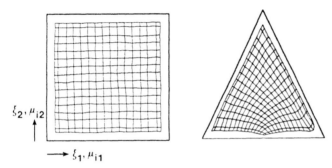

ξ_2, μ_{i2}

ξ_1, μ_{i1}

Fig. 3.5. Two-dimensional distributions of input vectors (framed areas), and the networks of reference vectors approximating them

The codebook vectors, while being ordered, also tend to approximate $p(x)$, the probability density function of x. This approximation, however, is not quite accurate, as will be seen later.

The examples shown in Fig. 3.5 represent the approximately *converged* state of the weight vectors. The different units have clearly become sensitized to different domains of input vectors in an orderly fashion. There is a boundary effect visible in Fig. 3.5, a slight contraction of the edges of the maps. We shall analyze this effect in Sect. 3.5.1 On the other hand, the density of weight vectors is correspondingly higher around the contraction. The relative contraction effect diminishes with increasing size of the array.

Examples of intermediate phases that occur during the self-organizing process are given in Figs. 3.6 and 3.7. The initial values $m_i(0)$ were selected at random from a certain (circular) support of values, and the structure of the network becomes visible after some time. Notice that the array can be, e.g., one-dimensional although the vectors are two-dimensional, as shown in

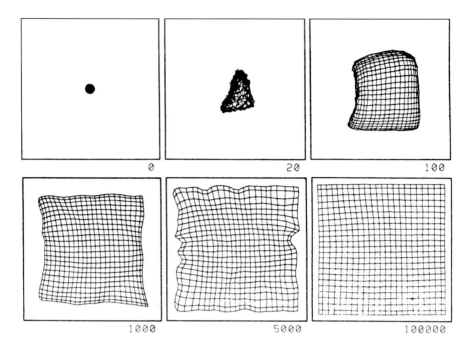

Fig. 3.6. Reference vectors during the ordering process, square array. The numbers at lower right-hand corner indicate learning cycles

Fig. 3.7. The "order" thereby created resembles a Peano curve, or fractal form.

Calibration. When a sufficient number of input samples $x(t)$ has been presented and the $m_i(t)$, in the process defined by (3.2) and (3.3), have converged to practically stationary values, the next step is *calibration* of the map, in order to locate images of different input data items on it. In practical applications for which such maps are used, it may be self-evident how a particular input data set ought to be interpreted and labeled. By inputting a number of typical, manually analyzed data sets, looking where the best matches on the map according to Eq. (3.2) lie, and labeling the map units correspondingly, the map becomes calibrated. Since this mapping is assumed to be continuous along a hypothetical "elastic surface", the unknown input data are approximated by the closest reference vectors, like in vector quantization.

Comment. An "optimal mapping" might be one that projects the probability density function $p(x)$ in the most "faithful" fashion, trying to preserve at least the local structures of $p(x)$ in the output plane. (You might think of $p(x)$ as a flower that is pressed!)

It has to be emphasized, however, that description of the *exact* form of $p(x)$ by the SOM is not the most important task. It will be pointed out that the SOM automatically finds those *dimensions* and *domains* in the signal

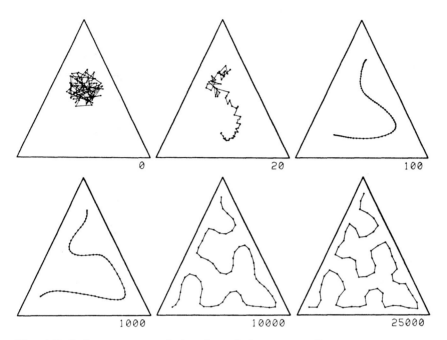

Fig. 3.7. Reference vectors during the ordering process, linear array

space where x has significant amounts of sample values, conforming to the usual philosophy in regression problems.

3.3 The "Dot-Product SOM"

It has sometimes been suggested that x be normalized before it is used in the algorithm. Normalization is not necessary in principle, but it may improve numerical accuracy because the resulting reference vectors then tend to have the same dynamic range.

Another aspect is that it is possible to apply many different metrics in matching; then, however, the matching and updating laws should be mutually *compatible* with respect to the same metric. For instance, if the dot-product definition of similarity of x and m_i is applied, the learning equations should read

$$x^{\mathrm{T}}(t)m_c(t) = \max_i\{x^{\mathrm{T}}(t)m_i(t)\} , \qquad (3.5)$$

$$m_i(t+1) = \begin{cases} \dfrac{m_i(t) + \alpha'(t)x(t)}{||m_i(t) + \alpha'(t)x(t)||} & \text{if } i \in N_c(t) , \\[2ex] m_i(t) & \text{if } i \notin N_c(t) , \end{cases} \qquad (3.6)$$

and $0 < \alpha'(t) < \infty$; for instance, $\alpha'(t) = 100/t$. This process automatically normalizes the reference vectors at each step. The normalization computations slow down the training algorithm. On the other hand, during matching, the dot product criterion applied during recognition is very simple and fast, and amenable to many kinds of simple analog computation, both electronic and optical. It also seems to have a connection to physiological processes, as later discussed in Chap. 4.

Besides the Euclidean distances and dot products, many other matching criteria can be used with the SOM. Sect. 3.12 points out how often other SOM algorithms may be derivable.

This author has frequently visualized the self-organizing process by means of a mechanical gadget, depicted in Fig. 3.8. A regular 4 by 4 array of rotary disks, each representing a neuron, has been pinned onto a substrate. An arrow, representing a normalized two-dimensional reference vector m_i, is painted on each disk. The initial orientations of the reference vectors can be randomized. The sequence of training vector values is selected at random, too, to represent samples of $p(x)$. The "winner" neuron is the one that has the smallest angle of its reference vector with the input vector; the reference vectors of all neurons in the neighborhood set (enclosed by the set line in Fig. 3.8) are corrected by a fraction of the angles between x and the m_i. After a few corrective steps the m_i values start looking smoothed and ordered, as seen from the series of pictures in Fig. 3.8.

3.4 Other Preliminary Demonstrations of Topology-Preserving Mappings

3.4.1 Ordering of Reference Vectors in the Input Space

The computer simulations given in this section will further illustrate the effect that the reference vectors tend to approximate various distributions of input vectors in an orderly fashion. In most examples, the input vectors were chosen two-dimensional for visual display purposes, and their probability density function was uniform over the area demarcated by its borderlines. Outside these borders the probability density function had the value zero. The vectors $x(t)$ were drawn from these density functions independently, at random, whereafter they caused adaptive changes in the reference vectors m_i.

The m_i vectors have mostly been illustrated as points in the same coordinate system where the $x(t)$ are represented; like earlier, in order to indicate to which unit each m_i belongs, the end points of the m_i have been connected by a lattice of lines that conforms to the topology of the processing unit array. A line connecting two weight vectors m_i and m_j is thus only used to indicate that the two corresponding units i and j are adjacent in the "neural" network.

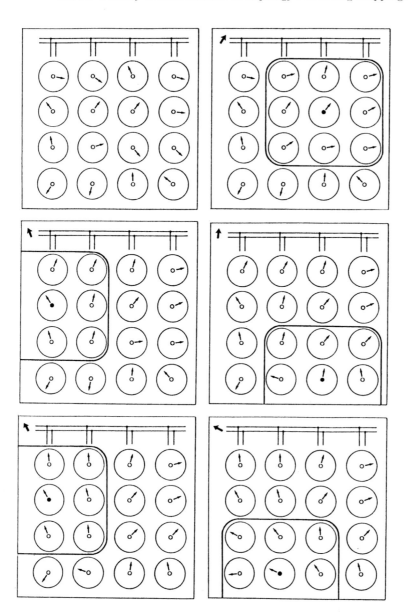

Fig. 3.8. This sequence of pictures illustrates successive phases in the operation of a two-dimensional dot-product SOM with two-dimensional input and reference vectors (arrows). The first picture represents the initial state. In the rest of the pictures, the small arrow in the upper left corner represents the input vector x, which specifies the best-matching reference vector, "winner". A set line drawn around the "winner" defines the neighborhood set N_c, in which (in this demonstration) the correction of each arrow was half of the angular difference of x and m_i. A smooth order of the m_i is seen to emerge

The results are still more interesting if the input vectors and the array have different dimensionalities: Fig. 3.9 illustrates a case in which the vectors were three-dimensional, but the array was two-dimensional. For clarity, $p(x)$ and the network formed of the m_i have been drawn separately, as explained in the caption of Fig. 3.9.

Comments. One remark ought to be made now in relation to Fig. 3.9. Some reference vectors seem to have remained outside $p(x)$. This, however, should not be interpreted as an error, as long as the SOM "net" is regarded to have a certain degree of "stiffness" and represent a nonparametric *regression*. In Sect. 5.4 we shall demonstrate how the SOM could be made to describe a structured $p(x)$ better, but is it then any longer a regression?

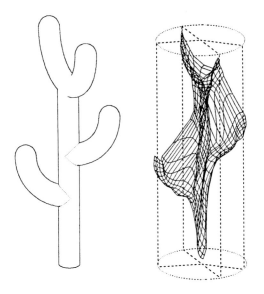

Fig. 3.9. SOM for a structured distribution of $p(x)$. For clarity, the three-dimensional $p(x)$, having uniform density value inside the "cactus" shape and zero outside it, is shown on the left, whereas the "net" of reference vectors is displayed on the right in a similar coordinate system

Before proceeding further one should also note that there is yet no factor present that would define a particular *orientation* of the map. Accordingly, the latter can be realized in the process in any mirror- or point-symmetric inversion. If a particular orientation had to be favored, the easiest way to reach this result would be to choose the initial values $m_i(0)$ asymmetrically. Since the symmetries may not be as interesting as the self-organizing behaviour itself, we shall ignore them in the preliminary examples at least. Practical questions of this kind will be discussed in Sect. 3.13.

The SOM as a Nonlinear, Adaptive Projection Screen. Figure 3.9 also illustrates another aspect of the SOM, namely, that of a *nonlinear projection*. One may regard the elastic network as a flexible *projection screen* that is first fitted through the distribution of the data points. After that, any point of $\Re^3$ (or in the general case, of an arbitrary-dimensional input space) will become projected onto the closest node of the "screen." This projection may not be orthogonal, though, because the screen is not continuous: it is only approximately orthogonal.

Automatic Selection of Feature Dimensions. There are two opposing tendencies in the self-organizing process. First, the set of weight vectors tends to describe the density function of the input vectors. Second, local interactions between processing units tend to preserve *continuity* in the double (two-dimensional) sequences of weight vectors. A result of these opposing "forces" is that the reference vector distribution, tending to approximate a smooth hypersurface, *also seeks an optimal orientation and form in the pattern space that best imitates the overall structure of the input vector density.*

A very important detail of the above reference-vector distribution is that it automatically tends to find those two dimensions of the pattern space where the input vectors have a high variance and which, accordingly, ought to be described in the map. As this effect might otherwise remain a bit obscure, the following extremely simple experiment is intended to illustrate what is meant. It is believed that the result that is here demonstrated using a one-dimensional topology (linear array of processing units) and a simple two-dimensional input density function, is easily generalizable for higher-order topology, arbitrary dimensionality of the input-vector density function, and structured distribution of the input samples.

Assume that the system consists of only *five* neurons connected as a linear open-ended array. Their reference vectors $m_i = [\mu_{i1}, \mu_{i2}]^{\mathrm{T}}, i = 1, 2, \ldots, 5$ and the components of the input vectors $x = [\xi_1, \xi_2]^{\mathrm{T}}$ are represented as an already familiar illustration in Fig. 3.10. The variances of ξ_1 and ξ_2 are now selected differently as shown by the borderlines in Fig. 3.10. As long as one of the variances is significantly higher, the weight vectors form an almost straight line that is aligned in the direction of the greater variance.

On the other hand, if the variances are almost equal, or if the length of the array is much greater than the range of lateral interaction, the straight form of the distribution is switched into a "Peano curve". The transition from straight to curved line is rather sharp, as shown in Fig. 3.11. Here the variances are fixed but the length of the array is varied; the borders of the Voronoi sets have also been drawn to the picture.

The next picture, Fig. 3.12 further illustrates what may happen when the input vectors have a higher dimensionality than the network topology (which in this case was two). As long as variance in the third dimension (ξ_3) is small enough, the map remains straight. However, with increasing variance and short lateral interaction range the map tends to become corrugated, and in

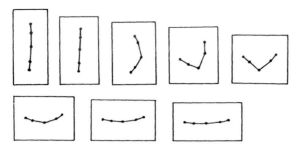

Fig. 3.10. Automatic selection of dimensions for mapping

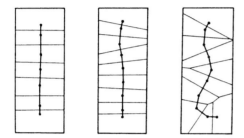

Fig. 3.11. Distribution of reference vectors with different length of a linear array

this connection one should note the "zebra stripes" that have been found in brain maps experimentally. *Here the "stripes" have a very simple and natural explanation, namely, they occur whenever a two-dimensional map tries to approximate a higher-dimensional signal distribution which has significant variance in more than two dimensions.*

The reader might be curious to know the mathematical explanation of this intriguing self-ordering effect. The phenomenon is actually rather delicate and needs a lengthy discussion, as will be seen in the rest of this book. We shall start the mathematical discussion in Sect. 3.5.

3.4.2 Demonstrations of Ordering of Responses in the Output Space

All simulations reported in this subsection were performed in our group in 1982, and most of them appeared in [2.37].

"The Magic TV". The following demonstration, Fig. 3.13 shows another, more concrete example of self-organization. It describes a hypothetical image-transferring system (dubbed here "The Magic TV") in which there are no control mechanisms for the encoding of the position of picture elements, but where the order of image points in the output display automatically results in the self-organizing learning process from their topological constraints. The result is a one-to-one mapping of the points of the input plane onto the points

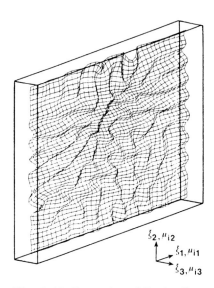

Fig. 3.12. Formation of "stripes"

in the output plane; in the transmission of the signals, however, this order was never specified explicitly. The system has to deduce the order gradually from the relations that are implicit in the transmitted signals. This system consists of a primitive "TV camera" and an adaptive system of the above type. The camera is thought to have a very poor optical system, such that whenever a spot of light appears on the input plane, it forms a very diffuse focus onto the photocathode. Assume that the cathode has three sectors, each producing a signal directly proportional to the area that is illuminated by the broad focus. Now let the light spot move at random in different points in the input plane, with a probability density that is uniform over a square area. The resulting signals ξ_1, ξ_2, and ξ_3 are then transmitted to the processing-unit array, where they cause adaptive changes. Let this process continue for a sufficient time, after which a test of the system is performed. This test is accomplished, e.g., by recording the output of each unit in turn and *locating that point in the input plane where the light spot must be in order to cause the best match in the unit under consideration.*

The resulting output map can be tested in either of the following ways: A) One can look at which processing unit each of the test vectors in turn has the best match, and call this unit the *image* of the test vector. The array is labeled accordingly. B) One can test to which training vector (with known classification) each of the units has become most sensitive, making the best match.

Obviously these tests are related, but the respective output maps look slightly different. With Test A, two or more test vectors may become mapped onto the same unit while the classification of some units may be left undefined.

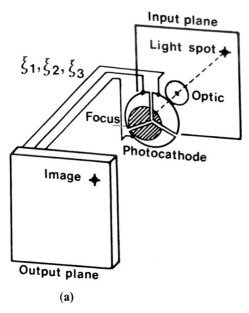

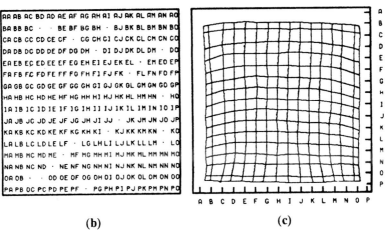

(b)   **(c)**

Fig. 3.13. "The Magic TV" **(a)** System **(b)** Output plane. Pairs of letters correspond to processing units, labeled by the images of test vectors (with coordinates
defined in **(c)**). **(c)** Input plane showing those points to which the various processing units (corresponding to nodes of the net) have become most sensitive

Test B always defines a unique matching input to every output unit but it
eventually happens that some input vectors, especially near the edges, do not
become mapped at all.

Mapping by a Feeler Mechanism. The purpose of this example is to
demonstrate that a map of the environment of a subject can be formed in

a self-organizing process whereby the observations can be mediated by very rude, nonlinear, and mutually dependent mechanisms such as arms and detectors of their turning angles. In this demonstration, two artificial arms, with two joints each, were used for the feeling of a planar surface. The geometry of the setup is illustrated in Fig. 3.14. The tips of both arms touched the same point on the plane, which during the training process was selected at random, with a uniform probability density over the framed area. At the same time, two signals, proportional to the bending angles, were obtained from each arm; these signals were led to a self-organizing array of the earlier type, and adaptation of its parameters took place.

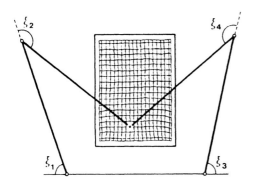

Fig. 3.14. Map of a feeler mechanism. The network of lines drawn onto the input plane shows a "virtual image" of the weight vectors: a crossing in this network is the input point to which the corresponding neuron in the SOM array is "tuned"

The lattice of lines that has been drawn onto the framed area in this picture represents a *virtual image* of the reference vectors, i.e., showing to which point on the plane each unit is most sensitive. When this point is touched, the corresponding processing unit makes the best match between x and m_i. One might also define the map so obtained as the map of the *receptive fields* of the array units. It can be tested for both arms separately, i.e., by letting each of them to touch the plane and looking at which point it had to be in order to cause the maximum response at a particular unit. These two maps coincide almost perfectly.

Formation of a Topographic Map of the Environment Through Many Different Channels. This example elucidates alternative mechanisms that are able to create the perception of space. In the following simulation, observations of the environment were mediated by one eye and two arms. Both arms touched some point on the input plane to which the gaze was also directed. The turning angles of the eye, and the bending angles of the arms were detected. Notice that no image analysis was yet performed. Every point on the input plane corresponds to six transmitted signals used for the ξ_i variables, $i = 1, \ldots, 6$.

Training was made by selecting the target point at random, and letting the resulting ξ_i signals affect the adaptive system. The asymptotic state was then tested *for each arm and the eye separately.* In other words, when the output of each processing unit was recorded in turn, one could, e.g., find the direction of gaze that caused the maximum match at this unit. (The other signals were thereby zero.) Figure 3.15 shows the results of this test, again projected on the target plane as a network of thin lines; their crossings correspond to processing units, and these networks have been determined as inverse mappings of the due weight vectors. It is remarkable that practically the same input-output mapping is obtained with very different nonlinear detectors.

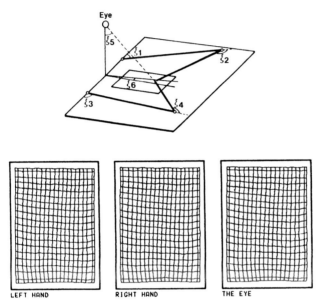

Fig. 3.15. Formation of a topographic map of the environment using one eye and two hands simultaneously

Construction of a Topographic Map of Stage by a Freely Moving Observer Using Vision Only. In this demonstration we show results of an experiment that was intended to illustrate how a complete image of space might be formed from its partial observations. Here the observatory mechanism was a very trivial one, consisting of a simple optic and a cylindrical retina with eight photosensitive segments. Consider Fig. 3.16 that shows the top view of a stage provided with walls, the scenery. Assume that the background is dark and the walls are white; a projection image of the scenery is formed onto the retina. The observers moves at random within a restricted area, and in this simple demonstration his orientation was fixed. The sig-

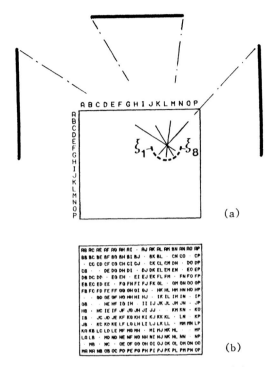

(a)

(b)

Fig. 3.16. Topographic map of a stage. (**a**) Thick lines: scenery. The framed area is where the observer moved: the coordinates are indicated by a pair of letters. (**b**) Output map, with pairs of letters corresponding to processing units. The units are labeled by coordinates of those points on the stage the images of which are formed on the units in question

nals obtained from the segments of the retina are led to a self-organizing processing unit array. After adaptation, the map was tested by letting the observer stand in a particular location and recording the coordinates of the corresponding image point in the map.

Tonotopic Map. The following effect was found already in the first SOM experiments, but it has later been ignored. Nonetheless it demonstrates a very important feature of self-organization, namely that the inputs to the map units need not be identical, not even having the same dimensionality, as long as some kind of metric-topological order of the various inputs to the different "neurons" is preserved. This property is necessary in the biological models, because the number of inputs to a neuron is never fixed. Let us demonstrate this effect with a simple model that refers to a possible mode of formation, or at least refinement, of the physiological *tonotopic map*. This experiment demonstrates that the inputs can be nonidentical as long as the signals to every unit are correlated, and the topological order of their input vectors in the respective signal subspaces is the same. Consider Fig. 3.17 that depicts

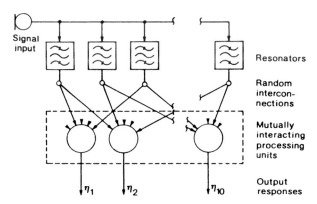

Fig. 3.17. System for tonotopic-map simulation

Table 3.1. Formation of frequency maps. There were twenty second-order filters with quality factor $Q = 2.5$ (where Q is inversely proportional to the relative width of resonance curve) and resonant frequencies distributed at random over the range $[1, 2]$. The training frequencies were drawn at random from the range $[0.5, 1]$. The numbers in the table indicate those test frequencies to which each processing unit became most sensitive.

Unit	1	2	3	4	5	6	7	8	9	10
Experiment 1, 2000 training steps	0.55	0.60	0.67	0.70	0.77	0.82	0.83	0.94	0.98	0.83
Experiment 2, 3500 training steps	0.99	0.98	0.98	0.97	0.90	0.81	0.73	0.69	0.62	0.59

a one-dimensional array of processing units. This system receives sinusoidal signals and becomes ordered according to their *frequency*. Assume a set of resonators or bandpass filters tuned at random. The filters may have a rather shallow resonance curve. The inputs to the array units (five to each) are now also picked up at random from the resonator outputs, different samples for different array units, so that there is no order or correlation in any structure or initial parameters. Next a series of adaptation operations is carried out, every time generating a new sinusoidal signal with a randomly chosen frequency. After a number of iteration steps, the units start to become sensitized to different frequencies in an ascending or descending order. Final results of two experiments are shown in Table 3.1.

Phonotopic Map. A more practical example is mapping of natural stochastic data, such as short-time spectra picked up from natural speech, onto the network. In this example the input vector x was 15-dimensional, and its components corresponded to the output powers of a 15-channel frequency filter

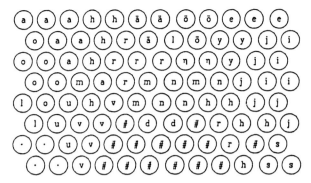

Fig. 3.18. The neurons, shown as circles, are labeled with the symbols of the phonemes with which they made the best match. Distinction of /k,p,t/ from this map is not reliable, and this phonemic class is indicated by symbol #. An analysis of the transient spectra of these phonemes by an auxiliary map is necessary

bank, averaged over 10 ms intervals, with the midfrequencies of the filters being selected from the range 200 Hz to 6400 Hz. The spectral samples were applied in their natural order of occurrence at the inputs of the SOM, and after learning the map was calibrated with known input samples and labeled according to the phonemic symbols of the samples. Details of the system used for the acquisition of speech samples will be given in Sects. 7.2 and 7.5. Figure 3.18 shows the result directly *on the neural network*, not in the signal space. The various neurons have become "tuned" to different categories of phonemes. (The speech used in this experiment was Finnish, phonetically rather similar to Latin.) The cells labeled with the same phonemic symbol actually correspond to somewhat different reference or codebook vectors, which like the various codebook vectors in VQ approximate the probability density function of a particular set of samples. As /k,p,t/ have a very weak signal power compared with the other phonemes, they are clustered together and represented by a broad phonetic class with symbol "#."

3.5 Basic Mathematical Approaches to Self-Organization

Although the basic principle of the above system seems simple, the process behaviour, especially relating to the more complex input representations, has been very difficult to describe in mathematical terms. The first approach made below discusses the process in its simplest form, but it seems that fundamentally similar results are obtainable with more complex systems, too. The other approaches made in this section are also simple, meant for basic understanding. More refined discussions can be found in the literature (Chap. 10).

3.5.1 One-Dimensional Case

We shall first try to justify the self-organizing ability analytically, using a very simple system model. The reasons for the self-ordering phenomena are actually very subtle and have strictly been proven only in the simplest cases. In this section we shall first delineate a basic Markov-process explanation that should help to understand the nature of the process.

In the first place we shall restrict our considerations to a *one-dimensional, linear, open-ended array* of functional units to each of which a *scalar-valued input signal* ξ is connected. Let the units be numbered $1, 2, \ldots, l$. Each unit i has a single scalar input weight or reference value μ_i, whereby the similarity between ξ and μ_i is defined by the absolute value of their difference $|\xi - \mu_i|$; the best match is defined by

$$|\xi - \mu_c| = \min_i \{|\xi - \mu_i|\} . \tag{3.7}$$

We shall define the set of units N_c selected for updating as follows:

$$N_c = \{\max(1, c-1), c, \min(l, c+1)\} . \tag{3.8}$$

In other words, unit i has the neighbors $i-1$ and $i+1$, except at the end points of the array, where the neighbor of unit 1 is 2, and the neighbor of unit l is $l-1$, respectively. Then N_c is simply the set of units consisting of unit c and its immediate neighbors.

The general nature of the process is similar for different values of $\alpha > 0$; it is mainly the speed of the process that is varied with α. In the *continuous-time formalism*, the equations read

$$\begin{aligned} d\mu_i/dt &= \alpha(\xi - \mu_i) &\text{for } i \in N_c , \\ d\mu_i/dt &= 0 &\text{otherwise} . \end{aligned} \tag{3.9}$$

Proposition 3.5.1. *Let ξ be a stochastic variable. Starting with randomly chosen initial values for the μ_i, these numbers will gradually assume new values in a process specified by (3.7)–(3.9), such that when $t \to \infty$, the set of numbers $(\mu_1, \mu_2, \ldots, \mu_l)$ becomes ordered in an ascending or descending sequence. Once the set is ordered, it remains so for all t. Moreover, the point density function of the μ_i will finally approximate some monotonic function of the probability density function $p(\xi)$ of ξ.*

The discussion shall be carried out in two parts: formation of ordered sequences of the μ_i, and their convergence to certain "fixed points", respectively.

Ordering of Weights.

Proposition 3.5.2. *In the process defined by (3.7)–(3.9), the μ_i become ordered with probability one in an ascending or descending order when $t \to \infty$.*

One might like to have a rigorous proof for that ordering occurs almost surely (i.e., with probability one). Following the argumentation presented by *Grenander* for a related problem [3.9], the proof of ordering can be delineated as indicated below. Let $\xi = \xi(t) \in \Re$ be a random (scalar) input that has the probability density $p(\xi)$ over a finite support, with $\xi(t_1)$ and $\xi(t_2)$ independent for any $t_1 \neq t_2$.

The proof follows from general properties of Markov processes, especially that of the *absorbing state* for which the transition probability into itself is unity. It can be shown [3.10] that if such a state, starting from arbitrary initial state, is reached by *some* sequence of inputs that has a *positive* probability, then allowing a random sequence of inputs, the absorbing state is reached almost surely (i.e., with probability one), when $t \to \infty$.

The absorbing state is now identified with any of the ordered sequences of the μ_i. (Notice that there are two different ordered sequences, which together constitute the absorbing state.) On the line of real numbers, select an interval such that ξ has a positive probability on it. By repeatedly choosing values of ξ from this interval, it is possible to bring all μ_i within it in a finite time. After that it is possible to repeatedly choose values of ξ such that if, e.g., μ_{i-2}, μ_{i-1}, and μ_i are initially disordered, then μ_{i-1} will be brought between μ_{i-2} and μ_i, while the relative order in the other subsequences is not changed. Notice that if unit i is selected, then units $i-1$ *and* $i+1$ will change; if there is disorder on both sides of i, we may consider that side which is ordered first, stopping the application of (3.9) at that point, and calling this an elementary sorting operation. For instance, if $\mu_{i-1} < \mu_{i-2} < \mu_i$, then selection of ξ from the vicinity of μ_i will bring μ_{i-1} between μ_{i-2} and μ_i (notice that μ_{i-2} is not changed). The sorting can be continued systematically along similar lines. An overall order will then result in a finite number of steps. Since the above ξ values are realized with positive probability, the proof of Proposition 3.5.2 is concluded.

Cottrell and *Fort* [3.11] have presented an exhaustive and mathematically stringent proof of the above ordering process in the one-dimensional case, but since it is very lengthy (forty pages), it will be omitted here. The reader might instead study the shorter constructive proof presented in Sect. 3.5.2 for an almost equivalent system.

Corollary 3.5.1. *If all the values* $\mu_1, \mu_2, \ldots, \mu_l$ *are ordered, they cannot become disordered in further updating.*

The proof follows directly from the observation that if all partial sequences are ordered, then process (3.9) cannot change the relative order of any pair $(\mu_i, \mu_j), i \neq j$.

Convergence Phase. After the μ_i have become ordered, their final convergence to the asymptotic values is of particular interest since the latter represent the image of the input distribution $p(\xi)$.

In this subsection it is assumed henceforth that the $\mu_i, i = 1, 2, \ldots, l$ are already ordered and, on account of Corollary 3.5.1, remain such in further updating processes. The aim is to calculate the asymptotic values of the μ_i. To be quite strict, asymptotic values are obtained in the sense of mean squares or almost sure convergence only if the "learning rate factor" $\alpha = \alpha(t)$ in (3.9) decreases to zero; the sequence $\{\alpha(t)|t = 0, 1, \ldots\}$ must obviously satisfy certain conditions similar to those imposed on the Robbins-Monro stochastic approximation process, cf. Sect. 1.3.3.

The convergence properties of the μ_i are discussed in this section in a less restricted sense, namely, only the dynamic behavior of the *expectation values* $E\{\mu_i\}$ is analyzed. These numbers will be shown to converge to unique limits. The variances of the μ_i can then be made arbitrarily small by a suitable choice of $\alpha(t), t \to \infty$.

It may be useful to refer to Fig. 3.19 that represents the values μ_i on a line of real numbers. As stated above, the μ_i shall already be in order; we may restrict ourselves to the case of increasing values. It is also assumed that $[\mu_1, \mu_l]$ is a proper subset of $[a, b]$, the support of $p(\xi)$, which is obviously due if ordering has occurred through a process described above.

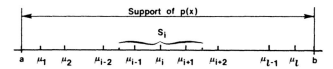

Fig. 3.19. Reference values after ordering

Since ordering of the μ_i was postulated, and because a selected node can only affect its immediate neighbours, it is obvious from (3.7)–(3.9) that any particular value μ_i can be affected only if ξ hits an interval S_i defined in the following way: assuming $l \geq 5$, we have

$$
\begin{array}{llll}
\text{for } 3 \leq & i \leq l-2 & : & S_i = [\frac{1}{2}(\mu_{i-2} + \mu_{i-1}), \frac{1}{2}(\mu_{i+1} + \mu_{i+2})] , \\
\text{for} & i = 1 & : & S_i = [a, \frac{1}{2}(\mu_2 + \mu_3)] , \\
\text{for} & i = 2 & : & S_i = [a, \frac{1}{2}(\mu_3 + \mu_4)] , \\
\text{for} & i = l-1 & : & S_i = [\frac{1}{2}(\mu_{l-3} + \mu_{l-2}), b] , \\
\text{for} & i = l & : & S_i = [\frac{1}{2}(\mu_{l-2} + \mu_{l-1}), b] .
\end{array}
\tag{3.10}
$$

The expectation values of the $d\mu_i/dt \stackrel{\text{def}}{=} \dot{\mu}_i$, conditional on the $\mu_1, \ldots, \mu_l$, according to (3.9) read

$$
\langle \dot{\mu}_i \rangle \stackrel{\text{def}}{=} E\{\dot{\mu}_i\} = \alpha(E\{\xi|\xi \in S_i\} - \mu_i)P(\xi \in S_i) ,
\tag{3.11}
$$

where $P(\xi \in S_i)$ is the probability for ξ falling into the interval S_i. Now $E\{\xi|\xi \in S_i\}$ is the center of gravity of S_i, see (3.10), which is a function of the μ_i when $p(\xi)$ has been defined. In order to solve the problem in simplified closed form, it is assumed that $p(\xi) \equiv \text{const}$ over the support $[a, b]$ and zero outside it, whereby one first obtains:

for $3 \leq i \leq l - 2$:

$$\langle \dot{\mu}_i \rangle = \frac{\alpha}{4}(\mu_{i-2} + \mu_{i-1} + \mu_{i+1} + \mu_{1+2} - 4\mu_i)P(\xi \in S_i) ,$$

$$\langle \dot{\mu}_1 \rangle = \frac{\alpha}{4}(2a + \mu_2 + \mu_3 - 4\mu_1)P(\xi \in S_1) ,$$

$$\langle \dot{\mu}_2 \rangle = \frac{\alpha}{4}(2a + \mu_3 + \mu_4 - 4\mu_2)P(\xi \in S_2) ,$$

$$\langle \dot{\mu}_{l-1} \rangle = \frac{\alpha}{4}(\mu_{l-3} + \mu_{l-2} + 2b - 4\mu_{l-1})P(\xi \in S_{l-1}) ,$$

$$\langle \dot{\mu}_l \rangle = \frac{\alpha}{4}(\mu_{l-2} + \mu_{l-1} + 2b - 4\mu_l)P(\xi \in S_l) . \tag{3.12}$$

Starting with arbitrary initial conditions $\mu_i(0)$, the most probable, "averaged" trajectories $\mu_i(t)$ are obtained as solutions of an equivalent differential equation corresponding to (3.12), namely,

$$dz/dt = P(z)(Fz + h) , \quad \text{where} \tag{3.13}$$

$$z = [\mu_1, \mu_2, \ldots, \mu_l]^{\mathrm{T}} ,$$

$$F = \frac{\alpha}{4}
\begin{bmatrix}
-4 & 1 & 1 & 0\ 0\ 0\ 0\ \cdots & & & & \\
0 & -4 & 1 & 1\ 0\ 0\ 0 & & & & \\
1 & 1 & -4 & 1\ 1\ 0\ 0 & & & & \\
0 & 1 & 1 & -4\ 1\ 1\ 0 & & & & \\
& \vdots & & & & & & \vdots \\
& & & & 0\ 1\ 1\ -4 & 1 & 1 & 0 \\
& & & & 0\ 0\ 1 & 1 & -4 & 1 & 1 \\
& & & & 0\ 0\ 0 & 1 & 1 & -4 & 0 \\
& & & & \cdots\ 0\ 0\ 0 & 0 & 1 & 1 & -4
\end{bmatrix} ,$$

$$h = \frac{\alpha}{2}[a, a, 0, 0, \ldots, 0, b, b]^{\mathrm{T}} , \tag{3.14}$$

and $P(z)$ is now a diagonal matrix with the $P(\xi \in S_i)$ its diagonal elements. The averaging, producing (3.13) from (3.12), could be made rigorous along the lines given by *Geman* [2.54]. Equation (3.13) is a first-order differential equation with constant coefficients. It has a fixed-point solution, a particular solution with $dz/dt = 0$, which is (taking into account that $P(z)$ is diagonal and positive definite)

$$z_0 = -F^{-1}h , \tag{3.15}$$

provided that F^{-1} exists; this has been shown strictly by the author [3.12] but the proof is too cumbersome and lengthy to be reproduced here. The general solution of (3.13) is difficult to obtain. Some recent studies, e.g. [3.11] indicate that convergence to the fixed point is generally true, and thus we restrict our considerations to showing the asymptotic solution only.

The asymptotic values of the μ_i, for uniform $p(\xi)$ over the support $[0, 1]$ have been calculated for a few lengths l of the array and presented in Table 3.2 as well as in Fig. 3.20.

Table 3.2. Asymptotic values for the μ_i, with $a = 0$ and $b = 1$

Length of array (l)	μ_1	μ_2	μ_3	μ_4	μ_5	μ_6	μ_7	μ_8	μ_9	μ_{10}
5	0.2	0.3	0.5	0.7	0.8	–	–	–	–	–
6	0.17	0.25	0.43	0.56	0.75	0.83	–	–	–	–
7	0.15	0.22	0.37	0.5	0.63	0.78	0.85	–	–	–
8	0.13	0.19	0.33	0.44	0.56	0.67	0.81	0.87	–	–
9	0.12	0.17	0.29	0.39	0.5	0.61	0.7	0.83	0.88	–
10	0.11	0.16	0.27	0.36	0.45	0.55	0.64	0.73	0.84	0.89

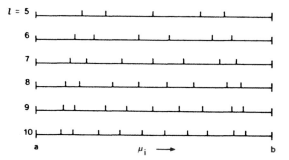

Fig. 3.20. Asymptotic values for the μ_i for different lengths of the array, shown graphically

It can be concluded that

- the outermost values μ_1 and μ_l are shifted inwards by an amount which is approximately $1/l$ whereas for uniform distribution this amount should be about $1/2l$; this is the "boundary effect" that occurs in the illustrations of Sect. 3.4, and vanishes with increasing l
- the values μ_3 through μ_{l-2} seem to be distributed almost evenly.

3.5.2 Constructive Proof of Ordering of Another One-Dimensional SOM

Since the ordering proof of the original SOM algorithm has turned out problematic although possible, we now carry out a simple proof for a somewhat modified but still logically similar system model. This proof appeared in

[3.13]. One of the main differences is redefinition of the "winner," because in the case of a tie, all "winners" are taken into account for learning in this modified model. The learning rule is then applied to the neighborhoods of all "winners," and the neighborhood set is modified in such a way that the "winner" itself is excluded from it. These modifications allow a *rigorous constructive proof of ordering* in the one-dimensional case.

The Problem. Assume an open-ended linear array of units named *nodes*. Let the nodes be indexed by $1, 2, \ldots, l$.

With every node, a real number $\mu_i = \mu_i(t) \in \Re$, which is a function of time, is associated. Let $\xi = \xi(t) \in \Re$ be a random input variable with a stationary probability density function $p(\xi)$ over a support $[a, b]$.

A self-organizing process is defined as follows: the degree of similarity of ξ with the numbers μ_i is at every instant t evaluated in terms of the distances $|\xi - \mu_i|$ that define one or several "winner" nodes c according to

$$|\xi - \mu_c| = \min_i \{|\xi - \mu_i|\} . \tag{3.16}$$

The numbers μ_i are continuously *updated* according to

$$\frac{d\mu_i}{dt} = \alpha(\xi - \mu_i) \quad \text{for } i \in N_c ,$$
$$\frac{d\mu_i}{dt} = 0 \quad \text{otherwise} , \tag{3.17}$$

where α is a "gain coefficient" (> 0), and N_c is a set of indices defined in the following way:

$$N_1 = \{2\} ,$$
$$N_2 = \{i - 1, i + 1\} \quad \text{for } 2 \le i \le l - 1 ,$$
$$N_l = \{l - 1\} . \tag{3.18}$$

In case c is not unique, the union of all sets N_c must be used in stead of N_c in (3.17), excluding all the selected nodes themselves. The non-uniqueness of c is commented in the proof of Theorem 3.5.1.

The following discussion aims at showing that with time, the set of numbers $(\mu_1, \mu_2, \ldots, \mu_l)$ becomes ordered almost surely if certain rather mild conditions are fulfilled.

The degree of ordering is conveniently expressed in terms of the *index of disorder D*,

$$D = \sum_{i=2}^{l} |\mu_i - \mu_{i-1}| - |\mu_l - \mu_1| , \tag{3.19}$$

which is ≥ 0.

Definition. *The numbers* $\mu_1, \ldots, \mu_l$ *are ordered if and only if* $D = 0$, *which is equivalent to either* $\mu_1 \geq \mu_2 \geq \ldots \geq \mu_l$ *or* $\mu_1 \leq \mu_2 \leq \ldots \leq \mu_l$.

Note that this definition also allows the case in which a subset of the μ_i, or in fact all of them, become equal. In the present process this is indeed possible unless the input attains all its values sufficiently often. In numerous computer simulations, however, under the assumptions of Theorem 3.5.1, the asymptotic values have invariably ordered in a strictly monotonic sequence.

Theorem 3.5.1. *Let* $\xi = \xi(t)$ *be a random process satisfying the following assumptions:*

(i) $\xi(t)$ *is almost surely integrable on finite intervals;*
(ii) the probability density $p(\xi)$ *of* $\xi(t)$ *is independent of* t *and strictly positive on* $[a, b]$ *and zero elsewhere, and* $\xi(t)$ *attains all values on* $[a, b]$ *almost surely during all time intervals* $[t, \infty)$;
(iii) the initial values for the μ_i *are randomly chosen from an absolutely continuous distribution on* $[a, b]$.

Then, in the process defined by (3.16), (3.17), and (3.18), the μ_i *will become almost surely ordered asymptotically.*

Proof. Under the assumptions on ξ, and starting from almost surely different initial values, two numbers $\mu_i(t)$ and $\mu_j(t)$, $i \neq j$, are almost surely different except in the following case: if c is the "winner," or one of the "winners," then μ_{c-1} or μ_{c+1} may become equal to μ_c at some instant t. Then the index, say $c - 1$, is added to the set of "winner" indices. Since now $d\mu_m/dt = d\mu_{m-1}/dt = 0$, the values of μ_c and μ_{c-1} stay equal as long as c remains one of the "winners." The result is that a set of values with consequent indices, say $\mu_{c-r}, \mu_{c-r+1}, \ldots, \mu_{c+p}$ with $r \geq 0$, $p \geq 0$, may be equal over time intervals of positive length. Then and only then c is non-unique with a positive probability.

Convention 3.5.1. In view of the above, the considerations will be reduced to the case of a unique c with always $\mu_{c-1} \neq \mu_c$ and $\mu_{c+1} \neq \mu_c$. If c is not unique, the indices must be redefined so that $c - 1$ stands for $c - r - 1$ and $c + 1$ stand for $c + p + 1$, and μ_c stands for all $\mu_{c-r}, \ldots, \mu_{c+p}$. This convention is followed throughout the proof without further mention.

Step 1. It will be shown first that $D = D(t)$ is a monotonically decreasing function of time, for all t, if $\xi = \xi(t)$ satisfies assumption (i).

The process (3.17) affects at most three (according to Convention 3.5.1) successive values of μ_i at a time; therefore it will suffice to inspect only those terms in D which may change. Let these constitute the partial sum S dependent on c in the following way (notice that $|\mu_l - \mu_1|$ may change only for $c = 2$ or $l - 1$):

$$\text{for } 3 \ \le \ c \le l-2, \quad S = \sum_{i=c-1}^{c+2} |\mu_i - \mu_{i-1}| \ ;$$

$$\text{for } c \ = \ 1, \ \ S = |\mu_3 - \mu_2| + |\mu_2 - \mu_1| \ ;$$

$$\text{for } c \ = \ 2, \ \ S = |\mu_4 - \mu_3| + |\mu_3 - \mu_2| + |\mu_2 - \mu_1| - |\mu_l - \mu_1| \ ;$$

$$\text{for } c \ = \ l-1,$$

$$S = |\mu_l - \mu_{l-1}| + |\mu_{l-1} - \mu_{l-2}| + |\mu_{l-2} - \mu_{l-3}| - |\mu_l - \mu_1| \ ;$$

$$\text{for } c \ = \ l, \ \ S = |\mu_l - \mu_{l-1}| + |\mu_{l-1} - \mu_{l-2}| \ . \tag{3.20}$$

Since the cases $c = l - 1$ and $c = l$ are obviously symmetric with respect to $c = 2$ and $c = 1$, respectively, they need not be discussed separately.

At any given instant t, and any c, the signs of the differences $\mu_i - \mu_{i-1}$ in S attain one of at most 16 combinations. For any particular sign combination, the expression of dS/dt is given below. Of all different cases, half are symmetric in the sense that they produce the same analytical expression; accordingly, only the cases listed below need to be discussed separately. Around c one has (denoting $\dot\mu_i = d\mu_i/dt$)

$$\dot\mu_{c-2} \ = \ 0 \ ,$$

$$\dot\mu_{c-1} \ = \ \alpha(\xi - \mu_{c-1}) \ ,$$

$$\dot\mu_c \ = \ 0 \ ,$$

$$\dot\mu_{c+1} \ = \ \alpha(\xi - \mu_{c+1}) \ ,$$

$$\dot\mu_{c+2} \ = \ 0 \ . \tag{3.21}$$

These linear differential equations have almost surely unique continuous solutions due to assumption (i).

We shall now consider the cases $3 \le c \le l-2$, $c = 1$, and $c = 2$ separately.

A. $3 \le c \le l - 2$; assume $\mu_{c-1} \ge \mu_{c-2}$.

Case	$\mu_c - \mu_{c-1}$	$\mu_{c+1} - \mu_c$	$\mu_{c+2} - \mu_{c+1}$
a0	> 0	> 0	≥ 0
a1	> 0	> 0	≤ 0
a2	> 0	< 0	≥ 0
a3	> 0	< 0	≤ 0
a4	< 0	> 0	≥ 0
a5	< 0	> 0	≤ 0
a6	< 0	< 0	≥ 0
a7	< 0	< 0	≤ 0

If the symbols a0 through a7 are used as subscripts for the labeling of S, and it is denoted $\dot S = dS/dt$, one obtains taking into account (3.21):

$$S_{a0} = -\mu_{c-2} + \mu_{c+2},$$
$$S_{a1} = -\mu_{c-2} + 2\mu_{c+1} - \mu_{c+2},$$
$$S_{a2} = -\mu_{c-2} + 2\mu_c - 2\mu_{c+1} + \mu_{c+2},$$
$$S_{a3} = -\mu_{c-2} + 2\mu_c - \mu_{c+2},$$
$$S_{a4} = -\mu_{c-2} + 2\mu_{c-1} - 2\mu_c + \mu_{c+2},$$
$$S_{a5} = -\mu_{c-2} + 2\mu_{c-1} - 2\mu_c + 2\mu_{c+1} - \mu_{c+2},$$

$$\dot{S}_{a0} = 0 \ ;$$
$$\dot{S}_{a1} = 2\alpha(\xi - \mu_{c+1}) < 0 \ ; \quad *)$$
$$\dot{S}_{a2} = 2\alpha(\mu_{c+1} - \xi) < 0 \ ;$$
$$\dot{S}_{a3} = 0 \ ;$$
$$\dot{S}_{a4} = 2\alpha(\xi - \mu_{c-1}) < 0 \ ;$$
$$\dot{S}_{a5} = 2\alpha[(\xi - \mu_{c-1})$$
$$+ (\xi - \mu_{c+1})] < 0 \ ;$$

$$S_{a6} = -\mu_{c-2} + 2\mu_{c-1} - 2\mu_{c+1} + \mu_{c+2}, \quad \dot{S}_{a6} = 2\alpha(\mu_{c+1} - \mu_{c-1}) < 0 \ ;$$
$$S_{a7} = -\mu_{c-2} + 2\mu_{c-1} - \mu_{c2}, \quad \dot{S}_{a7} = 2\alpha(\xi - \mu_{c-1}) < 0 \ .$$

*)   Notice that for c the "winner," $\frac{1}{2}(\mu_{c-1} + \mu_c) \le \xi \le \frac{1}{2}(\mu_c + \mu_{c+1})$

(3.22)

B. $c = 1$: The proof is similar for nodes 1 and l and will be carried out for 1 only. Now only four cases need be considered.

Case	$\mu_2 - \mu_1$	$\mu_3 - \mu_2$
b0	> 0	≥ 0
b1	> 0	≤ 0
b2	< 0	≥ 0
b3	< 0	≤ 0

$$S_{b0} = -\mu_1 + \mu_3, \qquad \dot{S}_{b0} = 0;$$
$$S_{b1} = -\mu_1 + 2\mu_2 - \mu_3, \quad \dot{S}_{b1} = 2\alpha(\xi - \mu_2) < 0;$$
$$S_{b2} = \mu_1 - 2\mu_2 + \mu_3, \quad \dot{S}_{b2} = 2\alpha(\mu_2 - \xi) < 0;$$
$$S_{b3} = \mu_1 - \mu_3, \qquad \dot{S}_{b3} = 0.$$

(3.23)

C. $c = 2$: It will suffice to consider the case $\mu_l \ge \mu_1$ only, since the case $\mu_l \le \mu_1$ is symmetric to the other one.

Case	$\mu_2 - \mu_1$	$\mu_3 - \mu_2$	$\mu_4 - \mu_3$
c0	> 0	> 0	≥ 0
c1	> 0	> 0	≤ 0
c2	> 0	< 0	≥ 0
c3	> 0	< 0	≤ 0
c4	< 0	> 0	≥ 0
c5	< 0	> 0	≤ 0
c6	< 0	< 0	≥ 0
c7	< 0	< 0	≤ 0

$$
\begin{aligned}
S_{c0} &= -\mu_l + \mu_4, & \dot{S}_{c0} &= 0; \\
S_{c1} &= -\mu_l + 2\mu_3 - \mu_4, & \dot{S}_{c1} &= 2\alpha(\xi - \mu_3) < 0; \\
S_{c2} &= -\mu_l + 2\mu_2 - 2\mu_3 + \mu_4, & \dot{S}_{c2} &= 2\alpha(\mu_3 - \xi) < 0; \\
S_{c3} &= -\mu_l + 2\mu_2 - \mu_4, & \dot{S}_{c3} &= 0; \\
S_{c4} &= -\mu_l + 2\mu_1 - 2\mu_2 + \mu_4, & \dot{S}_{c4} &= 2\alpha(\xi - \mu_1) < 0; \\
S_{c5} &= -\mu_l + 2\mu_1 - 2\mu_2 + 2\mu_3 - \mu_4, & \dot{S}_{c5} &= 2\alpha[(\xi - \mu_1) + (\xi - \mu_3)] < 0; \\
S_{c6} &= -\mu_l + 2\mu_1 - 2\mu_3 + \mu_4, & \dot{S}_{c6} &= 2\alpha(\mu_3 - \mu_1) < 0; \\
S_{c7} &= -\mu_l + 2\mu_1 - \mu_4, & \dot{S}_{c7} &= 2\alpha(\xi - \mu_1) < 0.
\end{aligned}
$$

$$(3.24)$$

Step 2. Since $D(t)$ is monotonically decreasing and nonnegative, it must tend to some limit $D^* = \lim_{t\to\infty} D(t)$. It is now shown that $D^* = 0$ with probability one if $\xi(t)$ satisfies the assumption (ii).

Since D^* is constant with respect to t, $dD^*/dt = 0$ for all t. Still, D^* satisfies the same differential equation as $D(t)$. Assume now that $D^* > 0$. Then there is disorder in the set $(\mu_1, \ldots, \mu_l)$; without loss of generality, assume that there is an index j, $2 \le j \le l-1$, such that $\mu_j > \mu_{j+1}$ and $\mu_j > \mu_{j-1}$. (The proof would be similar for the case in which $\mu_j < \mu_{j+1}$ and $\mu_j < \mu_{j-1}$. Note that here, too, it is possible that μ_j stands for a set of equal values with consequent indices.)

With probability one $\xi(t)$ will eventually attain a value such that $j = c$ is the "winner," due to assumption (ii). If $j = 2$ (or $j = l-1$), then we have either case c2 with $dD^*/dt < 0$ which would mean contradiction, or case c3 with $dD^*/dt = 0$.

The magnitude relations of the μ_i values corresponding to case c3 are: $\mu_2 > \mu_1$, $\mu_3 < \mu_2$, and $\mu_4 \le \mu_3$. Due to assumption (ii), with probability one $\xi(t)$ will eventually also attain a value such that $c = 3$ is the "winner," and $\xi < \mu_3$. If the μ_i values are as above, with $c = 3$ they can be written as $\mu_{c-1} > \mu_{c-2}$, $\mu_c < \mu_{c-1}$, and $\mu_{c+1} \le \mu_c$. Then they correspond to case a6 or a7, with the difference that equality is possible in $\mu_{c+1} \le \mu_c$. However, both $\dot{S}_{a6}$ and $\dot{S}_{a7}$ in (3.22) are now negative. This implies $dD^*/dt < 0$ which then again leads to contradiction.

If the "winner" c is equal to j with $3 \le j \le l-2$, then either case a2 is due, and $dD^*/dt < 0$, which means contradiction, or case a3 is due whereby $dD^*/dt = 0$. The magnitude relations of the μ_i values corresponding to case a3 are: $\mu_j > \mu_{j-1}$, $\mu_{j+1} < \mu_j$, and $\mu_{j+2} \le \mu_{j+1}$. Again, due to assumption (ii), with probability one $\xi(t)$ will eventually also attain a value such that $j + 1 = c$ is the "winner" and $\xi < \mu_{j+1}$. If the μ_i values are as above, with $c = j + 1$ they can be written as $\mu_{c-1} > \mu_{c-2}$, $\mu_c < \mu_{c-1}$, and $\mu_{c+1} \le \mu_c$. Then they correspond to case a6 or a7, with the difference that equality is possible in $\mu_{c+1} \le \mu_c$. However, again both $\dot{S}_{a6}$ and $\dot{S}_{a7}$ in (3.22) are negative, leading to the contradiction $dD^*/dt < 0$.

From the above, we can conclude that D^* must be equal to zero, or otherwise a contradiction is bound to occur. The asymptotic state must therefore be an ordered one.

3.6 The Batch Map

It will be useful to understand what the convergence limits m_i^* in the sequence defined by (3.3) actually represent.

Assuming that the convergence to some ordered state is true, the expectation values of $m_i(t+1)$ and $m_i(t)$ for $t \to \infty$ must be equal, even if $h_{ci}(t)$ were then selected nonzero. In other words, in the stationary state we must have

$$\forall i, \quad \mathrm{E}\{h_{ci}(x - m_i^*)\} = 0 . \tag{3.25}$$

In the simplest case $h_{ci}(t)$ was defined: $h_{ci} = 1$ if i belongs to some *topological neighborhood set* N_c of cell c in the cell array, whereas otherwise $h_{ci} = 0$. With this h_{ci} there follows

$$m_i^* = \frac{\int_{V_i} x p(x) dx}{\int_{V_i} p(x) dx} , \tag{3.26}$$

where V_i is the set of those x values in the integrands that are able to update vector m_i; in other words, the "winner" node c for each $x \in V_i$ must belong to the neighborhood set N_i of cell i.

Let us now exemplify (3.25) or (3.26) by Fig. 3.21. In this special case, the neighborhood of cell i consists of the cells $i - 1, i$, and $i + 1$, except at the ends of the array, where only one neighbor exists.

In the case described in Fig. 3.21 we have two-dimensional vectors as inputs, and the probability density function of $x \in \Re^2$ is uniform over the framed area (support of the x values) and zero outside it. The neighborhood set N_c has the simple form defined above. *In this case at least the equilibrium condition (3.25) or (3.26) means that each m_i^* must coincide with the*

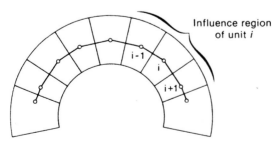

Fig. 3.21. Illustration for the explanation of the equilibrium state in self-organization and definition of "influence region"

centroid of the respective influence region. At least in this case it may be intuitively clear that the equilibrium represents the case whereby *the Voronoi sets around the m_i contact each other in the same order as the "topological links" between the nodes in the neuron array are defined.* In general, with more complicated network architectures, a similar topological correspondence has been discussed in [3.14].

The equilibrium condition (3.25) or (3.26) with a general probability density function $p(x)$ means that each m_i^* must coincide with the centroid of $p(x)$ over the respective influence region. This could then be taken for the definition of the ordered state. It may be intuitively clear that with general dimensionalities, such an equilibrium can only be valid with a particular configuration of the m_i.

Equation (3.26) is already in the form in which the so-called *iterative contractive mapping* used in the solving of nonlinear equations is directly applicable. Let z be an unknown vector that has to satisfy the equation $f(z) = 0$; then, since it is always possible to write the equation as $z = z + f(z) = g(z)$, the successive approximations of the root can be computed as a series $\{z_n\}$ where

$$z_{n+1} = g(z_n) . \tag{3.27}$$

We shall not discuss any convergence problems here. In the SOM context we have never encountered any, but if there would exist some, they could be overcome by the so-called Wegstein modification of (3.27):

$$z_{n+1} = (1 - \lambda)g(z_n) + \lambda z_n \quad , \text{where } 0 < \lambda \leq 1 . \tag{3.28}$$

The iterative process in which a number of samples of x is first classified into the respective V_i regions, and the updating of the m_i^* is made iteratively as defined by (3.26), can be expressed as the following steps. The algorithm dubbed *"Batch Map"* [3.15, 16] resembles the Linde-Buzo-Gray algorithm discussed in Sect. 1.5, where all the training samples are assumed to be available when learning begins. The learning steps are defined as follows:

1. For the initial reference vectors, take, for instance, the first K training samples, where K is the number of reference vectors.
2. For each map unit i, collect a list of copies of all those training samples x whose nearest reference vector belongs to unit i.
3. Take for each new reference vector the mean over the union of the lists in N_i.
4. Repeat from 2 a few times.

It is easy to see that this algorithm describes the same process that was described in the general setting in Sect. 3.1. If now a general neighborhood function h_{ji} is used, and $\bar{x}_j$ is the mean of the $x(t)$ in Voronoi set V_j, then we shall weight it by the number n_j of samples V_j and the neighborhood function. Now we obtain

$$m_i^* = \frac{\sum_j n_j h_{ji} \bar{x}_j}{\sum_j n_j h_{ji}} \ , \tag{3.29}$$

where the sum over j is taken for all units of the SOM, or if h_{ji} is truncated, over the neighborhood set N_i in which it is defined. For the case in which no weighting in the neighborhood is used,

$$m_i^* = \frac{\sum_{j \in N_i} n_j \bar{x}_j}{\sum_{j \in N_i} n_j} \ . \tag{3.30}$$

A discussion of the convergence and ordering of the Batch Map type algorithm has been presented in [3.8].

This algorithm is particularly effective if the initial values of the reference vectors are already roughly ordered, even if they might not yet approximate the distribution of the samples. It should be noticed that the above algorithm contains no learning-rate parameter; therefore it has no convergence problems and yields stabler asymptotic values for the m_i than the original SOM.

The size of the neighborhood set N_i above can be similar to the size used in the basic SOM algorithms. "Shrinking" of N_i in this algorithm means that the neighborhood radius is decreased while steps 2 and 3 are repeated. At the last couple of iterations, N_i may contain the element i only, and the last steps of the algorithm are then equivalent with the K-means algorithm, which guarantees the most accurate approximation of the density function of the input samples. A few iterations of this algorithm will usually suffice.

Elimination of Border Effects for Low-Dimensional Signals. Inspection of Fig. 3.22 may further clarify the border effects in the one-dimensional SOM and help to understand how they could be eliminated.

If every cell of the SOM has two neighbors, except one at the ends, the "influence region" of cell i ($i > 2$ and $i < k - 1$) (or the range of x values that can affect cell i) is defined $V_i = [\frac{1}{2}(m_{i-2} + m_{i-1}), \frac{1}{2}(m_{i+1} + m_{i+2})]$. In the asymptotic equilibrium, according to (6.18), every m_i must coincide with the centroid of $p(x)$ over the respective V_i. The definition of the "influence

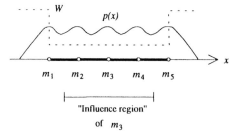

Fig. 3.22. One-dimensional SOM with five reference "vectors" m_i (scalars) that approximate the probability density function $p(x)$, and delineation of the weighting function W

regions" near the borders of the SOM is different, however, and therefore the m_i do not approximate $p(x)$ everywhere in the same way.

In computing the centroids, it is now possible to provide the x samples with *conditional weights* W that depend on index i and the relative magnitude of x and m_i. This weighting can be used both in the old stepwise SOM algorithm (with the given definition of the neighborhood set N_c), and with the Batch Map, too. In the former case, the weight should be applied to the learning-rate factor α, not to x. For to guarantee stability, one must then have $\alpha W < 1$, so this trick is not applicable during the first steps when α is still large. In the Batch Map algorithm, however, the x samples are always weighted directly, so no such restriction exists. Henceforth we assume that the Batch Map is used.

The following rules may first sound a bit complicated, but they are simple to program, and in practice they are also very effective and robust in eliminating the border effects to a large extent. Assume that the m_i values are already ordered.

Weighting Rule for the One-Dimensional SOM:

In updating, each x sample is provided with weight W. Normally $W = 1$, but $W > 1$ for the border (end) cells in the case that x is bigger than the biggest m_i or smaller than the smallest m_i, AND when updating of the border cell (but not of its neighbors) is due.

Consider the special case that $p(x)$ is *uniform* over some singly connected domain of x and zero outside it. It may then be easy to deduce on the basis of Fig. 3.22 and the above weighting rule that if we select for the special weight a value of $W = 9$, all the m_i will become equidistant in the asymptotic equilibrium; then they describe $p(x)$ in an unbiased way. Naturally, for other forms of $p(x)$, we should take other values for W. In many practical cases, however, the default value $W = 9$ compensates for the most part of the border effects in general.

It is possible to eliminate the border effects *totally*, if after a couple of Batch Map iterations the neighborhood set N_i is replaced by $\{i\}$, i.e., having a couple of simple K-means iterations at the end.

In a two-dimensional SOM, the weighting rules are slightly different. While we used, say, the value of $W = 9$ for the end cells in the one-dimensional array, in the updating of the two-dimensional array we must have a different weight W_1 for the corner cells, and another value W_2 for edge cells that are not in the corner. Inside the array, the weight is equal to unity.

Weighting Rules for the Two-Dimensional SOM:

The value W_1 is applied if both of the following two conditions are satisfied: A1. The value of x is in one of the four "outer corner sectors," i.e. outside the array and such that the m_i of some corner *cell is closest. A2. Updating of this selected m_i (but not of any other of its topological neighbors) is due.*

The value W_2 is applied if both of the following conditions are satisfied: B1. The value of x lies outside the m_i array, but the closest m_i does not belong to any corner cell. B2. Updating of the selected edge cell or any of its topological neighbors, which must be one of the edge cells (eventually even a corner cell) is due.

If $p(x)$ in the two-dimensional input space were uniform over a square domain and zero outside it, it would be easy to deduce, in analogy with the one-dimensional case, that for an equidistant equilibrium distribution of the m_i values we must have $W_1 = 81$, $W_2 = 9$. Again, for other $p(x)$ the compensation is not complete with these weights. Then, as earlier, the Batch Map process may be run for a couple of iterations, followed by a couple of K-means iterations. Such a combination of methods is again both robust and unbiased and follows the two-dimensional input states very effectively.

3.7 Initialization of the SOM Algorithms

Random Initialization. The reason for using random initial values in the demonstrations of the SOM was that the SOM algorithms *can* be initialized using *arbitrary* values for the codebook vectors $m_i(0)$. In other words it has been demonstrated that initially unordered vectors will be ordered in the long run, in usual applications in a few hundred initial steps. This does *not* mean, however, that random initialization would be the *best* or *fastest* policy and should be used in practice.

Linear Initialization. As the $m_i(0)$ can be arbitrary, one might reason that *any ordered initial state* is profitable, even if these values do not lie along the main extensions of $p(x)$. A method we have used with success is to first determine the two eigenvectors of the autocorrelation matrix of x that have the largest eigenvalues, and then to let these eigenvectors span a two-dimensional linear subspace. A rectangular array (with rectangular or hexagonal regular lattice) is defined along this subspace, its centroid coinciding with that of the mean of the $x(t)$, and the main dimensions being the same as the two largest eigenvalues. The initial values of $m_i(0)$ are then identified with the array points. If one wants to obtain an approximately uniform lattice spacing in the SOM, the relative numbers of cells in the horizontal and vertical directions of the lattice, respectively, should be proportional to the two largest eigenvalues considered above.

Since the $m_i(0)$ are now already ordered and their point density roughly approximates $p(x)$, it will be possible to directly start the learning with the convergence phase, whereby one can use values for $\alpha(t)$ that from the beginning are significantly smaller than unity, and a neighborhood function the width of which is close to its final value, to approach the equilibrium smoothly.

3.8 On the "Optimal" Learning-Rate Factor

The period during which a rough order in the SOM is obtained is usually relatively short, on the order of 1000 steps, whereas most of the computing time is spent for the final convergence phase, in order to achieve a sufficiently good statistical accuracy. It is not clear how the learning-rate factor should be optimized during the first phase, since the width of the neighborhood function is thereby changing, too, which complicates the situation.

On the other hand, in Sect. 3.7 we already pointed out that it is always possible to start the SOM algorithm with an already ordered state, e.g. with all the m_i lying in a regular array along a two-dimensional hyperplane. Also, since during the final convergence phase we usually keep the width of the neighborhood fixed, it seems possible to determine some kind of "optimal" law for the sequence $\{\alpha(t)\}$ during the convergence phase. Before deriving this law for the SOM we shall discuss a couple of simpler examples, from which the basic idea may be seen.

Recursive Means. Consider a set of vectorial samples $\{x(t)\}$, $x(t) \in \Re^n$, $t = 0, 1, 2, \ldots, T$. From the way in which the mean m of the $x(t)$ is computed recursively it may be clear that $m(t)$ is at all times correct with respect to samples taken into account up to step $t + 1$:

$$m = \frac{1}{T} \sum_{t=1}^{T} x(t) = m(T) ;$$

$$m(t + 1) = \frac{t}{t+1} m(t) + \frac{1}{t+1} x(t+1)$$

$$= m(t) + \frac{1}{t+1} [x(t+1) - m(t)] . \tag{3.31}$$

This law will now be reflected in the VQ processes, too.

"Optimized" Learning-Rate Factors for Recursive VQ. The steepest-descent recursion for the classical vector quantization was derived in Sect. 1.5 and reads

$$m_i(t + 1) = m_i(t) + \alpha(t)\delta_{ci}[x(t) - m_i(t)] , \tag{3.32}$$

where δ_{ci} is the Kronecker delta. Let us rewrite (3.32) as

$$m_i(t + 1) = [(1 - \alpha(t)\delta_{ci}]m_i(t) + \alpha(t)\delta_{ci}x(t) . \tag{3.33}$$

(Notice a shift in the index of $x(t)$ with respect to (3.31). It is a matter of convention how we label the $x(t)$.) If we consider $m_i(t+1)$ as a "memory" of all the values $x(t')$, $t' = 0, 1, \ldots, t$, we see that if m_i is the "winner", then a memory trace of $x(t)$, scaled down by $\alpha(t)$, is superimposed on it. If m_i was "winner" at step t, it has a memory trace of $x(t-1)$, scaled down by $[1-\alpha(t)]\alpha(t-1)$ through the first term of (3.33); if $m_i(t)$ was not winner, no memory trace from $x(t-1)$ was obtained. It is now easy to see what happens to earlier memory traces from $x(t')$; every time when m_i is a "winner", all $x(t')$ are scaled down by the factor $[1-\alpha(t)]$, which we assume < 1. If $\alpha(t)$ were constant in time, we see that in the long run the effects of the $x(t')$ on $m(t+1)$ will be *forgotten*, i.e., $m(t+1)$ only depends on relatively few last samples, the "forgetting time" depending on the value of α. If, on the other hand, the earlier values of $\alpha(t')$ are selected as bigger, we might be able to *compensate for* the "forgetting" and have a roughly equal influence of the $x(t')$ on $m(t+1)$.

Comment. Since the Voronoi tessellation in VQ is changed at every step, the supports from which the $x(t)$ values are drawn for a particular m_i are also changing, so this case is not as clear as for the simple recursive means; therefore the $\alpha(t)$ values we derive will only approximately be "optimal."

Consider now the two *closest* instants of time t_1 and t_2, $t_2 > t_1$, for which the same m_i, denoted by m_c, was the "winner." If we stipulate that the memory traces of $x(t_1)$ and $x(t_2)$ shall be equal in different neurons at all times, obviously we have to select an *individual learning-rate factor $\alpha_i(t)$* for each m_i. For steps t_1, and t_2 we must then have

$$[1-\alpha_c(t_2)]\alpha_c(t_1) = \alpha_c(t_2) . \tag{3.34}$$

Solving for $\alpha_c(t_2)$ from (3.34) we get

$$\alpha_c(t_2) = \frac{\alpha_c(t_1)}{1+\alpha_c(t_1)} . \tag{3.35}$$

Let us emphasize that $\alpha_i(t)$ is now *only* changed when m_i was the "winner"; otherwise $\alpha_i(t)$ remains *unchanged*. Since $\alpha_c(t_1)$ retains the same value up to step t_2-1, we can easily see that the following recursion of α_i can be applied at the same time when the "winner" m_c is updated:

$$\alpha_c(t+1) = \frac{\alpha_c(t)}{1+\alpha_c(t)} . \tag{3.36}$$

It must once again be emphasized that (3.36), as well as similar expressions to be derived for SOM and LVQ, do not *guarantee* absolutely optimal convergence, because the Voronoi tessellations in these algorithms are changing. Conversely, however, one may safely state that if (3.36) is *not* taken into account, the convergence is *less optimal* on the average.

"Optimized" Learning-Rate Factor for the SOM. It is now straightforward to generalize the above philosophy for the SOM. If we define an individual $\alpha_i(t)$ for each m_i and write

$$m_i(t+1) = m_i(t) + \alpha_i(t)h_{ci}[x(t) - m_i(t)] , \tag{3.37}$$

where h_{ci} (in the convergence phase) is time-invariant, *we may update* $\alpha_i(t)$ *whenever a correction to the* m_i *values is made, relative to that correction:*

$$\alpha_i(t+1) = \frac{\alpha_i(t)}{1 + h_{ci}\alpha_i(t)} \ . \tag{3.38}$$

The law for $\alpha_i(t)$ derived above was based on a theoretical speculation; it may not work in the best way in practice, due to the very different values for $\alpha_i(t)$ obtained in the long run for different i. Let us recall that the point density of the codebook vectors in the original SOM algorithm was some monotonic function of $p(x)$. This density is influenced by differences in learning-rate factors.

In vector quantization, especially in Learning Vector Quantization discussed in Chap. 6 this idea works reasonably well, though: no harmful deformations of point densities of the m_i have thereby been observed.

Semi-Empirical Learning-Rate Factor. For the above mentioned reasons it also seems justified to keep $\alpha(t)$ in the SOM identical for all the neurons and to look for an *average optimal rate. Mulier* and *Cherkassky* [3.17] have ended up with an expression of the form

$$\alpha(t) = \frac{A}{t+B} \ , \tag{3.39}$$

where A and B are suitably chosen constants. At least this form satisfies the stochastic-approximation conditions. The main justification for (3.39) is that earlier and later samples will be taken into account with approximately similar average weights.

3.9 Effect of the Form of the Neighborhood Function

As long as one starts the self-organizing process with a wide neighborhood function, i.e., with a wide radius of the neighborhood set $N_c(0)$, or a wide standard deviation of $h_{ci}(0)$ such a value being of the same order of magnitude as half of the largest dimension of the array, there are usually no risks for ending up in "metastable" configurations of the map (for which the average expected distortion measure or average expected quantization error would end up in a local minimum instead of the global one.) However, with time-invariant neighborhood function the situation may be quite different, especially if the neighborhood function is narrow.

Erwin et al. [3.18] have analyzed the "metastable states" in a one-dimensional array. They first defined the neighborhood function being *convex* on a certain interval $I \equiv \{0, 1, 2, \ldots, N\}$, if the conditions $|s - q| > |s - r|$ and $|s - q| > |r - q|$ imply that $[h(s, s) + h(s, q)] < [h(s, r) + h(r, q)]$ for all $s, r, q \in I$. Otherwise the neighborhood function was said to be concave.

The main results they obtained were that if the neighborhood function is convex, there exist no stable states other than the ordered ones. If the neighborhood function is concave, there exist metastable states that may slow down the ordering process by orders of magnitude. Therefore, if in the beginning of the process the neighborhood function is convex, like the middle part of the Gaussian $h_{ci}(t)$ at large standard deviation is, the ordering can be achieved almost surely; and after ordering the neighborhood function can be shrunk to achieve an improved approximation of $p(x)$.

The ordering conditions in general are most severe if the input signal space has the same dimensionality as the array; in practice, however, the dimensionality of the input signal space is usually much higher, whereby "ordering" takes place easier.

3.10 Does the SOM Algorithm Ensue from a Distortion Measure?

Since the SOM belongs to the category of vector quantization (VQ) methods, one might assume that the starting point in its optimization must be some kind of *quantization error in the vector space*. Assume that $x \in \Re^n$ is the input vector and the $m_i \in \Re^n, i \in$ {indices of neurons} are the reference vectors; let $d(x, m_i)$ define a generalized distance function of x and m_i. The quantization error is then defined as

$$d(x, m_c) = \min_i \{d(x, m_i)\} , \qquad (3.40)$$

where c is the index of the "closest" reference vector to x in the space of input signals.

An even more central function in the SOM, however, is the *neighborhood function* $h_{ci} = h_{ci}(t)$ that describes the interaction of reference vectors m_i and m_c during adaptation and is often a function of time t. To this end one might try to define the *distortion measure* in the following way. Denote the set of indices of all lattice units by L; the distortion measure e is defined as

$$e = \sum_{i \in L} h_{ci} d(x, m_i) , \qquad (3.41)$$

that is, as a sum of distance functions weighted by h_{ci}, whereby c is the index of the closest codebook vector to x. If we now form the *average expected distortion measure*

$$E = \int ep(x)dx = \int \sum_{i \in L} h_{ci} d(x, m_i) p(x)dx , \qquad (3.42)$$

one way of defining the SOM is to define it as the set of the m_i that globally minimizes E.

Exact optimization of (3.42), however, is still an unsolved theoretical problem, and extremely heavy numerically. The best approximative solution that has been found so far is based on the Robbins-Monro stochastic approximation (Sect. 1.3.3). Following this idea we consider stochastic samples of the distortion measure: if $\{x(t), t = 1, 2, \ldots\}$ is a sequence of input samples and $\{m_i(t), t = 1, 2, \ldots\}$ the recursively defined sequence of codebook vector m_i, then

$$e(t) = \sum_{i \in L} h_{ci}(t) d[x(t), m_i(t)] \tag{3.43}$$

is a stochastic variable, and the sequence defined by

$$m_i(t+1) = m_i(t) - \lambda \cdot \nabla_{m_i(t)} e(t) \tag{3.44}$$

is used to find an approximation to the optimum, as asymptotic values of the m_i. This would then define the SOM algorithm for a generalized distance function $d(x, m_i)$. It must be emphasized, however, that although the convergence properties of stochastic approximation have been thoroughly known since 1951, the asymptotic values of the m_i obtained from (3.44) only minimize E approximately. We shall see in Sect. 3.12.2 that at least the point density of the reference vectors obtained from (3.44) are differen from those derived directly on the basis of (3.42). *Then, on the other hand, (3.43) and (3.44) may be taken as still another definition of a class of SOM algorithms.*

Comment 3.10.1. For those readers who are not familiar with the Robbins-Monro stochastic approximation it may be necessary to point out that E need not be a *potential function*, and the convergence limit does not necessarily represent the *exact* minimum of E, only an approximation of it. Nonetheless the convergence properties of this class of processes are very robust and they have been studied thoroughly long ago in the theory of this method.

Comment 3.10.2. Above, the neighborhood function h_{ci} was assumed time-invariant, and it will be assumed such also in Sect. 3.11 where an optimization attempt will be made. It seems to be an important property of the SOM algorithms, however, that the kernels h_{ci} are time-variable.

Example 3.10.1. Let $d(x, m_i) = ||x - m_i||^2$. Then we obtain the original SOM algorithm:

$$m(t+1) = m_i(t) + \alpha(t) h_{ci}(t)[x(t) - m_i(t)]. \tag{3.45}$$

Example 3.10.2. In the so-called *city-block metric*, or *Minkowski metric of power one*,

$$d(x, m_i) = \sum_j |\xi_j - \mu_{ij}|. \tag{3.46}$$

Now we have to write the SOM algorithm in component form:

$$\mu_{ij}(t+1) = \mu_{ij}(t) + \alpha(t) h_{ci}(t) \operatorname{sgn}[\xi_j(t) - \mu_{ij}(t)], \tag{3.47}$$

where sgn[·] is the signum function of its argument.

3.11 An Attempt to Optimize the SOM

It has to be emphasized that we do not know any theoretical reason for which the recursive algorithm of the basic SOM *should* ensue from any objective function E that describes, e.g., the average expected distortion measure. The following facts only *happen* to hold true: 1. The usual stochastic-approximation optimization of E leads to the basic SOM algorithm, but this optimization method is only approximate, as will be seen below. 2. The (heuristically established) basic SOM algorithm describes a nonparametric regression that often reflects important and interesting topological relationships between clusters of the primary data.

In order to demonstrate what the "energy function" formalism pursued in this context actually may mean, we shall proceed one step deeper in the analysis of the average expected distortion measure [1.75].

The effect of the index c of the "winner," which is a discontinuous function of x and all the m_i, can be seen more clearly if the integral E in (3.42), with $d(x, m_i) = ||x - m_i||^2$ is expressed as a sum of partial integrals taken over those domains X_i where x is closest to the respective m_i (partitions in the Voronoi tessellation, cf. Fig. 3.23):

$$E = \sum_i \int_{x \in X_i} \sum_k h_{ik}||x - m_k||^2 p(x)dx . \tag{3.48}$$

When forming the true (global) gradient of E with respect to an arbitrary m_j, one has to take into account two different kinds of terms: first, when the integrand is differentiated but the integration limits are held constant, and second, when the integration limits are differentiated (due to changing m_j) but the integrand is held constant. Let us call these terms G and H, respectively:

$$\nabla_{m_j} E = G + H , \tag{3.49}$$

whereby it is readily obtained

$$G = -2 \cdot \sum_i \int_{x \in X_i} h_{ij}(x - m_j)p(x)dx$$
$$= -2 \cdot \int_{x \text{ space}} h_{cj}(x - m_j)p(x)dx . \tag{3.50}$$

In the classical VQ (Sect. 1.5), $H = 0$, due to the facts that $h_{ck} = \delta_{ck}$, and there, across the border of the Voronoi tessellation between X_i and X_j, the terms $||x - m_i||^2$ and $||x - m_j||^2$ were equal. With general h_{ck}, computation of H is a very cumbersome task, because the integrands are quite different when crossing the borders in the Voronoi tessellation.

To start with, I am making use of the fact that only those borders of the tessellation that delimit X_j are shifting in differentiation.

Consider Fig. 3.23 that delineates the partition X_j, and the shift of its borders due to dm_j. The first extra contribution to $\nabla_{m_j} E$, with the $||x - m_k||^2$

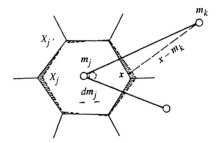

Fig. 3.23. Illustration of the border effect in differentiation with respect to m_j

evaluated over the topological neighborhood of m_j, is now obtained as the integral of the integrand of E taken over the shaded differential hypervolume. Because all the borders are segments of midplanes (hyperplanes) between the neighboring m_i, at least some kind of "average shift" seems to be estimable. As a matter of fact, if there were so many neighbors around m_j that X_j could be approximated by a hypersphere, and m_j were varied by the amount dm_j, then X_j would preserve its shape and be shifted by the amount $(1/2)dm_j$. In a general case with arbitrary dimensionality of the x space and constellation of the m_i, the shape of X_j would be changed; but my simplifying and averaging approximation is that this change of shape is not considered, in the first approximation at least, and X_j is only shifted by $(1/2)dm_j$. Another approximation that is needed to simplify the order-of-magnitude discussion is to assume $p(x)$ constant over the partition X_j, and this is justified if many m_i are used to approximate $p(x)$. If both of the above approximations are made, then the first contribution to H obtained from the variation of X_j, denoted H_1, is approximately the difference of two integrals: the integral over the displaced X_j, and the integral over the undisplaced X_j, respectively. This is equal to the integral of $\sum h_{jk}||x - (m_k - (1/2)dm_j)||^2$ *times* $p(x)$ *minus* the integral of $\sum h_{jk}||x - m_k||^2$ *times* $p(x)$. But one may think that this again is equivalent to the case where each of the m_k is differentiated by the same amount $-(1/2)dm_j$. Therefore this difference can also be expressed as the sum of the gradients of the integral of $\sum h_{jk}||x - m_k||^2$ *times* $p(x)$ taken with respect to each of the m_k, and multiplied by $-(1/2)dm_j$:

$$H_1 = -1/2 \cdot (-2) \cdot \int_{x \in X_j} \sum_{k \neq j} h_{jk}(x - m_k)p(x)dx$$
$$= \int_{x \in X_c} \sum_{k \neq c} h_{ck}(x - m_k)p(x)dx \ . \tag{3.51}$$

The case $k = c$ can be excluded above because this term corresponds to the basic Vector Quantization, and its contribution can be shown to be zero [1.75].

We must not forget that E is a sum over all i, and so the second extra contribution to H, named H_2, is due to the integral of the integrand of E

over all the partial differential hypervolumes (shaded segments in Fig. 3.23) bordering to $X_{j'}$ where j' is the index of any of the neighboring partitions of X_j. To compute H_2 seems to be even more cumbersome than to get the approximation of H_1. There are good reasons, however, to assume that on the average, $||H_2|| < ||H_1||$, since each of the integration domains in H_2 is only one segment of the differential of X_j, and the $x - m_k$ are different in each of these subdomains. In an order-of-magnitude analysis of the extra corrections at least, it is perhaps more interesting to concentrate on the major contribution H_1 quantitatively, whereby it is also possible to give an illustrative interpretation for it.

When trying to derive the results obtained into a recursive stepwise descent in the "E landscape", we may notice that in the expression of G, x runs over the whole x space, and $\nabla_{m_j} E$ has contributions from all those X_c for which m_j is a topological neighbor in the network. Conversely, in H_1, the integral is only over X_j, whereas the integrand contains terms that depend on the topological neighbors m_k of m_j. In analogy with Eqs. (1.166), (1.167), and (1.168) in Sect. 1.5 we can then write (neglecting terms due to H_2):

$$m_c(t+1) = m_c(t) + \alpha(t)\{h_{cc}[x(t) - m_c(t)]$$
$$- 1/2\sum_{k \neq c} h_{ck}[x(t) - m_k(t)]\} , \tag{3.52}$$

$$m_i(t+1) = m_i(t) + \alpha(t)h_{ci}[x(t) - m_i(t)] , \quad \text{for } i \neq c .$$

Notice in particular that the term $-1/2\sum_k h_{ck}[x(t) - m_k(t)]$ was derived under the assumption that $p(x)$ is constant over each partition in the Voronoi tessellation. For nodes at the edges of the tessellation, the corresponding partitions X_i extend to infinity, and this approximation is then no longer quite valid; so boundary effects, slightly different from those encountered with the basic SOM, are also here discernible.

Before reporting numerical experiments, we can give an interpretation to the extra terms due to the H_1 integral. Consider Fig. 3.24 that illustrates the old and the new algorithm. Assume that at some learning step, one of the m_i is offset from the rest of the parameter vector values. When x is closest to it, in the old algorithm this $m_i = m_c$ and its topological neighbors m_k would be shifted toward x. In the new algorithm, only the $m_k \neq m_c$ are shifted toward x, whereas m_c is "relaxed" in the opposite direction toward the center of the m_k neighborhood. This extra relaxation seems to be useful for self-organization, as demonstrated below.

A numerical experiment was performed to study the convergence properties of (3.52). A square array of nodes, with two-dimensional x and m_i, was used in the experiment. Both the old algorithm and (3.52) were run with identical initial state, values of $\alpha(t)$, and random number sequence. The $m_i(0)$ were selected as independent random values and $\alpha(t)$ decreased linearly from 0.5 to 0. The neighborhood kernel h_{ck} was constant for all nodes within one

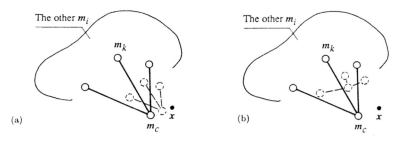

Fig. 3.24. Differences in correction: **(a)** The basic SOM, **(b)** Eq. (3.52)

lattice spacing from c in the horizontal and vertical direction, and $h_{ck} = 0$ otherwise. (It must be recalled that in practical applications, to make convergence faster and safer, the kernel is usually made to "shrink" over time.) In Fig. 3.25, the values of m_i for a square lattice as in Fig. 3.5 after 8000 steps of the process are displayed. The $p(x)$ had a constant value in the framed square and zero outside it. From many similar experiments (of which one is shown in Fig. 3.25), two conclusions transpire: 1. The new algorithm orders the m_i somewhat faster and safer; after 8000 steps (with the narrow constant kernel h_{ck}), none of the old-algorithm states was completely ordered, whereas about half of the new-algorithm states were. 2. The boundary effects in the new algorithm are stronger.

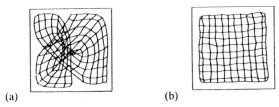

Fig. 3.25. Two-dimensional map after 8000 steps. **(a)** The basic SOM with small neighborhood, **(b)** The same neighborhood but with (3.52) used

Special Case. If the probability density function $p(x)$ of input data is *discrete-valued*, such as the case is in the famous traveling-salesman problem, the original SOM algorithm can be derived from the average expected distortion measure. (For a textbook account, see [1.40], Sect. 6.2 of that book.) In view of the above discussion this result may now be clear, because $p(x)$ at the borders of the Voronoi tessellation is then zero, and the extra term H, due to differentiation of the integration limits, then vanishes.

Modification of the Definition of the "Winner." It has been pointed out [3.19, 20] that if the matching criterion for the definition of the "winner" c is modified as

$$\sum_i h_{ci}||x - m_i||^2 = \min_j \left\{ \sum_i h_{ji}||x - m_i||^2 \right\} , \tag{3.53}$$

where h_{ci} is the same neighborhood function as that applied during learning, then the average expected distortion measure becomes an energy or potential function, the minimization of which can be carried out by the usual gradient-descent principle accurately. While this observation has certain mathematical interest, the following facts somewhat neutralize its value in neural-network modeling: 1. In the physiological explanation of the SOM (Chap. 4) the neighborhood function h_{ck} only defines the control action on the synaptic plasticity, not control of network acticity in the WTA function as implied by (3.53). 2. The h_{ji} above have to be normalized, $\sum_j h_{ji} = 1$, whereby in general $h_{ji} \neq h_{ij}$ for $i \neq j$.

For a practical ANN algorithm, computation of (3.53) for finding the "winner" is also more tedious compared with simple distance calculations.

Nonetheless, this kind of parallel theory, by analogy, may shed some extra light on the nature of the SOM process.

3.12 Point Density of the Model Vectors

3.12.1 Earlier Studies

In biological brain maps, the areas allocated to the representations of various sensory features are often believed to reflect the importance of the corresponding feature sets; the scale of such a map is somewhat loosely called the "magnification factor." As a matter of fact, different parts of a receptive surface such as the retina are transformed in the brain in different scales, almost like in mathematical *quasiconformal mappings*.

It is true that a higher magnification factor usually corresponds to a higher density of receptor cells, which might give a reason to assume that the magnification factor depends on the number of axons connecting the receptive surfaces to that brain area. However, in the nervous systems there also exist numerous "processing stations," called nuclei in the sensory pathways, whereby one cannot directly compare input-output point densities in such a mapping. In the light of the SOM theory it rather seems that the area allocated to the representation of a feature in a brain map is somehow *proportional to the statistical frequency of occurrence of that feature in observations*.

As the nervous pathways in the biological realms are very confused, we shall use the term "magnification factor" in the sequel only to mean *the inverse of point density of the m_i*. For instance, in the classical vector quantization with squared errors we had this density proportional to $[p(x)]^{\frac{n}{n+2}}$, where n is the dimensionality of x. The "magnification factor" would then be the inverse of this.

Ritter and Schulten [3.21] and *Ritter* [3.22] analyzed the point density for the linear map in the case that the map contained a very large number of codebook vectors over a finite area. If N neighbors on both sides of the "winner" and the "winner" itself were included in the neighborhood set, the asymptotic point density (denoted here M) can be shown to be $M \propto [p(x)]^r$, where the exponent is

$$r = \frac{2}{3} - \frac{1}{3N^2 + 3(N+1)^2} .$$ (3.54)

One thing has to be made clear first. The SOM process, although it may start with a wide neighborhood function, can have an arbitrary neighborhood function width at the end of the learning process (especially if the Gaussian kernel is used). Then at the last phases we may even have the zero-order topology case, i.e., no neighbors except the "winner" itself. However, with zero-order topological interaction the process no longer maintains the order of the codebook vectors, and this order may disappear in extensive learning with narrow neighborhoods. Therefore one has to make a compromise between the wanted approximation accuracy of $p(x)$, which is best in the VQ case, and the wanted stability of ordering, for which the neighborhood interactions are needed.

With $N = 0$ we get above the one-dimensional VQ (or the so-called scalar quantization) case, whereby $r = 1/3$. It seems that the low value of r (compared with unity) is often regarded as a handicap; it may be felt that approximation of probability density functions should be necessary in all statistical pattern recognition tasks, to which the neural networks are also supposed to bring the "optimal" solution. One has to note, however: 1. If classification is based on finding the Bayesian borders where the density functions of two classes have equal value, the same result can be obtained by comparing *any* (but the same) monotonic functions of densities. 2. Most practical applications have data vectors with very high dimensionality, say, dozens to hundreds. Then, e.g., in classical VQ, the exponent actually is $n/(n+2) \approx 1$, where n is the dimensionality of x. It is not yet quite clear what the corresponding exponent in two-dimensional SOMs with very high input density is, but it is certainly lower than unity.

In the work of Dersch and Tavan [3.23] the neighborhood function was Gaussian, and it is possible to see from Fig. 3.26 how the exponent α depends on the normalized second moment σ of the neighborhood function. (The integral values of σ correspond, very roughly, to N^2 above.)

Further works on the SOM point density can be found in [3.24, 25].

3.12.2 Numerical Check of Point Densities in a Finite One-Dimensional SOM

Strictly speaking, the scalar entity named the "point density" as a function of x has a meaning only in either of the following cases: 1. The number of

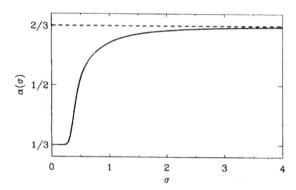

Fig. 3.26. Distortion exponent α for a Gaussian neighborhood interaction as the function of its normalized second moment σ [3.23]

points (samples) in any reasonable "volume" differential is large, or 2. The points (samples) are stochastic variables, and their differential probability of falling into a given differential "volume," i.e., the probability density $p(x)$ can be defined.

Since in vector quantization problems one aims at the minimum expected quantization error, the model or codebook vectors m_i tend to assume a more or less regular optimal configuration, and cannot be regarded as stochastic. Neither can one usually assume that their number in any differential volume is high.

Consider now the one-dimensional coordinate axis x, on which some probability density function $p(x)$ is defined. Further consider two successive points m_i and m_{i+1} on this same axis. One may regard $(m_{i+1} - m_i)^{-1}$ as the local "point density." However, to which value of x should it be related? The same problem is encountered if a functional dependence is assumed between the point density and $p(x)$. Below we shall define the point density as the inverse of the width of the Voronoi set, i.e., $[(m_{i+1} - m_i)/2]^{-1}$, and refer the density to the model m_i, which is only one choice, of course.

Asymptotic State of the One-Dimensional, Finite-Grid SOM Algorithm in Three Exemplary Cases. Consider a series of samples of the input $x(t) \in \Re$, $t = 0, 1, 2, \ldots$ and a set of k model (codebook) values $m_i(t) \in \Re$, $t = 0, 1, 2, \ldots$, whereupon i is the model index $(i = 1, \ldots, k)$. For convenience assume $0 \leq x(t) \leq 1$.

The original one-dimensional self-organizing map (SOM) algorithm with at most one neighbor on each side of the best-matching m_i reads:

$$
\begin{aligned}
m_i(t+1) &= m_i(t) + \varepsilon(t)[x(t) - m_i(t)] \text{ for } i \in N_c , \\
m_i(t+1) &= m_i(t) \text{ for } i \notin N_c , \\
c &= \arg\min_i\{|x(t) - m_i(t)|\} , \text{ and} \\
N_c &= \{\max(1, c-1), c, \min(k, c+1)\} ,
\end{aligned}
\qquad (3.55)
$$

where N_c is the neighborhood set around node c, and $\varepsilon(t)$ is a small scalar value called the learning-rate factor. In order to analyze the asymptotic values of the m_i, let us assume that the m_i are already ordered. Let the Voronoi set V_i around m_i be defined as

$$\text{for } 1 < i < k, \; V_i = \left[\frac{m_{i-1} + m_i}{2}, \frac{m_i + m_{i+1}}{2}\right],$$

$$V_1 = \left[0, \frac{m_1 + m_2}{2}\right], \; V_k = \left[\frac{m_{k-1} + m_k}{2}, 1\right], \text{ and denote}$$

$$\text{for } 1 < i < k, \; U_i = V_{i-1} \cup V_i \cup V_{i+1},$$

$$U_1 = V_1 \cup V_2, \; U_k = V_{k-1} \cup V_k. \tag{3.56}$$

In other words, U_i is the set of such $x(t)$ values that are able to modify $m_i(t)$ during one learning step. Following the simple case discussed in Sect. 3.5.1 one can write the condition for stationary equilibrium of the m_i for a constant ε as:

$$\forall i, \; m_i = \mathrm{E}\left\{x \mid x \in U_i\right\}. \tag{3.57}$$

This means that every m_i must coincide with the centroid of the probability mass in the respective U_i.

For $2 < i < k - 1$ we have for the limits of the U_i:

$$A_i = \frac{1}{2}(m_{i-2} + m_{i-1}),$$

$$B_i = \frac{1}{2}(m_{i+1} + m_{i+2}). \tag{3.58}$$

For $i = 1$ and $i = 2$ we must take B_i as above, but $A_i = 0$; and for $i = k - 1$ and $i = k$ we have A_i as above and $B_i = 1$.

Case 1: $p(x) = 2x$. The first case we discuss here is the one where the probability density function of x is linear, $p(x) = 2x$ for $0 \leq x \leq 1$ and $p(x) = 0$ for all the other values of x.

It is now straightforward to compute the centroids of the trapezoidal probability masses in the U_i:

$$\mathrm{E}\{x|x \in U_i\} = \frac{2(B_i^3 - A_i^3)}{3(B_i^2 - A_i^2)}. \tag{3.59}$$

The stationary values of the m_i are defined by the set of nonlinear equations

$$\forall i, \; m_i = \frac{2(B_i^3 - A_i^3)}{3(B_i^2 - A_i^2)} \tag{3.60}$$

and the solution of (3.60) is sought by the so-called *contractive mapping*. Let us denote

$$z = [m_1, m_2, \ldots, m_k]^T. \tag{3.61}$$

Then the equation to be solved is of the form

$$z = f(\mathbf{z}) \ . \tag{3.62}$$

Starting with the first approximation for $\mathbf{z}$ denoted $\mathbf{z}^{(0)}$, each improved approximation for the root is obtained recursively:

$$z^{(s+1)} = f(z^{(s)}) \ . \tag{3.63}$$

In this case one may select for the first approximation of the m_i equidistant values.

With a small number of grid points, (3.63) converges reasonably fast, but already with 100 grid points the required number of steps for the accuracy of, say, five decimal places may be about five thousand.

It may now be expedient to define the point density q_i around m_i as the inverse of the length of the Voronoi set, or $q_i = [(m_{i+1} - m_{i-1})/2]^{-1}$.

The problem expressed in a number of previous works, e.g., [3.21, 22, 23], is to find out whether q_i could be approximated by the functional form $const.[p(m_i)]^\alpha$. Previously this was only shown for the continuum limit, i.e. for an infinite number of grid points. The present numerical analysis allows us to derive results for finite-length grids, too. Assuming tentatively that the power law holds for the models m_i through m_j (leaving aside models near to the ends of the grid), we shall then have

$$\alpha = \frac{\log(m_{i+1} - m_{i-1}) - \log(m_{j+1} - m_{j-1})}{\log[p(m_j)] - \log[p(m_i)]} \ . \tag{3.64}$$

Naturally, more values of the m_i could be taken for improved accuracy. In Table 3.3, using $i = 4$ and $j = k - 3$, between which the border effects may be assumed as negligible, the exponent α has been estimated from (3.64) for 10, 25, 50, and 100 grid points, respectively.

Case 2: $p(x) = 3x^2$ (convex). Now we have the system of equations

$$\forall i, \quad m_i = \frac{3(B_i^4 - A_i^4)}{4(B_i^3 - A_i^3)} \tag{3.65}$$

and the approximations for α are in Table 3.3.

Case 3: $p(x) = 3x - \frac{3}{2}x^2$ (concave). The system of equations reads

$$\forall i, \quad m_i = \frac{8(B_i^3 - A_i^3) - 3(B_i^4 - A_i^4)}{12(B_i^2 - A_i^2) - 4(B_i^3 - A_i^3)} \tag{3.66}$$

and the approximations for α are also in Table 3.3.

These simulations show convincingly that for three qualitatively different $p(x)$ the exponent α, even for a reasonably small number of grid points, is fairly close to the value of $\alpha = 0.6$ as derived in the continuum limit in [3.22], in the case of one neighbor on both sides of the best-matching m_i.

Table 3.3. Exponent derived from the SOM algorithm

Grid points	Exponent α		
	Case 1	Case 2	Case 3
10	0.5831	0.5845	0.5845
25	0.5976	0.5982	0.5978
50	0.5987	0.5991	0.5987
100	0.5991	0.5994	0.5990

Numerically Accurate Optimum of the One-Dimensional SOM Distortion Measure with Finite Grids: Case 1. Equation 3.48 can also be written as

$$E = \sum_i \sum_j \int_{x \in V_i} h_{ij} \|x - m_j\|^2 p(x) dx \, , \tag{3.67}$$

where i and j run over all the values for which h_{ij} has been defined, and V_i is the Voronoi set around m_i.

In the simple one-dimensional case, when h_{ij} is defined as

$$h_{ij} = 1 \text{ if } |i - j| < 2 \, , \text{ and } h_{ij} = 0 \text{ otherwise} \, , \tag{3.68}$$

when we take Case 1, or $p(x) = 2x$ for $0 \le x \le 1$, $p(x) = 0$ otherwise, and when we assume the m_i as ordered in the ascending sequence, (3.67) becomes

$$
\begin{aligned}
E &= 2 \sum_i \sum_{j \in N_i} \int_{C_i}^{D_i} (x - m_j)^2 x \, dx \\
&= \sum_i \sum_{j \in N_i} m_j^2 (D_i^2 - C_i^2) - \frac{4}{3} m_j (D_i^3 - C_i^3) + \frac{1}{2} (D_i^4 - C_i^4) \quad (3.69)
\end{aligned}
$$

where the *neighborhood set of indices* N_i was defined in (3.55), and the borders C_i and D_i of the Voronoi set V_i are

$$
\begin{aligned}
C_1 &= 0 \, , \\
C_i &= \frac{m_{i-1} + m_i}{2} \quad \text{for } 2 \le i \le k \, , \\
D_i &= \frac{m_i + m_{i+1}}{2} \quad \text{for } 1 \le i \le k - 1 \, , \\
D_k &= 1 \, . \tag{3.70}
\end{aligned}
$$

When forming the accurate gradient of E, it must be noticed that index i is contained in N_{i-1}, N_i, and N_{i+1}, whereupon

$$
\begin{aligned}
\frac{\partial E}{\partial m_i} = \frac{\partial}{\partial m_i} \sum_{j \in N_{i-1}} \Bigg(& m_j^2 (D_{i-1}^2 - C_{i-1}^2) - \frac{4}{3} m_j (D_{i-1}^3 - C_{i-1}^3) \\
& + \frac{1}{2} (D_{i-1}^4 - C_{i-1}^4) \Bigg)
\end{aligned}
$$

$$+ \frac{\partial}{\partial m_i} \sum_{j \in N_i} \left(m_j^2(D_i^2 - C_i^2) - \frac{4}{3}m_j(D_i^3 - C_i^3) + \frac{1}{2}(D_i^4 - C_i^4) \right)$$

$$+ \frac{\partial}{\partial m_i} \sum_{j \in N_{i+1}} \left(m_j^2(D_{i+1}^2 - C_{i+1}^2) - \frac{4}{3}m_j(D_{i+1}^3 - C_{i+1}^3) \right.$$

$$\left. + \frac{1}{2}(D_{i+1}^4 - C_{i+1}^4) \right) . \tag{3.71}$$

The result of this differentiation is given as follows (notice that $C_i = D_{i-1}$):

$$\frac{\partial E}{\partial m_1} = 2m_1 D_2^2 - \frac{4}{3}D_2^3 - m_3^2 C_2 + 2m_3 C_2^2 - C_2^3 ,$$

$$\frac{\partial E}{\partial m_2} = m_1^2 D_2 - 2m_1 D_2^2 + D_2^3 + 2m_2 D_3^2 - \frac{4}{3}D_3^3 - m_3^2 C_2 + 2m_3 C_2^2$$
$$- C_2^3 - m_4^2 C_3 + 2m_4 C_3^2 - C_3^3 ,$$

$$\frac{\partial E}{\partial m_i} = m_{i-2}^2 D_{i-1} - 2m_{i-2}D_{i-1}^2 + D_{i-1}^3 + m_{i-1}^2 D_i - 2m_{i-1}D_i^2 + D_i^3$$
$$- m_{i+1}^2 C_i + 2m_{i+1}C_i^2 - C_i^3 - m_{i+2}^2 C_{i+1} + 2m_{i+2}C_{i+1}^2 - C_{i+1}^3$$
$$+ 2m_i(D_{i+1}^2 - C_{i-1}^2) - \frac{4}{3}(D_{i+1}^3 - C_{i-1}^3) \quad \text{for } 2 < i < k-1 ,$$

$$\frac{\partial E}{\partial m_{k-1}} = m_{k-3}^2 D_{k-2} - 2m_{k-3}D_{k-2}^2 + D_{k-2}^3 + m_{k-2}^2 D_{k-1}$$
$$- 2m_{k-2}D_{k-1}^2 + D_{k-1}^3 - m_k^2 C_{k-1} + 2m_k C_{k-1}^2$$
$$- C_{k-1}^3 + 2m_{k-1}(1 - C_{k-2}^2) - \frac{4}{3}(1 - C_{k-2}^3) , \quad \text{and}$$

$$\frac{\partial E}{\partial m_k} = m_{k-2}^2 D_{k-1} - 2m_{k-2}D_{k-1}^2 + D_{k-1}^3$$
$$+ 2m_k(1 - C_{k-1}^2) - \frac{4}{3}(1 - C_{k-1}^3) . \tag{3.72}$$

The question is whether one can obtain the optimal values of the m_i by the gradient-descent method, i.e.,

$$\forall i , \quad m_i(t+1) = m_i(t) - \lambda(t) \cdot \partial E/\partial m_i|_t , \tag{3.73}$$

where $\lambda(t)$ is a suitable small scalar factor. In the present problem E is of the fourth degree in the m_i and at least one kind of spurious local optimum has been found: for instance, when starting with the asymptotic m_i values obtained from the SOM algorithm and keeping $\lambda(t)$ at a value of the order of .001 or smaller, a very shallow local minimum of E has been reached, which has given the wrong value of about .6 for α. However, with $\lambda(t) > .01$ (even with $\lambda(t) = 10$) and starting with very different initial values for the m_i, the process robustly converges to a unique global minimum. After computation of the optimal values $\{m_i\}$, in order to facilitate a direct comparison with the values presented in Table 3.3 the exponent α of the tentative power law was computed from (3.64) of the previous section and presented in Table 3.4 for different lengths of the grid.

Table 3.4. Exponent derived from the SOM distortion measure

Grid points	Exponent α
10	0.3281
25	0.3331
50	0.3333
100	0.3331

Clearly the computed α is an approximation of the value of $1/3$, the same as the exponent in vector quantization for $n = 1$ and $r = 2$, rather than of $\alpha = .6$ of the simple SOM algorithm.

The result that transpired in this numerical check is that the point density of the model (codebook) vectors resulting as asymptotic values in the basic SOM algorithm is different from that ensuing as the parameter values of the SOM distortion measure at its minimum. Nonetheless the m_i in both cases can be regarded as the nodes of an "elastic" network that is regressed onto the manifold of the input samples in an orderly fashion. The conclusion is thus that the Robbins-Monro stochastic approximation does not exactly lead to the basic SOM algorithm, but the algorithm and the distortion measure may be two optional ways to define the self-organizing map.

3.13 Practical Advice for the Construction of Good Maps

Although it is possible to obtain some kind of maps without taking into account any precautions, nonetheless it will be useful to pay attention to the following advice in order that the resulting mappings be stable, well oriented, and least ambiguous. Let us recall that the initialization problem was already discussed in Sect. 3.7. The reader is now adviced to study the two extensive software packages SOM_PAK [3.26] and LVQ_PAK [3.27].

Form of the Array. For visual inspection, the *hexagonal lattice* is to be preferred, because it does not favor horizontal and vertical directions as much as the rectangular array. The *edges* of the array ought to be rectangular rather than square, because the "elastic network" formed of the reference vectors m_i must be oriented along with $p(x)$ and be stabilized in the learning process. Notice that if the array were, e.g., circular, it would have no stable orientation in the data space; so any oblongated form is to be preferred. On the other hand, since the m_i have to approximate $p(x)$, it would be desirable to find such dimensions for the array that roughly correspond to the major dimensions of $p(x)$. Therefore, visual inspection of the rough form of $p(x)$, e.g., by *Sammon's mapping* [1.34] (Sect. 1.3.2) ought to be done first.

Learning with a Small Number of Available Training Samples. Since for a good statistical accuracy the complete learning process may require an appreciable number, say, 100'000 steps, and the number of available samples is usually much smaller, it is obvious that the samples must be used *reiteratively* in training. Several alternatives then exist: the samples may be applied cyclically or in a randomly permuted order, or picked up at random from the basic set (so-called *bootstrap learning*). It has turned out in practice that ordered cyclic application is not noticeably worse than the other, mathematically better justifiable methods.

Enhancement of Rare Cases. It may also be obvious from the above that the SOM in one way or another tends to represent $p(x)$. However, in many practical problems important cases (input data) may occur with small statistical frequency, whereby they are not able to occupy any territory at all in the SOM. Therefore, such important cases can be enhanced in learning by an arbitrary amount by taking a higher value of α or h_{ci} for these samples, or repeating these samples in a random order a sufficient number of times during the learning process. Determination of proper enhancement in learning should be done in cooperation with the end users of these maps.

Scaling of the Pattern Components. This is a very subtle problem. One may easily realize that the orientation, or ordered "regression" of the reference vectors in the input space must depend on the scaling of the components (or dimensions) of the input data vectors. However, if the data elements have already been represented in different scales, there does not exist any simple rule to determine what kind of optimal rescaling should be used before entering the training data to the learning algorithm. The first guess, which is usually rather effective, especially with high input dimensionality, is to normalize the variance of each component over the training data. One may also try many heuristically justifiable rescalings and check the quality of the resulting maps by means of Sammon's mapping or average quantization errors.

Forcing Representations to a Wanted Place on the Map. Sometimes, especially when the SOM is used to monitor experimental data, it may be desirable to map "normal" data onto a specified location (say, into the middle) of the map. In order to force particular data into wanted places, it is advisable to use their copies for the initial values of reference vectors at these locations, and to keep the learning-rate factor α low for these locations during their updating.

Monitoring of the Quality of Learning. Different learning processes can be defined starting with different initial values $m_i(0)$, and applying different sequences of the training vectors $x(t)$ and different learning parameters. It is obvious that some optimal map for the same input data may exist. It may also be obvious that the best map is expected to yield the smallest average *quantization error*, approximately at least, because it is then fitted best to the same data. The mean of $||x - m_c||$, defined via inputting the training data once again after learning, is then a useful performance index. Therefore, an appreciable number (say, several dozens) of random initializations of the $m_i(0)$ and different learning sequences ought to be tried, and the map with the minimum quantization error might be selected.

The accuracy of the maps in *preserving the topology*, or neighborhood relations, of the input space has been measured in various ways. One approach is to compare the relative positions of the reference vectors with the relative positions of the corresponding units on the map [3.28]. For example, the number of times the Voronoi region of another map unit "intrudes" the middle of the reference vectors of two neighbor units can be measured [3.29]. A different approach is to consider, for each input vector, the distance of the best-matching unit and the second-best-matching unit on the map: If the units are not neighbors, then the topology is not preserved [3.30, 31] . When the distance between map units is defined suitably, such a measure can be combined with the quantization error to form a single measure of map goodness [3.32]. Although it is not at present possible to indicate the best measure of map quality, these measures may nevertheless be useful in choosing suitable learning parameters and map sizes.

3.14 Examples of Data Analyses Implemented by the SOM

This section contains demonstrations of the SOM, intended to visualize so-called *data matrices*. In the simplest cases we assume that all items possess values for all their attributes; later in Sect. 3.14.2 we consider another case in which almost all of the items lack some of their attributes.

3.14.1 Attribute Maps with Full Data Matrix

Abstract Hierarchical Data Structure. Consider a finite set of items, each one having a number of characteristics or *attributes*. The latter can be binary, or integer- or continuous-valued; at any rate the sets of attributes should be comparable in some metric.

If we first apply the SOM to abstract data vectors consisting of hypothetical attributes, we can thereby define very clear data structures. We shall consider an example with implicitly defined (hierarchical) structures in the primary data, which the map algorithm is then supposed to reveal and display. Although the SOM is a single-level network, it can produce a hierarchical representation of the relations implicit in the primary data.

In Table 3.5, 32 items, with five hypothetical attributes each, are recorded in a data matrix. (Let us recall that this example is completely artificial.) Each of the columns represents one item, and for later inspection the items are labeled "A" through "6", although these labels are not referred to during learning.

Table 3.5. Input data matrix

Attribute	A	B	C	D	E	F	G	H	I	J	K	L	M	N	O	P	Q	R	S	T	U	V	W	X	Y	Z	1	2	3	4	5	6
a_1	1	2	3	4	5	3	3	3	3	3	3	3	3	3	3	3	3	3	3	3	3	3	3	3	3	3	3	3	3	3	3	3
a_2	0	0	0	0	0	1	2	3	4	5	3	3	3	3	3	3	3	3	3	3	3	3	3	3	3	3	3	3	3	3	3	3
a_3	0	0	0	0	0	0	0	0	0	0	1	2	3	4	5	6	7	8	3	3	3	3	6	6	6	6	6	6	6	6	6	6
a_4	0	0	0	0	0	0	0	0	0	0	0	0	0	0	0	0	0	0	1	2	3	4	1	2	3	4	2	2	2	2	2	2
a_5	0	0	0	0	0	0	0	0	0	0	0	0	0	0	0	0	0	0	0	0	0	0	0	0	0	0	1	2	3	4	5	6

The attribute values $(a_1, a_2, \ldots, a_5)$ were defined artificially and they constitute the pattern vector x that acts as a set of signal values at the inputs of the network of the type of Fig. 3.3. During training, the vectors x were selected from Table 3.5 at random. Sampling and adaptation was continued iteratively until one could regard the asymptotic state of the SOM as stationary. Such a "learned" network was then calibrated using the items from Table 3.5 and labeling the map units according to the best match with the different items. Such a labeled map is shown in Fig. 3.27. It is discernible that the "images" of different items are related according to a taxonomic graph where the different branches are visible. For comparison, Fig. 3.28 illustrates the so-called *minimal spanning tree* (where the most similar pairs of items are linked) that describes the similarity relations of the items in Table 3.5.

Map of a Binary Data Matrix. Attributes are usually variables with scalar-valued discrete or continuous values, but they may also attain qualitative properties such as "good" or "bad". If the property of being "good" or

```
B C D E * Q R * Y Z
A * * * * P * * X *
  * F * N O * W * * 1
* G * M * * * * 2 *
  H K L * T U * 3 * *
* I * * * * * * 4 *
* J * S * * V * 5 6
```

Fig. 3.27. Self-organized map of the data matrix in Table 3.5

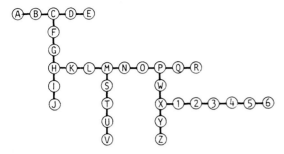

Fig. 3.28. Minimal spanning tree of the data matrix in Table 3.5

"bad", respectively, should be describable by a numerical attribute, it would be simplest to assume that such an attribute has the binary value, say 1 or 0, depending on the presence vs. absence of that attribute, respectively. Then the (unnormalized) similarity between two (binary) attribute sets may be defined in terms of the number of attributes common to both sets, i.e., as the dot product of the respective attribute vectors. It might seem more effective to use the value +1 to indicate the presence of an attribute, and -1 for its absence, respectively; however, if we *normalize* the input vectors, in their subsequent comparison using the dot product the attribute values 0 have a qualitatively similar effect as negative components in a comparison on the basis of vectorial differences. Euclidean distances can naturally be used directly for comparison, too.

To illustrate the self-organizing result with a concrete model simulation [2.73], consider the data given in Table 3.6. Each column is a schematic description of an animal, based on the presence (= 1) or absence (= 0) of some of the 13 different attributes given on the left. Some attributes, such as "feathers" and "2 legs" are correlated, indicating more significant differences than the other attributes, but we shall not take this correlation into account in learning in any way. In the following, we will take each column for the input vector of the animal indicated at the top. The animal name itself does not belong to the vector but instead specifies the label of the animal in the calibration of the map.

The members of the data set were presented iteratively and in a random order to a SOM of 10×10 neurons subject to the adaptation process described above. The initial connection strengths between the neurons and their $n = 29$ input lines were chosen to be small random values, i.e. no prior order was imposed. However, after a total of 2000 presentations, each neuron became more or less responsive to one of the occuring attribute combinations and simultaneously to one of the 16 animal names, too. Thus we obtain the map shown in Fig. 3.29 (the dots indicate neurons with weaker responses). It is very apparent that the spatial order of the responses has captured the essential "family relationships" among the animals. Cells responding to, e.g.,

Table 3.6. Animal names and their attributes

		dove	hen	duck	goose	owl	hawk	eagle	fox	dog	wolf	cat	tiger	lion	horse	zebra	cow
is	small	1	1	1	1	1	1	0	0	0	0	1	0	0	0	0	0
	medium	0	0	0	0	0	0	1	1	1	1	0	0	0	0	0	0
	big	0	0	0	0	0	0	0	0	0	0	0	1	1	1	1	1
has	2 legs	1	1	1	1	1	1	1	0	0	0	0	0	0	0	0	0
	4 legs	0	0	0	0	0	0	0	1	1	1	1	1	1	1	1	1
	hair	0	0	0	0	0	0	0	1	1	1	1	1	1	1	1	1
	hooves	0	0	0	0	0	0	0	0	0	0	0	0	0	1	1	1
	mane	0	0	0	0	0	0	0	0	0	1	0	0	1	1	1	0
	feathers	1	1	1	1	1	1	1	0	0	0	0	0	0	0	0	0
likes to	hunt	0	0	0	0	1	1	1	1	0	1	1	1	1	0	0	0
	run	0	0	0	0	0	0	0	0	1	1	0	1	1	1	1	0
	fly	1	0	1	1	1	1	0	0	0	0	0	0	0	0	0	0
	swim	0	0	1	1	0	0	0	0	0	0	0	0	0	0	0	0

"birds" occupy the left part of the lattice, "hunters" such as "tiger", "lion" and "cat" are clustered toward the right, and more "peaceful" species such as "zebra", "horse" and "cow" are situated in the upper middle. Within each cluster, a further grouping according to similarity is discernible.

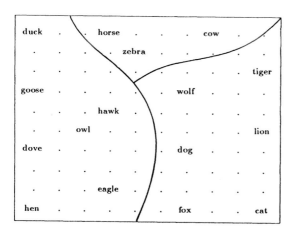

Fig. 3.29. After the network had been trained with inputs describing attribute sets from Table 3.6, the map was calibrated by the columns of Table 3.6 and labeled correspondingly. A grouping according to similarity has emerged

3.14.2 Case Example of Attribute Maps Based on Incomplete Data Matrices (Missing Data): "Poverty Map"

It has turned out that the SOM is a very robust algorithm, compared with many other neural models. One of the frequently encountered problems in practical (especially statistical) applications is caused by missing data [3.33, 34]. However, if the number of attributes taken into account is appreciable, say, at least on the order of hundreds, an appreciable fraction of data may be missing without making the similarity comparison impossible.

To start with, it is advisable to *normalize each attribute scale* such that its variance taken over all the items is unity. Similarly one may subtract the mean from each attribute: thus the scales are said to be $(0, 1)$ -normalized.

Comparison of an item with the codebook vectors is best made in the Euclidean metric, unless some better metric is deducible from the nature of the problem. Then, however, only the known components of x, and the corresponding components of each m_i, can be taken into account in the comparison: during this step the vectors are redimensioned correspondingly.

The data used in this case example were the same as in Sect. 1.3.2, namely, from the statistics published by *World Bank* [1.26]. The 39 indicators describe the poverty of the countries, or their citizens. All indicators are relative to population.

For 126 countries of the world listed, 12 or more indicator values of 39 possible ones were given to 78 countries and these countries were then taken to the data matrix used in training the SOM process. In Fig. 3.30, these countries are labeled by capital letters. For the countries labeled by lower-case symbols, more than 11 attribute values were missing, and they were mapped onto the SOM after training.

The "poverty map" of the countries is presented in Fig. 3.30. The symbols used in the map have been explained in Table 3.7. It has to be noted that horizontal or vertical directions have no explicit meaning in this map: only the local geometric relations are important. One might say that countries that are mapped close to each other in the SOM have a similar state of development, expenditure pattern, and policy.

3.15 Using Gray Levels to Indicate Clusters in the SOM

From the above map and the previous ones, however, we still do not see any boundaries between eventual *clusters*. That problem will be discussed in this section.

A graphic display called the *U-matrix*, to illustrate the clustering of code-book vectors in the SOM has been developed by *Ultsch and Siemon* [3.35], as well as *Kraaijveld et al.* [3.36]. They suggested a method in which the average distances between neighboring codebook vectors are represented by shades in a gray scale (or eventually pseudocolor scales might be used). If the

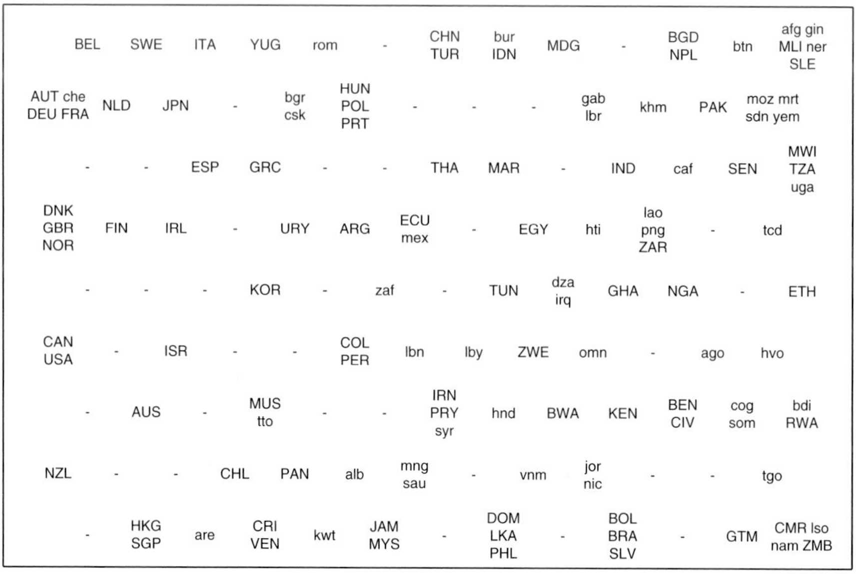

BEL	SWE	ITA	YUG	rom	-	CHN TUR	bur IDN	MDG	-	BGD NPL	btn	afg gin MLI ner SLE
AUT che DEU FRA	NLD	JPN	-	bgr csk	HUN POL PRT	-	-	-	gab lbr	khm	PAK	moz mrt sdn yem
-	-	ESP	GRC	-	-	THA	MAR	-	IND	caf	SEN	MWI TZA uga
DNK GBR NOR	FIN	IRL	-	URY	ARG	ECU mex	-	EGY	hti	lao png ZAR	-	tcd
-	-	-	KOR	-	zaf	-	TUN	dza irq	GHA	NGA	-	ETH
CAN USA	-	ISR	-	-	COL PER	lbn	lby	ZWE	omn	-	ago	hvo
-	AUS	-	MUS tto	-	-	IRN PRY syr	hnd	BWA	KEN	BEN CIV	cog som	bdi RWA
NZL	-	-	CHL	PAN	alb	mng sau	-	vnm	jor nic	-	-	tgo
-	HKG SGP	are	CRI VEN	kwt	JAM MYS	-	DOM LKA PHL	-	BOL BRA SLV	-	GTM	CMR lso nam ZMB

Fig. 3.30. "Poverty map" of 126 countries of the world. The symbols written in capital letters correspond to countries used in the formation of the map; the rest of the symbols (in lower case) signify countries for which more than 11 attribute values were missing, and which were mapped to this SOM after the learning phase

average distance of neighboring m_i is small, a light shade is used; and vice versa, dark shades represent large distances. A "cluster landscape" formed over the SOM then clearly visualizes the classification.

The case example of Sect. 3.14.2, with the gray-level map superimposed on it, is now shown in Fig. 3.31, and it clearly shows "ravines" between clusters. The interpretation is left to the reader.

3.16 Interpretation of the SOM Mapping

The Self-Organizing Map combines nonlinear projection (Sect. 1.3.2) and clustering (Sect. 1.3.4) methods in an ordered vector quantization graph. Therefore the mapping that it produces is expected to be explainable in terms of some classical concepts of statistics.

3.16.1 "Local Principal Components"

The main difference between the SOM and the principal component analysis (PCA) is that the latter describes the global statistical properties of the data distribution: the first "principal axis" defines the direction in which the variance of the distribution is largest, the second "principal axis" that

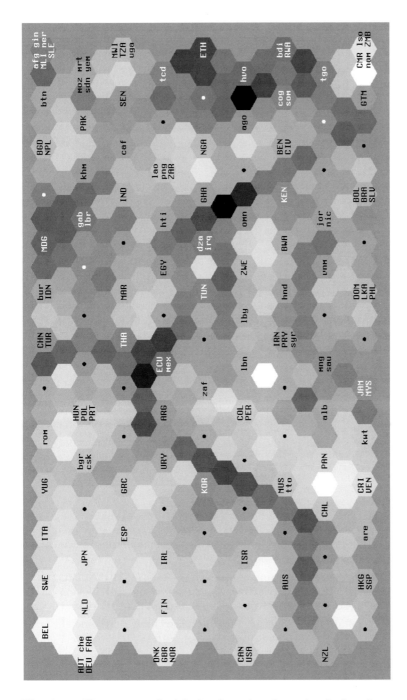

Fig. 3.31. "Poverty map," with the clustering shown by shades of gray

Table 3.7. Legend of symbols used in Figs. 3.29 and 3.30:

AFG	Afghanistan	GRC	Greece	NOR	Norway
AGO	Angola	GTM	Guatemala	NPL	Nepal
ALB	Albania	HKG	Hong Kong	NZL	New Zealand
ARE	United Arab Emirates	HND	Honduras	OAN	Taiwan, China
ARG	Argentina	HTI	Haiti	OMN	Oman
AUS	Australia	HUN	Hungary	PAK	Pakistan
AUT	Austria	HVO	Burkina Faso	PAN	Panama
BDI	Burundi	IDN	Indonesia	PER	Peru
BEL	Belgium	IND	India	PHL	Philippines
BEN	Benin	IRL	Ireland	PNG	Papua New Guinea
BGD	Bangladesh	IRN	Iran, Islamic Rep.	POL	Poland
BGR	Bulgaria	IRQ	Iraq	PRT	Portugal
BOL	Bolivia	ISR	Israel	PRY	Paraguay
BRA	Brazil	ITA	Italy	ROM	Romania
BTN	Bhutan	JAM	Jamaica	RWA	Rwanda
BUR	Myanmar	JOR	Jordan	SAU	Saudi Arabia
BWA	Botswana	JPN	Japan	SDN	Sudan
CAF	Central African Rep.	KEN	Kenya	SEN	Senegal
CAN	Canada	KHM	Cambodia	SGP	Singapore
CHE	Switzerland	KOR	Korea, Rep.	SLE	Sierra Leone
CHL	Chile	KWT	Kuwait	SLV	El Salvador
CHN	China	LAO	Lao PDR	SOM	Somalia
CIV	Cote d'Ivoire	LBN	Lebanon	SWE	Sweden
CMR	Cameroon	LBR	Liberia	SYR	Syrian Arab Rep.
COG	Congo	LBY	Libya	TCD	Chad
COL	Colombia	LKA	Sri Lanka	TGO	Togo
CRI	Costa Rica	LSO	Lesotho	THA	Thailand
CSK	Czechoslovakia	MAR	Morocco	TTO	Trinidad and Tobago
DEU	Germany	MDG	Madagascar	TUN	Tunisia
DNK	Denmark	MEX	Mexico	TUR	Turkey
DOM	Dominican Rep.	MLI	Mali	TZA	Tanzania
DZA	Algeria	MNG	Mongolia	UGA	Uganda
ECU	Ecuador	MOZ	Mozambique	URY	Uruguay
EGY	Egypt, Arab Rep.	MRT	Mauritania	USA	United States
ESP	Spain	MUS	Mauritius	VEN	Venezuela
ETH	Ethiopia	MWI	Malawi	VNM	Viet Nam
FIN	Finland	MYS	Malaysia	YEM	Yemen, Rep.
FRA	France	NAM	Namibia	YUG	Yugoslavia
GAB	Gabon	NER	Niger	ZAF	South Africa
GBR	United Kingdom	NGA	Nigeria	ZAR	Zaire
GHA	Ghana	NIC	Nigaragua	ZMB	Zambia
GIN	Guinea	NLD	Netherlands	ZWE	Zimbabwe

direction of all directions orthogonal to the first principal axis in which the residual variance is largest, and so on.

The SOM, however, can be characterized as a two-dimensional, finite-element "elastic surface" or network that is fitted to the distribution of the input samples. This network has *locally*, at every node of it, two principal directions that are found considering the differences between the neighboring vectors. For instance, the subset of reference vectors $\{m_j\}$ in the closest neighborhood of reference vector m_i and including the latter may be regarded as a smoothed set of samples that is trying to represent a *local two-dimensional hyperplane*. The principal axes of this hyperplane, and the two largest principal components of this subset can be computed in the normal way from the subset of neighboring reference vectors.

As the reference vectors resulting in the smoothing process carried out by the SOM may not accurately represent the original statistics of the input samples, we may only qualitatively regard the two principal components of the neighborhood set as the "local principal components." If we want more than two of them, we must fit a higher-dimensional SOM to the input data. Nonetheless, as the "local principal components" can be computed readily and in an almost unique way (depending only on the SOM process and the definition of the neighborhood), they allow one to get an illustrative insight into complex and "noisy" data distributions.

One may further apply the classical *factor analysis* (Sect. 1.3.1) to each neighborhood set of the reference vectors, in order to find the *factor loadings* of the input variables in each local area of the SOM. Notice that the local factor loadings may be totally different in different domains of the input data.

Notice that it would be absurd to carry out a local PCA, or to compute the local factors on the basis of the subset of input samples that are mapped to neighboring Voronoi sets. Consider, for instance, Fig. 3.11 and look what distribution the samples in the union of any three neighboring Voronoi sets have: no "principal axes" have any sense. On the other hand, the neighboring SOM reference vectors lie neatly in a direction that takes into account the global form of the distribution, but still the local directions are sensitive to local statistics. Therefore it is more sensible to define the "local factors" on the basis of the reference vectors. In other words, like in the mathematical discipline called differential geometry, the local directions must fulfill certain "compatibility conditions" that the SOM automatically takes into account.

3.16.2 Contribution of a Variable to Cluster Structures

This analysis is related to the local PCA and local factor analysis but is computationally simpler and more directly describes the discriminatory power of an input variable in the mapping.

Consider the vectorial differences between neighboring reference vectors. Similary one can form the respective differences of each vector component separately. If the correlation between the vectorial differences and the differences of some component in a local area of the SOM is large, this component (variable) has a significant contribution to the cluster structure and a large explanatory power in that domain of values.

The discriminatory power of a variable is manifested most strongly at the cluster borders, where one is looking for variables that make the biggest difference between the neighboring clusters.

Example. In Fig. 3.32, the animal example defined in Table 3.6 and depicted in Fig. 3.29 has been interpreted by both of the above methods.

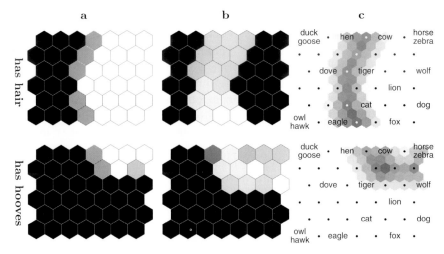

Fig. 3.32. Illustration of the two interpretation methods applied to the animal data. The top row visualizes the variable "has hair" and the bottom row "has hooves," respectively. **(a)** The component planes. The shade of gray describes the value of the respective component of the reference vectors (white: large, black: small). **(b)** The contribution of the variables in the two local factors (white: maximal contribution, black: minimal contribution). **(c)** The (spatially smoothed) contribution of the variables in the local cluster structures (dark: large contribution, white: minimal contribution)

3.17 Speedup of SOM Computation

3.17.1 Shortcut Winner Search

If there are M map units (neurons) in the SOM, and one stipulates that for a certain statistical accuracy the number of updating operations per unit shall be some constant (say, on the order of 100), then the total number of comparison operations to be performed during learning by an exhaustive search of the winners is $\sim M^2$.

By a tree-structured multilayer SOM architecture to be discussed in Sect. 5.3 it will be possible to reduce the number of searching operations to $\sim M \log M$. However, since the winner is thereby determined in a multistep decision process in which the later decisions depend on the earlier ones, the partition of the input signal space is not exactly the same as the Voronoi tessellation discussed earlier.

We will now show that the total number of comparison operations can be made $\sim M$, provided that the training vectors have been given in the beginning, i.e., their set is finite and closed. *Koikkalainen* [3.37, 38] has used a somewhat similar idea in the tree-structured SOM, but the scheme presented below [3.39] can be used in connection with any traditional SOM. Moreover, in principle at least, the decision can also be regarded as a one-level process, comparable to the basic SOM.

Assume that we are somewhere in the middle of an iterative training process, whereby the last winner corresponding to each training vector has been determined at an earlier cycle. If the training vectors are expressed as a linear table, a *pointer* to the corresponding *tentative winner location* can be stored with each training vector (Fig. 3.33).

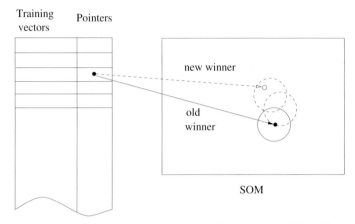

Fig. 3.33. Finding the new winner in the vicinity of the old one, whereby the old winner is directly located by a pointer. The pointer is then updated

Assume further that the SOM is already smoothly ordered although not yet asymptotically stable. This is the situation, e.g., during the lengthy fine-tuning phase of the SOM, whereupon the size of the neighborhood set is also constant and small. Consider that updating of a number of map units is made before the same training input is used again some time later. Nonetheless it may be clear that if the sum of the corrections made during this period is not large, the new winner is found at or in the vicinity of the old one. Therefore, in searching for the best match, it will suffice to locate the map unit corresponding to the associated pointer, and then to perform a local search for the winner in the neighborhood around the located unit. This will be a significantly faster operation than an exhaustive winner search over the whole SOM. The search can first be made in the immediate surround of the said location, and only if the best match is found at its edge, searching is continued in the surround of the preliminary best match, until the winner is one of the middle units in the search domain. After the new winner location has been identified, the associated pointer in the input table is replaced by the pointer to the new winner location.

For instance, if the array topology of the SOM is hexagonal, the first search might be made in the neighborhood consisting of the seven units at or around the winner. If the tentative winner is one of the edge units of this neighborhood, the search must be continued in the new neighborhood of seven units centered around the last tentative winner. Notice that in each new neighborhood, only the three map units that have not yet been checked earlier need to be examined.

This principle can be used with both the usual incremental-learning SOM and its batch-computing version.

A benchmark with two large SOMs relating to our recent practical experiments has been made. The approximate codebook vector values were first computed roughly by a traditional SOM algorithm, whereafter they were fine-tuned using this fast method. During the fine-tuning phase, the radius of the neighborhood set in the hexagonal lattice decreased linearly from 3 to 1 units equivalent to the smallest lattice spacings, and the learning-rate factor at the same time decreased linearly from 0.02 to zero. There were 3645 training vectors for the first map, and 9907 training vectors for the second map, respectively. The results are reported in Table 3.8.

Table 3.8. Speedup due to shortcut winner search

Input dimensionality	Map size	Speedup factor in winner search	Speedup factor in training
270	315	43	14
315	768	93	16

The theoretical maximum of speedup in winner search is: 45 for the first map, and 110 for the second map, respectively. The training involves the winner searches, codebook updating, and overhead times due to the operating system and the SOM software used. The latter figures may still be improved by optimization of the management of the tables.

3.17.2 Increasing the Number of Units in the SOM

Several suggestions for "growing SOMs" (cf., e.g. [3.40–54]) have been made. The detailed idea presented below has been optimized in order to make very large maps, and is believed to be new. The basic idea is to estimate good initial values for a map that has plenty of units, on the basis of asymptotic values of a map with a much smaller number of units.

As the general nature of the SOM process and its asymptotic states is now fairly well known, we can utilize some "expert knowledge" here. One fact is that the asymptotic distribution of codebook vectors is generally smooth, at least for a continuous, smooth probability density function (pdf) of input, and therefore the lattice spacings can be smoothed, interpolated, and extrapolated locally.

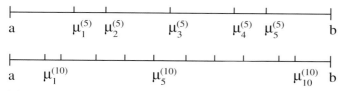

Fig. 3.34. Asymptotic values for the μ_i for two lengths of the array

As an introductory example consider, for instance, the one-dimensional SOM and assume tentatively a uniform pdf of the scalar input in the range $[a, b]$. Then we have the theoretical asymptotic codebook values for different numbers of map units that approximate the same pdf, as shown in Fig. 3.34.

Assume now that we want to estimate the locations of the codebook values for an *arbitrary* pdf and for a 10-unit SOM on the basis of known codebook values of the 5-unit SOM. A linear *local* interpolation-extrapolation scheme can then be used. For instance, to interpolate $\mu_5^{(10)}$ on the basis of $\mu_2^{(5)}$ and $\mu_3^{(5)}$, we first need the *interpolation coefficient* λ_5, computed from the two ideal lattices with uniform pdf:

$$\mu_5^{(10)} = \lambda_5 \mu_2^{(5)} + (1 - \lambda_5)\mu_3^{(5)} , \tag{3.74}$$

from which λ_5 for $\mu_5^{(10)}$ can be solved. If then, for an arbitrary pdf, the *true* values of $\mu_2'^{(5)}$ and $\mu_3'^{(5)}$ have been computed, the estimate of the true $\hat{\mu}_5'^{(10)}$ is

$$\hat{\mu}_5'^{(10)} = \lambda_5 \mu_2'^{(5)} + (1 - \lambda_5)\mu_3'^{(5)} . \tag{3.75}$$

Notice that a similar equation can also be used for the *extrapolation* of, say, $\mu_1^{(10)}$ on the basis of $\mu_1^{(5)}$ and $\mu_2^{(5)}$.

Application of local interpolation and extrapolation to *two-dimensional* SOM lattices (rectangular, hexagonal, or other) is straightforward, although the expressions become a little more complicated. Interpolation and extrapolation of a codebook vector in a two-dimensional lattice must be made on the basis of vectors defined at least in *three* lattice points. As the maps in practice may be very nonlinear, the best estimation results are usually obtained with the three closest reference vectors.

Consider two similarly positioned overlapping "ideal" two-dimensional lattices in the same plane, with the codebook vectors $m_h^{(d)} \in \Re^2, m_i^{(s)} \in \Re^2, m_j^{(s)} \in \Re^2$, and $m_k^{(s)} \in \Re^2$ its nodes, where the superscript d refers to a "dense" lattice, and s to a "sparse" lattice, respectively. If $m_i^{(s)}, m_j^{(s)}$, and $m_k^{(s)}$ do not lie on the same straight line, then in the two-dimensional signal plane any $m_h^{(d)}$ can be expressed as the linear combination

$$m_h^{(d)} = \alpha_h m_i^{(s)} + \beta_h m_j^{(s)} + (1 - \alpha_h - \beta_h)m_k^{(s)} , \tag{3.76}$$

where α_h and β_h are interpolation-extrapolation coefficients. This is a two-dimensional vector equation from which the two unknowns α_h and β_h can be solved.

Consider then two SOM lattices with the same topology as in the ideal example, in a space of arbitrary dimensionality. When the true pdf may also be arbitrary, we may not assume the lattices of true codebook vectors as planar. Nonetheless we can perform a *local linear estimation* of the true codebook vectors $m_h'^{(d)} \in \Re^n$ of the "dense" lattice on the basis of the true codebook vectors $m_i'^{(s)}, m_j'^{(s)}$, and $m_k'^{(s)} \in \Re^n$ of the "sparse" lattice, respectively.

In practice, in order that the linear estimate be most accurate, we may stipulate that the respective indices h, i, j, and k are such that in the ideal lattice $m_i^{(s)}, m_j^{(s)}$, and $m_k^{(s)}$ are the three codebook vectors *closest* to $m_h^{(d)}$ in the signal space (but not on the same line). With α_h and β_h solved from (3.76) for each node h separately we obtain the wanted interpolation-extrapolation formula as

$$\hat{m}_h'^{(d)} = \alpha_h m_i'^{(s)} + \beta_h m_j'^{(s)} + (1 - \alpha_h - \beta_h) m_k'^{(s)} . \tag{3.77}$$

Notice that the indices h, i, j, and k refer to *topologically identical lattice points* in (3.76) and (3.77). The interpolation-extrapolation coefficients for two-dimensional lattices depend on their topology and the neighborhood function used in the last phase of learning. For the "sparse" and the "dense" lattice, respectively, we should first compute the ideal codebook vector values in analogy with Fig. 3.34. As the closed solutions for the reference lattices may be very difficult to obtain, the asymptotic codebook vector values may be approximated by simulation. If the ratio of the horizontal vs. vertical dimension of the lattice is $H : V$, we may assume tentatively that two-dimensional input vectors are selected at random from a uniform, rectangular pdf, the width of which in the horizontal direction is H and the vertical width of which is V. These inputs are then used to train two-dimensional SOMs that approximate the ideal two-dimensional lattices.

Initialization of the Pointers for a Larger Map. When the size (number of grid nodes) of the maps is increased stepwise during learning using the estimation procedure, the initial pointers for all data vectors after each increase can be estimated quickly by utilizing the formula that was used in increasing the map size, equation (3.76). The winner is the map unit for which the inner product with the data vector is the largest, and so the inner products can be computed rapidly using the expression

$$x^T m_h^{(d)} = \alpha_h x^T m_i^{(s)} + \beta_h x^T m_j^{(s)} + (1 - \alpha_h - \beta_h) x^T m_k^{(s)} . \tag{3.78}$$

Expression (3.78) can be interpreted as the inner product between two *three-dimensional* vectors, $[\alpha_h; \beta_h; (1 - \alpha_h - \beta_h)]^T$ and $[x^T m_i^{(s)}; x^T m_j^{(s)}; x^T m_k^{(s)}]^T$, *irrespective of the dimensionality of* x. If necessary, the winner search can still be speeded up by restricting the winner search to the area of the dense map that corresponds to the neighborhood of the winner on the sparse map.

This is especially fast if only a subset (albeit a subset that covers the whole map) of all the possible triplets (i, j, k) is allowed in (3.76) and (3.78).

3.17.3 Smoothing

It may not always be desirable or possible to use an excessive computing time just to guarantee very high accuracy or correctness of the asymptotic state of the map. Consider, for instance, that the number of available training samples is so small that they anyway do not approximate the pdf well enough. However, as the smooth form of the map is necessary for good resolution in comparing matching at the adjacent units, one might rather stop the training when there are still appreciable statistical errors in the codebook vectors, and apply a *smoothing* procedure to achieve this fine resolution.

Smoothing and *relaxation* are related concepts, although with quite different scopes. In smoothing one aims at the reduction of stochastic variations, whereas in relaxation a typical task is to solve a differential equation by difference approximations. An example of the latter is the *Dirichlet problem*, computation of the values of a harmonic function inside a domain, when the boundary conditions have been given.

Smoothing also differs from relaxation because it is usually only applied a few times differentially; in relaxation problems the "smoothing" steps are iterated until the asymptotic state can be regarded as stable, whereby the final state may look very different from the initial one.

Consider, for example, that we stop the SOM algorithm when the codebook vectors have achieved the values m'_h. A smoothing step looks similar to (3.76) and (3.77), except that the lattice is now the same: the new value of a codebook vector is obtained from, say, three old codebook vectors that are closest to the one to be smoothed in the two-dimensional signal space of the ideal lattice, but do not lie on the same straight line. In analogy with (3.76) and (3.77), let m_h (of the ideal lattice) first be expressed as the linear combination

$$m_h = \gamma_h m_i + \delta_h m_j + (1 - \gamma_h - \delta_h) m_k \ . \tag{3.79}$$

From this vector equation, γ_h and δ_h can be solved. The smoothed value of m'_h in the corresponding true SOM lattice with arbitrary pdf reads

$$S(m'_h) = \varepsilon[\gamma_h m'_i + \delta_h m'_j + (1 - \gamma_h - \delta_h) m'_k] + (1 - \varepsilon) m'_h \ , \tag{3.80}$$

where the degree of smoothing has been further *moderated* using the factor ε, $0 < \varepsilon < 1$.

This scheme is similar for units both in the inside and at the edges of the SOM lattice. In smoothing, the new values of all codebook vectors must first be computed on the basis of the old ones and buffered, whereafter they are made to replace the old values simultaneously.

3.17.4 Combination of Smoothing, Lattice Growing, and SOM Algorithm

It now seems to be the most reasonable strategy to first perform an *equal number* of *identical* smoothing steps *for both the ideal and the real lattice*, respectively, after which (3.76), (3.77), and (3.80) are applied:

$$m_h^{(d)} = \alpha_h S^k(m_i^{(s)}) + \beta_h S^k(m_j^{(s)}) + (1 - \alpha_h - \beta_h)S^k(m_k^{(s)}) , \quad (3.81)$$

$$\hat{m}_h'^{(d)} = \alpha_h S^k(m_i'^{(s)}) + \beta_h S^k(m_j'^{(s)}) + (1 - \alpha_h - \beta_h)S^k(m_k'^{(s)}) , (3.82)$$

where $S^k(\cdot)$ means k successive smoothing operations.

Finally, some fine-tuning steps of the map can be made with the SOM, eventually using the shortcut winner search at its later cycles.

Comment. The results of this Sect. 3.17 have been utilized in the application to be described in Sect. 7.8, whereby very large SOMs are computed.

4. Physiological Interpretation of SOM

It may be recalled (Sect. 2.15) that three types of neuronal organization can be called "brain maps": sets of feature-sensitive cells, ordered projections between neuronal layers, and ordered anatomical maps of *abstract features*. The latter reflect the most central properties of the organism's experiences and other occurrences. Such feature maps are probably learned in a process that involves parallel input to neurons in a brain area and adaptation of neurons in the neighborhood of the cells that respond most strongly to this input.

Since it might take days, weeks, months, or years for the biological SOM processes in neural realms to form or alter neural maps, the organizing effects are not immediately measurable and may be disguised by other, stronger changes in short-term signal activities. Changes in the m_i and their effects on signal transmission occur gradually, probably only intermittently during short periods of "learning" when some specific plasticity-control factor is on.

Nonetheless it has now become possible to demonstrate theoretically that a SOM process is implementable by quite usual biological components, and there may even exist many candidates for the latter. The purpose of this chapter is to devise general principles according to which physiological SOM models may be built. It is based on [2.45].

4.1 Conditions for Abstract Feature Maps in the Brain

It were unrealistic to expect that maps as clearly organized as in the simulation studies should exist in the brain. The biological sensory preprocessing is already much more complex than that used in simulations. Nonetheless there exist many kinds of feature maps in the brain, either orderly representations of the skin, retina, or cochlea, or more abstract maps. For instance, there are acoustic maps associated with sound to which the organism is most frequently exposed to. The tonotopic maps are known to exist both in the auditory pathways and in the areas of hearing on the cortex [4.1, 2] They are often thought to represent the order of the acoustic resonances on the basilar membrane of the inner ear, and preservation of the same order in the pathways originating from the hair cells and ending up at the cortex. Nonetheless similar maps have been produced succesfully by the SOM algorithm directly [2.36]. Other feature maps were discussed in Sect. 2.15.

Materialization of the SOM principle in the nervous systems would first require a mechanism that distributes essentially the same or strongly correlated signals to the neurons or neuron groups of a certain region. Spatial diameters of the maps would then be determined by the longest distances reachable by common (or highly correlated) input, as well as the ranges of the lateral interactions. The SOM principle might work well in the thalamocortical system, where the diameter of the map would be determined primarily by the ramification of the thalamocortical axons [4.3], and by the diameters of the apical dendritic trees [4.4]. Thus, a SOM that would describe a single type of feature would be maximally four to five millimeters.

Second, one needs a mechanism that selects the winner neurons, that is, the center around which adaptation will take place. We will see in this chapter that lateral interconnectivity with excitatory and inhibitory connections [2.11–18, 4.5] may play a central role in this selection.

Third, the restriction of learning to the neighborhood of the winner(s) might simply result from the spreading of triggering activity, but one can also assume that some type of chemical "learning factor" that controls modification locally, but does not activate the neurons, is diffusing from the active neurons [2.45]. This idea comes close to the concept of *metaplasticity*, that is, active control of synaptic plasticity [4.6].

Essentially every level of the nervous system exhibits plasticity under certain circumstances [4.7] and, thus, feature maps can also be expected at all levels.

4.2 Two Different Lateral Control Mechanisms

Simulations performed during many years by a great number of scientists have convincingly shown that the best self-organizing results are obtained if the following two partial processes are implemented in their purest forms: 1. Decoding of that m_i, denoted by m_c ("winner") that has the best match with x. 2. Adaptive improvement of the match in the neighborhood of neurons centered around the "winner."

The former operation is signified as the *winner-take-all (WTA)* function. Traditionally, the WTA function has been implemented in neural networks by lateral-feedback circuits [2.11–18]. The following kind of control of the neighborhood, however, as propounded by this author, represents a new direction in neural modeling: the "winner" *modulates the synaptic plasticity directly* in the lateral direction, without really enhancing the activity of all cells in this neighborhood. Accordingly, for modeling of the physiological SOM process we need to define *two separate* interaction kernels (Fig. 4.1): 1. The *activation kernel*, usually the so-called "Mexican hat" function (positive feedback in the center, negative in the surround); and 2. The *plasticity control kernel*, which

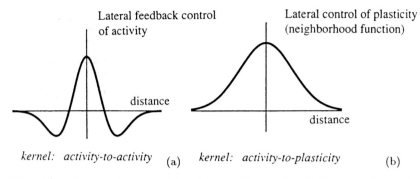

Lateral feedback control of activity

distance

kernel: activity-to-activity   (a)

Lateral control of plasticity (neighborhood function)

distance

kernel: activity-to-plasticity   (b)

Fig. 4.1. The two types of lateral interaction: **(a)** activity-control kernel, **(b)** plasticity-control kernel

defines how local activity determines the learning rate in its neighborhood. This kernel is nonnegative and may take on the Gaussian form. We shall next discuss these two types of lateral control separately.

4.2.1 The WTA Function, Based on Lateral Activity Control

This discussion is related to [2.11, 12, 14–16], with the main difference being that the new nonlinear dynamic model occurs in the description of the neurons. Another refinement is the local reset function that uses slow inhibitory interneurons of the same type but with greater time constants. A network model based on these model neurons is intended to make the discussion more lucid and the simulations economical.

Consider Fig. 4.2 that delineates the cross section of a two-dimensional neural network, where each principal cell receives input from some external sources, and the cells are further interconnected by abundant lateral feedback. In the simplest model the same set of afferent input signals is connected to all principal cells.

The output activity η_i (spiking frequency) of every neuron i in the network is described [2.40, 45] by a law of the general form

$$d\eta_i/dt = I_i - \gamma(\eta_i) , \quad \eta_i \geq 0 , \tag{4.1}$$

where I_i is the combined effect of all inputs, e.g., afferent (external) inputs as well as lateral feedbacks, on cell i embedded in the layered network. In modeling, without much loss of generality, I_i may be thought to be proportional to the dot product of the signal vector and the synaptic transmittance vector, or some monotonic function of it. Let $\gamma(\eta_i)$ describe all loss or leakage effects that oppose I_i. This is an abbreviated way of writing: since $\eta_i \geq 0$, (4.1) only holds when $\eta_i > 0$, or when $\eta_i = 0$ and $I_i - \gamma(\eta_i) \geq 0$, whereas otherwise $d\eta_i/dt = 0$. We have also found that for stable convergence, $\gamma(\eta_i)$ must be convex, i.e., $\gamma''(\eta_i) > 0$. Here $\gamma''(\cdot)$ means the second derivative with respect to the argument of γ.

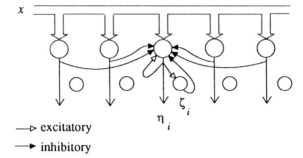

x

—▷ excitatory

—▶ inhibitory

Fig. 4.2. Simplified model of a distributed neural network (cross section of a two-dimensional array). Each location consists of an excitatory principal input neuron and an inhibitory interneuron that feeds back locally. The lateral connections between the principal input neurons may or may not be made via interneurons

For the principal neurons the input I_i consists of two parts I_i^e and I_i^f, respectively: $I_i = I_i^e + I_i^f$, where the superscript e means "external" or afferent input, and f the lateral feedback, respectively. In the simplest case these terms read

$$I_i^e = m_i^{\mathrm{T}} x = \sum_j \mu_{ij} \xi_j , \tag{4.2}$$

$$I_i^f = \sum_k g_{ik} \eta_k . \tag{4.3}$$

Here $x = [\xi_1, \xi_2, \ldots, \xi_n]^{\mathrm{T}} \in \Re^n$ means the afferent input data vector, or the vector of signal activities on a set of axons that are assumed to be connected in parallel to all principal cells of this network, whereas $m_i = [\mu_{i1}, \mu_{i2}, \ldots, \mu_{in}]^{\mathrm{T}} \in \Re^n$ is redefined to be the corresponding vector of synaptic strengths of cell i. The $g_{ik} \in \Re$ describe effective lateral connection strengths of cells. For simplicity, it is assumed that g_{ii} is independent of i, and the g_{ik}, $k \neq i$ are mutually equal.

In another study [4.5] we have been able to show that the lateral feedback can also be made via interneurons that are describable by Eq. (4.1), whereupon, for stable convergence, the time constants must be significantly smaller than those of the principal cells in order that the interneurons converge faster than the principal cells.

In the above system of equations, by far the only and rather general restrictions are $\mu_{ij} > 0$, $\xi_j > 0$ (i.e., the external inputs are excitatory), $g_{ii} > 0$, $g_{ik} < 0$ for $k \neq i$, $|g_{ik}| > |g_{ii}|$, $\gamma > 0$, and $\gamma''(\eta_i) > 0$.

Starting with arbitrary nonnegative, different initial values $m_i(0)$ and with $\eta_i(0) = 0$, the output η_c of that cell for which $m_i^{\mathrm{T}} x$ is maximum ("winner") can be shown to converge to an asymptotic high value, whereas the other η_i, $i \neq c$ tend to zero [2.45]. This convergence is very robust.

We shall present a formal proof for this convergence.

The following discussion relates to a WTA function with respect to persistent external inputs I_i^e, whereby the I_i^e are assumed constant in time, and the difference between the largest I_i^e and all the other inputs is not infinitesimally small. We shall restrict our discussion to the case in which all g_{ii} are equal and all g_{ik}, $k \neq i$ are mutually equal. For the convenience of mathematical treatment we introduce the new notations g^+ and g^- defined in the following way:

$$
\begin{aligned}
g_{ik} &= g^- \quad \text{for } k \neq i , \\
g_{ii} &= g^+ + g^- ,
\end{aligned}
\tag{4.4}
$$

whereby the system equations read

$$
H = I_i^e + g^+ \eta_i + g^- \sum_k \eta_k - \gamma(\eta_i) ,
$$

$$
\frac{d\eta_i}{dt} = H \quad \text{if } \eta_i > 0 , \text{ or if } \eta_i = 0 \text{ and } H \geq 0 ,
$$

$$
\frac{d\eta_i}{dt} = 0 \quad \text{if } \eta_i = 0 \text{ and } H < 0 .
\tag{4.5}
$$

The sum over k, now including the term $k = i$, does not depend on i.

If $\gamma(0) = 0$ and we start with all the $\eta_i(0) = 0$, the initial derivatives are $d\eta_i/dt = I_i^e$, and the η_i start growing in the correct order. Let us derive a lemma that describes the further development of two arbitrary outputs η_i and η_j. Let the value η_0 be defined by

$$
\gamma'(\eta_0) = g^+ ,
\tag{4.6}
$$

where γ' is the first derivative of γ.

Lemma 4.2.1. *If $I_i^e - I_j^e > 0$, $0 < \eta_i < \eta_0$, $0 < \eta_j < \eta_0$, $\eta_i - \eta_j > 0$, $g^+ > 0$, and $\gamma''(\eta) > 0$, then the difference $\eta_i - \eta_j$ grows.*

The proof is obvious almost immediately. Let us denote $e = \eta_i - \eta_j > 0$. Then

$$
\frac{de}{dt} = I_i^e - I_j^e + g^+ e - (\gamma(\eta_i) - \gamma(\eta_j)) ,
\tag{4.7}
$$

and from $\gamma''(\eta) > 0$ there follows that $\gamma(\eta_i) - \gamma(\eta_j) < \gamma'(\eta_0) \cdot e = g^+ e$, whereby $de/dt > 0$.

From this point on we shall always assume that $g^+ > 0$, $g^- < 0$, and $\gamma(\eta) > 0$, $\gamma''(\eta) > 0$ for all $\eta > 0$.

Denote $d\eta_i/dt = \dot{\eta}_i$. Next we show that η_c, the activity of the "winner," grows continually for all $\eta_c \leq \eta_0$; in the proof we allow the existence of discrete points at which $\dot{\eta}_c = 0$, but it is not likely that such points occur in the operation. Assume tentatively that we had $\dot{\eta}_c(t_0) = 0$ at some time $t = t_0$ and at least one $\eta_i > 0$, $i \neq c$. At time $t_0 + \Delta t$, where Δt is infinitesimally small, we would then have (noting that η_c and $\gamma(\eta_c)$ are constant during Δt in the first approximation)

$$\dot{\eta}_c(t_0 + \Delta t) = \dot{\eta}_c(t_0) + g^- \sum_{k \neq c} \dot{\eta}_k(t_0) \cdot \Delta t + \text{terms of the order of}(\Delta t)^2 . \quad (4.8)$$

According to Lemma 4.2.1, we would now have $\dot{\eta}_k(t_0) < 0$ with $k \neq c$, for the cell or cells that have $\eta_k > 0$. Let us recall that $g^- < 0$. If the differences between I_c^e and the I_k^e, and thus those between $\dot{\eta}_c$ and the $\dot{\eta}_k$ are not infinitesimal, then we can neglect the second-order terms, whereby it is seen that $\dot{\eta}_c(t + \Delta t) > 0$. In other words, $\dot{\eta}_c$ has thereby been shown to grow continually up to $\eta_c = \eta_0$.

If all η_i, $i \neq c$ have reached zero, the proof is trivial, as shown later by (4.10).

From the above it is also clear that η_c will reach η_0 in a finite time, whereby all the other η_i are still smaller.

In practice it is always possible to select the system parameters relative to the input signal values such that the "winner" is unique. We shall show that if the following conditions are satisfied, all the other activities η_i except that of the "winner" will converge to zero. Recalling that $g^- < 0$, assume that

(i) $g^+ > -g^-$ (or in original notations, $g_{ii} > 0$)   and
(ii) $I_i^e < (-2g^- - g^+)\eta_0$ for $i \neq c$,

which also implies that $-2g^- > g^+$ (or in original notations that the inhibitory couplings are stronger than the excitatory ones). At least when $\eta_c \geq \eta_0 > \eta_i > 0$, we now have

$$\frac{d\eta_i}{dt} < (-2g^- - g^+)\eta_0 + g^+\eta_i + g^-(\eta_0 + \eta_i)$$
$$= (\eta_0 - \eta_i)(-g^- - g^+) < 0 . \quad (4.9)$$

Using this result, it can now also be proved, in analogy with the discussion prior to (4.8), that $\dot{\eta}_c > 0$ for $\eta_c > \eta_0$, whereby η_c stays $> \eta_0$. Thus all $\dot{\eta}_i$, $i \neq c$, stay < 0 and the η_i will reach zero. After they have reached zero, the "winner" is described by

$$\frac{d\eta_c}{dt} = I_c^e + (g^+ + g^-)\eta_c - \gamma(\eta_c) ; \quad (4.10)$$

since $\gamma''(\eta_c) > 0$, η_c can be seen to have an upper limit to which it converges. ∎

A biological neural network, however, must be able to respond to all new inputs, and therefore the output state $\{\eta_i\}$ must be *reset* before application of the next input. There are many ways to reset the activity [2.45]. In most solutions one needs some kind of parallel control of all neurons of the network, which does not seem very plausible in biology. In this model resetting is done automatically and *locally*, by slow inhibitory interneurons with output variable ζ_i (Fig. 4.2). These interneurons could be described by (4.1), too, but for further simplification, to facilitate more general mathematical proofs, the loss term is linearized piecewise:

$$d\zeta_i/dt = b\eta_i - \theta ,\qquad(4.11)$$

where b and θ are scalar parameters. Also here we must state that (4.11) only holds if $\zeta_i > 0$, or if $\zeta_i = 0$ and $b\eta_i - \theta \geq 0$, whereas otherwise $d\zeta_i/dt = 0$. The complete equation corresponding to (4.1) then reads

$$d\eta_i/dt = I_i - a\zeta_i - \gamma(\eta_i) ,\qquad(4.12)$$

where a is another scalar parameter.

This WTA circuit can be seen to operate in *cycles*, where each cycle can be thought to correspond to one discrete-time phase of the SOM algorithm. Normally the input would be changed at each new cycle; however, if the input is steady for a longer time, the next cycle selects the "runner-up", after which the "winner" is selected again, etc.

The cyclic operation of this WTA circuit is illustrated in Fig. 4.3.

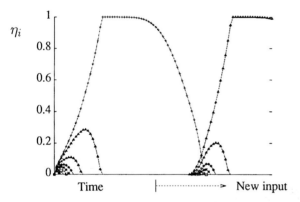

Fig. 4.3. Demonstration of the WTA function provided with automatic reset. The first inputs were applied at time zero. New inputs were applied as indicated by the dotted arrow. The network consisted of 20 cells, and the inputs $I_i^e = m_i^T x$ were selected as random numbers from the interval $(0,1)$. The g_{ii} were equal to 0.5 and the $g_{ij}, i \neq j$, equal to -2.0, respectively. The loss function had the form $\gamma(\eta) = \ln\frac{1+\eta}{1-\eta}$; other simpler laws can also be used. The feedback parameters were $a = b = 1$, $\theta = 0.5$. The network operates as follows: The first "winner" is the cell that receives the largest input; its response will first stabilize to a high value, while the other outputs tend to zero. When the activity of the "winner" is temporarily depressed by the dynamic feedback, the other cells continue competing. The solution was obtained by the classical Runge-Kutta numerical integration method, using a step size of 10^{-5}

4.2.2 Lateral Control of Plasticity

In certain early works that demonstrated the self-organizing map effect [2.32, 34–36], the learning law was assumed essentially Hebbian, i.e., the synaptic plasticity was independent of location, and learning was thus *proportional to local output activity level.* If the lateral activity control ("Mexican hat") has positive feedback to a reasonably wide area, the output activity, due to this kernel form, becomes clustered into finite-size "bubbles" of activity. The neighborhood function h_{ci} in the basic SOM model would then be proportional and mediated by these activity "bubbles."

It is possible to implement some kind of SOM processes based on these "bubbles." The ordering power, however, thereby is not particularly good, and formation of ordered maps is unsure, sensitive to parameter values. The border effects at the edges of the map can be large.

A recent idea of this author [2.45] was to assume that *the plasticity control is mediated by another set of biological variables, eventually some kind of diffuse chemical agents or even special chemical transmitters or messengers, the amount of which in the neighborhood is proportional to local activity, but which does not control activity itself in the neighborhood; the local neural activity in one place only controls the plasticity (learning-rate factor) in another location, the strength of control being some function of distance between these locations.* It may not be necessary to specify the control agent, and many alternatives for it may indeed exist in the neural realms, chemical as well neural (synaptic); if the control is synaptic, we have to assume that *these synapses only control the plasticity of other synapses but do not contribute to the output activity.* The essential idea is thus a separate non-activating plasticity control.

It has been observed quite recently that lateral control of synaptic plasticity is often mediated by gaseous NO (nitric oxide) molecules that are formed at the synapse, on the postsynaptic side of it. These light molecules can be diffused freely through the medium to nearby locations, where they modify synaptic plasticity (for a review, cf. [2.50]). *We shall not claim, however, that the control we need is implemented just by NO molecules; they are only mentioned here for to demonstrate a theoretical possibility.* For instance, CO and CN molecules have been found to have similar effects, but there may exist many other agents that have not yet been observed.

It must also be noted that the WTA function analyzed theoretically above, with one cell becoming a "winner", would be too weak to produce a sufficient amount of chemical control agents of the above kind. It seems more plausible that *the WTA function activates a whole "bubble" of activity* whereby, say, 1000 nearby neurons become active and are able to elicit a sufficient amount of the control agent.

4.3 Learning Equation

The biologically inspired neural models usually involve, in the simplest cases at least, that neuron i is activated in proportion to the dot product $m_i^T x$ where x is the presynaptic signal vector and m_i represents the synaptic strengths. Therefore we shall take the "dot-product SOM," and the Riccati-type learning equation as discussed in Sect. 2.13.2 for a basis of the present physiological SOM model.

Let us first see what follows from the SOM equations (3.5) and (3.6) that were originally suggested by this author [2.36] in 1982. Assume tentatively that the $m_i(t)$ were already normalized to the value $||m_i(t)|| = 1$. If h_{ci} in (3.5) is a small factor, the Taylor expansion in the two lowest-order terms reads

$$m_i(t + 1) \approx m_i(t) + h_{ci}(t)[x(t) - m_i(t)m_i^T(t)x(t)] . \tag{4.13}$$

Since h_{ci} is the learning-rate factor of m_i, corresponding to P in (2.12) or (2.13), this expression (in the continuous-time limit) has a close correspondence with the Riccati-type learning law (2.14). In the latter there holds that starting with *arbitrary* norm $||m_i(t)||$, the norm always converges to a constant value, here equal to unity.

Conversely, if we start with the Riccati-type learning law expressed for the external (afferent) signal connections, and P in (2.13) is identified with the "mean field" type (spatially integrated) lateral plasticity-control effect equal to $h_{ci}(t)$, and if we use the discrete-time approximation of (2.14), we indeed end up with the learning law

$$m_i(t + 1) = m_i(t) + h_{ci}(t)[x(t) - m_i(t)m_i^T(t)x(t)] . \tag{4.14}$$

Thus we see a close mutual correspondence between the dot-product-SOM learning law and the Riccati-type learning law.

4.4 System Models of SOM and Their Simulations

The next step is to combine the WTA function, the learning law, and some description of the neighborhood function. We can do this by using a continuous-time system model, where the neurons, described by (4.1), (4.2) and (4.3), form a regular array; in the simulations reported below this array was a two-dimensional and hexagonally organized layer.

The output activities are first made to oscillate as demonstrated in Fig. 4.3. A lateral control effect of plasticity, in proportion to $\eta_j(t)$, is made to spread from each neuron j. There is still considerable freedom in describing this effect analytically, depending on the type of plasticity control assumed in the model. This may be a diffuse chemical effect, or a neural effect contributing to the lateral control of plasticity according to essentially similar dynamics.

However, in order to forestall a potential misinterpretation of the WTA circuit of Fig. 4.2 it will be necessary to emphasize that each formal neuron i of the simplified model may in reality stand for a group of, say, hundreds of tightly coupled biological neurons. Therefore it would be more accurate to talk of *site i*. It is not necessary to specify whether such a site is a microanatomical aggregate, or whether the activity of an active neuron simply triggers the activity in nearby cells. The most important condition for the chemical-interaction assumption is that a sufficient amount of the "learning factor" is released from the "winner site," the total activity of which we simply denote by η_i in simulations.

Let us assume that the active site constitutes a thin cylindrical local source, perpendicular to the cortex, and the control agent for plasticity is diffused from this source proportional to its activity η_j. Consider now the dynamics of cylindrical diffusion from a line source of strength $\alpha(t)\eta_j(t)$; this model need not be quite accurate. The slowly decreasing coefficient $\alpha(t)$ describes temporal variations in the source strength. In a homogeneous medium, the density of the agent at time t and at cylindrical distance r from the source is

$$D_j(t,r) = \int_0^t \alpha(\tau)\eta_j(\tau)\frac{1}{4\pi(t-\tau)}e^{-r^2/4(t-\tau)}d\tau. \tag{4.15}$$

For better readability, the time constants and scaling coefficients needed in the actual simulations have been omitted above. In simulations we describe the neighborhood function by

$$h_{ij}(t) = D_j(t, r_{ij}) , \tag{4.16}$$

where r_{ij} denotes the lateral distance between sites i and j.

In the simulations the control factor describing the plasticity, (4.16), was complemented with two additional details: first, the possible mechanisms responsible for the inactivation of the diffusing substance were modelled simply by using a finite integration length in the discrete-time approximation of (4.15). Second, the plasticity was required to exceed a small threshold before taking effect.

It should be noted that the equation (4.15) can be approximated by the expression $D_j(t,r) \approx \alpha(t)h(r)\eta_j(t)$, where h is a coefficient which depends neither on η_j nor t, if the activity $\eta_j(t)$ changes slowly compared to the time rate of the diffusion.

The learning law was already postulated by (4.14).

A simulation of the system behavior defined by the above system model was performed and its solution was illustrated by an animation film. Like in many earlier simulations, the input $x(t)$ was defined as a series of successive samples picked up at random from the framed rectangular areas in Fig. 4.4. Now $x(t)$ has to be approximated by steady values during WTA cycles. As the hexagonal topology was used in the neuron array, the auxiliary lines drawn to Fig. 4.4 also mean that the distance r_{ij} between neighboring nodes, used in (4.16), is evaluated at the discrete nodes of the hexagonal lattice only.

Three frames from this film are shown in Fig. 4.4. It took about three weeks to run this simulation on a modern workstation, and even so the final asymptotic state was not yet achieved. Most of the time was spent for the computation of (4.15) and also for the Runge-Kutta computation of the time functions $\eta_i(t)$. A similar simulation based on the dot-product SOM would have only taken less than one second on the same computer. This simulation then proves that the general behavior of both the physiological SOM model and the dot-product SOM is qualitatively similar, and therefore *the dot-product SOM can henceforth be regarded as a fast physiological SOM model.*

Just for a reference, we also include here another simulation that is otherwise similar to the above except that the slower diffusion effect was significantly simplified, making the simulation much faster. Now the spread of the learning factor was simply assumed to be of the form

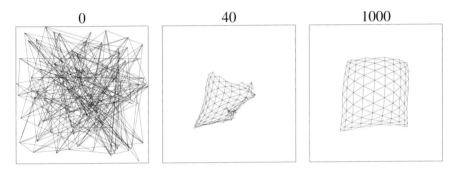

Fig. 4.4. Three frames from an animation film (learning steps 0, 40, and 1000, respectively) that illustrates the operation of a complete physiological SOM system, after which the simulation was stopped

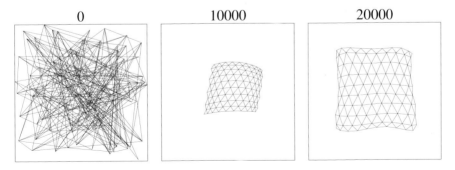

Fig. 4.5. Three frames from an animation film based on the precomputed diffusion solution (4.17)

$$D_j(t,r) = \alpha(t)\eta_j(t)e^{-r^2/2[\sigma(t)]^2} \ , \tag{4.17}$$

where $\alpha(t)$ and $\sigma(t)$ are (linearly) decreasing parameters like in the dot-product SOM. Then again, (4.16) can be used to define $h_{ij}(t)$ as a sum of contributions from the different neurons.

A complete simulation based on (4.17) was also performed. Again, an animation film was made, and three frames from it are shown in Fig. 4.5.

4.5 Recapitulation of the Features of the Physiological SOM Model

The most important message of this chapter is that certain assumptions of the SOM theory seem to have biological counterparts: 1. A natural, robust implementation of the *WTA function* based on a simple nonlinear dynamic model for a neuron and a simple laterally interconnected neural network. 2. Automatic resetting of the WTA function by a special kind of integrating inhibitory interneuron, resulting in *cyclic operation* of the network. 3. A realistic interpretation of the law of *synaptic plasticity*. 4. Automatic *normalization* of *subsets* of synaptic vectors, in particular, normalization of the subset of synapses corresponding to the afferent signals. 5. A natural explanation of the *lateral control of plasticity* (neighborhood function) between neurons in learning.

Although it now has been demonstrated that the SOM models, both the simpler algorithmic (discrete) and the more complex distributed ones, can be regarded as genuine physiological models, nonetheless *it is advisable to use only the simpler ones in practical applications. The approach taken in this chapter was strictly meant for a study of dynamic phenomena that may occur in a network made of nonlinear dynamic neuron models.*

4.6 Similarities Between the Brain Maps and Simulated Feature Maps

The abstract feature maps may be more common in the brain than it is generally thought. One difficulty in finding them is their extent, which is probably very small, only a few millimeters. Therefore, simulated SOMs and real brain maps have not been compared frequently. However, their similarities suggest that the same simple principle might underlie, at least in an abstract form, the emergence of feature maps in the living brain at scales of less than a few millimeters. It is not yet known whether larger-scale maps could be formed according to the same principle or perhaps by a multilevel organization.

4.6.1 Magnification

The cortical representations occupy a territory, the area of which is proportional to some measure of occurrence of that feature in experiences. In the simulated feature maps, the area occupied by a feature value is proportional to some power of the probability of occurrence of the feature. Theoretically, the exponent can assume values between one third and one.

4.6.2 Imperfect Maps

The simulated SOMs sometimes contain typical unintended features. For example, if self-organization starts from different random initial states, the orientation of the resulting maps can vary or form a mirror image. If the self-organization process is not perfect, artificial networks may be fragmented into two or several parts with opposite order.

The "initial values" of brain maps are most probably determined genetically, but reverse ordering is still possible in them.

4.6.3 Overlapping Maps

In different situations, the same brain cells may respond to signals of different modality. The neurons of the model network merge signals of different modalities. If only one modality is used at a time, several independent maps can be superimposed on the same network. However, if, for example. two types of input are used simultaneously, a common map is formed, with representations for both modalities (3.15). The cells respond similarly to inputs from either modality.

5. Variants of SOM

In order to create spatially ordered and organized representations of input occurrences in a neural network, the most essential principle seems to be *to confine the learning corrections to subsets of network units that lie in the topological neighborhood of the best-matching unit*. There seems to exist an indefinite number of ways to define the matching of an input occurrence with the internal representations, and even the neighborhood of a unit can be defined in many ways. It is neither necessary to define the corrections as gradient steps in the parameter space: improvements in matching may be achieved by batch computation or evolutionary changes. Consequently, all such cases will henceforth be regarded to belong to the broader category of the Self-Organizing Map (SOM) algorithms. This category may also be thought to contain both supervised and unsupervised learning methods.

5.1 Overview of Ideas to Modify the Basic SOM

The basic SOM defines a *nonparametric regression solution* to a class of vector quantization problems and in that sense does not need any modification. Nonetheless, there exist other related problems where the SOM philosophy can be applied in various modified ways. Examples of such lines of thought are given below.

Different Matching Criteria. It seems that the matching criteria in defining the "winner" can be generalized in many ways, using either different metrics or other definitions of matching. Generalization of metric was already made in Sect. 3.8; Sect. 5.2 contains another idea.

Nonidentical Inputs. A straightforward structural generalization would be to compose the input vector x of subsets of signals that are connected to different, eventually intersecting areas of the SOM, whereby the dimensionalities of x and the m_i may also depend on location. Very special ordering patterns, kind of abstractions, may be obtained in cells that lie at the intersections of input connection sets. We will not present any examples of such maps in this book.

Accelerated Searching by Traditional Optimization Methods. Abundant knowledge that the ANN community may need is available in the literature of related areas. For instance, there exists a plethora of textbooks on numerical optimization methods; some of the latter might be used to implement an accelerated search in the ANN and in particular SOM algorithms, especially when the initial order in the latter has already been formed. The following is a list, without explanation, of the most important standard optimization algorithms:

- linear programming (LP)
  - simplex method
  - interior-point method (Karmarkar method)
- integer programming (IP) and mixed integer programming (MIP)
- quadratic programming (QP)
  and sequential quadratic programming (SQP)
- unconstrained nonlinear programming
  - steepest-descent methods
  - Newton's method
  - quasi-Newtonian and secant method
  - conjugate gradient method
- nonlinear least-square methods
  - Gauss-Newton method
  - Levenberg-Marquardt method
- constrained nonlinear programming
  - sequential linear programming
  - sequential quadratic programming
  - methods using cost and constraint functions
  - feasible-direction methods
  - augmented Lagrangian method
  - projection methods
- statistical (heuristic) search

Numerical algorithms for many of these methods can be found in [5.1].

Hierarchical Searching. Two particular searching methods to find the "winner" in the SOM very quickly will be discussed in this book, namely, the *tree-search SOM*, and the *Hypermap*. The "winner" m_c is thereby defined by a sequential search process. For the tree-search SOM, see Sect. 5.3. In the principle of "Hypermap" (Sect. 6.10), the central idea is that by means of one subset of input signals only a *candidate set* of best-matching nodes is defined, eventually using coarse and fast methods, after which the "winner" is selected from this subset by means of other inputs. After decisions made at the earlier step, the exact signal values thereby applied are no longer taken into account at the next steps or in making the final selection of the "winner." This principle can be generalized for an arbitrary number of levels.

We shall revert to the Hypermap in Sect. 6.10, in connection with developments of LVQ.

Dynamically Defined Network Topology. Several authors [3.40–54] have suggested that the size or structure of the SOM network could be made dependent on intermediate results (e.g., quantization error during the process). There is one case where changing the architecture of the network according to tentative ordering results is profitable, and that is when the SOM is used to find *clusters* or *filaments* in the distribution of primary data. In the latter idea, one is adding new neurons to the network (i.e., making it grow) or deleting them upon need to describe the probability density function of input more accurately.

Definition of Neighborhoods in the Signal Space. If description of density functions is the primary goal, then in *nonordered vector quantization* [3.5], [5.2] one can achieve significantly accelerated learning by defining the neighborhoods in the *signal space*, as shown in Sect. 5.5. Somewhat similar interactions are defined in the "neural gas" approach [3.6, 5.3, 5.4].

Hierarchical Maps. The basic SOM is already able to represent hierarchical structures of data, as demonstrated in Sect. 3.15. However, one objective in SOM research has been to construct structured maps using elementary SOMs as modules. Such constructs are still at an elementary stage of development.

Acceleration of Learning in SOM. One of the most effective means to guarantee fast convergence is to define the initial reference vector values properly. As stated in Sect. 3.7, they may be selected, for instance, as a two-dimensional regular (rectangular) array along the two-dimensional linear subspace that is fitted to $p(x)$, and the dimensions of the array may then be made to correspond to the two largest eigenvalues of the correlation matrix of x.

One problem posed long ago concerns the optimal sequence of the learning-rate parameter. In the original Robbins-Monro stochastic approximation (Sect. 1.3.3) the two necessary and sufficient conditions were

$$\sum_{t=1}^{\infty} \lambda^2(t) < \infty, \quad \sum_{t=1}^{\infty} \lambda(t) = \infty . \tag{5.1}$$

It seems that at least with very large maps, the effect of the learning-rate parameters on the convergence must be taken into account. It seems that a time function of the type of (3.39) should be used in $h_{ci}(t)$ or $\alpha(t)$.

In Sect. 3.6 we have introduced a fast computing scheme, the Batch Map, in which the problem about the learning rate has been eliminated (although the problem about the neighborhood size still exists).

SOM for Sequential Signals. Natural data are often dynamic (changing smoothly in time) and statistically dependent, whereas the basic SOM was designed to decode pattern sequences the members of which are essentially statistically independent of each other. There exist simple ways to take the sequential relations of signals into account in the SOM, as discussed in Sect. 5.6.

SOM for Symbol Strings. The SOM can be defined for string variables, too (Sect. 5.7). The algorithm is then expressed as batch version, where the average, the so-called generalized median, over a list of symbol strings is defined to be the string that has the smallest sum of generalized distance functions from all the other strings.

Operator Maps. If a finite sequence $X_t = \{x(t-n+1), x(t-n+2), \ldots, x(t)\}$ of input samples is regarded as a single input entity to the SOM, each cell might be made to correspond to an operator that analyzes X_t; then matching of x with the neurons might mean finding the operator with parameters m_i such that its response to x would be maximum. Other interesting cases are obtained if the cell function is an *estimator* G_i that defines the *prediction* $\hat{x}_i(t) = G_i(X_{t-1})$ of the signal $x(t)$ by unit i at time t on the basis of X_{t-1}, and matching means finding the neuron that is the best estimator of $x(t)$. This will be discussed in Sect. 5.8. Cf. also [5.5].

Evolutionary-Learning SOM. Although no distance function over the input data is definable, it is still possible to implement the SOM process using evolutionary-learning operations. The process can be made to converge rapidly when the probabilistic trials of conventional evolutionary learning are replaced by averaging using the Batch Map version of the SOM. No other condition or metric than a *fitness function* between the input samples and the models need be assumed, and an order in the map that complies with the "functional similarity" of the models can be seen to emerge.

Supervised SOM. Normally one would recommend that for a supervised statistical pattern recognition problem the LVQ algorithms should be used. It may be of historical interest to note that the basic SOM too can be modified to perform supervised classification, as shown in Sect. 5.10.

Adaptive-Subspace SOM. A new line of thinking, not only in the SOM theory but in the ANN field in general, is description of neural units as *manifolds*. By this approach it is possible to create *invariant-feature detectors*. The adaptive-subspace SOM (ASSOM) described in Sects. 5.11 and 5.12 is the first step in that direction.

Systems of SOMs. A far-reaching goal in self-organization is to create autonomous systems, the parts of which control each other and learn from each other. Such control structures may be implemented by special SOMs; the main problem thereby is the interface, especially automatic scaling of interconnecting signals between the modules, and picking up relevant signals to interfaces between modules. We shall leave this idea for future research.

5.2 Adaptive Tensorial Weights

The Euclidean metric is a natural choice for matching, if no special assumptions about the input vectors can be made. If there are significant differences

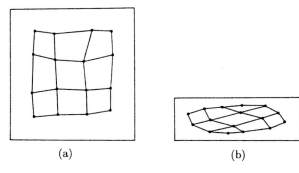

(a)   (b)

Fig. 5.1. An example of oblique orientation of the SOM

in variances of the components of x, however, an oblique orientation of the map may thereby result, as demonstrated in Fig. 5.1(b).

A better orientation can be guaranteed by the introduction of a weighted Euclidean distance in the matching law [3.5], [5.2]. Its square is defined as

$$d^2[x(t), m_i(t)] = \sum_{j=1}^{N} \omega_{ij}^2 [\xi_j(t) - \mu_{ij}(t)]^2 , \tag{5.2}$$

where the ξ_j are the components of x, the μ_{ij} are the components of the m_i, and ω_{ij} is the weight of the jth component associated with cell i, respectively. The central idea is to estimate such ω_{ij} values recursively in the course of the unsupervised learning process that the effects of errors (variance disparity) are balanced. To that end, each cell i is first made to store the backwards exponentially weighted averages of the absolute values of the errors $|\xi_j(t) - \mu_{ij}(t)|$ in each component. Denoting these values at step t by $e_{ij}(t)$, we first obtain

$$e_{ij}(t + 1) = (1 - \kappa_1)e_{ij}(t) + \kappa_1 \omega_{ij} |\xi_j(t) - \mu_{ij}(t)| , \tag{5.3}$$

where the effective backward range of averaging is defined by parameter κ_1 (a small scalar).

By the same token as in the basic learning process, averages of only those cells need to be updated that at each time belong to the learning neighborhood.

Next we define the mean of the $e_{ij}(t)$ over the inputs of each cell:

$$e_i(t) = \frac{1}{N} \sum_{j=1}^{N} e_{ij}(t) . \tag{5.4}$$

Within each cell, we then try to maintain the same average level of weighted errors over all the inputs:

$$\forall j , \quad E_t\{\omega_{ij}|\xi_j - \mu_{ij}|\} = e_i . \tag{5.5}$$

The following simple stabilizing control has been found adequate to approximate (5.5):

$$\omega_{ij}(t+1) = \kappa_2 \omega_{ij}(t),\ 0 < \kappa_2 < 1 \ \text{if} \ \omega_{ij}|\xi_j(t) - \mu_{ij}(t)| > e_i(t)\ , \quad (5.6)$$
$$\omega_{ij}(t+1) = \kappa_3 \omega_{ij}(t),\ 1 < \kappa_3 \ \text{if} \ \omega_{ij}|\xi_j(t) - \mu_{ij}(t)| < e_i(t)\ . \quad (5.7)$$

Suitable, experimentally found control parameters are: $\kappa_1 = 0.0001$, $\kappa_2 = 0.99$, and $\kappa_3 = 1.02$.

To further force each cell to "win" approximately as often as the others, it is possible to stipulate that

$$\forall i\ ,\ \prod_{j=1}^{N} \frac{1}{\omega_{ij}} = \text{const.}\ . \quad (5.8)$$

Introduction of the above weighted Euclidean norm can be justified by simulations, as illustrated in Fig. 5.2. Two comparative series of experiments were carried out, one with the unweighted Euclidean norm, and the other with the adaptively weighted Euclidean norm, respectively. The self-organizing maps contained 16 cells arranged in a rectangular grid. The learning vectors were selected randomly from a uniform rectangular distribution, and relatively few learning steps were used. The variance along the vertical dimension versus the horizontal one was varied (1:1, 1:2, 1:3 and 1:4, respectively). The

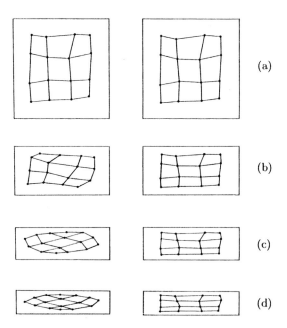

(a)

(b)

(c)

(d)

Fig. 5.2. Rectified orientation of the SOM when tensorial weights are used. The Euclidean distance was used in the left counterparts of Figs. **(a)** through **(d)**, and tensorial weights in the right counterparts, respectively

results with the unweighted metric are shown on the left of Fig. 5.2. and the result with the weighted metric on the right of Fig. 5.2, respectively.

If the variances along the dimensions are the same, both norms result in a similar alignment of the m_i vectors, as shown in Fig. 5.2a. However, an optimum array to approximate the distribution of Fig. 5.2d would have been a 2 by 8 lattice. Since we had preselected the dimensions of the array, as well as the neighborhood function, with widely different variances of x in the different coordinate dimensions only a suboptimal vector quantization thereby results, as indicated by its oblique orientation. If now the tensorial weights are applied in the norm calculation, each neighborhood easily adjusts itself to the form of the distribution of x in an acceptable way, as will be seen.

In real-world applications, the form of $p(x)$, however, is never so regular as exemplified above. It seems that a nice orientation of the SOM along the structured form of $p(x)$ may even not be desirable in all cases; it seems to be a "cosmetic" improvement in "toy examples." It is far more important that the average expected quantization error defined by the m_i is minimized. Evaluation of the quantization error is more complicated if tensorial weights are used.

5.3 Tree-Structured SOM in Searching

Sequential search for the "winner" in software implementations of SOM can be very time consuming. To speed up this search, *Koikkalainen and Oja* [5.6], *Koikkalainen* [5.7]. and *Truong* [5.8] have introduced a hierarchical searching scheme that involves several SOMs describing the same $p(x)$, organized as a pyramid structure (Fig. 5.3).

The basic idea is to start at the root level (top) of the tree and form a one-neuron SOM, after which its codebook vector is fixed. Thereafter the lower but larger SOMs are trained, one level at a time, and their codebook vectors thereafter fixed. *In locating the "winner" in a SOM that is under formation, the previous, already fixed levels are being used as a search tree. Searching, relative to x, starts at the root proceeding to lower levels, and at a lower level*

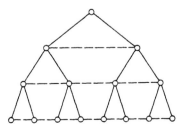

Fig. 5.3. A pyramid of SOMs for fast determination of the "winner"

the "winner" is only searched for among the "descendants" defined in the following way.

Figure 5.3 shows the search tree in the one-dimensional case. The links between units on the same level define the neighborhood to be used on that level *indirectly*; no "shrinking" is needed. Unlike in the normal tree search where the result on the lower level is searched for among the candidates linked to the upper level, searching on the lower level is here performed actually *in all units that are linked to the "winner" and its neighbors on the next-higher level*. Eventual false ordering on a higher level is thus compensated for by this indirect definition of neighborhood.

Searching starts at the root level, descending to the next-lower level, on which the "winner" of that level is determined; on the still-next-lower level only a subset of units is subordinate to the "winner" and need be studied, etc. As the global ordering of the map system starts at the top, the organization of the upper level effectively guides the ordering of the lower-level nodes.

5.4 Different Definitions of the Neighborhood

Disjoint Topologies. This author, as early as 1982, experimented with SOM structures consisting of several disjoint maps. Although a disjoint topology was defined, a unique "winner" neuron, for a particular input x, was selected from all the reference vectors. As a matter of fact this case is already a *hierarchical* vector quantization. We may think that the simple codebook vectors are replaced by separate, interacting SOMs, whereby in the set of SOMs, each set may be regarded as a codebook vector in VQ. It has turned out, as shown by the example of Fig. 5.4, that the clusters are thereby identified more accurately, even in difficult cases.

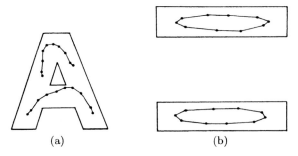

(a)   (b)

Fig. 5.4. Demonstrations of disjoint-topology one-dimensional SOMs, performed in 1982 by the author; there were open-ended topologies on the left, but closed-loop topologies on the right. A unique "winner" over all the nodes was determined in comparison

Recently, e.g. in [5.9, 10] this idea has been applied to clustered data, in land surveying from satellite pictures.

Dynamically Defined Topologies. It has also been clear for a long time that the neighborhood function h_{ci} can be made more complicated than just symmetric, and its *structure* can be *adaptive*. The idea of the author around 1982 was to define h_{ci}, or actually the neighborhood set N_c as a set of *address pointers to the neighboring neurons*, the former being stored at each location of the map. If after tentative convergence of the map the vectorial difference of the codebook vector m_i from m_c was too large, the corresponding pointer was *inactivated*, and learning was continued. Figure 5.5 demonstrates how the dynamically altered topology adapts better to $p(x)$. It turned out later that absolute values of the differences are not very good for a decision on the inactivation of the pointers, because the point densities of the m_i in different parts of the map can be very different. In order to describe pure structures,

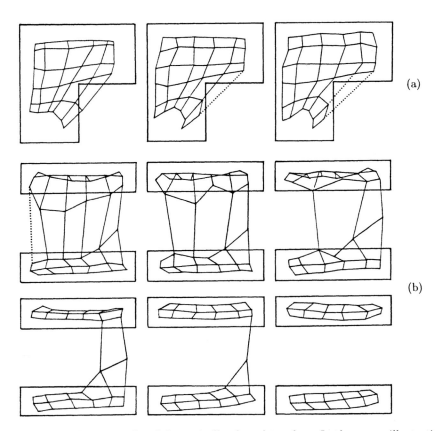

Fig. 5.5. Two examples of dynamically altered topology. In the upper illustration, inactivated pointers are shown dotted. In the lower picture they are left out for clarity

the decision on inactivation must be based on the *relative* values of the above vectorial differences, referring to the averages taken over the differences, the criterion being the average distance around neuron i.

As mentioned earlier, other adaptive methods to change the h_{ci} have been published in [3.40–54].

It must be realized, however, that the original regular-topology SOM and the dynamic-topology versions represent quite different philosophies of thinking. The SOM was originally conceived for a *nonparametric regression*, whereupon it was considered more important to find the main dimensions, the "local principal axes" in the signal space along which the samples are distributed locally, and in Sect. 3.16 it was indeed shown how such a SOM can be interpreted in classical terms of statistics. It seems, however, that the dynamic-topology SOM has been introduced to describe *shapes* of the sample distribution. At least the ability of the SOM to represent *abstractions* is largely lost with the dynamical topologies and growing of the network, because the branches then tend to describe every detail of $p(x)$ and not to generalize it. Neither can one easily interpret the structured SOMs statistically.

One should also notice that in almost all practical problems the dimensionality n of x and the m_i is very high, and, for instance, to describe a "hyperrectangle" one needs 2^n vertices. It seems very difficult to define any more complicated forms of regions in high-dimensional spaces by a restricted number of codebook vectors and training samples.

It seems that dynamical-topology SOMs are at their best in artificially constructed nonstatistical examples, while with natural stochastic data they have difficulties in deciding when to make a new branch.

Hypercube Topology. In addition to one- and two-dimensional maps, one may consider any dimensionality of the array. A special case is the $\{0,1\}^N$ SOM, where the nodes form the vertices of an N-dimensional hypercube. In this case the neighborhood of a vertex point consists of itself and all the neighboring vertices sharing the same edges [5.11, 12].

Cyclic Maps. Also cyclic, e.g. toroidal two-dimensional arrays may sometimes be used, especially if it can be concluded that the data are cyclic, too.

5.5 Neighborhoods in the Signal Space

When the input vector distribution has a prominent shape, the results of the best-match computations tend to be concentrated on a fraction of cells in the map. It may easily happen that the reference vectors lying in zero-density areas are affected by input vectors from all the surrounding parts of the nonzero distribution. This causes a statistical instability, which ceases when the neighborhoods are shrunk, but a residual effect from the rigid neighborhoods may be that some cells remain *outliers*. If the input vector distribution is more uniform, the reference vector set neatly adapts to the input data.

These observations led us [3.5], [5.2] to look for new mechanisms in which the local neighborhood relationships could be defined adaptively in the course of learning. We abandoned the definition of topologies on the basis of spatial adjacency relations of the network, and we reverted to the original vector quantization problem. It seemed interesting and advantageous to define the neighborhoods *according to the relative magnitudes of the vectorial differences of the m_i vectors, i.e, in the input space.* In the new learning algorithm, the neighborhood relationships were defined along the so-called *minimal spanning tree (MST)*. As is well known from the literature [5.13], the MST algorithm assigns arcs between the nodes such that all nodes are connected through single linkages and, furthermore, the total sum of the lengths of the arcs is minimized. Here, the lengths of the arcs are defined to be the nonweighted Euclidean norms of the vectorial differences of the corresponding reference vectors. The neighborhood of a cell in such an MST topology is defined along the arcs emanating from that cell (Fig. 5.6).

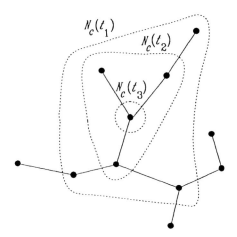

Fig. 5.6. Examples of neighborhoods of a cell in the minimal-spanning-tree topology

Like in the original algorithm, learning here starts with wide neighborhoods, which means traversing more MST arcs off the selected cell, in order to make up the neighborhood. Later on in the learning process the neighborhoods are shrunk and the type of the topology becomes less critical. Since the adaptation in general is temporally smooth, it is not necessary to compute the MST after each successive learning vector $x(t)$. We have found in the simulations that recomputation of the topological neighborhood every 10 to 100 steps will suffice. It has to be pointed out that this self-organizing process usually does not lead to spatially ordered mappings; on the other hand, it makes *vector quantization* significantly faster and stabler.

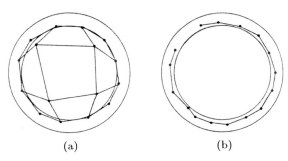

Fig. 5.7. Final adaptation results of a 16-cell self-organizing network to a spherical distribution in **(a)** rectangular topology and **(b)** MST topology

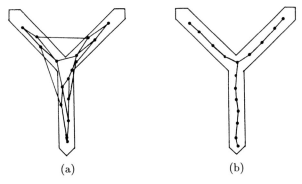

Fig. 5.8. Final adaptation of a 16-cell self-organizing network to a "Y-shaped" input distribution in **(a)** rectangular topology and **(b)** MST topology

The virtues of the dynamic MST topology were demonstrated by two series of simulations. Certain twodimensional distributions of the x, as shown in Figs. 5.7 through 5.9, were used. On the left in the illustrations, the asymptotic SOM configurations referring to the rectangular topology (4-neighbors) are shown. On the right of the illustrations, the final alignment of the minimal spanning tree is depicted showing a better fit to the input data. For a reference, Fig. 5.9 depicts the corresponding asymptotic map vectors with square distributions. In such cases, the MST topology seems to generate poorer results than the rectangular network topology.

Another virtue of the MST topology can be seen from Figs. 5.10 and 5.11. In these simulations, the actual input distribution consisted of two clusters. We wanted to see how well the m_i would follow the input samples in time if the latter were first drawn from one cluster only and then, later on, from both clusters. Learning was thus carried out in two phases. In the first phase, the input vectors were drawn from the lower cluster only. The configurations of reference vectors in both topologies after the first phase are shown on the left in Figs. 5.10 and 5.11. In the second phase, the input vectors were drawn

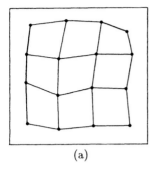 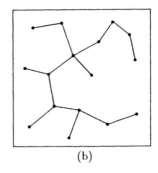

(a)   (b)

Fig. 5.9. Final adaptation results of a 16-cell self-organizing network to a square input distribution in **(a)** rectangular topology and **(b)** MST topology

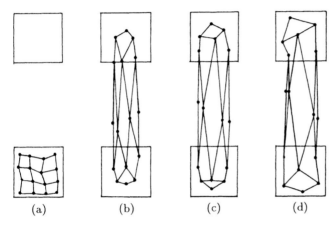

(a)   (b)   (c)   (d)

Fig. 5.10. (a) Final adaptation results of a 4 by 4 cell self-organizing network in rectangular topology when the input samples were drawn from the lower cluster only. **(b)–(d)** Snapshots of the adaptation during phase 2, when input samples were drawn from the upper cluster, too

from the entire distribution. Different phases of the second-phase adaptation are shown in Figs. 5.10 and 5.11. The MST topology exhibited a remarkable flexibility when adapting to changes in the input distribution. It is an essential improvement that now almost all nodes are finally situated within the different subsets of the distributions.

Compared with the classical VQ this MST method converges extremely fast and finds the hierarchical clusters effectively. It is also possible to trace the map along the links, thereby eventually being able to discover structures of hierarchical relations like in the growing maps discussed earlier.

The so-called "neural gas" idea of *Martinetz* [3.6], [5.3, 4] also defines the interactions in the input space.

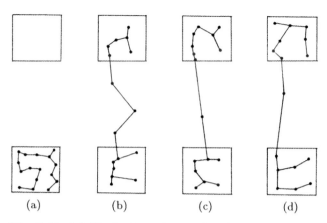

Fig. 5.11. (a) Final adaptation results of a 4 by 4 cell self-organizing network in the MST topology when the input samples were drawn from the lower cluster only. **(b)–(d)** Snapshots of the adaptation during phase 2, when input samples were also drawn from the upper cluster

5.6 Dynamical Elements Added to the SOM

The original SOM idea was based on the matching of *static* signal patterns only, and its asymptotic state is not steady unless the topological relationships between the different input patterns are steady. The input patterns may, however, occur in a sequence, like the spectral samples picked up from the natural speech do. Statistical dependence between the successive samples can now be taken into account in various ways. The simplest one is definition of a "time window" in the sequence of time samples, and collection and concatenation of successive spectral or other samples into a pattern vector of correspondingly higher dimensionality. Time dependence of successive samples is then reflected in the order of the concatenated elements in the pattern vector. In digital computation such a concatenation is extremely easy, but in analog computation (and especially in the biological networks) concatenation would imply a very special serial-to-parallel conversion of the signal values.

For the reasons mentioned above, attempts to implement "sequential SOMs" using delayed feedbacks from the outputs to input terminals have never been successful.

The simplest and most natural method of converting sequential information into static form is to use low-pass filters, such as resistor-capacitor elements of electrical circuits, as short-term memories (integrators). Such elements can be added to either inputs or outputs of the neurons, or both. In speech recognition they have had a positive effect on recognition accuracy [5.14–18]. They constitute the first step toward dynamic networks and a more general definition of the SOM that we shall discuss in the next section.

5.7 The SOM for Symbol Strings

The Self-Organizing Maps are usually defined in metric vector spaces. The SOM is a *similarity diagram of complex entities*. Many different kinds of vectorial entities are amenable to SOM representation. In the first place one may think of ordered sets of numbers that stand for sets of signals, measurements, or statistical indicators. For instance, textual documents, as we shall see in Sect. 7.8 can be described by statistical vectors, if the latter represent the usage of words; the word histograms or their compressed versions can be regarded as real vectors.

It can be shown, however, that the entity to be ordered can be much more general. If x and y are any entities, a sufficient condition for them to be mapped into a SOM diagram is that some kind of symmetric *distance function* $d = d(x, y)$ is definable for all pairs (x, y).

We shall demonstrate the organization of *symbol strings* on a SOM array, whereby the relative locations of the "images" of the strings (points) on the SOM ought to reflect, e.g., some distance measure, such as the *Levenshtein distance (LD)* or *feature distance (FD)* between the strings (Sect. 1.2.2) If one tries to apply the SOM algorithm to such entities, the difficulty immediately encountered is that *incremental learning laws cannot be expressed for symbol strings*, which are discrete entities. Neither can a string be regarded as a vector. The author has shown [5.19] that the SOM philosophy is nonetheless amenable to the construction of ordered similarity diagrams for string variables, if the following ideas are applied:

1. The *Batch Map* principle (Sect. 3.6) is used to define learning as a succession of certain generalized *conditional averages over subsets of selected strings*.
2. These "averages" over the strings are computed as *generalized medians* (Sect. 1.2.3) of the strings.

5.7.1 Initialization of the SOM for Strings

It is possible to initialize a usual vector-space SOM by random vectorial values. We have been able to obtain organized SOMs for string variables, too, starting with random reference strings. However, it is of a great advantage if the initial values are already ordered, even roughly, along the SOM array. Then one can use a smaller, even fixed-size neighborhood set, eventually consisting of only the closest neighbors of the node and the node itself. Now we can make use of the fact that the set median is one of the input samples. It will be possible to define ordered strings for the initial values, if one first forms *Sammon's mapping* (Sect. 1.3.2) of the input samples. From the projection of a sufficient number of representative input samples it is then possible to pick up manually a subset of samples that seem to be ordered two-dimensionally.

If the symbol strings are long enough, one can define a rough tentative order for them by considering the histograms of symbols in each string, over the alphabet of the symbols. This means that one tentatively constructs a reference vector for each map unit of the SOM, with the same dimensionality as the histograms, too. During this first tentative phase of initialization, a traditional SOM is first constructed. Its map units could be labeled by all strings whose histograms are mapped to the corresponding units, but in order to obtain a unique label for each unit, a majority voting over the labels of each unit is then carried out. After one has obtained a tentative, roughly ordered set of symbol strings that label the map units of the SOM, one can regard them as the initial values of the reference strings and ignore the vectorial histogram models. The more accurate self-organizing process of the strings then proceeds as described in the following.

5.7.2 The Batch Map for Strings

The conventional Batch Map computing steps of the SOM are applicable to string variables almost as such:

1. Select the initial reference strings by some method described in Sect. 5.7.1.
2. For each map unit i, collect a list of those sample strings to whom the reference string of unit i is the nearest reference string.
3. For each map unit i, take for the new reference string the generalized median over the union of the lists that belong to the topological neighborhood set N_i of unit i, as described in Sect. 1.2.3.
4. Repeat from 2 a sufficient number of times, until the reference strings are no longer changed in further iterations.

It may sometimes occur that the corrections lead to a "limit cycle" whereupon the reference strings oscillate between two alternatives, the latter usually forming a tie in winner search. In such a case the algorithm must be terminated somehow, e.g., by a random choice of either alternative.

5.7.3 Tie-Break Rules

Notice that all strings are discrete-valued entities. Therefore ties in various kinds of comparison operations may occur frequently, and the convergence to the final values in learning may not be particularly fast if the map is big and unless certain precautions are taken. First of all, ties may occur 1. In the winner search, 2. In finding the median.

In the winner search, a tie may occur easily if the strings are very short. Then the distances from an input to the closest reference string may be equal even if the reference strings are very different. If the weighted Levenshtein distance is used, the number of ties of this kind remains small, and a random choice between equal winners can be made. If, during the formation of the map, *set medians* are used for new values of the reference strings (in order to speed up computations), and if a tie occurs, one might speculate that the best candidate might be selected according to the lengths of the strings. In our experiments, however, there were no benefits for favoring either shorter or longer strings; a random choice was sufficient.

After the tentative organization of the histograms and subsequent labeling of the map units by all samples mapped to that unit, the majority voting may also result in a tie. One might then select the label according to its length, or alternatively, from the ties at random. In our experiments random selection has given even slightly better results compared to favoring either short or long labels.

5.7.4 A Simple Example:
The SOM of Phonemic Transcriptions

We have made SOMs of strings, e.g., for phonemic transcriptions produced by the speech recognition system similar to that to be reported later in Sect. 7.5. The string classes represented 22 Finnish command words; 1760 phoneme strings used in the following experiment were collected from 20 speakers (15 male speakers and 5 female speakers). Finnish is pronounced almost like Latin. The speech recognizer was a vocabulary-free phoneme recognizer that had no grammatical restrictions to its output: the system and the details of the experiment have been reported in [5.20]. Figure 5.12(a) shows Sammon's projection of some of the produced strings, from which two-dimensionally ordered samples were picked up for the initialization of the SOM as shown in Fig. 5.12(b). After ten batch training cycles, the SOM looks like shown in Fig. 5.12(c).

5.8 Operator Maps

One of the more difficult problems associated with a simple SOM has indeed been implementation of selective responses to *dynamic phenomena*. As stated above, some trivial but still workable solutions have been to buffer subsequent samples of dynamic signals in static registers, or to provide the input and output ports with a memory for previous signal values.

There exists no reason, however, for which the cells in the SOM could not be made to decode dynamic patterns directly. The generalization is to regard the cells as *operators* denoted by G, eventually representing some kind of filters for dynamic patterns. Such operators may contain tunable parameters, which then correspond to the weight parameters μ_i of the simple models. The central problem is to define the tuning of an operator to an input sequence $X = \{x(t)\}$. The ordering of the operators may then mean that some functionals $G(X)$ of the sequences (or some kind of their time integrals) are ordered, and this order is then relative to the statistics of X.

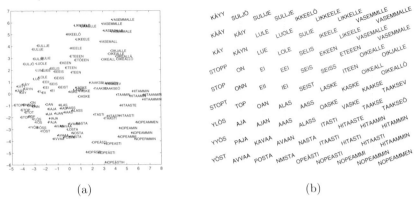

(a)   (b)

(c)

Fig. 5.12. (a) Sammon's projection of phoneme strings. **(b)** Initialization of the SOM from Sammon's projection. **(c)** The reference strings of the SOM after 10 batch rounds

Estimator Map. Consider first a parametric operator G_i associated with cell i; let w_i be its tuning parameter vector. Let $X_{t-1} = \{x(t-n), x(t-n+1), \ldots, x(t-1)\}$ be a finite sequence of eventually vectorial stochastic input samples available up to time $t-1$. Assume first in the simplest case that G_i is an *estimator* that defines the *prediction* $\hat{x}_i(t) = G_i(X_{t-1})$ of the signal $x(t)$ in unit i at time t on the basis of X_{t-1}. Each cell shall make its own prediction; let the prediction error $x(t) - \hat{x}_i(t)$ define the degree of matching at unit i. Examples of such operators are: the recursive expression that defines the *autoregressive (AR)* process, the generally known *Kalman filter* [5.21] for signals corrupted by time-variable colored noise, etc.

Denote the best-matching unit ("winner") by subscript c, and let it be defined by

$$||x(t) - \hat{x}_c(t)|| = \min_i \{||x(t) - \hat{x}_i(t)||\} . \tag{5.9}$$

In analogy with the earlier discussion of the SOM, define the strength of control exercised by the best-matching cell c on an arbitrary cell i during learning in terms of the *neighborhood function* h_{ci}. Then, *the average expected squared prediction error that is locally weighted with respect to the winner*, may be defined by the functional

$$E = \int \sum_i h_{ci} ||x - \hat{x}_i||^2 p(X) dX \ , \tag{5.10}$$

where $p(X)$ means the joint probability density function of the whole input sequence over which the estimator is formed, and dX is a "volume differential" in the space in which X is defined (this notation also means that the Jacobi functional determinant is eventually taken into account when needed). It is also understood that the expectation value is taken over an infinite sequence of stochastic samples $x = x(t)$.

By the same token as when deriving the original SOM algorithm, we shall resort to the stochastic approximation for the minimization of E, i.e., to find optimal parameter values. The basic idea is again to try to decrease the *sample function* $E_1(t)$ of E at each new step t by descending in the direction of the negative gradient of $E_1(t)$ with respect to the present parameter vector w_i. The sample function of E reads

$$E_1(t) = \sum_i h_{ci} ||x - \hat{x}_i||^2 \ , \tag{5.11}$$

where the unknown parameter vectors w_i occur in both the estimates $\hat{x}_i$ and the subscript c. The recursive formula for the parameter vector w_i reads

$$w_i(t+1) = w_i(t) - (1/2)\alpha(t)\frac{\partial E_1(t)}{\partial w_i(t)} \ , \tag{5.12}$$

whereby we assume that c is not changed in a gradient step (because the correction of the "winner" is towards input) and $\alpha(t)$ is a scalar that defines the size of the step. In this discussion, too, $\alpha(t)$ $(0 < \alpha(t) < 1)$ should be decreasing in time and satisfy the usual restrictions relating to stochastic approximation.

Example [5.5]: Let the prediction of a scalar variable $x(t) \in \Re$ be defined by

$$\hat{x}_i(t) = \sum_{k=1}^n b_{ik}(t)x(t-k) \ , \tag{5.13}$$

which describes an AR process. The coefficients $b_{ik}(t)$ are also called *linear predictive coding (LPC) coefficients*. This estimator has been used widely, e.g., in speech synthesis and recognition, because the time-domain signal values can thereby be applied without prior frequency analysis. The LPC coefficients will be "learned" adaptively upon application of stochastic training sequences, whereby

$$\frac{dE_1(t)}{db_{ik}} = -2h_{ck}[x(t) - \hat{x}_i(t)] \cdot \frac{\partial \hat{x}_i(t)}{\partial b_{ik}(t)}$$

$$= -2h_{ck}[x(t) - \sum_{j=1}^{n} b_{ij}(t)x(t-j)]x(t-k) , \qquad (5.14)$$

where the summation index has been changed for clarity, and

$$b_{ik}(t+1) = b_{ik}(t) + \alpha(t)h_{ck}[x(t) - \hat{x}_i(t)]x(t-k) . \qquad (5.15)$$

More General Filter Map. If the degree of "tuning" of the different G_i to X is defined in terms of the magnitudes of the *output responses* $\eta_i = G_i(X)$, then, first of all, the G_i must somehow be kept normalized all the time to facilitate their comparison. For instance, one might stipulate that the energy

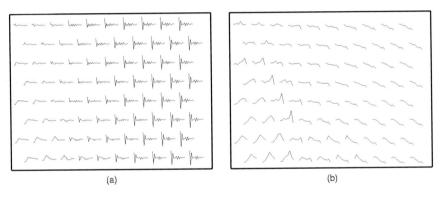

(a)   (b)

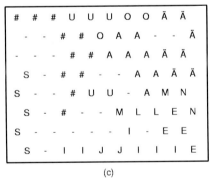

(c)

Fig. 5.13. The 16th-order LPC coefficient vectors have been organized in this SOM, and shown graphically at each map unit position. **(a)** The ordinate values of the plots drawn at each neuron position of the map represent the LPC coefficients are shown in each map position: b_{i1} on the left, $b_{i,16}$ on the right. **(b)** Frequency responses computed from the LPC coefficients are shown as a plot in each map position: low frequencies (300 Hz) on the left, high frequencies (8 kHz) on the right. **(c)** Labeling of the map units by phonemic symbols. (# here means all the weak phonemes such as /k/, /p/, and /t/)

of the impulse response of G_i must be the same for all cells. After that, the parameters of G_i can be varied under the normalization restriction, eventually ending up with a learning law like (5.15).

Application: Phoneme Map. The LPC coefficients have often been used for the extraction of features from the speech signal, because they contain the same information as the amplitude spectra, but are computationally lighter than the FFT. For this reason we will demonstrate in the following what kind of phoneme maps are created when the *time-domain signal samples* are directly mapped onto the LPC-coefficient SOM.

The input data were discrete-time sequences of samples from natural Finnish speech, picked up from the stationarity regions of the phonemes. The sampling frequency was 16 kHz, and the order of the LPC process was $n = 16$. No attempts were made to optimize either.

The ordering result has been illustrated in three different ways in Figs. 5.13a, 5.13b, and 5.13c. It has to be emphasized that this is still a provisional example only; carefully constructed phoneme maps formed of, e.g., amplitude spectra have been more accurate. This example has mainly been meant to illustrate that the extension of the original SOM idea is possible and to prove that the operator maps can learn.

5.9 Evolutionary-Learning SOM

5.9.1 Evolutionary-Learning Filters

The SOM philosophy can actually be much more general than discussed so far. For instance, even when the operators of Sect. 5.8 are not parametric, or the parameters are not well-defined or directly observable or controllable, it is possible to run the same input sequences $\{x(t)\}$ through each filter and compare either the prediction errors or eventually the *integrated responses*. But how can we make such cells learn, to develop an ordered SOM?

The solution to this problem might be a kind of "evolution" or "natural choice" that resembles the genetic algorithms. Let us again denote the operator associated with cell i in the array by G_i. Denote the input information, eventually a sequence, by X. The "winner" with respect to X is now denoted by G_c. Assume that new candidates or versions for the operators can be generated easily: for instance, if they belong to some category that is characterizable in some stochastic way, by a random choice one can generate a new candidate G. However, it is also possible to define a mixture of categories in each of which the operators have a different structure (e.g., different order of filters). During one "learning" step, the same input process X that was operated by G_c can also be operated by one or a number of tentative G, and among these tests one eventually finds the operator G_b that is best or at least better than G_c. Then G_c is replaced by G_b with a probability P, where P corresponds to the learning-rate factor $\alpha(t)h_{cc}$ in the usual SOM models. Moreover, operators in the neighborhood around cell c will similarly be replaced by G_b with the probability $\alpha(t)h_{ci}$. These replacements should be statistically independent, i.e., the decisions concerning replacements in the neighboring cells must drawn independently, but with the given probabilities.

5.9.2 Self-Organization According to a Fitness Function

Although no distance function over the input data or the models is definable, it is still possible to implement the SOM process using evolutionary-learning operations. The process can be made to converge more rapidly when the probabilistic trials of conventional evolutionary learning are replaced by averaging using the Batch Map version of the SOM. Although no other condition or metric than a *fitness function* between the input samples and the models is assumed, an order in the map that complies with the "functional similarity" of the models can be seen to emerge. Any two neighboring models, say, M_1 and M_2 are regarded as *functionally similar* if their fitness-function values $f(X, M_1)$ and $f(X, M_2)$ with respect to the same X input, which selects either one of them for the winner, are approximately the same. If such an order over the array can be produced, it will become possible to find functionally similar models around each location. The following procedure produces such a "local functional order."

Modification of the models is easier if they can be defined by a set of parameters, numerical or other (e.g., codes). However, we can start with the general formulation of the process in which even the parametrization of the models is not necessary. The minimum requirement is that there exists some set, open or closed, of possible models.

It may be advisable to refer to Fig. 3.1 in the beginning of Chapter 3. Fast evolutionary learning can be based on the batch-type SOM and is defined in the general form by the following steps:

1. Initialize the models M_i, e.g., by a random choice from a set of possible models.
2. Input a number of X items and list each of them under the respective winner unit (i.e. that M_i for which some fitness function $f(X, M_i)$ is maximum). In case there is a tie, i.e., two or more M_i have the same fitness to X, select one of them randomly for the effective unique "winner" under which the listing is made.
3. Find a new value M_i' for each M_i such that if U_i is the union of lists relating to model M_i in the same way as in the Batch Map algorithm, the sum of the fitness-function values $f(X, M_i')$, $X \subset U_i$ is increased.
4. Repeat from step 2.

The addends in the sum of fitness values can be weighted by the neighborhood function h_{ij} centered at model M_i.

The modification of M_i at step 3 is a "natural choice" process, and if the models form a closed set and there is sufficient computing capacity available, one may perform an exhaustive search over all the possible models, in order to find the model for which the sum of the $f(X, M_i')$ in U_i is maximum. If the set is open or very large so than the exhaustive search is not possible, a large number of random trials can be made. The replacement of the M_i by the M_i' must be made as the so-called two-rank operation, first computing the candidate for each M_i' on the basis of the old lists, and then making the substitution of all models in the array in one operation.

In case the models can be parametrized, we may try genetic-algorithmtype operations. The parameters of model M_i' may be varied one at a time ("mutation"), or a whole subset of parameters can be exchanged with another possible model ("crossover").

A Very Simple Example. We shall clarify the fast evolutionary-learning algorithm using a very simple example. Since the latter is based totally on fiction, it need not comply with natural conditions in any aspects.

Let us consider twelve possible fictive animals that are defined by three parameters describing their size, type of skin, and type of legs, respectively. Let there be only two choices for the size, "small" and "big," respectively. Let the skin be "hairless" or "haired." Let there be three choices for the legs, "short," "long," and "legs with webfeet," respectively.

Since this is a fictive example, we could further assume for simplicity that any of the types of animals can freely mate for crossing.

Let these fictive animals live in a mixture of environments where one can distinguish three types of territories: "fen" (that is, very wet moss), "moss" (not so wet), and "moor." These territories shall exist in either summer or winter conditions. In the latter case everything is frozen and eventually covered with snow. Since the example is fictive, we need not consider the intermediate seasons of autumn and spring.

Next we assume that the fitness of each of the twelve possible species to the different environmental conditions can be tabulated. In doing so we do not consider that these "animals" might eat each other: they are vegetarians. However, if we assign a model M_i to each animal, we must assume in this simulation that the number of animals, or the number of models is fixed. Again, this assumption is somewhat artificial.

The *total fitness* of an animal to the environmental conditions shall be described by three factors F_1, F_2, and F_3. Here F_1 is the "fitness of size," F_2 is the "fitness of skin," and F_3 is the "fitness of legs," respectively. The total fitness F we express in the model as the product $F = F_1 \cdot F_2 \cdot F_3$. We have chosen the product form here, because it emphasizes the fact that if any of the partial fitness factors has a very low value, the situation will be fatal to the survival of the animal. Let Table 5.1 define F_1, F_2 and F_3 in a scale of 0 to 10.

Table 5.1. Fitness factors for fictive animals

		F_1 Size		F_2 Skin			F_3 Legs	
		Small	Big	Hairless	Hairy	Short	Long	Webfeet
Parameter values		(0)	(1)	(0)	(1)	(0)	(1)	(2)
	Fen	10	1	10	5	1	1	10
Summer	Moss	10	5	10	5	3	5	10
	Moor	10	5	10	5	10	10	3
	Fen	10	5	2	10	10	10	3
Winter	Moss	10	5	2	10	10	10	3
	Moor	10	5	2	10	10	10	3

Examples of parametric coding of animals:
"Small hairy animal with long legs"   =   011
"Small hairless animal with webfeet"   =   002

We have selected the fitness values based on intuition and fixed them to arbitrary values in this fictive example. We have punished the big size, because bigger animals need more resources. The webfeet are very good on "fen" and "moss" in summer conditions, especially since other animals are not interfering, and somewhat worse in other conditions, hence the "10s" and "3s" in the last column. All the other values may be self-explanatory.

The simulation program consists of the following steps:

1. Initialize the models M_i using random choices for the parameters.
2. Select the environment (one of the six rows in Table 5.1) randomly with a probability that can be controlled according to the assumed relative length of the winter, and fraction of "fen," "moss," and "moor." The selected environment is identified with X.
3. List X under that SOM location that corresponds to the maximum value of $F = F_1 \cdot F_2 \cdot F_3$.
4. Repeat steps 2 and 3 an appreciable number of times.
5. For each model M_i, generate variations, by randomly replacing parameters of M_i with those of the models in the closest neighborhood of M_i in the SOM array. Compute the total fitness values for each variation, relating each of them to all X in the union of lists, and find that variation whose sum of total fitness values with respect to all X in the union of lists is maximum.
6. Replace M_i by the best-fitting variation.
7. Repeat from step 2 an indefinite number of times.

A five-by-five rectangular SOM array was used. In its initialization, the parameters of a random choice from the twelve possible "animals" in Table 5.1 were associated with each SOM location. With such a small array dimensionality it is possible to restrict the neighborhood set N_i to the immediately adjacent neighbors of unit i in the horizontal, vertical, and diagonal directions, and unit i shall belong to N_i, too.

Selection of any of the three territories "fen," "moss," and "moor" was made randomly with equal probability, whereas the length of "winter" vs. "summer" was varied. In total 1000 inputs X were applied. In the beginning of the process the probability for exchanging a parameter was set equal to .5, and this value decreased to zero linearly with the number of iterations used in the Batch Map principle.

Figure 5.14 illustrates the results after a certain number of iterations in the evolutionary process. The three-digit codes written to the map locations represent codes of the animal types as defined in Table 5.1.

(a)

002	002	002	002	002
002	002	002	002	002
002	002	002	002	001
001	001	001	001	001
001	001	001	001	001

(b)

002	011	011	011	010
011	011	011	011	011
011	011	011	010	010
011	011	011	011	011
011	011	011	011	011

(c)

010	010	010	010	010
010	010	010	010	010
010	011	010	010	010
011	011	010	011	011
011	011	011	011	011

Fig. 5.14. The models in the map array (their three-digit codes referring to Table 5.1) after having been computed in ten iterations. **(a)** Length of winter zero, **(b)** Length of winter 6 months, **(c)** Length of winter 12 months

5.10 Supervised SOM

It is generally held that the SOM is formed in an *unsupervised* process, i.e., like classical clustering methods are traditionally regarded as *unsupervised classification*. As we originally tried to use the SOM for statistical pattern recognition tasks (recognition of phonemes from natural speech), it turned out that the class-separation of codebook vectors, and thus also the classification accuracy could be improved by a significant amount if information about the class-identity could be taken into account in the learning phase. This idea led to the so-called *supervised SOM* idea that was applied to speech recognition, where it already, around 1984, yielded an almost as good accuracy as the Learning Vector Quantization methods to be discussed in the next chapter yield nowadays [5.22].

The above "neural" pattern recognition principle was applied to our speech recognition system up to 1986 [5.23]. In order to make the SOM supervised, the input vectors were formed of two parts x_s and x_u, where x_s was a 15-component short-time acoustic spectrum vector computed over 10 milliseconds, and x_u corresponded to a unit vector with its components assigned to one of the 19 phonemic classes that were taken into account. (In reality, during training, a value of .1 was used for each "1" position of the "unit" vector. During recognition, x_u was not considered.) The concatenated 34-dimensional vectors $x = [x_s^T, x_u^T]^T$ were then used as inputs to the SOM. Notice that since x_u is the same for vectors of the same class but different for different classes, clustering of the vectors x along with the classes is enhanced, leading to improved class-separation. The weight vectors $m_i \in \Re^{34}$ then also tended to approximate to the density of the concatenated x, not of the signals x_s.

Supervised learning here means that whereas the classification of each x_s in the training set is known, the corresponding x_u value must be used during training. During recognition of an unknown x, only its x_s part is compared with the corresponding part of the weight vectors.

The unsupervised SOM constructs a topology-preserving representation of the statistical distribution of all input data. The supervised SOM tunes this representation to discriminate better between pattern classes. The weight vectors of differently labeled cells in the x_u part define decision borders between classes. *This special supervised training was used in our original speech recognition system known as the "Phonetic Typewriter" [5.23], and it indeed made use of topological ordering of the SOM, contrary to the newer architectures that are based on Learning Vector Quantization.*

5.11 The Adaptive-Subspace SOM (ASSOM)

5.11.1 The Problem of Invariant Features

A long-standing problem in the theory of perception has been *invariance* of sensory experiences. For instance, we can create a steady visual perception of an object, in spite of its image on our retinas being in motion.

But notice that the image of a moving object is in motion on the cortex, too, because the visual signals are mapped to brain areas *retinotopically*, that is, preserving their geometric relations.

It may be clear that the traditional neural-network models do not tolerate this kind of displacements of the input patterns, because they are directly matching signal patterns with synaptic patterns ("template matching").

In practical object recognition applications, the conventional solution for the achievement of invariances in perception, such as the invariance with respect to movements of the objects, is to provide a simple classifier with a heuristically designed *preprocessing stage* that extracts a set of *invariant features* from the primary signals. In contemporary pattern recognition, certain *local features* such as pieces of sinusoidal waveforms called the *wavelets* have become popular as invariant features. The wavelets are usually found by transformation techniques, as will be discussed in Sect. 5.11.2 below. Classification, for instance by neural networks, is then based on these features. The results are invariant in spite of the observations being subject to certain elementary transformations, of which their *translation* (in space or time) is the most important one. Invariances with respect to rotations, scalings, and different illuminations of the objects can be achieved by other transformations, to a limited extent at least. The analytical, mathematical forms of these filters have so far been postulated, and only their parameters have been adjusted.

We now have to forestall eventual misinterpretations by telling already at this point that such feature-filter based invariant matching is only possible for certain elementary features, and each filter can be invariant to one transformation group only. Then we should take into account threefacts: 1. At a certain accuracy, an arbitrary pattern can be decomposed into a very large, although finite number of *different kinds* of elementary features. 2. The *mixture* of the different features, i.e., different transformations can be fitted to the occurring images or other signals optimally. 3. There can be a mixture of different types of filters for different features, and this mixture can be different in different parts of, say, the field of vision.

The Self-Organizing Map has also been called the "Self-Organizing Feature Map," and many attempts have been made to use it for the optimal extraction of features from primary signals. However, although the map units often become sensitive to some kind of elementary patterns, these cannot yet be regarded as *invariant* features.

We shall show below that there exists a special kind of SOM, called the *Adaptive-Subspace SOM (ASSOM)*, in which the various map units adaptively develop into filters of many basic invariant features. *The mathematical forms of these filters need not be fixed a priori; the filters and their mixture find their forms automatically in response to typical transformations that occur in observations.* The ASSOM principle was introduced for the first time in the first edition of this book in 1995, and some improvements to it were suggested in [5.24–26].

If, instead of referring to weight vectors of neurons as templates for patterns, a "neural" unit of the SOM is made to represent a *manifold* such as a *linear subspace*, the SOM units will then be able to match certain elementary patterns *under some of their basic transformations* such as translation, rotation, and scaling.

Let us recall from Sect. 1.1.1 that a linear subspace is a manifold, defined as the general linear combination of its *basis vectors*. We shall show below that a new kind of SOM, the ASSOM, can be constructed of neural units that represent linear subspaces, and invariant feature detectors for various wavelets will emerge at the map units during learning. This discussion is conceptually already more sophisticated and the underlying details more subtle than of any of the SOM variants discussed so far.

The ASSOM is related to the subspace methods of classification, discussed in Sect. 1.4.

If the basis vectors of the subspaces are chosen in a suitable way, for instance, as certain kinds of elementary patterns (representations of features), the different linear subspaces can be made to represent different *invariance groups* of these elementary patterns. As already mentioned, we shall further show that the basis vectors can be learned completely automatically, without assumption of any mathematical forms for them, in response to pieces of sequences of input signals that contain these transformations. In the ASSOM, every map unit is made to represent *its own subspace* with a small dimensionality, say, two, and the basis vectors are determined adaptively, following the idea of the Adaptive-Subspace Theorem discussed in Sect. 1.4.2. The fundamental difference with respect to the latter is that we now have *many* subspaces instead of one that *compete* on the same input; the "winner subspace" and its neighbors will learn the present sequence, whereas other sequences may be learned by other "winners" at other times.

As emphasized before, another fundamental difference between the basic SOM and the ASSOM algorithm is thus that a map unit is not described by a single weight vector, but *the unit is supposed to represent a linear subspace spanned by the adaptive basic vectors.* Matching thereby does not mean comparison of dot products or Euclidean differences of the input and weight vectors, but *comparison of orthogonal projections of the input vector on the different subspaces represented by the different map units. Accordingly, the learning step must modify all the basis vectors that define the selected subspaces.*

The most essential function in the ASSOM, however, is *competitive learning of episodes*, whereupon it is not a pattern but a *sequence of patterns* on which the map units compete, and to which they are adapting. This scheme does not occur in any other neural-network model.

The ASSOM algorithm may also be thought to differ from all the other neural-network algorithms in the respect that it does not learn particular patterns, but rather transformation kernels.

5.11.2 Relation Between Invariant Features and Linear Subspaces

Wavelet and Gabor Transforms. In the recognition of acoustic signals and images, the *wavelet transforms* and in particular the *Gabor transforms* are gaining importance in preprocessing, in the definition of translation-invariant feature filters.

In order to understand what the ASSOM is actually doing, it will first be necessary to relate these transforms to the linear-subspace formalism.

Consider first a scalar-valued time function $f(t)$, such as an acoustic waveform described in the time domain. It can be expanded in terms of basis functions called the *self-similar wavelets*. Like in Fourier transformation, there exist cosine and sine wavelets, denoted by ψ_c and ψ_s, respectively; if they are centered around instant t_0, they are of the form

$$
\begin{aligned}
\psi_c(t - t_0, \omega) &= e^{-\frac{\omega^2(t-t_0)^2}{2\sigma^2}} \cos(\omega(t - t_0)) , \\
\psi_s(t - t_0, \omega) &= e^{-\frac{\omega^2(t-t_0)^2}{2\sigma^2}} \sin(\omega(t - t_0)) .
\end{aligned}
\tag{5.16}
$$

The *wavelet transforms* of $f(t)$ evaluated at t are defined as

$$
\begin{aligned}
F_c(t, \omega) &= \int_\tau \psi_c(t - \tau, \omega) f(\tau) d\tau , \\
F_s(t, \omega) &= \int_\tau \psi_s(t - \tau, \omega) f(\tau) d\tau .
\end{aligned}
\tag{5.17}
$$

The square of the *wavelet amplitude transform* A is

$$
A^2 = F_c^2(t, \omega) + F_s^2(t, \omega) .
\tag{5.18}
$$

It will be easy to show that the sinusoidal oscillation is eliminated from A^2, and this expression is then almost invariant to shifts of $f(x)$ in the time scale, provided that these shifts are at most of the order of magnitude of σ/ω.

A set of wavelet amplitude transforms A, taken for different ω and σ, can then serve as a *feature vector that has an approximate invariance with respect to limited time shifts*.

In image analysis, the two-dimensional wavelets are called the *Gabor functions* [2.56, 5.27, 28], and the corresponding translationally invariant features are the *Gabor amplitude transforms*. In the following we denote the image intensity at location r by $I(r)$.

It is possible to express the cosine and and sine kernels (5.16) and transforms (5.17) as functions of a complex variable.

For instance, the two-dimensional (unnormalized) cosine and sine wavelets can be combined into the *complex Gabor functions*

$$
\psi(r, k) = \exp[i(k \cdot r) - (k \cdot k)(r \cdot r)/2\sigma^2] .
\tag{5.19}
$$

The *complex Gabor transform* can then be written

$$
G = G(r_0, k) = \int_{r \text{ space}} \psi(r - r_0, k) I(r) dr ,
\tag{5.20}
$$

where the integration is carried out over the image points located by their position vectors r. The modulus of G, or $|G(r_0, k)|$ has then an approximate invariance with respect to limited shifts of the image.

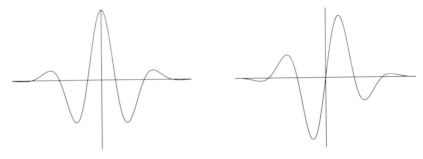

Fig. 5.15. Example of the one-dimensional cosine and sine wavelet, respectively, with $\sigma = 1$; arbitrary scales

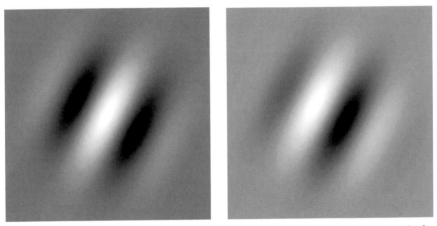

Fig. 5.16. Examples of Gabor functions of the cosine and sine type, respectively, shown in a gray scale

Fig. 5.15 exemplifies the one-dimensional cosine and sine wavelets, and Fig. 5.16 the corresponding two-dimensional wavelets, respectively.

We shall now show that the invariance property of the wavelet and Gabor transforms can also be expressed by the *subspace formalism*.

Wavelets as a Linearly Dependent Set. In this subsection we shall show in what ways the wavelets can satisfy linear constraints, or form linear subspaces.

Consider a scalar signal $f(t)$. The *sample vector* of it may be defined as

$$x = [f(t_k), f(t_{k+1}), \ldots, f(t_{k+n-1})]^{\mathrm{T}} \in \Re^n , \tag{5.21}$$

where for simplicity the $t_k \ldots t_{k+n-1}$ are regarded here as equidistant discrete-time coordinates.

Consider then tentatively that we had selected $f(t_k) = A \sin \omega t_k = S_k$, where A and ω are some constants, and let the C_k below denote samples of the corresponding cosine function. Using the familiar formulae for trigonometric functions of sum angles we directly find out that

$$S_{k+p} = \alpha_1 S_k + \alpha_2 C_k , \qquad (5.22)$$
$$C_{k+p} = \alpha_1 C_k - \alpha_2 S_k ,$$

where $\alpha_1 = \cos \omega(t_{k+p} - t_k)$ and $\alpha_2 = \sin \omega(t_{k+p} - t_k)$. Here α_1 and α_2 are just two constants that may take arbitrary values. Notice that $t_{k+p} - t_k$ is a constant displacement of p sampling intervals, but in principle it need not even be a multiple of the interval.

If the sides of the first equation (5.22) are multiplied by some arbitrary scalar a_0 and similar equations are expressed for $p + 1$, $p + 2$, etc. and for different multipliers a_i, the corresponding vector equation becomes

$$\begin{bmatrix} a_0 S_{k+p} \\ a_1 S_{k+p+1} \\ \vdots \\ a_n S_{k+p+n-1} \end{bmatrix} = \alpha_1 \begin{bmatrix} a_0 S_k \\ a_1 S_{k+1} \\ \vdots \\ a_n S_{k+n-1} \end{bmatrix} + \alpha_2 \begin{bmatrix} a_0 C_k \\ a_1 C_{k+1} \\ \vdots \\ a_n C_{k+n-1} \end{bmatrix} \qquad (5.23)$$

$$= \alpha_1 b_1 + \alpha_2 b_2 .$$

Notice that the set of multipliers $[a_0, a_1, \ldots, a_r]$ may correspond to the Gaussian envelope like in (5.16), but (5.23) allows even more general envelopes.

The two column vectors b_1 and $b_2 \in \Re^n$ on the right-hand side of (5.23) can be interpreted as *the two basis vectors of some two-dimensional subspace* $\mathcal{L} \subset \Re^n$. Their components $a_i S_{k+i}$ and $a_i C_{k+i}$, respectively, can be regarded as *samples of the sine and cosine wavelet with general amplitude modulation.* If the scalar constants α_1 and α_2 take on all possible values, the basis vectors b_1 and b_2 can be seen to span *a linear subspace of similar wavelets with arbitrary phase,* whereby all wavelets of this subspace are mutually linearly dependent.

Related linear dependencies can be expressed for many other pattern vectors, for instance for two-dimensional patterns represented as column vectors. They are defined in the following way. Assume that $I(r)$ is the scalar intensity of the pattern element at pixel $r \in \Re^2$, whereby the input samples form the pattern vector

$$x = [I(r_k), I(r_{k+1}), \ldots, I(r_{k+n-1})]^{\mathrm{T}} \in \Re^n . \qquad (5.24)$$

Here the $r_k, \ldots, r_{k+n-1}$ usually form a two-dimensional *sampling lattice.* The intensity pattern $I(r)$ can be decomposed into two-dimensional wavelets in different ways.

For instance, assume tentatively that $I(r) = A \sin (k \cdot r)$, where $k \in \Re^2$ is the two-dimensional *wave-number vector,* A is a scalar amplitude, and we define the samples of the sine and cosine functions as $S_k = I(r_k) = A \sin (k \cdot r_k)$ and $C_k = A \cos (k \cdot r_k)$, respectively. Then it holds for an arbitrary pixel at $r_{k'}$ that

$$S_{k'} = \alpha_1 S_k + \alpha_2 C_k \, , \tag{5.25}$$

with $\alpha_1 = \cos\left[k_0(r - r_0)\right]$ and $\alpha_2 = \sin\left[k_0(r - r_0)\right]$. The two-dimensional vectorial linear-constraint equation is similar to (5.23).

5.11.3 The ASSOM Algorithm

The various operations discussed in this section constitute the complete *Adaptive-Subspace SOM (ASSOM)* algorithm.

Let us recall that the purpose of the ASSOM is to learn a number of various invariant features, usually pieces of elementary one- or two-dimensional waveforms with different frequencies called the "wavelets," independent of their phase. These waveforms will be represented and analyzed by special *filters*, resembling the Gabor filters, and phase-invariant filtering is usually implemented by summing up the squares of the outputs of two associated orthogonal linear filters. This pair of linear filters carries out convolution-type integral transforms, and the kernels of the transformation pair constitute matched filters to the "wavelets," one of them usually corresponding to the cosine transform, and the other to the sine transform, respectively. In principle it is also possible to use more than two orthogonal filters to describe a wavelet, but for simplicity we shall restrict to filter pairs in this discussion.

The formalism used to implement these filters adaptively is that of the *learning-subspace classifier* discussed in Sect. 1.4.3. The basis vectors of the subspaces will be learned from input data and they then correspond to the transformation kernels; since we are mainly interested in the cosine- and sine-type transformation, the number of orthogonal basis vectors in all the subspace classifiers discussed in this section will also be restricted to two.

The subspaces that represent these filters are associated with the neural units of a special SOM architecture, they are made to compete on the same inputs during learning; in this way they learn how they shall mutually partition the input signal space. *The Adaptive-Subspace (AS) Theorem* (Theorem 1.4.1) derived in Sect. 1.4.2 already described how a single "neural" subspace can be made to converge to a subspace spanned by a small number of adjacent input signal samples, i.e., to become a subset of the latter. If there are several "neural" subspaces, they will thus competitively partition the signal space into a number of subspaces, each one describing its own signal manifold (wavelet set).

In the concrete examples discussed below in this section the subspaces, as mentioned before, are assumed as two-dimensional, in order to compare the results with, say, the two-dimensional orthogonal wavelet pairs. This restriction, however, is not essential; many other interesting sets of nonorthogonal basis vectors and subspaces spanned by them will also be formed by the same algorithm.

A natural observation is usually a mixture of signal components or elementary waveforms. Some of the latter may satisfy certain simple linear constraints, while the rest does not. The ASSOM will be adapted to those signal components that satisfy the *linear* constraints, whereas the rest of the components will be regarded as noise and smoothed out.

"Episode" and "Representative Winner". An essential new idea we need in the sequel is the concept of *a "representative winner" for a set of subsequent sequences, i.e., a set of $x(t)$ vectors that occur adjacent in time*. In a traditional SOM we can, of course, define the usual "winner" node $c = c(t)$ for every $x(t)$, but what we actually want a particular map unit and its neighborhood in the ASSOM array to do is *to learn the general linear combination of adjacent sequences $x(t)$*. If a particular neural unit has to become sensitive *to a whole set of temporally adjacent sequences independently of their phases (or shift in the time scale)*, i.e., to a subset of input vectors, then obviously *the "winner" must be defined for that whole subset of temporally adjacent sequences of samples*, which we shall call the *episode*. The winner shall be fixed for the whole episode during a learning step. All the adjacent $x(t)$ will then be learned by the "representative winner" unit and its neighboring units in the array during this step.

It may help to understand this idea if we consider that during a short *episode $\mathcal{S}$*, equivalent with a set of successive sampling instants $\{t_p\}$, we have collected a set of input vectors $\{x(t_p)\}$, each member of which represents a sample of a sequence and all these sequences are linearly dependent. (There may also exist other, linearly independent components in the input, but they will be regarded as noise and smoothed out in the learning process.) In particular, if $\mathcal{S}$ is finite and eventually rather small, the $x(t_p)$ during it may be thought to span a *signal subspace $\mathcal{X}$* of a much lower dimensionality than n. Consider also a set of operational units that will be identified with the SOM (ASSOM) units a little later. With each unit we can associate its own subspace $\mathcal{L}^{(i)}$. If we then identify the "winner" subspace $\mathcal{L}^{(c)}$ to which $\mathcal{X}$ is closest, $\mathcal{L}^{(c)}$ may be regarded to approximate $\mathcal{X}$ better than any other $\mathcal{L}^{(i)}$.

The main problem in matching signal subspaces $\mathcal{X}$ with the "neural" subspaces $\mathcal{L}^{(i)}$ is that since the number of samples in the episodes $\mathcal{S}$ can be taken as arbitrary, the *dimensionality* of the subspaces $\mathcal{X}$ spanned by the $x(t_p)$ becomes arbitrarily defined, too. Therefore this author has suggested a simpler and more robust method for the matching of an episode with the "neural" subspaces. Consider the "energy of the projections" of the $x(t_p), t_p \in \mathcal{S}$, on the different $\mathcal{L}^{(i)}$. This "energy" is defined as the sum of squared projections over the episode on each of the $\mathcal{L}^{(i)}$. Then the maximum of these entities shall define the *"representative winner"* (over the episode), denoted by c_r:

$$c_r = \arg \max_i \left\{ \sum_{t_p \in \mathcal{S}} ||\hat{x}^{(i)}(t_p)||^2 \right\}. \tag{5.26}$$

This "representative winner" is fixed for the whole $\mathcal{S}$, and the units at or around index c are made to learn $\mathcal{X}$ as described by the *Adaptive-Subspace Theorem* discussed in Sect. 1.4.2.

There exist several possibilities for the definition of the "representative winner" for a set of adjacent sequences, for instance, the extremum of all projections during the episode:

$$c_r = \arg\max_p \{||\hat{x}_{c_p}(t_p)||\} , \quad \text{or}$$
$$c_r = \arg\min_p \{||\tilde{x}_{c_p}(t_p)||\} . \tag{5.27}$$

Still another definition of the "representative winner" is the majority of matches over the episode. Consider the sequences $x(t_1), x(t_2), \ldots, x(t_k)$ where each vector x is defined as a different set of samples taken from the same signal waveform. Let the corresponding projections on subspace $\mathcal{L}^{(i)}$ be $\hat{x}_i(t_1), \ldots, \hat{x}_i(t_k)$. For each t_p, $p = 1, 2, \ldots, k$ we may now find the "winners" in either of the following ways:

$$c_p = \arg\max_i \{||\hat{x}_i(t_p)||\} , \quad \text{or}$$
$$c_p = \arg\min_i \{||\tilde{x}_i(t_p)||\} . \tag{5.28}$$

Then the "representative winner" c for this whole set of sequences may be defined as

$$c = \text{maj}_p \{c_p\} , \tag{5.29}$$

where maj_p means the majority of symbols over the index p.

All the above definitions have yielded roughly similar ASSOMs in practical experiments. The learning law is independent of the definition of the "representative winner."

To recapitulate, *if a phase-insensitive feature detector has to be formed at a particular node of the SOM and different features shall become represented by different locations of the SOM, then the "representative winner" and the whole neighborhood N_c in which learning takes place ought to be defined in some way for the whole set of adjacent sequences, and this "winner" must be held steady for the set of adjacent sequences during a learning step.*

Stabilization of Learning Rates of the ASSOM. In many traditional learning processes we will not encounter a problem that, however, is present in many projective methods, in particular in the learning subspace methods such as the ASSOM. Notice (Sect.1.4.2) that the basis vectors b of a subspace are updated multiplying them by matrix operators of the form $P = (I + \lambda x x^T)$. But notice too that

$$(I + \lambda x x^T)b = b + \lambda(x^T b)b , \tag{5.30}$$

whereby the rate of adaptation is proportional to the dot product of x and b. This rate becomes very small if x and b are even approximately orthogonal.

We should also notice that in the subspace methods, the projection error $||\tilde{x}^{(i)}||$ is defined by the whole subspace and not by its individual basis vectors. However, even then, the corrections become *smaller* with *increasing* angle between input x and the subspace to be adapted. When this angle approaches $\pi/2$, the corrections become zero, which certainly would be wrong. On the contrary, it might be self-evident that in general *the corrections to be made on neural units shall be monotonically increasing functions of the error.* In other words, we should stipulate that *any correction shall be a monotonically increasing function of* $||\tilde{x}^{(i)}||$ *or a monotonically decreasing function of* $||\hat{x}^{(i)}||$.

The λ factor can now be made a *function* of x and the basis vectors, such that it guarantees monotonic corrections of the $||\tilde{x}^{(i)}||$ or $||\hat{x}^{(i)}||$. One of the simplest methods to warrant this is to multiply the learning-rate factor by $||x||/||\hat{x}^{(i)}||$. Let us denote the new learning-rate parameter by α; then the projection operator that guarantees smooth learning in the above sense is

$$R = \left(I + \alpha \frac{xx^{\mathrm{T}}}{||\hat{x}^{(i)}|| \, ||x||} \right) . \tag{5.31}$$

Even if x is orthogonal to all the basis vectors $b_h^{(i)}$ of the subspace, whereby $||\hat{x}^{(i)}|| = 0$, R is nonzero, as wanted.

With $\alpha > 0$, the rotation is always toward x. Notice that the complete scalar factor that multiplies xx^{T} still just defines an effective learning rate.

Competitive Learning of the Episodes. The Self-Organizing Map (SOM) is in general a regular (usually one or two dimensional) array of neural cells or nodes, and every cell (in the simplest case) receives the same input. The learning rule in any SOM architecture consists of the following steps:

1. Locate the cell ("winner"), the parametric representation of which matches best (in the sense of some metric criterion) with the representation of input data.
2. Change the parameters of the winner and its neighbors in the array to improve matching with input data.

The *Adaptive-Subspace SOM (ASSOM)* also relates to an array of "neural" units, but each unit in it describes a subspace. Such a unit may be composed physically of several neurons, as we have seen in Fig. 1.9. Following the subspace classifier principle, the ASSOM algorithm is defined in detail as:

1. Locate the unit ("representative winner"), on which the "projection energy" $\sum_{t_p \in \mathcal{S}} ||\hat{x}^{(i)}(t_p)||^2$ is maximum.
2. Rotate the basis vectors $b_h^{(i)}$ of the winner unit and units in its neighborhood in the neural-unit array.

The basis vectors $b_h^{(i)}$ of unit i are rotated for each sample vector $x(t_p)$ of the episode $\mathcal{S}$ as:

$$b'^{(i)}_h = \prod_{t_p \in \mathcal{S}} \left[I + \alpha(t_p) \frac{x(t_p)x^{\mathrm{T}}(t_p)}{||\hat{x}^{(i)}(t_p)|| \, ||x(t_p)||} \right] b_h^{(i)}(t_p) . \tag{5.32}$$

This "rotation" must be restricted to the units $i \in N^{(c_r)}$, where $N^{(c_r)}$ is the *neighborhood* of the "representative winner" in the ASSOM network.

The samples $x(t)$ should preferably be normalized prior to application of (5.32); this can be made for each sample separately, or for some larger set of samples (i.e., dividing by the average norm). As mentioned earlier, below we shall always take $h \in \{1, 2\}$ corresponding to two-dimensional sub-spaces. In principle the $b_h^{(i)}$ need not be orthogonalized or normalized, but orthonormalization, even after some tens or hundreds of learning periods, is recommendable for the stabilization of the process. Also, if we want to iden-tify the $b_h^{(i)}$ with the pairs of wavelets or Gabor filters discussed next, then orthogonality of these components would be desirable. Henceforth we shall always give the basis vectors obtained in this process in the orthonormalized form.

Comment. There now seems to exist a paradox that whereas the "repre-sentative winner" was already determined by one set of samples that formed the episode, these samples are no longer available for learning at a late time. This paradox, however, is more apparent than real. In many practical cases, e.g., when analyzing natural signal waveforms, the correlation of having the same "representative winner" during the closest subsequent episodes is high, at least when the waveforms are changing slowly. Thus one could use one period of time for determining the representative "winner" and the next one for learning, respectively. In technological and biological systems there might exist some kinds of short-term memories for the *buffering* of the training samples, to be used for learning after determination of the "representative winner." *In the simulations reported below, learning was simply based on the same set of samples that was used for locating the "representative winner" and the corresponding neighborhood $N^{(c_r)}$ in the array.*

5.11.4 Derivation of the ASSOM Algorithm by Stochastic Approximation

We shall now derive the ASSOM algorithm on the basis of *the "energy" of the relative projection errors* of the input samples.

Assume first that no neighborhood interactions between the modules are taken into account, and the input vectors are normalized. The average dis-tance between the input subspace and the subspace of the winning module, computed over the space of all possible Cartesian products X of input sam-ples collected during the episodes $\mathcal{S}$, is expressed in terms of the average expected value of the relative projection error, as the objective function

$$E = \int \sum_{t_p \in \mathcal{S}} \|\tilde{x}^{(c_r)}(t_p)\|^2 p(X) dX \ . \tag{5.33}$$

Here $p(X)$ is the probability density of the X, and dX is a shorthand notation for a volume differential in the Cartesian product space of all the samples of the episode. The index of the winning module c_r, "representative winner," depends on the whole episode, as well as on the subspaces $\mathcal{L}^{(i)}$ of all modules i.

Exact minimization of (5.33) by traditional optimization methods may be a very complicated task. Therefore, as in the derivation of the SOM algorithm, we shall again resort to the *Robbins-Monro stochastic approximation* (Sect. 1.3.3) in minimizing (5.33). It has to be emphasized that the objective function thereby need not be an energy function; on the other hand, the optimum is then only computed approximately.

In order to create an *ordered* set of subspaces in analogy with the basic SOM, the integrand in E can further be multiplied by the *neighborhood kernel* $h_{c_r}^{(i)}$, which is a decreasing function of the distance between the modules c_r and i in the ASSOM array, and summed up over all modules i of the array. The new objective function then reads

$$E_1 = \int \sum_i h_{c_r}^{(i)} \sum_{t_p \in \mathcal{S}} \|\tilde{x}^{(i)}(t_p)\|^2 p(X) dX \ . \tag{5.34}$$

In stochastic approximation, the gradient of E_1 is approximated by the gradient of the *sample function*, which now reads

$$E_2(t) = \sum_i h_{c_r}^{(i)} \sum_{t_p \in \mathcal{S}(t)} \|\tilde{x}^{(i)}(t_p)\|^2 \ . \tag{5.35}$$

Consider that the set of values $\{x(t_p)\}$ during the whole episode $\mathcal{S} = \mathcal{S}(t)$ can be regarded to constitute a "sample" in the stochastic approximation. Instead of optimizing the original objective function E_1, in this approximation a step into the direction of the negative gradient of $E_2(t)$ with respect to the *last* values of the basis vectors $b_h^{(i)}$ is taken.

Earlier we pointed out that the "representative winner" c_r must be the same for the whole episode during learning. Thus, for the "sample" of the stochastic-approximation step, the index c_r shall also be used.

If the basis vectors are orthonormalized after each episode, the gradient of the sample function with respect to basis vector $b_h^{(i)}$ is readily found after a few substitutions:

$$\frac{\partial E_2}{\partial b_h^{(i)}}(t) = -2h_{c_r}^{(i)} \sum_{t_p \in \mathcal{S}(t)} \left[x(t_p)x^{\mathrm{T}}(t_p)\right] b_h^{(i)}(t) \ . \tag{5.36}$$

Comment 1. As index c_r will be changed abruptly when the signal subspace passes the border between its two closest subspaces $\mathcal{L}^{(i)}$, it will be necessary to stipulate that the signal subspace does not belong to the infinitesimal neighborhood of such a border in (5.36). For continuous, stochastic signals, this possibility can be neglected. Cf. also a similar discussion in [1.75].

Comment 2. As the basis vectors that span a subspace are not unique, the optimum is not unique, although the optimal value of E_2 is. Notwithstanding it may be clear that *any* of these optima (for which E_2 is minimized) is an equivalent solution, so it will suffice to find *one* optimal basis.

With a step of length $\lambda(t)$ in the direction of the negative gradient we obtain the delta-type rule

$$
\begin{aligned}
b_h^{(i)}(t+1) &= b_h^{(i)}(t) - \frac{1}{2}\lambda(t)\frac{\partial E_2}{\partial b_h^{(i)}}(t) \\
&= (I + \lambda(t)h_{c_r}^{(i)} \sum_{t_p \in \mathcal{S}(t)} x(t_p)x^{\mathrm{T}}(t_p))b_h^{(i)}(t) .
\end{aligned}
\tag{5.37}
$$

If the input vectors are not normalized, the objective function (5.34) should be modified by considering the *normalized projection error* on the winning subspace, or $\|\tilde{x}^{(c_r)}(t_p)\|/\|x(t_p)\|$. Then the learning rule becomes

$$
b_h^{(i)}(t+1) = (I + \lambda(t)h_{c_r}^{(i)} \sum_{t_p \in \mathcal{S}(t)} \frac{x(t_p)x^{\mathrm{T}}(t_p)}{\|x(t_p)\|^2})b_h^{(i)}(t) .
\tag{5.38}
$$

In Sect. 5.9.3 we stipulated that the product of unnormalized rotation operators should be used for learning. Using $h_{c_r}^{(i)}$ to denote the neighborhood interactions, the learning law was

$$
b'_h^{(i)} = \prod_{t_p \in \mathcal{S}} \left[I + \alpha(t_p)h_{c_r}^{(i)} \frac{x(t_p)x^{\mathrm{T}}(t_p)}{\|\hat{x}^{(i)}(t_p)\|\,\|x(t_p)\|} \right] b_h^{(i)}(t_p) .
\tag{5.39}
$$

When $\alpha(t_p)$ is small, and we define $\lambda(t_p) = \alpha(t_p)\|x(t_p)\|/\|\hat{x}^{(i)}(t_p)\|$, the corrections ensuing from (5.38) and (5.39) can be found equivalent, and (5.38) may be regarded as a batch version of (5.39), where the whole episode is used in one operation.

5.11.5 ASSOM Experiments

Input Patterns to the ASSOM. It may be necessary to emphasize that the biological visual systems contain a vast number of *local processors*, each one analyzing only a rather narrow *receptive field* of the complete field of vision. If the visual system is simulated by artificial neural networks, *we would need, in principle at least, a separate neural network for each of the receptive fields.* Since the computing speed of artificial neural networks, or their simulation by contemporary computers is usually very high compared with the time constants of the optic patterns, it is a normal practice in engineering to have only one image processing algorithm or hardware processor to which the signals from different receptive fields are *multiplexed.*

In the rest of this discussion it may thus suffice to study what kind of adaptive processes there occur in a single neural network such as the ASSOM, when it receives input signals from a single receptive field (often called the *window* in engineering). We shall define each input x to the SOM as a set of discrete *samples* taken from a continuous signal pattern. If the signal process is a time function $f(t)$, then the input vector could be defined as

$$x = x(t) = [f(t - t_1), f(t - t_2), \ldots, f(t - t_n)]^T \in \Re^n , \qquad (5.40)$$

where the $t_1, \ldots, t_n$ are constant displacements in time relative to t (not necessarily equidistant). If, on the other hand, we study two-dimensional images with intensity $I(r)$ at the image coordinate $r \in \Re^2$, the sampled pattern vector can be defined as

$$x = x(r) = [I(r - r_1), I(r - r_2), \ldots, I(r - r_n)]^T \in \Re^n, \qquad (5.41)$$

where now the $r_1, \ldots, r_n \in \Re^2$ form a two-dimensional *sampling lattice*. The spacing in this lattice can be equidistant, but alternatively, some more "biological" spacing such as the one shown in Fig. 5.17 can be used. In the latter case, in order to avoid too high density of the sampling points near the center, the middle sampling points can be taken equidistant. One might use a similar nonuniform sampling for time-domain signals, too, with the highest density in the middle of the window.

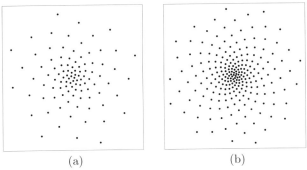

(a)   (b)

Fig. 5.17. Nonuniformly distributed two-dimensional sampling lattices. **(a)** 100 sampling points, **(b)** 225 sampling points

ASSOMs for Speech. In the experiments reported next we generated wavelet filters for time-domain speech waveforms using data from the generally known standard TIMIT database that contains a vast number of samples from U.S. speakers. These signals have been sampled at 12.8 kHz. The input vectors consisted of 64 successive samples of the speech waveform. Each learning episode began at a randomly chosen time instant and consisted of eight vectors, each displaced by a random amount (eight samples on the average) from the previous one. Thus, the transformation group inherent in the data consisted of translations in time. A segment of the original signal and the eight input vectors forming an episode are exemplified in Fig. 5.18.

The training process consisted of 30000 steps. At each step the episode consisted of eight windows displaced randomly over 64 adjacent sampling

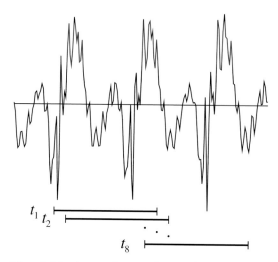

Fig. 5.18. A sample speech waveform. The locations of the eight input vectors (with starting times $t_1, \ldots, t_8$) belonging to an episode are identified with the line segments under the speech waveform.

instants. All displacements were different. The "representative winners" were computed from (5.26).

In the following we describe some auxiliary details of the experiment. They may seem a bit arbitrarily chosen, but actually there exist very clear reasons behind them.

First, the time-domain signal data in the TIMIT database were preprocessed by taking the differences of successive values (high-pass filtering). The data vectors x were then normalized.

Second, the b_{im} vectors were normalized after each projection operation performed according to (5.32). This normalization could have been made more seldom, say, after every 100 projection operations.

Third, if the sampling lattice would have been equidistant, the wavelet pattern might have been formed freely anywhere within the window; so, in order to stabilize the wavelets into the center of the window, the middle samples were provided with higher *weights*. A Gaussian weighting function centered in the window was used. The weighting was started using a narrow Gaussian having the full width at half maximum, $\mathrm{FWHM}(t)$, of eight sampling points (equivalent to 0.7 ms). In the self-organizing process, the narrow Gaussian can be found to stabilize the high-frequency filters first, whereby also all the filters are roughly ordered. When the Gaussian is thereafter let to broaden linearly up to 50 sampling points (equivalent to 4 ms) during the learning phase, the low-frequency filters will acquire their asymptotic values.

A similar centering effect would have been obtained without weighting, if the sampling points would have been distributed unequally within the window, having the highest density in its middle point.

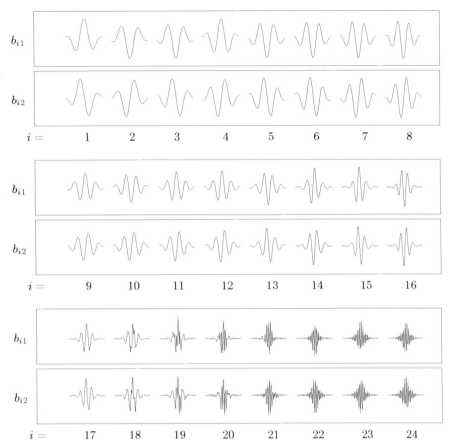

Fig. 5.19. The 24 wavelet filters (basis vectors of the subspaces) that were formed automatically in the ASSOM process. The 64 abscissae points of each curve correspond to the 64 components of the basis vectors, and the ordinate points their values, respectively

Figure 5.20 recapitulates and illustrates the selection of parameters relating to the experiment, the results of which are given in Fig. 5.19.

Figure 5.19 shows the two basis vectors for each subspace formed in the self-organizing process. These vectors have clearly assumed the wavelet form, and their distribution has been "optimized" automatically in the competitive-learning process.

It is possible to obtain qualitatively correct-looking filters easily, but for their good *distribution* the process has to be controlled very closely, as seen next.

A Characteristic Instability. The time-dependent learning parameters described in Fig. 5.20 were selected such that the self-organization process converged quickly, the wavelets acquired simple, smooth forms, and the mid-

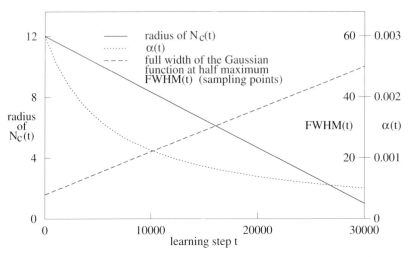

Fig. 5.20. The learning parameters vs. number of learning steps. The learning-rate factor $\alpha(t)$ was of the form $A/(B+t)$, where $A = 18$ and $B = 6000$. In 30000 steps the radius of $N_c(t)$ decreased linearly from 12 lattice spacings to one spacing, and $\mathrm{FWHM}(t)$ increased linearly from eight to 50 sampling points, respectively

frequencies of the filters corresponding to them were fairly smoothly distributed over the range of the speech frequencies. However, upon closer examination of this process it transpired that the waveforms shown in Fig. 5.19 were not asymptotically stable. A characteristic instability was almost always encountered with an extended learning period: the asymptotic distribution of the mid-frequencies of the filters usually had a steep step somewhere in the middle of the frequency range, as exemplified in Fig. 5.21. It also turned out that although the filters at both ends of the array had a single passband around their midfrequency, the filters around the discontinuous "step" usually had *two* clearly separated passbands; also their wavelet form was more complex.

In looking for an explanation to this instability, it seemed obvious that if no competitive learning were used, a linear filter of this order (number of input samples and parameters) could freely be tuned to a rather complicated frequency spectrum. However, since competitive learning tends to partition the signal space in such a way that the neighboring filters of the array get an equitable representation of the signal domain, each filter effectively only learns a narrow "slice" of the frequency band and tends to acquire the single-passband form. Nonetheless there remains this tendency to form multiple passbands, especially in the middle of the filter array where the ordering effects of both halves of the array meet. If a severe instability has already been formed it seems to remain stable.

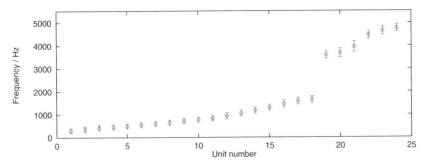

Fig. 5.21. Distribution of the mean frequencies of the filters. The vertical line segments at the experimental points describe the half-widths of the respective filters. There is a steep step, due to instability in the formation of the basis vectors, from filter no. 18 to filter no. 19

Without further speculation we would like to mention that this explanation seemed to be correct, since the problem was solved in the following simple way.

Dissipation. The following very effective practical remedy was found experimentally, although it also had an intuitive theoretical motivation. If, during the learning steps, we set such components of the $b_h^{(i)}$ to zero that already have a small absolute value, then *we are in fact reducing the effective order of the linear filter*, i.e., eliminating those degrees of freedom where we have "nuisance information" in the description of the $b_h^{(i)}$. The $b_h^{(i)}$ will then better become tuned to the *dominant frequency bands* by means of the *larger components* of the $b_h^{(i)}$.

A very simple *dissipation* of this type can be implemented computationally after each rotation and before normalization in the following way. If we denote $b_h^{(i)} = [\beta_{h1}^{(i)}, \ldots, \beta_{hn}^{(i)}]^{\mathrm{T}} \in \Re^n$, and the corrected value of $\beta_{hj}^{(i)}$ is denoted $\beta'^{(i)}_{hj}$, the value after dissipation could be selected as

$$\beta''^{(i)}_{hj} = \mathrm{sgn}(\beta'^{(i)}_{hj}) \max(0, |\beta'^{(i)}_{hj}| - \varepsilon) , \qquad (5.42)$$

where ε $(0 < \varepsilon < 1)$ is a very small term.

Summary of the Practical ASSOM Learning Algorithm. The details of the ASSOM algorithm, including the nonlinear operation necessary for robust ordering, are now incorporated into the following rules.

For each learning episode $\mathcal{S}(t)$ consisting of successive time instants $t_p \in \mathcal{S}(t)$ do the following:

– Find the winner, indexed by c:

$$c = \arg \max_i \left\{ \sum_{t_p \in \mathcal{S}(t)} ||\hat{x}^{(i)}(t_p)||^2 \right\} .$$

– For each sample $\mathbf{x}(t_p)$, $t_p \in \mathcal{S}(t)$:

1. Rotate the basis vectors of the modules:

$$b_h^{(i)}(t+1) = \left[I + \lambda(t)h_c^{(i)}(t) \frac{x(t_p)x(t_p)^{\mathrm{T}}}{\|\hat{x}^{(i)}(t_p)\| \|x(t_p)\|} \right] b_h^{(i)}(t) .$$

2. Dissipate the components $b_{hj}^{(i)}$ of the basis vectors $b_h^{(i)}$:

$$b'^{(i)}_{hj} = \mathrm{sgn}(b_{hj}^{(i)}) \max(0, |b_{hj}^{(i)}| - \varepsilon) ,$$

where

$$\varepsilon = \varepsilon_h^{(i)}(t) = \alpha |b_h^{(i)}(t) - b_h^{(i)}(t-1)| .$$

3. Orthonormalize the basis vectors of each module.

The orthonormalization can also be made more seldom, say, after every hundred steps. Above, $\lambda(t)$ and α are suitable small scalar parameters.

Improved Speech Filters. In the following experiment, the dissipation effect was used to stabilize self-organization and to produce smooth, asymptotically stable, single-peaked bandpass filter with a distribution of their midfrequencies that was continuous over the range of speech frequencies. After each orthonormalization of the basis vectors, a small constant value (equivalent to $\alpha/50$) was subtracted from the *amplitude* of each component of the basis vectors (when they were orthonormalized at each step according to (5.42). Figure 5.22 shows the pairwise basis vectors formed in the self-organizing process at each "neural" unit. *Notice that the* $a_i^2 = b_1^{(i)2} + b_2^{(i)2}$ *correspond to the widths of the wavelets; the forms of these "optimized" wavelets can be seen to depend on frequency!*

Automatic Formation of Gabor-Type Filters. In the following experiments, sampling of *images* was performed using sampling lattices with 225 points. At each learning cycle, this lattice was randomly shifted in a few (say, four) closeby positions, and the pattern vectors $x(t)$ thereby obtained formed the two-dimensional "episodes" over which the "representative winner" in the SOM, and updating of the corresponding $N^{(c_r)}$ were made. This learning cycle was repeated for other parts of the image area, or for different images.

In the first experiment, sampling was nonuniform (Fig. 5.17 b). A two-dimensional monochromatic *sinusoidal wave* with randomly varying wavelength and direction was presented in a number of images. Figure 5.23 shows the basis vectors thereby formed in a gray scale, for each of the SOM nodes. It is discernible that the Gaussian amplitude modulation of the wavelets was formed automatically, in spite of the waves having constant amplitude. This is due to interference of the waves near the borders of the receptive fields. In Fig. 5.23(c) the sum of squares of the respective components in the basis vectors has been displayed, showing the envelope typical to Gabor filters [2.56, 5.27]. The attenuation at the edges of this filter discernible already in Figs. 5.23(a) and (b) is due to the fact that in learning, the waves match best in the center of the receptive fields, but as there are learning samples with slightly different direction and wavelength, they interfere near the edges. The Fourier transforms of the filters have also been shown.

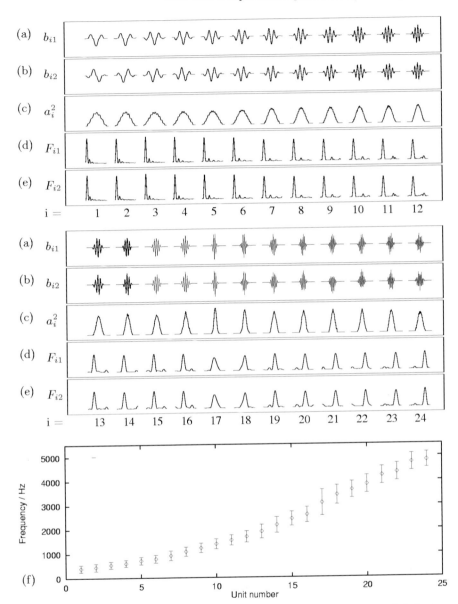

Fig. 5.22. Notice that the diagrams (a) through (e) are in two parts. (a) Cosine-type wavelets (b_{i1}), (b) Sine-type wavelets (b_{i2}), (c) Sums of squares of the respective components of b_{i1} and b_{i2} (a_i^2), (d) Fourier transforms (F_{i1}) of the b_{i1}, (e) Fourier transforms (F_{i2}) of the b_{i2}, (f) Distribution of the mean frequencies of the filters. The steeper part around 2500 Hz corresponds to a broad minimum of the speech spectrum. The vertical line segments at the experimental points describe the half-widths of the receptive filters

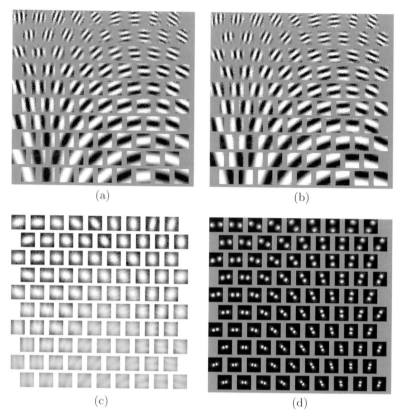

(a)   (b)

(c)   (d)

Fig. 5.23. Automatic formation of Gabor-type spatial feature filters in the adaptive-subspace-SOM (ASSOM), in response to artificial, two-dimensional, randomly oriented random-frequency sinusoidal waves. The square sampling lattice consisted of 15 by 15 contiguous pixels. These SOM images show the basis vectors at each map node, in a gray scale. **(a)** The first counterparts of the basis vectors. **(b)** The second counterparts of the basis vectors, phase-shifted by 90 degrees with respect to the first ones. White: positive value. Black: negative value. **(c)** The sums of squares of respective components of the basis vectors as shown here can be seen to resemble the envelopes of Gabor filters. **(d)** Fourier transforms of the basis vectors

In the second experiment, shown in Fig. 5.24, photographic images were used as training data. High frequencies were emphasized by subtracting local averages from the pixels. Sampling was thereby equidistant. The sampling lattice was further weighted by a Gaussian function for better concentration of the filters to their middle points. The Gaussian was flattened during learning.

Filters for Different Invariances. The ASSOM filters are not restricted to translations. For instance, we have formed two-dimensional filters that reflect different kinds of invariances, although the algorithm and the basic pattern type were the same. *What was only different was the transformation of the basic pattern over the episode.* The following kinds of filters thereby emerged:

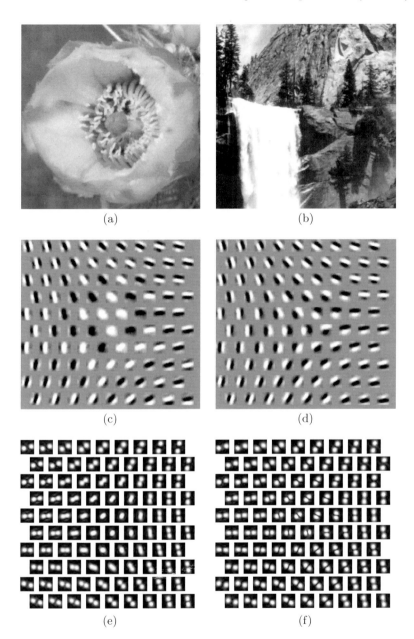

Fig. 5.24. Photographs were used in this ASSOM experiment. The sampling lattice consisted of 15 by 15 contiguous pixels. **(a)**, **(b)** Parts of two original images (300 by 300 pixels). **(c)** One basis vector is shown at each ASSOM position. **(d)** The associated basis vectors that are 90 degrees off the phase with respect to their first counterparts. **(e)**, **(f)** Fourier transforms of the basis vectors. (Notice that pictures (c)–(f) are magnified with respect to (a) and (b).)

Fig. 5.25. Colored noise (second-order Butterworth-filtered white noise with cut-off frequency of 0.6 times the Nyquist frequency of the lattice) used as input data. The receptive field is demarcated by the white circle

translation-invariant filters, rotation-invariant filters, and approach- (zoom-) invariant filters.

The receptive field of the input layer consisted of a circular sampling lattice (equivalent to 316-dimensional pattern vectors), shown in Fig. 5.25.

Over the input field we generated patterns consisting of colored noise (white noise, low-pass filtered by a second-order Butterworth filter with cut-off frequency of 0.6 times the Nyquist frequency of the sampling lattice). The input episodes for learning were formed by taking samples from this data field. The mean of the samples was always subtracted from the pattern vector.

Exemplary episodes of translated, rotated, and scaled colored noise are shown in Fig 5.26.

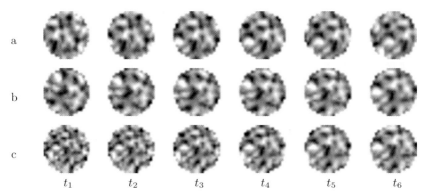

Fig. 5.26. Samples of transformed colored noise patterns. **(a)** Translated, **(b)** rotated, and **(c)** scaled inputs. Each partial figure forms an episode

In the *translation-invariant filter* experiment, the ASSOM was a hexagonal, 9 by 10 unit array, with two basis vectors at each unit. The episodes were formed by shifting the receptive field randomly into five nearby locations, the average shift thereby being ±2 pixels in both dimensions. The next episodes were taken from completely different places. In order to symmetrize the filters with respect to the center of the receptive field (because they might easily be formed as eccentric) we weighted the input samples by a Gaussian function that was symmetrically placed over the receptive field, and had a full width at half maximum that varied linearly during the learning process from one to 15 sampling lattice spacings. The radius of the circular neighborhood function $N^{(c_r)}$ decreased linearly from five to one ASSOM array spacings during the process, and the learning-rate factor was of the form $\alpha(t) = T/(T + 99t)$, where t was the index of the learning episode, and T the total number of episodes, respectively; in the present experiment T was 30000. The initial values of the basis vectors were selected as random but normalized vectors. The basis vectors were orthonormalized at each learning step, but before orthogonalization, a dissipation of $\alpha/10000$ was imposed on each component of both basis vectors.

Figure 5.27 shows the basis vectors b_{i1} and b_{i2} graphically in a gray scale at each array point. One should notice that the spatial frequencies of the basis vectors are the same, but the b_{i1} and b_{i2} are mutually 90 degrees out of phase. (The absolute phase of b_{i1} can be zero or 180 degrees, though.)

In the *rotation-invariant filter* experiment, the ASSOM lattice was linear, consisting of 24 units, whereas the receptive fields were essentially the same as in the first experiment. The episodes were formed by *rotating* the input field at random five times in the range of zero to 60 degrees, the rotation center coinciding with the center of the receptive field. Each new episode was

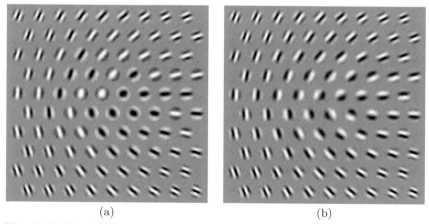

(a)   (b)

Fig. 5.27. The ASSOM that has formed Gabor-type filters: **(a)** The b_{i1}, **(b)** The b_{i2}

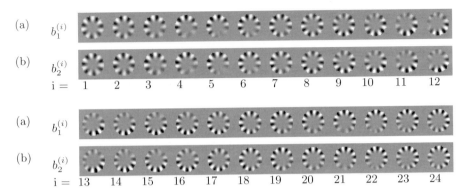

Fig. 5.28. One-dimensional rotation-invariant ASSOM. **(a)** Cosine-type "azimuthal wavelets" (b_{i1}), **(b)** Sine-type "azimuthal wavelets" (b_{i2}). Notice that the linear array has been shown in two parts

taken from a completely different place of the input field. In this experiment, no Gaussian weighting of the samples was necessary. Figure 5.28 shows the rotation filters thereby formed at the ASSOM units; clearly they are sensitive to azimuthal optic flow.

The *scale-invariant filters* were formed by *zooming* the input pattern field, with the center of the receptive field coinciding with the zooming center. No Gaussian weighting was used. The zooming range was $1 : 2$, from which five patterns were picked up to an episode; the next episode was formed from a quite different place of the input field. The filters thereby formed, shown in Fig. 5.29, have clearly become sensitive to radial optic flow, corresponding to approaching or withdrawing objects.

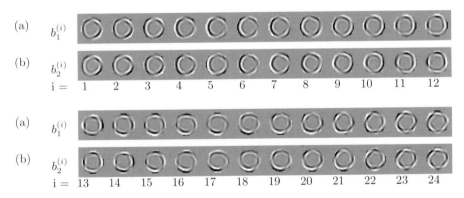

Fig. 5.29. One-dimensional zoom-invariant ASSOM. **(a)** Cosine-type "radial wavelets" (b_{i1}), **(b)** Sine-type "radial wavelets" (b_{i2}). Notice that the linear array has been shown in two parts

Competition on Different Functions. Finally we will see what happens if *different transformations* occur in the input samples. We have formed a mixture of episodes, some of which were formed by translations, other by rotations and scalings of the basic pattern defined in Fig. 5.25, respectively. As expected, some of the filters of the ASSOM array become translation-invariant, whereas others rotation- and scale-invariant, respectively (Fig. 5.30). In other words, the ASSOM can also compete on various *transformation types*, too.

On the Biological Connection of the ASSOM. It is an intriguing question what possible biological counterparts the various neuron types in the ASSOM architecture might have. In a series of interesting studies, *Daugman* [5.30, 31], *Marcelja* [5.31], and *Jones and Palmer* [5.32] advanced the hypothesis that the receptive fields of the *simple cells* in the mammalian visual cortex are describable by Gabor-type functions. What expired in the present study is that such functions in fact emerge in the *input layer* of the ASSOM structure, whereby the cells of this layer act in the similar way as the simple cells. If the output of the quadratic neuron were recorded, this neuron might have been seen to operate like a cortical neural cell called the *complex cell.*

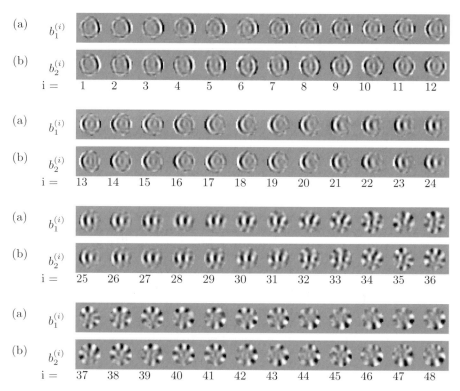

Fig. 5.30. An ASSOM that became tuned to a mixture of transformations (translation, rotation, and scaling)

It may be possible to create something like the so-called *hypercomplex cells*, by continuing the layered architectures, and using more complex input data.

5.12 Feedback-Controlled Adaptive-Subspace SOM (FASSOM)

Now we are ready to introduce a very important pattern-recognition architecture, namely, a combination of output-directed and input-directed pattern-recognition functions, both of which can be made adaptive. It consists of an ASSOM, in which the adaptive learning process that takes place in the subspaces is further controlled by higher-level information. This information may constitute of, e.g., classification results obtained by some pattern-recognition algorithm (such as LVQ) to which the ASSOM only delivers the input features. Also favorable reactions obtained from the environment can control learning in the ASSOM.

It is a question subject to much speculation how the processing results obtained in the biological central nervous system could be propagated back to the peripheral circuits such as sensory detectors. One possibility would be some kind of *retrograde* control, i.e., information sent back through the sensory pathways, eventually based on chemical messengers. Another possibility is the so-called *centrifugal* control, i.e., neural connections projecting back to the same places where afferent pathways came from. Examples of both types of control exist in the neural realms. Sometimes information about successful performance is communicated chemically via circulation.

Figure 5.31 delineates the general organization of the *feedback-controlled adaptive-subspace SOM (FASSOM)* architecture. The central idea is that the learning-rate factor $\alpha(t)$ of the "unsupervised" algorithm such as (5.32) is made a function of classification results: $\alpha(t)$ may, for instance, attain a high value if the classification is correct, and a low value if the classification is

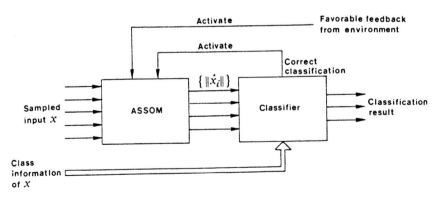

Fig. 5.31. The FASSOM architecture

wrong. In this way we only apply the reward strategy. Naturally, the sign of $\alpha(t)$ might also be taken negative when the training classification is wrong; this would then correspond to punishment like in supervised learning in general. Even without the punishment strategy, decision-controlled competitive learning that takes place in the feature-extraction stage would allocate the computing resources optimally in relation to various conflicting situations. When the preprocessing stage has settled down to its optimal configuration, the classification algorithm (eventually LVQ) may still be fine-tuned using supervised learning.

It should be noticed that the ASSOM stage can be learning all the time, whereby the "winners" are only used to define the neighborhood set for learning, but they are not propagated to the classification stage. The ASSOM stage computes the projections $\hat{x}_i$ that constitute the feature vector $f = [||\hat{x}_1||, ||\hat{x}_2||, \ldots]^{\mathrm{T}}$ to be used as input to the classification stage.

Caution. Special care should be taken for that all frequencies of the primary signals become processed in the preprocessing stage, without leaving "absorption bands" in the frequency domain. In other words, it has to be checked that the transmission bands of the filters overlap by a sufficient amount. Unless this is the case, the number of filters must be increased, or their bandwidths broadened by some means.

6. Learning Vector Quantization

Closely related to VQ and SOM is *Learning Vector Quantization (LVQ)*. This name signifies a class of related algorithms, such as LVQ1, LVQ2, LVQ3, and OLVQ1. While VQ and the basic SOM are *unsupervised* clustering and learning methods, LVQ describes *supervised learning*. On the other hand, unlike in SOM, no neighborhoods around the "winner" are defined during learning in the basic LVQ, whereby also no spatial order of the codebook vectors is expected to ensue.

Since LVQ is strictly meant for a *statistical classification or recognition method*, its only purpose is to define *class regions* in the input data space. To this end a subset of similarly labeled codebook vectors is placed into each class region; even if the class distributions of the input samples would overlap at the class borders, the codebook vectors of each class in these algorithms can be placed in and shown to stay within each class region for all times. The quantization regions, like the Voronoi sets in VQ, are defined by midplanes (hyperplanes) between neighboring codebook vectors. An additional feature in LVQ is that for class borders one can only take such borders of the Voronoi tessellation that separate Voronoi sets into different classes. The class borders thereby defined are piecewise linear.

6.1 Optimal Decision

The problem of *optimal decision* or *statistical pattern recognition* is usually discussed within the framework of the Bayes theory of probability (Sect. 1.3.3). Assume that all samples of x are derived from a finite set of classes $\{S_k\}$, the distributions of which usually overlap. Let $P(S_k)$ be the *a priori* probability of class S_k, and $p(x|x \in S_k)$ be the conditional probability density function of x on S_k, respectively. Define the *discriminant functions*

$$\delta_k(x) = p(x|x \in S_k)P(S_k) . \tag{6.1}$$

Let it be recalled that on the average, the unknown samples are classified optimally (i.e., the rate of misclassifications is minimized) if a sample x is determined to belong to class S_c according to the decision

$$\delta_c(x) = \max_k \{\delta_k(x)\} . \tag{6.2}$$

Let us recall that the traditional method in practical statistical pattern recognition was to first develop approximations for the $p(x|x \in S_k)P(S_k)$ and then to use them for the $\delta_k(x)$ in (6.1). The LVQ approach, on the other hand, is based on a totally different philosophy. Consider Fig. 6.1. We first assign *a subset of codebook vectors to each class* S_k and then search for that codebook vector m_i that has the smallest Euclidean distance from x. The sample x is thought to belong to the same class as the closest m_i. The codebook vectors can be placed in such a way that those ones belonging to different classes are not intermingled, although the class distributions of x overlap. Since now only the codebook vectors that lie closest to the class borders are important to the optimal decision, obviously a good approximation of $p(x|x \in S_k)$ is not necessary everywhere. It is more important to place the m_i into the signal space in such a way that the nearest-neighbor rule used for classification minimizes the average expected misclassification probability.

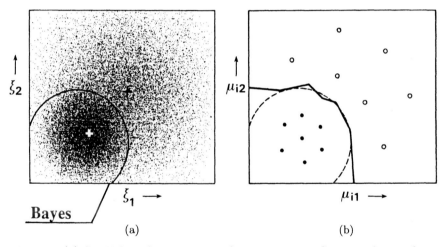

(a)   (b)

Fig. 6.1. (a) Small dots: Superposition of two symmetric Gaussian density functions corresponding to classes S_1 and S_2, with with their centroids shown by the white and dark cross, respectively. Solid curve: Bayes decision border. **(b)** Large black dots: reference vectors of class S_1. Open circles: reference vectors of class S_2. Solid curve: decision border in the learning vector quantization. Broken curve: Bayes decision border

6.2 The LVQ1

Assume that several codebook vectors are assigned to each class of x values, and x is then determined to belong to the same class to which the nearest m_i belongs. Let

$$c = \arg \min_i \{||x - m_i||\} \tag{6.3}$$

define the index of the nearest m_i to x.

Notice that c, the index of the "winner", depends on x and all the m_i. If x is a natural, stochastic, continuous-valued vectorial variable, we need not consider multiple minima: the probability for $||x - m_i|| = ||x - m_j||$ for $i \neq j$ is then zero.

Let $x(t)$ be an input sample and let the $m_i(t)$ represent sequential values of the m_i in the discrete-time domain, $t = 0, 1, 2, \ldots$. Values for the m_i in (6.3) that approximately minimize the rate of misclassification errors are found as asymptotic values in the following learning process [2.38], [2.40], [2.41]. Starting with properly defined initial values (as will be discussed in Sect. 6.9), the following equations define the basic Learning Vector Quantization process; this particular algorithm is called LVQ1.

$$
\begin{aligned}
m_c(t+1) &= m_c(t) + \alpha(t)[x(t) - m_c(t)] \\
&\quad \text{if } x \text{ and } m_c \text{ belong to the same class,} \\
m_c(t+1) &= m_c(t) - \alpha(t)[x(t) - m_c(t)] \\
&\quad \text{if } x \text{ and } m_c \text{ belong to different classes,} \\
m_i(t+1) &= m_i(t) \text{ for } i \neq c .
\end{aligned} \tag{6.4}
$$

Here $0 < \alpha(t) < 1$, and $\alpha(t)$ (learning rate) is usually made to decrease monotonically with time. It is recommended that α should already initially be rather small, say, smaller than 0.1. The exact law $\alpha = \alpha(t)$ is not crucial, and $\alpha(t)$ may even be made to decrease linearly to zero, as long as the number of learning steps is sufficient; see, however, Sect. 6.3. If also only a restricted set of training samples is available, they may be applied cyclically, or the samples presented to (6.3)-(6.4) may be picked up from the basic set of training samples at random.

The general idea underlying all LVQ algorithms is *supervised learning*, or the reward-punishment scheme. However, it is extremely difficult to show what the exact convergence limits are. The following discussion is based on the observation that the classical VQ tends to approximate $p(x)$ (or some monotonic function of it). Instead of $p(x)$, we may also consider approximation of *any* other (nonnegative) density function $f(x)$ by the VQ. For instance, let the optimal decision borders, or the Bayesian borders (which divide the signal space into class regions B_k such that the rate of misclassifications is minimized) be defined by (6.1) and (6.2); *all such borders together* are defined by the condition $f(x) = 0$, where

for $x \in B_k$ and $h \neq k$,

$$f(x) = p(x|x \in S_k)P(S_k) - \max_h \{p(x|x \in S_h)P(S_h)\} . \tag{6.5}$$

Let Fig. 6.2 now illustrate the form of $f(x)$ in the case of scalar x and three classes S_1, S_2, and S_3 being defined on the x axis by their respective distributions $p(x|x \in S_k)P(S_k)$. In Fig. 6.2(a) the optimal Bayesian borders have been indicated with dotted lines. The function $f(x)$ has zero points at these borders according to (6.5), as shown by Fig. 6.2(b); otherwise $f(x) > 0$ in the three "humps."

If we then use VQ to define the point density of the m_i that approximates $f(x)$, *this density also tends to zero at all Bayesian borders.* Thus VQ and (6.5) together define the Bayesian borders with arbitrarily good accuracy, depending on the number of codebook vectors used.

The optimal values for the m_i in the classical VQ were found by minimizing the average expected quantization error E, and in Sect. 1.5.2 its gradient was found to be

$$\nabla_{m_i} E = -2 \int \delta_{ci} \cdot (x - m_i) p(x) dx ; \tag{6.6}$$

here δ_{ci} is the Kronecker delta and c is the index of the m_i that is closest to x (i.e., the "winner"). The gradient step of vector m_i is

$$m_i(t+1) = m_i(t) - \lambda \cdot \nabla_{m_i(t)} E , \tag{6.7}$$

where λ defines the step size, and the so-called sample function of $\nabla_{m_i} E$ at step t is $\nabla_{m_i(t)} E = -2\delta_{ci}[x(t) - m_i(t)]$. One result is obvious from (6.6): only the "winner" $m_c(t)$ should be updated, while all the other $m_i(t)$ are left intact during this step.

If $p(x)$ in E is now replaced by $f(x)$, the gradient steps must be computed separately in the event that the sample $x(t)$ belongs to S_k, and in the event that $x(t) \in S_h$, respectively.

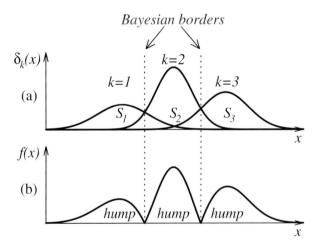

Fig. 6.2. (a) Distribution of scalar samples in three classes $S_1, S_2,$ and S_3. (b) Illustration of the respective function $f(x)$

The gradient of E, with $p(x)$ replaced by $f(x)$, is

$$
\begin{aligned}
\nabla_{m_i} E \ &= \ -2 \int \delta_{ci}(x - m_i) f(x) dx \\
&= \ -2 \int \delta_{ci}(x - m_i)[p(x|x \in S_k)P(S_k) \\
&\quad - \max_h \{ p(x|x \in S_h)P(S_h)\}] dx \ .
\end{aligned}
\tag{6.8}
$$

In the event that $x(t) \in S_k$ we thus obtain for the sample function of $\nabla_{m_i} E$ *with the a priori probability* $P(S_k)$:

$$
\nabla_{m_i(t)} E = -2\delta_{ci}[x(t) - m_i(t)] \ .
\tag{6.9}
$$

If the class with $\max_h \{ p(x|x \in S_h)P(S_h)\}$ is signified by index r meaning the "runner up" class, and in the event that $x(t) \in S_r$, the sample function of $\nabla_{m_i} E$ is obtained with the *a priori* probability $P(S_r)$ and reads

$$
\nabla_{m_i(t)} E = +2\delta_{ci}[x(t) - m_i(t)] \ .
\tag{6.10}
$$

Rewriting $\alpha(t) = 2\lambda$, there results

$$
\begin{aligned}
m_c(t+1) \ &= \ m_c(t) + \alpha(t)[x(t) - m_c(t)] \ \ \text{if } x(t) \in B_k \text{ and } x(t) \in S_k \ , \\
m_c(t+1) \ &= \ m_c(t) - \alpha(t)[x(t) - m_c(t)] \ \ \text{if } x(t) \in B_k \text{ and } x(t) \in S_r \ , \\
m_c(t+1) \ &= \ m_c(t) \ \ \text{if } x(t) \in B_k \text{ and } x(t) \in S_h \ , \ h \neq r \ , \\
m_i(t+1) \ &= \ m_i(t) \ \ \text{if } i \neq c.
\end{aligned}
\tag{6.11}
$$

If the m_i of class S_k were already within B_k, and we take into account the humped form of $f(x)$ (Fig. 6.2b), the $m_i \in S_k$ would further be attracted by VQ to the hump corresponding to the B_k, at least if the learning steps are small.

Near equilibrium, close to the borders at least, (6.4) and (6.11) can be seen to define almost similar corrections; notice that in (6.4), *the classification of x was approximated by the nearest-neighbor rule*, and this approximation will be improved during learning. Near the borders the condition $x \in S_r$ is approximated by looking for which one of the m_i is second-closest to x; in the middle of B_k this method cannot be used, but there the exact values of the m_i are not important. However, notice that in (6.4), the minus-sign corrections were made every time when x was classified incorrectly, whereas (6.11) only makes the corresponding correction if x is exactly in the runner-up class. This difference may cause a small bias to the asymptotic values of the m_i in the LVQ1. As a matter of fact, the algorithms called LVQ2 and LVQ3 that will be discussed below are even closer to (6.11) in this respect. *Two* codebook vectors m_i and m_j that are the nearest neighbors to x are eventually updated in them at every learning step; the one that is classified correctly is provided with the plus sign, whereas in the case of incorrect classification the correction is opposite.

6.3 The Optimized-Learning-Rate LVQ1 (OLVQ1)

The basic LVQ1 algorithm will now be modified in such a way that an individual learning-rate factor $\alpha_i(t)$ is assigned to each m_i, whereby we obtain the following learning process [3.15]. Let c be defined by (6.3). Then we assume that

$$
\begin{aligned}
m_c(t+1) &= m_c(t) + \alpha_c(t)[x(t) - m_c(t)] \text{ if } x \text{ is classified correctly,} \\
m_c(t+1) &= m_c(t) - \alpha_c(t)[x(t) - m_c(t)] \text{ if } x \text{ is classified incorrectly,} \\
m_i(t+1) &= m_i(t) \text{ for } i \neq c. \quad\quad (6.12)
\end{aligned}
$$

The problem is whether the $\alpha_i(t)$ can be determined optimally for fastest convergence of (6.12). We express (6.12) in the form

$$
m_c(t+1) = [1 - s(t)\alpha_c(t)]m_c(t) + s(t)\alpha_c(t)x(t) , \quad\quad (6.13)
$$

where $s(t) = +1$ if the classification is correct, and $s(t) = -1$ if the classification is wrong. It may be obvious that the *statistical accuracy* of the learned codebook vector values is approximately optimal if all samples have been used with equal weight, i.e, if the effects of the corrections made at different times, when referring to the end of the learning period, are of approximately equal magnitude. Notice that $m_c(t+1)$ contains a trace of $x(t)$ through the last term in (6.13), and traces of the earlier $x(t'), t' = 1, 2, \ldots, t-1$ through $m_c(t)$. In a learning step, the magnitude of the last trace of $x(t)$ is scaled down by the factor $\alpha_c(t)$, and, for instance, during the same step the trace of $x(t-1)$ has become scaled down by $[1 - s(t)\alpha_c(t)] \cdot \alpha_c(t-1)$. Now we first stipulate that these two scalings must be identical:

$$
\alpha_c(t) = [1 - s(t)\alpha_c(t)]\alpha_c(t-1) . \quad\quad (6.14)
$$

If this condition is made to hold for all t, by induction it can be shown that the traces collected up to time t of all the earlier $x(t')$ will be scaled down by an equal amount at the end, and thus the 'optimal' values of $\alpha_i(t)$ determined by the recursion

$$
\alpha_c(t) = \frac{\alpha_c(t-1)}{1 + s(t)\alpha_c(t-1)} . \quad\quad (6.15)
$$

A precaution is necessary, however: since $\alpha_c(t)$ can also increase, it is especially important that it shall not rise above the value 1. This condition can be imposed in the algorithm itself. For the initial values of the α_i one may then take 0.5, but it is almost as good to start with something like $\alpha_i = 0.3$.

It must be warned that (6.15) is not applicable to the LVQ2 algorithm, since thereby the α_i, on the average, would not decrease, and the process would not converge. The reason for this is that LVQ2 is only a *partial* approximation of (6.11), whereas LVQ3 will be more accurate and probably could be modified like LVQ1. If LVQ3 is modified, then it should be called "OLVQ3."

6.4 The Batch-LVQ1

The basic LVQ1 algorithm can be written in a compressed form as

$$m_i(t+1) \quad = \quad m_i(t) + \alpha(t)s(t)\delta_{ci}[x(t) - m_i(t)] ,$$

$$\text{where } s(t) \quad = \quad +1 \text{ if } x \text{ and } m_c \text{ belong to the same class,}$$
$$\text{but } s(t) \quad = \quad -1 \text{ if } x \text{ and } m_c \text{ belong to different classes.} \qquad (6.16)$$

Here δ_{ci} is the Kronecker delta ($\delta_{ci} = 1$ for $c = i$, $\delta_{ci} = 0$ for $c \neq i$).

The LVQ1 algorithm, like the SOM, can be expressed as a batch version. In a similar way as with the Batch Map (SOM) algorithm, the equilibrium condition for the LVQ1 is expresssed as

$$\forall i, \ \mathrm{E}_t \{s\delta_{ci}(x - m_i^*)\} = 0 . \qquad (6.17)$$

The computing steps of the so-called *Batch-LVQ1* algorithm (in which at steps 2 and 3, the class labels of the nodes are redefined dynamically) can then be expressed, in analogy with the Batch Map, as follows:

1. For the initial reference vectors take, for instance, those values obtained in the preceding unsupervised SOM process, where the classification of $x(t)$ was not yet taken into account.
2. Input the $x(t)$ again, this time listing the $x(t)$ *as well as their class labels* under each of the corresponding winner nodes.
3. Determine the labels of the nodes according to the majorities of the class labels of the samples in these lists.
4. Multiply in each partial list all the $x(t)$ by the corresponding factors $s(t)$ that indicate whether $x(t)$ and $m_c(t)$ belong to the same class or not.
5. At each node i, take for the new value of the reference vector the entity

$$m_i^* = \frac{\sum_{t'} s(t')x(t')}{\sum_{t'} s(t')} , \qquad (6.18)$$

   where the summation is taken over the indices t' of those samples that were listed under node i.
6. Repeat from 2 a few times.

Comment 1. For stability reasons it may be necessary to check the sign of $\sum_{t'} s(t')$. If it becomes negative, no updating of this node is made.

Comment 2. Unlike in usual LVQ, the labeling of the nodes was allowed to change in the iterations. This has sometimes yielded slightly better classification accuracies than if the labels of the nodes were fixed at first steps. Alternatively, the labeling can be determined permanently immediately after the SOM process.

6.5 The Batch-LVQ1 for Symbol Strings

Consider that $\mathcal{S} = x(i)$ is the fundamental set of strings $x(i)$ that have been assigned to different classes. Let m_i denote one of the reference strings. The identity vs. nonidentity of the classes of $x(i)$ and m_i shall be denoted by $s(i)$, as in Sect. 6.4. Then the equilibrium condition that corresponds to (6.17) in 6.4 is assumed to read

$$\forall i, \quad \sum_{x(i)\in\mathcal{S}} s(i)d[x(i), m_i^*] = \min! , \qquad (6.19)$$

where d is some distance measure defined over all possible inputs and models, $s(i) = +1$ if $x(i)$ and m_i^* belong to the same class, but $s(i) = -1$ if $x(i)$ and m_i^* belong to different classes.

   In accordance with the Batch-LVQ1 procedure introduced in Sect. 6.4 we obtain the *Batch-LVQ1 for strings* by application of the following computational steps:

1. For the initial reference strings take, for instance, those strings obtained in the preceding SOM process.
2. Input the classified sample strings once again, listing the strings as well as their class labels under the winner nodes.
3. Determine the labels of the nodes according to the majorities of the class labels in these lists.
4. For each string in these lists, compute an expression equal to the left side of (6.19), where the distance of the string from every other string in the same list is provided with the plus sign, if the class label of the latter sample string agrees with the label of the node, but with the minus sign if the labels disagree.
5. Take for the set median in each list the string that has the smallest sum of expressions defined at step 4 with respect to all other strings in the respective list. Compute the generalized median by systematically varying each of the symbol positions in the set median by replacement, insertion, and deletion of a symbol, accepting the variation if the sum of distances (provided with the same plus and minus signs that were used at the previous step) between the new reference string and the sample strings in the list is decreased.
6. Repeat steps 1 through 5 a sufficient number of times.

6.6 The LVQ2 (LVQ2.1)

The classification decision in this algorithm is identical with that of the LVQ1. In learning, however, *two* codebook vectors m_i and m_j that are the nearest neighbors to x are now updated simultaneously. One of them must belong to the correct class and the other to a wrong class, respectively. Moreover, x

must fall into a zone of values called a 'window' that is defined around the midplane of m_i and m_j. Assume that d_i and d_j are the Euclidean distances of x from m_i and m_j, respectively; then x is defined to fall in a 'window' of relative width w if

$$\min\left(\frac{d_i}{d_j}, \frac{d_j}{d_i}\right) > s, \text{ where } s = \frac{1-w}{1+w}. \tag{6.20}$$

A relative 'window' width w of 0.2 to 0.3 is recommended. The version of LVQ2 called LVQ2.1 below is an improvement of the original LVQ2 algorithm [2.39] in the sense that it allows *either* m_i *or* m_j be the closest codebook vector to x, whereas in the original LVQ2, m_i had to be closest.

Algorithm LVQ2.1:

$$
\begin{aligned}
m_i(t+1) &= m_i(t) - \alpha(t)[x(t) - m_i(t)], \\
m_j(t+1) &= m_j(t) + \alpha(t)[x(t) - m_j(t)],
\end{aligned} \tag{6.21}
$$

where m_i and m_j are the two closest codebook vectors to x, whereby x and m_j belong to the same class, while x and m_i belong to different classes, respectively. Furthermore x must fall into the 'window.'

6.7 The LVQ3

The LVQ2 algorithm was based on the idea of *differentially* shifting the decision borders toward the Bayesian limits, while no attention was paid to what might happen to the location of the m_i in the long run if this process were continued. Therefore it seems necessary to introduce corrections that ensure that the m_i continue approximating the class distributions, or more accurately the $f(x)$ of (6.5), at least roughly. Combining the earlier ideas, we now obtain an improved algorithm [6.1–3] that may be called LVQ3:

$$
\begin{aligned}
m_i(t+1) &= m_i(t) - \alpha(t)[x(t) - m_i(t)], \\
m_j(t+1) &= m_j(t) + \alpha(t)[x(t) - m_j(t)],
\end{aligned}
$$

where m_i and m_j are the two closest codebook vectors to x, whereby x and m_j belong to the same class, while x and m_i belong to different classes, respectively; furthermore x must fall into the 'window';

$$m_k(t+1) = m_k(t) + \epsilon\alpha(t)[x(t) - m_k(t)], \tag{6.22}$$

for $k \in \{i, j\}$, if x, m_i, and m_j belong to the same class.

In a series of experiments, applicable values of ϵ between 0.1 and 0.5 were found, relating to $w = 0.2$ or 0.3. The optimal value of ϵ seems to depend on the size of the window, being smaller for narrower windows. This algorithm seems to be self-stabilizing, i.e., the optimal placement of the m_i does not change in continued learning.

Comment. If the idea were only to approximate the humps of $f(x)$ in (6.5) as accurately as possible, we might also take $w = 1$ (no window at all), whereby we would have to use the value $\varepsilon = 1$.

6.8 Differences Between LVQ1, LVQ2 and LVQ3

The three options for the LVQ-algorithms, namely, the LVQ1, the LVQ2 and the LVQ3 yield almost similar accuracies in most statistical pattern recognition tasks, although a different philosophy underlies each. The LVQ1 and the LVQ3 define a more robust process, whereby the codebook vectors assume stationary values even after extended learning periods. For the LVQ1 the learning rate can approximately be optimized for quick convergence (as shown in Sect. 6.3). In the LVQ2, the relative distances of the codebook vectors from the class borders are optimized whereas there is no guarantee of the codebook vectors being placed optimally to describe the forms of the class distributions. Therefore the LVQ2 should only be used in a *differential* fashion, using a small value of learning rate and a restricted number of training steps.

6.9 General Considerations

In the LVQ algorithms, vector quantization is not used to approximate the density functions of the class samples, but to directly define the class borders according to the nearest-neighbor rule. The accuracy achievable in any classification task to which the LVQ algorithms are applied and the time needed for learning depend on the following factors:

– an approximately optimal number of codebook vectors assigned to each class and their initial values,
– the detailed algorithm, a proper learning rate applied during the steps, and a proper criterion for the stopping of learning.

Initialization of the Codebook Vectors. Since the class borders are represented piecewise linearly by segments of midplanes between codebook vectors of neighboring classes (a subset of borders of the Voronoi tessellation), it may seem to be a proper strategy for optimal approximation of the borders that the average distances between the adjacent codebook vectors (which depend on their numbers per class) should be the same on both sides of the borders. Then, at least if the class distributions are symmetric, this means that the average shortest distances of the codebook vectors (or alternatively, the medians of the shortest distances) should be the same everywhere in every class. Because, due to unknown forms of the class distributions, the final placement of the codebook vectors is not known until at the end of the learning process, their distances and thus their optimal numbers cannot be determined before that. This kind of assignment of the codebook vectors to the various classes must therefore be made *iteratively*.

In many practical applications such as speech recognition, even when the a priori probabilities for the samples falling in different classes are very different, a very good strategy is thus to start with the same number of codebook vectors in each class. An upper limit to the total number of codebook vectors is set by the restricted recognition time and computing power available.

For good piecewise linear approximation of the borders, the medians of the shortest distances between the codebook vectors might also be selected somewhat smaller than the standard deviations (square roots of variances) of the input samples in all the respective classes. This criterion can be used to determine the minimum number of codebook vectors per class.

Once the tentative numbers of the codebook vectors for each class have been fixed, for their initial values one can use first samples of the real training data picked up from the respective classes. Since the codebook vectors should always remain inside the respective class domains, for the above initial values too one can only accept samples that are not misclassified. In other words, a sample is first tentatively classified against all the other samples in the training set, for instance by the K-nearest-neighbor (KNN) method, and accepted for a possible initial value only if this tentative classification is the same as the class identifier of the sample. (In the learning algorithm itself, however, no samples must be excluded; they must be applied independent of whether they fall on the correct side of the class border or not.)

Initialization by the SOM. If the class distributions have several modes (peaks), it may be difficult to distribute the initial values of the codebook vectors to all modes. Recent experience has shown that it is then a better strategy to first form a SOM, thereby regarding all samples as unclassified, for its initialization. After that the map units are labeled according to the class symbols by applying the training samples once again and taking their labels into account like in the calibration of the SOM discussed in Sect. 3.2.

The labeled SOM is then fine-tuned by the LVQ algorithms to approximate the Bayesian classification accuracy.

Learning. It is recommended that learning always be started with the optimized LVQ1 (OLVQ1) algorithm, which converges very fast; its asymptotic recognition accuracy will be achieved after a number of learning steps that is about 30 to 50 times the total number of codebook vectors. Other algorithms may continue from those codebook vector values that have been obtained in the first phase.

Often the OLVQ1 learning phase alone may be sufficient for practical applications, especially if the learning time is critical. However, in an attempt to ultimately improve recognition accuracy, one may continue with the basic LVQ1, the LVQ2.1, or the LVQ3, using a low initial value of learning rate, which is then the same for all the classes.

Stopping Rule. It often happens that the neural-network algorithms 'over-learn'; e.g., if learning and test phases are alternated, the recognition accuracy is first improved until an optimum is reached. After that, when learning is continued, the accuracy starts to decrease slowly. A possible explanation of this effect in the present case is that when the codebook vectors become very specifically tuned to the training data, the ability of the algorithm to generalize for new data suffers. It is therefore necessary to stop the learning process after some 'optimal' number of steps, say, 50 to 200 times the total number of the codebook vectors (depending on particular algorithm and learning rate). Such a stopping rule can only be found by experience, and it also depends on the input data.

Let us recall that the OLVQ1 algorithm may generally be stopped after a number of steps that is 30 to 50 times the number of codebook vectors.

6.10 The Hypermap-Type LVQ

The principle discussed in this section could apply to both LVQ and SOM algorithms. The central idea is selection of candidates for the "winner" in sequential steps. In this section we only discuss the Hypermap principle in the context of LVQ algorithms.

The idea suggested by this author [6.4] is to recognize a pattern that occurs in the context of other patterns (Fig. 6.3). The context around the pattern is first used to select a *subset of nodes* in the network, from which the best-matching node is then identified on the basis of the pattern part. The contexts need not be specified at high accuracy, as long as they can be assigned to a sufficient number of descriptive clusters; therefore, their representations can be formed by unsupervised learning, in which unclassified raw data are used for training. On the other hand, for best accuracy in the final classification of the pattern parts, the representations of the latter should be formed in a supervised learning process, whereby the classification of the training data must be well-validated. *This kind of context level hierarchy can be continued for an arbitrary number of levels.*

The general idea of the Hypermap principle is thus that if different sources of input information are used, then each of the partial sources of information only defines a set of possible *candidates* for best-matching cells; the final clas-

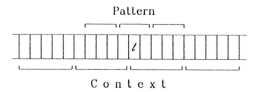

Fig. 6.3. Definition of pattern and context parts in a time series of samples

sification decision is suspended until all the sources have been utilized. Each of the neural cells is assumed to receive inputs from a number of different kinds of sources, but these subsets of inputs are not used simultaneously. Instead, after application of one particular signal set to the respective inputs, the neural cell is only left in a state of *preactivation*, in which it will not yet elicit an output response. Preactivation is a kind of binary memory state or bias of the neural cell, in which the latter is facilitated to be triggered by further inputs. Without preactivation, however, later triggering would be impossible. There may be one or more phases in preactivation. When a particular set of inputs has been used, information about its accurate input signal values will be "forgotten", and only the facilitatory (binary) information of the cell is memorized; so preactivation is not a bias for "priming" like in some other neural models. One can thereby discern that no linear vector sum of the respective input signal groups is formed at any phase; combination of the signals is done in a very nonlinear but at the same time robust way. Only in the last decision operation, in which the final recognition result is specified, the decision operation is based on continuous-valued discriminant functions.

Example: Two-Phase Recognition of Phonemes. The operation of the LVQ-Hypermap is now exemplified in detail by the following phoneme recognition experiment. It must be emphasized, however, that the principle is much more general. Consider Fig. 6.3, which illustrates a time sequence of samples, such as spectra. For higher accuracy, however, it is better to use in this experiment another feature set named the *cepstrum*, Sects. 7.2 and 7.5. We have used such features, so-called cepstral coefficients (e.g., 20 in number) instead of simple spectra. Assume that the phonemic identity of the speech signal shall be determined at time t. A pattern vector x_{patt} can be formed, e.g., as a concatenation of three parts, each part being the average of three adjacent cepstral feature vectors, and these nine samples are centered around time t. In order to demonstrate the Hypermap principle, x_{patt} is now considered in the context of a somewhat wider context vector x_{cont}, which is formed as a concatenation of four parts, each part being the average of five adjacent cepstral feature vectors. The context "window" is similarly centered around time t.

Figure 6.4 illustrates an array of neural cells. Each cell has two groups of inputs, one for x_{patt} and another for x_{cont}, respectively. The input weight vectors for the two groups, for cell i, are then denoted $m_{i,\mathrm{patt}}$ and $m_{i,\mathrm{cont}}$, respectively. We have to emphasize that the cells in this "map" were not yet ordered spatially: they must still be regarded as a set of unordered "codebook vectors" like in classical Vector Quantization.

Classification of the Input Sample. For simplicity, the classification operation is explained first, but for that we must assume that the input weights have already been formed, at least tentatively. The matching of x and m_i is determined in two phases.

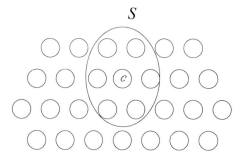

Fig. 6.4. Set of nodes (unordered) and the subset S from which the result is picked up

Matching Phase 1. Find the subset S of all cells for which

$$||x_{\text{cont}} - m_{i,\text{cont}}|| \le \delta \ , \tag{6.23}$$

where δ is a free, experimentally found parameter. Subset S now corresponds to one domain of context. The threshold δ, and thus the size of the context domain can be defined, e.g., relative to the best match at the Matching Phase 1:

$$\delta \ = \ r \cdot ||x_{\text{cont}} - m_{c,\text{cont}}|| \ , \tag{6.24}$$

$$\text{where } r \text{ is a suitable scalar parameter}(r > 1), \text{ and}$$

$$c \ = \ \arg\min_i \{||x_{\text{cont}} - m_{i,\text{cont}}\}||\} \ . \tag{6.25}$$

Matching Phase 2. In the subset of cells S found in Matching Phase 1, find the best match:

$$c_1 = \arg\min_{i \in S}\{||x_{\text{patt}} - m_{i,\text{patt}}\}|| \ . \tag{6.26}$$

Notice that in the comparisons, the pattern and context parts are never used simultaneously: they are separated sequentially in time, and no Euclidean metric is thus applied to the complete set of input signals.

The Learning Algorithm. Although quite natural and understandable, learning in this algorithm is an even more sophisticated process than classification. It consists of two different periods: 1. Adaptation of the context input weights and their subsequent fixation to constant values, 2. Adaptation of the pattern input weights. During the first period, only the x_{cont} are input, and learning consists of single phases and is unsupervised. During the second period, both x_{cont} and x_{patt} are used, and learning is two-phased. The second period is further divided into two subperiods. During the first subperiod, learning is still unsupervised. In order to improve classification accuracy, learning can be made supervised during the second subperiod.

For the above learning scheme, the weight vectors can be topologically organized or unordered as in the present example. Topological organization might allow certain geometric shortcuts in the selection of the subsets to be used, resulting in speedup of convergence. For clarity and simplicity, only nontopological vector quantization is used below.

Let us now write the learning scheme into algorithmic form. Like in usual vector quantization, all the input weights must first be initialized properly. The method we have used in this experiment is to choose for the whole set of cells values of $m_i = [m_{i,\text{patt}}^{\text{T}}, m_{i,\text{cont}}^{\text{T}}]^{\text{T}}$ that are copies of a number of the first inputs $x = [x_{\text{patt}}^{\text{T}}, x_{\text{cont}}^{\text{T}}]^{\text{T}}$. These values can be assigned to the cells in any order.

Learning Period 1 (Adaptation of context input weights):

$$m_{c,\text{cont}}(t+1) = m_{c,\text{cont}}(t) + \alpha(t)[x_{\text{cont}}(t) - m_{c,\text{cont}}(t)] ,$$
$$\text{where} \quad c = \arg\min_i\{||x_{\text{cont}} - m_{i,\text{cont}}||\} . \qquad (6.27)$$

Continue until the quantization error $||x_{\text{cont}} - m_{c,\text{cont}}||$ has settled down to an acceptable value. From now on fix all the weights $m_{i,\text{cont}}$.

Learning Period 2 (Adaptation of pattern input weights):
Subperiod 2A (unsupervised): First find the best-matching cell in two phases as in (6.23)–(6.26): denote this cell by index $c1$. Now perform the updating of the pattern part of m_{c1}:

$$m_{c1,\text{patt}}(t+1) = m_{c1,\text{patt}}(t) + \alpha(t)[x_{\text{patt}}(t) - m_{c1,\text{patt}}(t)] . \qquad (6.28)$$

Continue until the quantization error $||x_{\text{patt}} - m_{c1,\text{patt}}||$ has settled down to an acceptable value.
Subperiod 2B (supervised): Since no identifiers are assigned to the context part, a training sample $x = [x_{\text{patt}}^{\text{T}}, x_{\text{cont}}^{\text{T}}]^{\text{T}}$ is always labeled according to the pattern part. The different cells must now be provided with labels, too. The problem of what and how many labels shall be assigned to the set of cells is generally a tricky one and may be treated in different ways. The method used in the present study was to have a calibration period in the beginning of Subperiod 2B such that a number of labeled training samples is first input, and then the cells are labeled according to the majority of labels in those samples by which a certain cell is selected.

The supervised learning algorithm selected for this phase was the LVQ1, which is reasonably robust and stable. When (6.23)–(6.26) are again applied to find cell $c1$, then

$$m_{c1,\text{patt}}(t+1) = m_{c1,\text{patt}}(t) + \alpha(t)[x_{\text{patt}}(t) - m_{c1,\text{patt}}(t)]$$
$$\text{if the labels of } x_{\text{patt}} \text{ and } m_{c1,\text{patt}} \text{ agree} ,$$
$$m_{c1,\text{patt}}(t+1) = m_{c1,\text{patt}}(t) - \alpha(t)[x_{\text{patt}}(t) - m_{c1,\text{patt}}(t)]$$
$$\text{if the labels of } x_{\text{patt}} \text{ and } m_{c1,\text{patt}} \text{ disagree} . \qquad (6.29)$$

The training samples are input iteratively a sufficient number of times, until the weight vectors can be regarded as asymptotically stable.

Application Example: Recognition of Phonemes. Recently we have always used cepstral features for the description of speech segments: each speech sample taken every 10 ms thus corresponds to 20 cepstral coefficients. The pattern vector $x_{\text{patt}} \in \Re^{60}$ that describes a phoneme is formed as the concatenation of three parts, each part being the average of three adjacent cepstral feature vectors, picked up from the middle part of the phoneme. The context vector $x_{\text{cont}} \in \Re^{80}$ is formed as a concatenation of four parts, each part being the average of five adjacent cepstral feature vectors. The context "window" is similarly centered around the phoneme in this experiment. Thus the pattern vector contains samples from a period of 90 ms; the context vector corresponds to a period of 200 ms. Both 1000 and 2000 nodes were used in the network, whereby almost comparable results were obtained. The cepstra were picked up from Finnish speech spoken by two male Finns. Initialization of the nodes was made by the samples of Mr. J.K., while learning and testing were performed with samples of Mr. K.T. In the present experiment, the number of training samples was 7650, and that of the independent test samples 1860, respectively. The vocabulary used in training and testing was the same. Throughout this experiment the constant values $\alpha(t) = .1$ in unsupervised learning and $\alpha(t) = .01$ in supervised learning were used. The number of training steps was 100000 in each period. The effect of context is seen from the following experiments with 2000 nodes. When in (6.24) $r \gg 1$, the subset S contains all the nodes, whereby the result is the same as in one-level recognition. With $r = 2.5$ the context domains had approximately optimal sizes; this optimum was rather shallow.

LVQ1 classification:
> Accuracy in the one-level recognition of single cepstra: 89.6%
> Accuracy in the one-level recognition of $x_{\text{patt}} \in \Re^{60}$:   96.8%

Hypermap classification:
> Accuracy in the two-phase recognition: 97.7 %

It must be emphasized, however, that accuracy is not the only indicator to be considered; in the Hypermap principle, the subset of candidates selected at the first level can be compared by a more coarse principle, whereas more refined comparison methods need to be applied only to this smaller subset, which saves computing time.

Conclusions and Further Developments. As stated above, for the neural cell array in the previous example, no topology was yet defined. On the other hand, if the recognition operation in a SOM were based on a function like in (6.3), one might even ask what the purpose of any topological ordering in the array thereby would be. The reason is rather subtle: due to the smoothing effect of the neighborhood relations during learning, the parameter vectors m_i will be distributed into the input signal space more efficiently, and learning is indeed faster than without this smoothing effect. As a matter of fact, it may happen in the usual LVQ algorithms at very high input dimensionality and with a large number of codebook vectors that a fraction of the weight vectors becomes "trapped" into interstitial locations between the other vectors so that they are never selected for "winners" and remain useless. If neighborhood effects are used during learning, practically all vectors become updated, and recognition accuracy will be increased due to the higher number of *active* weight vectors thereby always applied.

6.11 The "LVQ-SOM"

It will be necessary to point out that LVQ and SOM may be combined in a straightforward way. Consider the basic learning equation of the (unsupervised) SOM:

$$m_i(t+1) = m_i(t) + h_{ci}(t)[x(t) - m_i(t)] . \tag{6.30}$$

The following supervised learning scheme can be used if every training sample $x(t)$ is known to belong to a particular class, and the $m_i(t)$ have been assigned to respective classes, too. Like in LVQ, if $x(t)$ and $m_i(t)$ belong to the same class, then in the "LVQ-SOM" $h_{ci}(t)$ should be selected positive. On the other hand, if $x(t)$ and $m_i(t)$ belong to different classes, then the sign of $h_{ci}(t)$ should be reversed. Notice that this sign rule is now applied individually to *every* $m_i(t)$ in the neighborhood of the "winner." It may be advisable to use the "LVQ-SOM" scheme only after an unsupervised SOM phase, when the neighborhood has shrunk to its final value.

7. Applications

The neural networks are meant to interact with the natural environment, and information about the latter is usually gathered through very noisy but redundant sensory signals. On the other hand, in the control of effectors or actuators (muscles, motors, etc.) one often has to coordinate many mutually dependent and redundant signals. In both cases, neural networks can be used to implement a great number of implicitly or otherwise poorly defined transformations between the variables.

In recent years, when enough computing capacity has become available, researchers in various fields have begun to apply artificial-neural-network (ANN) algorithms to very different problems. Many results of the ANN theory, however, have only been used as simple mathematical methods, such as fitting nonlinear functional expansions to experimental data.

The *SOM algorithm* was developed in the first place for the *visualization of nonlinear relations of multidimensional data*. It has turned out quite surprisingly, as we shall see, that also more abstract relations such as the contextual roles of symbols in symbol sequences may become clearly visible in the SOM.

The *LVQ methods*, on the other hand, were intentionally developed for *statistical pattern recognition*, especially when dealing with very noisy high-dimensional stochastic data. The main benefit thereby achieved is radical reduction of computing operations when compared with the more traditional statistical methods, while at the same time an almost optimal (Bayesian) recognition accuracy can be reached.

In many practical applications, SOM and LVQ methods are combined: for instance, one may first form a SOM array in an unsupervised learning process, for optimal allocation of neurons to the problem, whereupon lots of raw input data (without manual identification or verification of their classes) can be used for teaching. After that, the codebook vectors of the various classes or clusters can be fine-tuned using LVQ or other supervised training, whereby a much smaller number of data with well-validated classification will suffice.

The three major practical application areas in which ANNs could be used effectively are: 1. Industrial and other instrumentation (monitoring and control). 2. Medical applications (diagnostic methods, prostheses, modeling, and profiling of the patients). 3. Telecommunications (allocation of resources to networks, adaptive demodulation and transmission channel equalization.

Ordering in the SOM depends on the configuration of samples in the signal space, and the form of this configuration depends on the scaling of the dimensions. In statistical problems, especially in *exploratory data analysis* of which we had examples in Sect. 3.14, very different scales may have been used for different attributes (indicators). In the SOM algorithms, it is often a good strategy, especially when the number of attributes is high, to use scales in which the mean of all values of a particular attribute is zero, and the variance of these values is normalized to unity. This simple preprocessing of the variables is often sufficient before their application to the SOM algorithm.

For most *real-world computations*, however, no artificial neural networks are sufficient as such. For instance, in engineering applications the neural networks usually constitute the interface between the real-world situations and machines, whereupon the measurements may be made using very sophisticated instruments. The real-world measurements contain variations for very different reasons: due to movements of the objects, differing illuminating conditions, heavy disturbances and noise, aging of the devices and their faults, etc., which then should be eliminated or compensated for. This is frequently done by non-neural *preprocessing* methods. The share of preprocessing of all computations may be much larger than that of decision making.

In this chapter we shall first discuss the preprocessing problem that is common to most applications. Sections 7.3 through 7.11 then exemplify the main trends in SOM applications. A complete guide to the extensive SOM literature and an exhaustive listing of the detailed applications can be found in Chap. 10.

7.1 Preprocessing of Optic Patterns

Most of the basic ANN algorithms compare input pattern vectors with weight vectors using dot products or Euclidean differences, whereupon they are not meant to tolerate even insignificant amounts of transformation in elementary operations such as translation, rotation, or scale changes of the patterns. Consider Fig. 7.1; the two binary patterns are completely orthogonal, although looking very similar to us, and would be interpreted as totally dissimilar by such a system. Therefore, one should never use simple ANN methods for the classification of line figures without *preprocessing*.

In other words, the basic ANN is only supposed to act as a *statistical decision-making machine* that works very well if the input data elements represent *static* properties. Natural signals, which are often dynamic, and the elements of which are mutually dependent, must therefore be preprocessed before their presentation to the ANN algorithms. One of the purposes of such preprocessing is to select a set of *features* that are *invariant with respect to some basic transformation groups* of the input patterns. Sometimes, however,

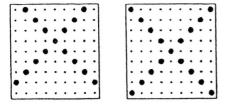

Fig. 7.1. Two line figures that are orthogonal, in spite of looking similar

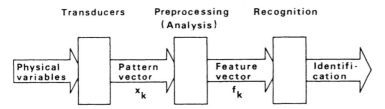

Fig. 7.2. Organization of a pattern recognition system

some features will be functions of the position and orientation of the pattern, whereby this information may be needed for the correct interpretation.

Figure 7.2 delineates a simple *pattern recognition system* showing the different stages and variables associated with it.

In this section we shall mention a few basic preprocessing methods for optic patterns; others are explained in connection with the applications.

7.1.1 Blurring

For line figures or photographs with a very high contrast at their contours, one of the simplest preprocessing methods is *blurring*, or linear convolution of the original pattern with some *point spread function* (Fig. 7.3). There is ample freedom in the choice for the latter. The dot product of blurred images is to some extent tolerant with respect to the translation of the patterns.

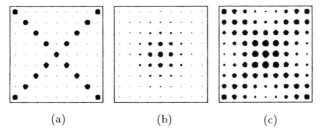

(a)   (b)   (c)

Fig. 7.3. Convolution of an input pattern with a blurring kernel. **(a)** Pattern **(b)** Point spread function (kernel) **(c)** Convolution

7.1.2 Expansion in Terms of Global Features

The original image can also be expanded in terms of some known two-dimensional *basis functions*, and the coefficients of such an expansion can be used as feature values for the ANN algorithm.

A typical task in engineering is to identify various distinct *objects* from their observations. As they may appear in different variations, direct identification of their images on the basis of the exact constellations of their pattern elements is usually not advisable. One principle, for instance in the recognition of faces or written characters well-positioned in the input field, is to describe the input data in terms of some *eigenvectors*. For instance, as discussed in Sect. 1.3.1, if the vector formed of all pixels of an image is regarded as a real vector $x \in \Re^n$, then one might first compute the correlation matrix

$$C_{xx} = \frac{1}{N} \sum_{t \in S} x(t)x^{\mathrm{T}}(t) \tag{7.1}$$

using a large set of sample images (here S is the set of image indices and N is the number of elements in S). If the eigenvectors of C_{xx} are denoted $u_k \in \Re^n, k = 1, 2, \ldots, n$, then any image vector x' can be expanded in terms of the eigenvectors,

$$x' = \sum_{k=1}^{p} (u_k^{\mathrm{T}} x') u_k + \varepsilon , \tag{7.2}$$

where $p \leq n$ and ε is a residual. A restricted set of the largest terms $u_k^{\mathrm{T}} x'$ called the *principal components* can be used in (7.2) to describe the approximate appearance of x'. This is an example of a *feature set* that is a compressed description of x' at some statistical accuracy. There also exist numerous other eigenfunctions of the input data, in terms of which the feature set can be defined (cf. Sect. 7.1.5).

When using features that refer to the whole input field, the images have to be *standardized*: for instance, their centroids are made to coincide and each image is scaled properly. Frequently, especially in the recognition of moving targets, this is not possible, however. The signals emitted by the object are transformed into observations in a complex way. In biology as well as artificial perception the solution is then to extract from the primary observations a set of *local features* that are more or less *invariant* with respect to the natural transformations.

7.1.3 Spectral Analysis

In order to make the representation of an image totally invariant to *translation* in any direction, one can use the elements of the two-dimensional Fourier amplitude-spectrum as features. (Naturally, by computational means, a simpler approach would be to shift every image into a position where the centroid of its intensity distribution coincides with the center of the picture frame; but this only works under carefully standardized conditions, not, e.g., under varying illuminating conditions that would change the centroid.) To achieve total *scale invariance*, the modulus of the Mellin transformation can be taken (whereby the picture must already be centered, etc.). If one tries to combine several successive transformations of the above types on the same pattern, numerical errors are accumulated and noise is enhanced. For simple patterns such as optical characters, computational normalization of their major dimensions, orientation, and position seems to be the best method in practice.

7.1.4 Expansion in Terms of Local Features (Wavelets)

For instance, in image and signal analysis, the *wavelets* (Sect. 5.11.2) have recently gained much importance as local features. The two-dimensional *self-similar wavelet* or the *Gabor function* is defined over the image field as the complex function

$$\psi_k(x) = \exp[i(k \cdot x) - (k \cdot k)(x \cdot x)/2\sigma^2] , \tag{7.3}$$

where $x \in \Re^2$ is the location vector on the image plane, and $k \in \Re^2$ is the wave vector, respectively. The wavelet transform F of function $f(x)$ centered around $x = x_0$ is

$$F = F(k, x_0) = \int \psi_k(x - x_0)f(x)dx . \tag{7.4}$$

Its modulus $|F(k, x_0)|$ is almost invariant with respect to displacements of $f(x)$ around location x_0 in a domain where the radius is of the order of σ. To the neural-network inputs one can select a set of such local features $|F(k, x_0)|$ with different values of k and x_0. Referring to the same x_0, such a set is also called the *Gabor jet* [7.1].

The ASSOM filter bank discussed in Sect. 5.11 seems to be a viable alternative for the mathematically defined Gabor functions. The ASSOM can also learn other transformations in addition to translation, and even mixtures of transformations (Fig. 5.30).

7.1.5 Recapitulation of Features of Optic Patterns

Many basis functions are useful for preprocessing. All of them do not possess invariance with respect to arbitrary shifts, rotations, etc.; they only reduce the effect of misalignments and deformations. The following is a short list of such generally known mathematical functions:

- Fourier transform
- Mellin transform
- Green's functions (e.g., Gaussian times Hermite's polynomials)
- eigenvectors of the input correlation matrix
- wavelets
- Gabor functions
- oblate spheroidal functions
- Walsh functions (binary)
  etc.

Let us recall that filters for wavelet-type and Gabor-type features are formed automatically in certain kinds of self-organizing maps (Sect. 5.11).

Heuristically Chosen Features. Especially in industrial robotics, the environmental conditions such as illumination can usually be standardized, so detection of features of the following kinds from targets will be easy: maximal and minimal diameters, lengths of perimeters, number of parts, holes, and cusps, and geometric moments, which are all invariant properties in translation and rotation of the targets, thereby allowing easy identification of the targets. A little more developed set of invariant features, relating to planar patterns, ensues from the *chain code* or *Freeman code* [7.2] that describes the form of the contour, starting from a reference point and detecting quantized changes in direction when traversing the contour.

In this chapter we shall also discuss the analysis of textural information in images. There exist numerous texture models and choices for textural features. In connection with the SOM we have found the so-called *co-occurrence matrix* [7.3] useful; its elements, and in particular statistical measures of the elements, can be used for the characterization of texture.

7.2 Acoustic Preprocessing

Acoustic signals may be picked up from speech, animal voices, machine and traffic noise, etc. A very special area is underwater acoustics. The microphones or other transducers can be very different and provided with different preamplifiers and filters. The analog-to-digital conversion of the signals can be performed at different sampling frequencies (we have used for speech 10 to 13 kHz) and different accuracies, say 12 to 16 bits. Higher numerical accuracy may be advantageous since it allows sounds with different intensities to be analyzed in the same dynamic range.

In the recognition of acoustic signals such as speech, local-feature extraction can be made in a computationally effective way by taking fast discrete Fourier transforms over specified time windows. In many practical systems the digital Fourier transform of the signal is formed at 256 points (we have used 10 ms time windows for speech). Logarithmization of the amplitude spectrum, and subtraction of the mean of all outputs makes detection of sounds intensity-invariant. The spectrum can still be filtered digitally, whereafter, in order to reduce dimensionality, the components of the input vector can be formed as averages over adjacent Fourier components (we have used, e.g., 15 averages, i.e. 15-dimensional speech vectors).

Normalization of the length of the acoustic feature vector before its presentation to the ANN algorithm is not always necessary, but we have found that the overall numerical accuracy of statistical computations, in connection with LVQ and SOM algorithms, is noticeably improved by that.

For the recognition of speech and, for instance, vehicles on the basis of their noise, it is usually advantageous to form the so-called *cepstrum*. The *cepstral coefficients* are then used for features. The cepstrum is defined as two successive Fourier amplitude transformations performed on the same signal, with the first transform usually logarithmized. If the signal is generated as a set of acoustic resonances driven by pulsed excitation, like the speech signal is, then the driving function can be eliminated by the cepstral analysis, using a mask for the driving function after the second transformation.

7.3 Process and Machine Monitoring

An industrial plant or machine is a very complicated system; it ought to be described in terms of state variables, the number of which may exceed the number of available measurements by orders of magnitude. The state variables may also be related in a highly nonlinear way. In the sense of classical systems theory, the analytical model of the plant or machine would then not be identifiable from measurements. Nonetheless there may exist much fewer characteristic states or state clusters in the system that determine its general behavior and are somehow reflected in the measurements. As the SOM is a nonlinear projection method, such characteristic states or clusters can often be made visible in the self-organized map, without explicit modeling of the system.

An important application of the SOM is in fault diagnosis. The SOM can be used in two ways: to detect the fault or to identify it. In practical engineering applications, we can then distinguish two different situations; either we have no prior measurements of a fault, or we have also been able to record or at least to define the fault.

7.3.1 Selection of Input Variables and Their Scaling

Consider a set of simple industrial measurements, whereby invariances of signal patterns are not yet taken into account. The measurements may have been taken with very different devices and in different scales. In statistical decision making, the set of measurements is often regarded as a *metric vector* that may be analyzed relating to the Euclidean space. The various elements of the vectors should therefore represent information in as balanced way as possible. This is particularly important with the SOM algorithm.

If the number of measuring channels or variables to be considered is very large, say, thousands, it will be necessary to select a much smaller number of representative variables to the inputs of the neural network. To this end one might analyze the information-theoretic entropy measure of the different channels to decide which variables should be selected.

To equalize the contributions of the different variables to the classification results, the simplest method is to *normalize* all variables into scales in which, e.g., the mean of each variable is zero and its variance is the same.

In the general case the variables are mutually dependent, i.e., their joint distribution is structured and cannot be expressed as the product of marginal distributions. If there is any doubt about the SOM mapping producing good projections, it is advisable to first make a clustering analysis of the measurements, e.g., by the *Sammon mapping method* (Sect. 1.3.2) in order to deduce whether simple rescaling is sufficient to separate the clusters, or whether the primary measurements should be somehow combined to eliminate their interdependence.

In Sect. 7.3.2 we will discuss two particular examples that are characteristic of measurements made in engineering, and for which a simple rescaling has been sufficient.

7.3.2 Analysis of Large Systems

Understanding and modeling of complex relationships between variables in large systems is often problematic. Automated measurements produce masses of data that may be very hard to interpret. In many practical situations, even minor knowledge about the characteristic properties of the system would be helpful. Therefore, the online measurements ought to be converted into some simple and easily comprehensible display which, despite the dimensionality reduction, would preserve the relationships between the system states. With this kind of transformation available, 1) it would be possible for the operating personnel to visually follow the development of the system state, 2) data understanding would facilitate estimation of the future behavior of the system, 3) abnormalities in the present or predicted behavior of the process would make identification of fault situations possible, and also 4) the control of the system could be based on the state analysis.

The SOM represents the most characteristic structures of the input density function in a low-dimensional display. Together with proper visualization methods the SOM is thus a powerful tool for discovering and visualizing general structures of the state space. Therefore the SOM is an efficient tool for visualizing the system behaviour, too.

Consider a physical system or device, from which several measurements are taken. The measurements are then normalized to make their dynamic ranges correspond to the same range of real numbers. Denote the measurement vector by $x = [\xi_1, \xi_2, \ldots, \xi_n]^{\mathrm{T}} \in \Re^n$. In addition to the system variables, control variables may be included in the measurement vector.

The device studied in the following case example was a power-engineering oil transformer, from which ten measurements (electrical variables, temperatures, gas pressures etc.) were taken [7.4]. The state of the system can be made to depend on the feeding and loading powers, and variations and faults in transducers can be induced experimentally or simulated by computer programs. Neural-network models usually need very many training examples, whereby destructive faults in the device itself cannot be made in large quantities.

Measurement vector values x can thus be varied in the laboratory or by simulation. If all state types are equally important from the point of view of monitoring the device, an equal number of samples of each state should be taken. To emphasize a particular state type, the learning rate α (Sect. 3.2) can be varied accordingly. The SOM is formed in the usual way, by iterating the x values in learning.

Assume that enough measurements are available for the SOM algorithm that forms the weight vectors $m_i = [\mu_{i1}, \mu_{i2}, \ldots, \mu_{in}]^{\mathrm{T}}$. The map array so created is then fixed prior to its use in monitoring.

Any component plane j of the SOM, i.e., the array of scalar values μ_{ij} representing the jth components of all the weight vectors m_i and having the same format as the SOM array, can be displayed separately, and the values of the μ_{ij} can be represented on it using shades of gray.

The SOM can then be used in two ways: 1) the *trajectory* consisting of successive images of operating points can be displayed *on the original SOM*, on which some labels may indicate general information about the system states, or 2) the same trajectory can also be drawn on the display of any of the *component planes*, whereby the shades of gray along the trajectory directly indicate the values of that process variable over time.

The practical example shown in Fig. 7.4, as mentioned, is a power transformer. In this example, the states relating to the measurements of ten variables have been followed during a period of 24 hours.

In Fig. 7.4, the trajectory has been drawn on the component plane that displays the oil temperature at the top of the transformer. Dark gray tones correspond to small loads and light tones to heavy loads, respectively. It can be seen that the operating point has moved from dark to light areas and back again corresponding to the daytime operation of the system, when the load was switched on and off. In this specific example, the trajectories of the successive days were very similar during normal operation.

Fault Identification. In the first place, fault detection can be based on the *quantization error*: when the feature vector corresponding to the measurements is compared with the weight vectors of all map units, and the smallest difference exceeds a predetermined threshold, the process is probably in a fault situation. This conclusion is based on the assumption that a

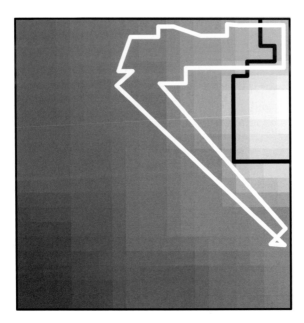

Fig. 7.4. The trajectory (in white) on the SOM indicates state changes of a power transformer during one day. The first component of the reference vector that represents the oil temperature at the top of the transformer was displayed in gray scale, showing high values in bright shade and low values in dark, respectively. Dangerous states are framed in the map with a dark line

large quantization error corresponds to the operation point belonging to the space not covered by the training data. Therefore, the situation is new and something is possibly going wrong.

In addition to visualizing the normal operating conditions by the SOM, however, it would be desirable to be able to visualize fault states, too. The main problem thereby is where to get abnormal but still typical measurements from. To observe all possible faults is obviously impossible. The faults may be rare and true measurements thus not available. In some cases, fault situations can be produced during testing, but especially in the case of industrial processes and big machines, production of severe faults might be too costly. Then the faults must be *simulated*.

If fault situations and their reasons are known well enough, simulated data for SOM training can be produced easily. In practical engineering systems, the different types of faults or errors most usually encountered are: 1) A break in signal lines resulting in abrupt drop of a signal, 2) heavy disturbance with a large sporadic change in signals, 3) jamming of a measurement value, often caused by a mechanical fault in measuring equipment, 4) slow drift in measured value due to aging of the device, and 5) sudden change in some regulated parameter value, indicating a fault in the control mechanism. All

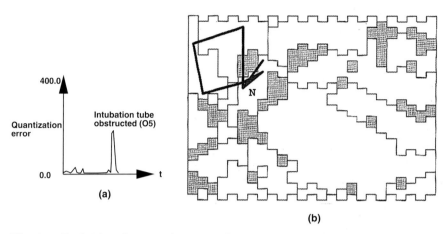

Fig. 7.5. Fault identification of an anaesthesia system tested in a true situation: N = normal state. The position of the patient was changed and the intubation tube was obstructed for a short period of time. The increase in the quantization error shown in figure (**a**) indicates that a fault has been detected. The trajectory of the operating point was moving from the area corresponding to the normal situation to the area that corresponded to an obstruction in the specific part of the system. The trajectory is depicted in figure (**b**)

the mentioned fault types can be simulated easily and independently. Thus, the feature vector x used in training can be varied artificially (by causing mechanical and electric disturbances), or by simulation.

If measurements of most of the typical fault types are available or can be simulated, mechanically or by modeling, the SOM can be used as a monitor of the operating conditions of the system. Figure 7.5(b) describes the SOM computed for an *anaesthesia system* [7.5], and different areas on it correspond to obstruction of tubes, hoses, or pumps, their breakage, wrong gas mixtures, wrong positioning of the intubation tube etc.

However, if the operating conditions of the system vary within wide limits, or if it is not possible to define "normal" situations such as a "normal" patient, the SOM must be used on two levels for the analysis of faults or alarms. On the first level, the so called *fault-detection map* is used. The fault detection is based on the *quantization error*. On the second level, a more detailed map is used to identify the reason of the fault. By storing a sequence of feature vectors, the behavior of the process before and during the occurence of the fault can be analyzed in more detail. Examples of fault detection and identification in an anaesthesia system have been described in [7.5].

For very different system states or conditions one has to compute several maps, one for each type of condition (an "atlas" of maps) and the most suitable one, based on some criteria, is then selected.

7.4 Diagnosis of Speech Voicing

The phoneme map example shown in Chap. 3, Fig. 3.18, suggests that the SOM might be used to detect statistical differences between *speakers*, too. To this end a special SOM, the "Phonotopic Map" [5.22] must be made using samples collected from many different speakers. It is advisable to make separate maps for male and female speakers, because their glottal frequencies differ approximately by an octave, which can easily be detected, whereas no large systematic differences in the spectral properties of their vocal tracts otherwise occur.

One of the first problems in speech disorder analysis is to develop objective indicators for the quality of voice, which mainly reflects the condition of the glottal cords. In the clinical evaluation of voice, attributes such as "sonorant", "strained", "breathy", "hoarse", and "rough" are generally used; but such diagnoses tend to be unreliable and at least subjective, differing from one physician to another. The original objective [7.6, 7] was to develop objective indicators that could be computed automatically from recorded data, such as speech spectra.

One of the central tasks thereby is to restrict the studies to the pronounciation of a representative test word, such as /sa:ri/, a Finnish word meaning "island." A Phonotopic Map (SOM) can be made for all the phonemes collected from several speakers, but only the area in the neighborhood of the representation of /a:/ is thereby utilized (Fig. 7.6).

For normal speakers, the trajectory representing speech is stationary around /a:/; the area of the ellipse drawn at a particular map unit is proportional to time, i.e., the number of subsequent speech samples selecting this unit for a "winner." Differences between test objects are clearly discernible. If one analyzes the lengths of transitions on the map during the pronounciation of /a:/, one can use their distribution as an objective indicator of the quality of voicing.

7.5 Transcription of Continuous Speech

Most devices developed for speech recognition try to spot and identify isolated words, whereupon the vocabulary is limited. A different approach altogether is to recognize *phonemes* from continuous speech. Some languages (French, Chinese, English) present thereby bigger problems than others (Italian, Spanish, Finnish, Japanese).

The level of ambition should also be specified. Interpretation of free speech from an arbitrary speaker, without prior speech samples given is impossible with the present techniques. Some experiments aim at recognition of command words in a very noisy environment, such as a helicopter cockpit. The special problem studied by us since the 1970s has been to transcribe care-

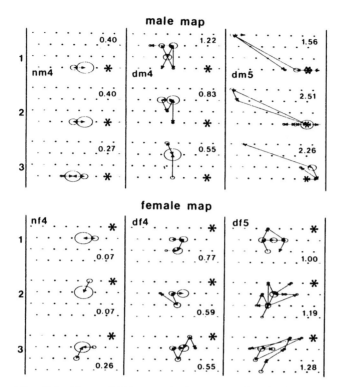

Fig. 7.6. Maps of 150-msec /a:/ segments during three repetitions of a test word (separated by seven other words) by 2 normal (nm4, nf4) and 4 dysphonic voices (dm4-5, df4-5). The value of the mean length of shifts is given on each map. The dysphonic voices were categorized as follows: dm4 mildly dysphonic, breathy, and strained; dm5 severely dysphonic, rough, breathy, and strained; df4 mildly dysphonic and breathy; and df5 severely dysphonic, rough, breathy, and strained

fully articulated but general speech into a unique phonemic transcription; in Finnish, like some other languages, this transcription is then reasonably well convertible into the official orthography of the written language.

Preprocessing. Spectral analysis for the preprocessing of speech signals has been almost standard in speech processing, whereas there exist several choices for speech features derivable from it. For the present, the two main alternatives for feature sets based on speech spectra are the short-time Fourier amplitude spectra as such, and the nonlinear combination of two cascaded Fourier transformations, which results in another feature set called the *cepstrum*.

Frequency analysis has been traditional in acoustics, and speech analysis has almost exclusively been based on so-called sonograms (also called sonagrams or spectrograms), i.e., sequences of short-time spectra plotted vs. time. In the automatic identification of these spectra and their division into phonemic classes there exists at least one principal problem: the speech voice is mainly generated by the vocal cords, and the pulsed flow of air causes characteristic acoustic resonances in the vocal tract. According to generally known system-theoretic principles, the vocal tract can be described by a linear transfer function, and the operation of the vocal chords by a system input that consists of a periodic train of impulses. The latter has a broad spectrum of frequency components spaced periodically apart in the frequency scale at multiples of the fundamental (glottal) frequency. The output voice signal thus has a spectrum that is the product of the transforms of the glottal air flow and that of the vocal tract transfer function, respectively. It may then be clear that the voice spectrum is modulated by all the harmonics of the fundamental frequency (Fig. 7.7(a)). This modulation is not stable; its depth, and the locations of its maxima and minima in the frequency scale vary with articulation and the pitch of speech. In direct template comparison between two spectra, severe mismatch errors may thus be caused due to this variable modulation.

In the cepstral analysis, the harmonics of the speech signal can be eliminated by a series of operations. Each component of the Fourier amplitude spectrum of the speech signal is usually logarithmized, and when a second Fourier amplitude spectrum is then taken, the periodicity of the modulation is converted into a single peak on the new transform (cepstrum) scale (Fig. 7.7(b)). This peak is easily filtered out.

The 256-point transforms, for computational reasons, are compressed by combining their components into 20 pattern elements. Details of the filtering operations of cepstral analysis can be found in [7.8].

Converting Cepstra into Quasiphonemes. Each of the feature vectors obtained above is classified into one of the 21 phonemic classes of the Finnish language. In the simplest version of implementation, each feature vector is classified individually; slight improvement results if several adjacent feature vectors (say, 3 to 8) are concatenated into a correspondingly longer pattern vector to encode more temporal information.

There exist many classical pattern recognition algorithms for this subtask. In our earlier system [5.23] we used the Supervised SOM discussed in Sect. 5.10. In order to achieve the highest possible speed without losing accuracy, we later have used Learning Vector Quantization (LVQ). This approximation is comparable in accuracy to the best nonlinear statistical decision methods. For the classification of stop consonants we have used auxiliary "transient maps."

Classification of a pattern vector is now a very fast, computationally light task: the unknown vector is simply compared with the list of reference vectors, and labeled according to the most similar one. With fast signal-processor chips, comparison with even 1000 reference vectors can be made every 10 ms. Careful prior optimization of the reference vectors usually results in higher classification accuracy.

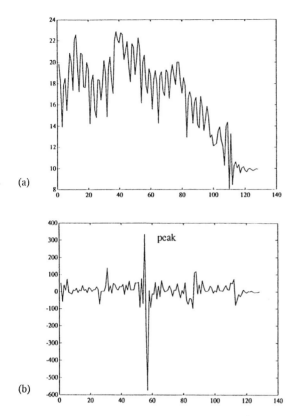

Fig. 7.7. **(a)** Spectrum (in arbitrary units) picked up from the diphthong /ai/, showing its periodicity due to the harmonics, **(b)** the corresponding cepstrum. Horizontal scale: channel number in FFT

What we obtain after the previous steps is still only a sequence of symbols, each of them identifying one (phonemic) state described by the corresponding short-time spectrum. These symbols, called *quasiphonemes*, are used to assign an acoustic label to the speech waveform every 10 ms. Another operation, as demanding as the previous ones, is needed to merge the quasiphonemes into segments, each one representing only one phoneme of speech. Since the quasi-phonemes contain a certain fraction of identification errors, this *segmentation* (also called *decoding*) into phonemes must involve statistical analyses.

Decoding Quasiphoneme Sequences into Phonemes. The simplest decoding method that we originally used was based on a kind of voting [5.22]. Consider n successive labels (a typical value being $n = 7$; however, n can be selected individually for each class). If m out of them are equal (say, $m = 4$), then this quasiphoneme segment is replaced by a single phoneme label of the type of majority. Some extra heuristic rules are needed to avoid overlappings of this rule and to obtain an approximately correct number of phonemes in a continuous scan.

Recently we have used for decoding another, widely known method called the *Hidden Markov Model (HMM)* [7.9]. Its basic idea is to regard the quasiphonemes as states of a stochastic model that undergo transitions according to certain conditional probabilities. The main problem is to estimate these probabilities from available examples of output sequences. By means of the HMM it is possible to decide optimally how many quasiphonemes correspond to one phoneme, and what the most probable phoneme is.

Grammatical Methods for the Correction of Symbol Strings. One problem, not yet discussed above, is connected with the *coarticulation effects* in speech. It has been stated that speech is an "overlearned" skill, which means that we are generally not trying to reproduce the standard phonemes, but complete utterances, thereby making all sorts of "shortcuts" in our articulation. Some of these transformations are just bad habits, which the machine then ought to learn, too. However, even in the speech of careful speakers, the intended phonemic spectra, due to dynamical dependencies, are transformed differently within different *contexts* of other phonemes. Often a phoneme is spectrally transformed into another phoneme. Any recognition system must then be able to take the neighboring phonemic context into account.

A novel method for the compensation of coarticulation effects is to perform this compensation at the symbolic level of the quasiphonemes, or even correcting errors from the final transcriptions (Sect. 1.6). One of the main benefits when working on the symbolic level is that information is already highly compressed, thereby allowing complex operations at very high speed, using relatively slow computing devices. The second, very significant advantage when working with discrete data is that the context can be defined *dynamically*; if some form of coarticulation is very common and consequent, even a short context may be sufficient for the description of the coarticulation effect. Consider the Finnish word /hauki/ (meaning pike), where /au/ is almost invariably pronounced (and detected by the speech recognizer) as /aou/; one might then use the substitution /aou/ ← /au/ to correct this part into orthographic form. More often, however, for the description of a particular coarticulation effect uniquely, one has to take into account a context that extends over several phonemes. It would therefore be advantageous if the width of the context could be different from case to case. In practice, due to the very great number of combinations of phonemes that may occur in speech, the number of different rules must be large; in our system we have had tens of thousands of rules (that are learned completely automatically from examples by our method). This kind of grammatical rule can already be applied at the quasiphoneme level, to make the phonemic segments more uniform (by averaging out random symbolic errors); the corresponding grammar developed for this task is called *Dynamically Focusing Context (DFC)* [7.10]. After decoding quasiphonemes into phonemes, another symbolic method named *Dynamically Expanding Context (DEC)* (Sect. 1.6) can be used to correct a significant portion of the remaining phonemic errors, as well as to convert the phonemic transcriptions into orthographic text. The DEC, invented by this author in 1986, is somewhat related to *Rissanen's minimum description length (MDL) principle* [1.78] published in the same year.

Benchmarking Experiments. As mentioned earlier, our original system already worked in real time. However, because appreciable changes were made later, it may be more interesting to report only the recognition accuracies relating to the latest system, whereby the speed of operation is of no primary concern; for instance, with contemporary workstations we are already very close to real time anyway, even when using software methods.

The scheme of the experimental setup that delineates interrelations of the processing modules is shown in Fig. 7.8.

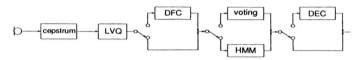

Fig. 7.8. Simplified configuration of the experimental system

The average figures given in Table 7.1 refer to an extensive series of experiments with three male Finns that represented average speakers. The results have been reported in more detail in [7.8]. For each speaker, four repetitions of a set of 311 words were used. Each set contained 1737 phonemes. Three of the repetitions were used for training, and the remaining one for testing. Four independent experiments were made by always leaving a different set out for testing (the *leave-one-out* principle); in this way, while maximally utilizing the statistics, one can be sure that the data used for training, and those used for testing are *statistically independent*.

Table 7.1. Benchmarking of different combinations of algorithms (with eight cepstra concatenated into a pattern vector)

DFC included	Decoding HMM	Voting	DEC included	Correctness %
	V			94.3
		V		90.6
V		V		93.9
	V		V	96.2
V		V	V	95.6

Notice that the DFC was only used in combination with the simple majority-voting decoding.

We have implemented the on-line, continuous-speech recognition system by various hardware architectures. The version of 1994 uses no other hardware than a Silicon Graphics Indigo II-workstation, whereby a software-based real-time operation has been achieved.

In a new phase of development [7.11–13] significantly wider contexts than mentioned above were used. Our last speech recognition experiments have been carried out using Markov models to describe the phoneme states. The SOM and LVQ algorithms were then used to approximate the state density functions [7.14–23].

7.6 Texture Analysis

Of all image-analysis applications of SOM and LVQ, those concentrating on *texture analysis* have been used longest in practice [2.71], [7.24–35]. In paper and woodmill industries, analysis of the surface quality is an important part of production. It can be performed by practically the same procedure that is discussed next, namely, the method used to segment and classify cloud types from satellite images (NOAA-satellite).

Preprocessing. The picture area in a satellite image may be divided into 9 by 9 groups of multispectral pixels, corresponding, e.g., to 4 by 4 km^2 or 1 by 1 km^2 each. There exist many approaches to the modeling of textures. With stochastic patterns such as the cloud images in which no natural contours occur, the *co-occurrence matrix* approach [7.3] has been used with success. To illustrate what a co-occurrence matrix is, we shall use the following very simple example. Consider Fig. 7.9a that delineates a pixel group with shade values 0, 1, and 2. These pixels lie on horizontal rows. Consider for simplicity only pairs of pixels with relative distance of $d = 1$ *from left to right*. There are two such pairs with values (0,0) and (0,1), and one with (2,1) and (2,2), respectively; no other combinations are found. The value pairs (0,0), (0,1) etc. identify the locations of elements in matrix $M_d(d = 1)$ in Fig. 7.9b, whereby

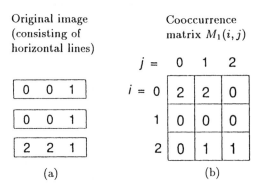

Fig. 7.9. (a) Rows of pixel values. (b) The corresponding co-occurrence matrix M_1

the value of the left pixel in the pair is the row index of the matrix running left to right, and the value of the right pixel is the column index running from top to bottom, respectively. The elements of M_d represent occurrences of the said pairs of values in the pixel groups.

The co-occurrence matrices M_d are in general constructed for several relative displacements d of the pixel pairs; in two-dimensional pixel groups they may be picked up in any direction.

The elements of M_d represent certain statistics of the shade values. For further compression of information, statistical indicators of the elements $M_d(i, j)$ themselves can be used as texture features: such are

$$\text{``Energy''} \quad = \quad \sum_i \sum_j M_d^2(i, j) \; ,$$

$$\text{``Momentum''} \quad = \quad \sum_i \sum_j (i - j)^2 M_d(i, j) \; , \quad \text{and} \tag{7.5}$$

$$\text{``Entropy''} \quad = \quad \sum_i \sum_j M_d(i, j) \ln M_d(i, j) \; .$$

Texture Map and Its Calibration. Each of these three indicators, representing the 9 by 9 pixel groups, can be computed for several values of d, whereafter the set of such indicators is defined to be the input vector x to the SOM. A clustering of the cloud types was first formed by a SOM. Labeling of the clusters must be made using known samples. The map vectors are fine tuned using LVQ and calibration data. Classification results from the map are then used to define the pseudocolor coding of a pixel group in the final image. Fig. 7.10 shows an example of cloud segmentation by this method.

7.7 Contextual Maps

Symbols in symbol strings may be regarded as "pixels" of a one-dimensional pattern, the domain of their values consisting of the *alphabet*. Similarly, words in sentences or clauses are "pixels" with values picked up from a *vocabulary*. In both cases the "pixels" can be encoded numerically, e.g., by unit vectors or random vectors (in which cases such a vector represents one "pixel"). In the case of using random vectors, the only requirement is that a metric is definable in which the distance of a "pixel" from itself (in the input space) is zero, whereas the distance between different "pixels" is nonzero and as independent of the symbols as possible.

Simple *"context patterns"* consist of groups of contiguous symbols. For instance, when scanning a string from left to right, symbol by symbol, *pairs* or *triplets* of symbols can be regarded as the feature patterns used as components of the input x to a SOM.

Very meaningful "symbol maps" are obtained if, say, the triplets are labeled by their middle symbol, and the SOM formed of the triplets is calibrated according to these labels. Such SOMs we call *contextual maps*; earlier [7.36]

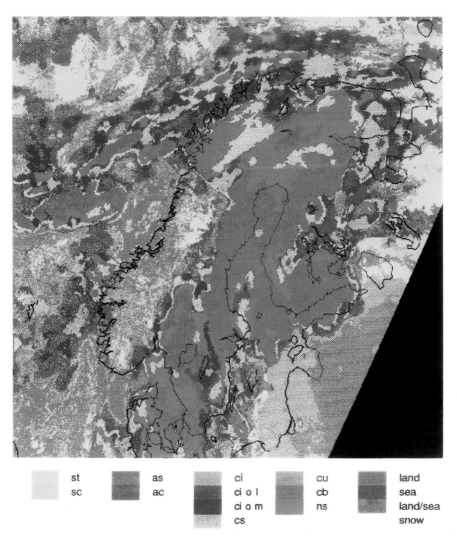

Fig. 7.10. Pseudocolor image of clouds over Scandinavia, segmented according to their texture by SOM. Legend: st = stratus, sc = stratocirrus, as = altostratus, ac = altocumulus, ci = cirrus, ci o l = cirrus over land, ci o m = cirrus over middle clouds, cs = cumulostratus, cu = cumulus, cb = cumulonimbus, ns = nimbostratus (rain cloud)

we also used the name "semantic map," but the only semantics thereby meant is defined by the *statistical occurrence of items in certain contexts*. Since we want to apply the same principle to many other similar but nonlinguistic problems, too, it may be expedient to call these SOMs henceforth also "contextual maps". Such maps are used to visualize the relevant relations between symbols, according to their roles in their use. In linguistics this would mean illustrating the emergence of word categories from the statistical occurrences of words in different contexts.

7.7.1 Artifically Generated Clauses

To illustrate the basic principle of contextual maps, we start with the "semantic maps" described in [7.36]. The symbols, the words, are defined by an artifically constructed vocabulary (Fig. 7.9a), which is further subdivided into categories of words, each of them shown on a separate row. Words in the same category are freely interchangeable in order to make meaningful clauses, whereas words of different categories have different roles in verbal expressions; all combinations of words do not make sense.

We only consider clauses of the type (noun, verb, noun) or (noun, verb, adverb) and it is believed that below all *meaningful* clauses of such types are defined by the sentence patterns (the numerals referring to word categories, i.e. rows) as listed in Fig. 7.11b.

As the codes used for the words should be as independent as possible, each word in the vocabulary was encoded by a seven-dimensional, unit-norm random vector $x_i \in \Re^7$ drawn from the surface of a seven-dimensional unit sphere. It was shown by analyses in [2.73] that such x_i and x_j, $i \neq j$ can be regarded as approximately orthogonal and statistically independent.

It may be necessary to advise how random vectors with a uniform distribution of their directions and unity length can be made. It is first necessary to make a spherically symmetric distribution around the origin. This follows automatically if each component ξ_{ij} of x_i has an independent Gaussian distribution with zero mean. Such a Gaussian distribution can be approximated with high accuracy by adding several, say, ten statistically independent

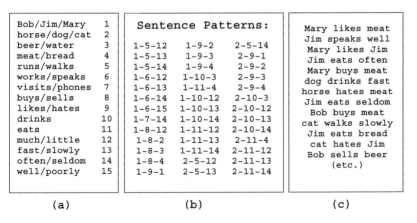

Fig. 7.11. (a) List of used words (nouns, verbs, and adverbs), (b) Legitimate sentence patterns, and (c) Some examples of generated three-word clauses

random numbers that have a constant probability density function on some interval $[-a, a]$ and zero outside it. After all the ξ_{ij} have been calculated, the vector x_i is normalized to unit length.

A great number (300 000) of meaningful clauses was generated by randomly selecting a sentence pattern from Fig. 7.11b and randomly substituting words to it from the vocabulary. Samples of meaningful clauses are exemplified in Fig. 7.11c. *Theses clauses were concatenated into a single source string without any delimiters.*

Input Patterns. The input patterns to the SOM could be formed, e.g., by concatenating triplets of successive words into a *context pattern* $x = [x_{i-1}^{T}, x_i^{T}, x_{i+1}^{T}]^{T} \in \Re^{21}$ (whereupon x and x_i are column vectors). We have also used pairs of words, without much difference in the resulting SOMs. The source string formed first was scanned left-to-right, symbol by symbol, and all triplets accepted for x. If the latter was formed of words in adjacent, independent sentences, such components relative to each other behaved like noise.

Accelerated and Balanced Learning. To speed up learning significantly and to make the labeling of the SOM more consequent, as we shall see soon, we actually used a batch process. Let a word in the vocabulary be indexed by k and represented by a unique random vector r_k. Let us then scan all occurrences of word k in the text in the positions $j(k)$, and construct for word k its "average context vector"

$$x_k = \begin{bmatrix} \mathrm{E}\{r_{j(k)-1}\} \\ \varepsilon r_{j(k)} \\ \mathrm{E}\{r_{j(k)+1}\} \end{bmatrix}, \tag{7.6}$$

where E means the average over all $j(k)$, $r_{j(k)}$ is the random vector representing word k in position $j = j(k)$ of the text, and ε is a scaling (balancing) parameter, e.g., equal to 0.2. Its purpose is to enhance the influence of the context part over the symbol part in learning, i.e., to enhance topological ordering according to context.

It is often possible to make $\varepsilon = 0$, and then for the average contect vector one can take $[\mathrm{E}\{r_{j(k)-1}^{T}\}, \mathrm{E}\{r_{j(k)+1}^{T}\}]^{T}$.

Notice that expression (7.6) has to be computed only once for each different word, because the $r_{j(k)}$ for all the $j = j(k)$ are identical. This usually also results in a more illustrative SOM than if the representations were weighted according to their statistical frequency of occurrence in the strings (especially in natural-language sentences, where these frequencies may be very different).

Calibration. In the learning process, an approximation of the x_i vectors will be formed into the middle field of the m_i vectors, corresponding to the middle field of the x vector in (7.6). The map can be calibrated using samples of the type $x = [\emptyset, x_i^{T}, \emptyset]^{T}$, where $\emptyset$ means "don't care" (these components not being involved in comparison) and locating and labeling those m_i vectors

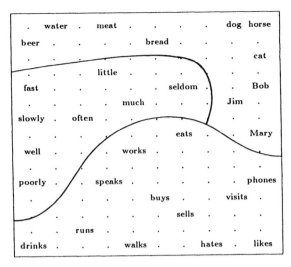

Fig. 7.12. "Semantic map" obtained on a network of 10 × 15 cells after 2000 presentations of word-context-pairs derived from 10,000 random sentences of the kind shown in Fig. 7.11. Nouns, verbs and adverbs are segregated into different domains. Within each domain a further grouping according to aspects of meaning is discernible

whose middle fields match best with these samples. Many SOM nodes may remain unlabeled, whereas the labeled nodes describe the relationships of the selected items in a clear geometric way.

The SOM formed in this example is shown in Fig. 7.12. A meaningful geometric order of the various word categories is clearly discernible.

The lines dividing nouns, verbs, and adverbs into contiguous regions have been drawn manually. A gray-scale shading as discussed in Sect. 3.15 would have distinguished these classes automatically.

7.7.2 Natural Text

For this experiment we collected about 5000 articles that appeared during the latter half of 1995 and up to June, 1996, in the Usenet newsgroup *comp.ai.neural_nets.* on the Internet.

Before application of the methods described in the previous section, we removed some non-textual information (e.g., ASCII drawings and automatically included signatures). Numerical expressions and special codes were replaced by special symbols using heuristic rules. To reduce the computational load, the words that occurred less than 50 times in the data base were neglected and treated as empty slots. The size of the vocabulary, after discarding the rare words, was 2500. The 800 most common words that were not supposed to have any specific "semantic" interest were still discarded manually.

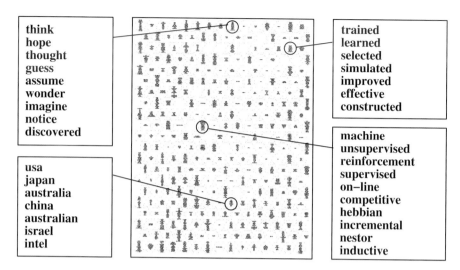

think		trained
hope		learned
thought		selected
guess		simulated
assume		improved
wonder		effective
imagine		constructed
notice		
discovered		**machine**
		unsupervised
usa		reinforcement
japan		supervised
australia		on–line
china		competitive
australian		hebbian
israel		incremental
intel		nestor
		inductive

Fig. 7.13. Examples of some clear "categories" of words on the word category map of the size of 15 by 21 units

The word category map contained 315 neurons with 270 inputs each. This particular choice of the dimensionalities was determined by the memory resources of the parallel CNAPS neurocomputer, initially intended to be used for document classification (Sect. 7.8). Similar words tended to occur in the same or nearby map nodes, forming "word categories" in the nodes, illustrated in Fig. 7.13.

7.8 Organization of Large Document Files

It is possible to form similarity graphs of *text documents* by the SOM principle [7.37–7.43], when models that describe collections of words in the documents are used. The models can simply be weighted histograms of the words regarded as real vectors, but usually some dimensionality reduction of the very-high dimensional histograms is carried out, as we shall see below.

A document organization, searching and browsing system called the WEBSOM is described in this section. The original WEBSOM [7.44] was a two-level SOM architecture, but we later simplified it as described here; at the same time we introduced several speed-up methods to the computation and increased the document map size by an order of magnitude.

7.8.1 Statistical Models of Documents

The Primitive Vector Space Model. In the basic *vector space model* [7.45] the stored documents are represented as real vectors in which each component corresponds to the frequency of occurence of a particular word in the document. The main problem of the vector space model is the large vocabulary in any sizable collection of free-text documents, which means a huge dimensionality of the model vectors.

Weighting of Words by Their Entropies. Experience has shown that if the documents stem from several clearly different groups such as from several newsgroups in the Internet, clustering is improved if one uses weighting of the words by the information-theoretic *entropy* (Shannon entropy) before forming the histograms. This kind of entropy-based weighting of the words is straightforward when the documents can be naturally and easily divided into groups such as the different newsgroups. If no natural division exists, word entropies might still be computed over individual documents (which then must be large enough for sufficient statistical accuracy) or clusters of documents.

Denote by $n_g(w)$ the frequency of occurrence of word w in group i, say newsgroup i ($i = 1, \ldots, N$). Denote by $P_g(w)$ the probability density for word w belonging to group g. The entropy H of this word is traditionally defined as

$$H(w) = -\sum_g P_g(w) \log P_g(w) \approx -\sum_g \frac{n_g(w)}{\sum_{g'} n_{g'}(w)} \ln \frac{n_g(w)}{\sum_{g'} n_{g'}(w)} , \quad (7.7)$$

and the weight $W(w)$ of word w is defined to be

$$W(w) = H_{\max} - H(w) , \quad (7.8)$$

where $H_{\max} = \ln N$.

For the weighting of a word according to its significance, one can also use the inverse of the number of documents in which the word occurs ("inverse document frequency").

Latent Semantic Indexing (LSI). In an attempt to reduce the dimensionality of the document vectors, one often first forms a matrix in which each column corresponds to the word histogram of a document, and there is one column for each document. After that the factors of the space spanned by the column vectors are computed by a method called the singular-value decomposition (SVD), and the factors that have the least influence on the matrix are omitted. The document vector formed from the histogram of the remaining factors has then a much smaller dimensionality. This method is called *latent semantic indexing (LSI)* [7.46].

Randomly Projected Histograms. We have shown experimentally that the dimensionality of the document vectors can be reduced radically by a much simpler method than the LSI, by a simple random projection method [7.47, 48], without essentially losing the power of discrimination between the documents. Consider the original document vector (weighted histogram) $n_i \in \Re^n$ and a rectangular random matrix R, the elements in each column of which are assumed to be normally distributed vectors having unit length. Let us form the document vectors as the *projections* $x_i \in \Re^m$, where $m \ll n$:

$$x_i = Rn_i \ . \tag{7.9}$$

It has transpired in our experiments that if m is at least of the order of 100, the similarity relations between arbitrary pairs of projection vectors (x_i, x_j) are very good approximations of the corresponding relations between the original document vectors (n_i, n_j), and the computing load of the projections is reasonable; on the other hand, with the decreasing dimensionality of the document vectors, the time needed to classify a document is radically decreased.

Construction of Random Projections of Word Histograms by Pointers. Before describing the new encoding of the documents [7.49] used in this work, some preliminary experimental results that motivate its idea are presented. Table 7.2 compares a few projection methods in which the model vectors, except in the first case, were always 315-dimensional. For the material in this smaller-scale preliminary experiment we used 18,540 English documents from 20 Usenet newsgroups of the Internet. When the text was preprocessed as explained in cf. Sect. 7.7.2, the remaining vocabulary consisted of 5,789 words or word forms. Each document was mapped onto one of its grid points. All documents that represented a minority newsgroup at any grid point were counted as classification errors.

The classification accuracy of 68.0 per cent reported on the first row of Table 7.2 refers to a classification that was carried out with the classical vector-space model with full 5789-dimensional histograms as document vectors. In practice, this kind of classification would be orders of magnitude too slow.

Random projection of the original document vectors onto a 315-dimensional space yielded, within the statistical accuracy of computation, the same figures as the basic vector space method. This is reported on the second row. The figures are averages from seven statistically independent tests, like in the rest of the cases.

Consider now that we want to simplify the projection matrix R in order to speed up computations. We do this by thresholding the matrix elements, or using sparse matrices. Such experiments are reported next. The following rows have the following meaning: Third row, the originally random matrix elements were thresholded to $+1$ or -1; fourth row, exactly 5 randomly distributed ones were generated in each column, whereas the other elements were zeros; fifth row, the number of ones was 3; and sixth row, the number of ones was 2, respectively.

These results are now supposed to give us the idea that if we, upon formation of the random projection, would reserve a memory array like an accumulator for the document vector x, another array for the weighted histogram n, and *permanent address pointers* from all the locations of the n array to

Table 7.2. Classification accuracies of documents, in per cent, with different projection matrices R. The figures are averages from seven test runs with different random elements of R

	Accuracy	Standard deviation due to different randomization of R
Vector space model	68.0	—
Normally distributed R	68.0	0.2
Thresholding to $+1$ or -1	67.9	0.2
5 ones in each column	67.8	0.3
3 ones in each column	67.4	0.2
2 ones in each column	67.3	0.2

all such locations of the x array for which the matrix element of R is equal to one, we could form the product very fast by following the pointers and summing up to x those components of the n vector that are indicated by the ones of R.

In the method that is actually being used we do not project ready histograms, but the pointers are already used with each word in the text in the construction of the low-dimensional document vectors. When scanning the text, the hash address for each word is formed, and if the word resides in the hash table, those elements of the x array that are found by the (say, three) address pointers stored at the corresponding hash table location are incremented by the weight value of that word. The weighted, randomly projected word histogram obtained in the above way may be optionally normalized.

The computing time needed to form the histograms in the above way was about 20 per cent of that of the usual matrix-product method. This is due to the fact that the histograms, and also their projections, contain plenty of zero elements.

Histograms on the Word Category Map. In our original version of the WEBSOM [7.44], the reduction of the dimensionality of the document vectors was carried out by letting the words of free natural text be *clustered* onto neighboring grid points of another special SOM. This *category histogram* then acted as the input pattern to the second-level SOM, the document map itself.

The old two-level SOM architecture thus consisted of two hierarchically organized Self-Organizing Maps, and is depicted in Fig. 7.14. The *word category map* (Fig. 7.14 a) is a "semantic SOM" of the type discussed in Sect. 7.7.2 that describes relations of words based on their averaged short contexts.

The input to such a "word category map" [7.44] consisted of triplets of adjacent words in the text taken over a moving window, whereupon each word in the vocabulary was represented by a unique random vector.

The word-category SOM is calibrated after its training process by inputting the preprocessed text of each document once again and labeling the best-matching units according to symbols corresponding to the $r_{j(k)}$ parts of

a)

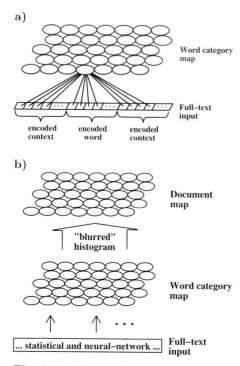

Word category
map

Full–text
input

encoded   encoded   encoded
context   word   context

b)

Document
map

"blurred"
histogram

Word category
map

↑   ↑   • • •

... statistical and neural–network ...   Full–text
input

Fig. 7.14. The basic two-level WEBSOM architecture. (**a**) The word category map first learns to represent relations of words based on their averaged contexts. This map is used to form a word histogram of the text to be analyzed. (**b**) The histogram, a "fingerprint" of the document, is then used as input to the second SOM, the document map

the averaged context vectors. It is to be noted that in this method a unit may become labeled by *several symbols*, often synonymic or forming a closed attribute set. Usually interrelated words that have similar contexts appear at the same node or close to each other on the map.

In the second phase the location of each document on the *document map* is determined. The word category map was constructed using all the documents of the data base. When one of these documents, as a new document is to be mapped, the text of a document is first mapped onto the word category map whereby a histogram of the "hits" on it is formed. To reduce the sensitivity of the histogram to small variations in the content, it can be slightly "blurred," e.g., using a convolution with a Gaussian convolution kernel having the full width at half maximum of 2 map spacings. Such "blurring" is a commonplace method in pattern recognition, and is also justified here, because the map is ordered. The document map is formed using the histograms, "fingerprints" of the documents, for its inputs. To speed up the computations, the positions of the word labels on the word category map may be looked up directly by *hash coding* (Sect. 1.2.2).

Case Study: A Usenet Newsgroup. We first organized the collection of the 5000 neural-network articles, using the old WEBSOM method. The word category map was depicted in Fig. 7.14. As we originally used the CNAPS neurocomputer, its local memory capacity restricted our computations to 315-input, 768-neuron SOMs. As explained in Sect. 3.17, however, we have been able to multiply the sizes of the SOMs by two solutions, but we then used a general-purpose computer: 1. Good initial values for a much larger map can be *estimated* on the basis of the asymptotic values of a smaller map, like the one computed with the CNAPS, by a local interpolation procedure. There is room for a much larger map in a general-purpose computer, and the number of steps needed for its fine tuning is quite tolerable. 2. In order to accelerate computations, the winner search can be speeded up by storing with each training sample an address pointer to the old winner location. During the next updating cycle, the approximate location of the winner can be found directly with the pointer, and only a *local search* around it need to be performed. The pointer is then updated. In order to guarantee that the asymptotic state is not affected by this approximation, updating with a full winner search was performed intermittently, after every 30 training cycles.

Later in Sect. 7.8.2 we shall show how really large maps can be made by the random-projection method, whereby yet other computational tricks are introduced.

The *document map* in this demonstration was clearly found to reflect relations between the newsgroup articles; similar articles occur near each other on the map. Several interesting clusters containing closely related articles have been found; sample nodes from a few of them are shown in Fig. 7.15. All nodes, however, were not well focused on one subject. While most discussions seem to be confined into rather small areas on the map, the discussions may also overlap. The clustering tendency or density of the documents in different areas of the "digital library" can be visualized using the gray scale, light areas representing high density of documents.

As the category-specific histograms have a strong correlation to the semantic contents of the texts, the various nodes in this second SOM can be found to contain closely related documents, such as discussions on the same topics, answers to the same questions, calls for papers, publications of software, related problems (such as financial applications, ANNs and the brain), etc.

The *WEBSOM* scheme may be regarded as a genuine *content-addressable memory*, because it does not only refer to exact or approximate replica of pieces of texts, but their more abstract contents or meanings. When the text of a sample document or a piece of it, or even a free-form quotation or reference of it is analyzed statistically as described in this section and input to the system, the best match on the document map then locates the document the contents of which resemble best the sample. Other, similar documents

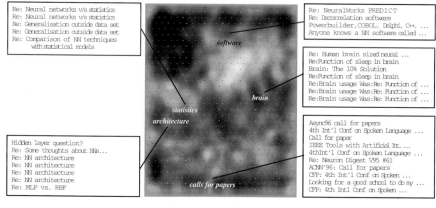

Fig. 7.15. A map of 4600 documents. The small dots denote the units on the map of the size of 24 by 32 units, and the clustering tendency has been indicated by shades of gray. White denotes a high degree of clustering whereas black denotes long distances, "ravines", between the clusters. The contents of five document map nodes have been illustrated by listing the subjects of their articles. Five labels, based on the sample nodes, have been written to the map

can then be found at the same node or its immediate surroundings in the document map.

Browsing Interface. To use the WEBSOM most effectively, it can be combined with standard document browsing tools. The document space has been presented at three basic levels of the system hierarchy: the map, the nodes, and the individual documents (Fig. 7.16). Any subarea of the map can be selected and zoomed by "clicking." One may explore the collection by following the links from hierarchy level to another. It is also possible to move to neighboring areas of the map, or to neighbors at the node level directly. This hierarchical system has been implemented as a set of WWW pages. They can be explored using any standard graphical browsing tool. A complete demo is accessible in the Internet at the address http://websom.hut.fi/websom/.

7.8.2 Construction of Very Large WEBSOM Maps by the Projection Method

After the experiments reported in Sect. 7.8.1 we abandoned the word category map since an even better accuracy of document classification was achieved by the straightforward random projection of the word histograms, explained in Sect. 7.8.1, and the computation of the latter is significantly faster, especially if the random pointer method of projection is used.

With large maps, both winner search and updating (especially taking into account the large neighborhoods in the beginning of the SOM process) are time-consuming tasks. The SOM algorithm is capable of organizing even a randomly initialized map. However, if the initialization is regular and closer

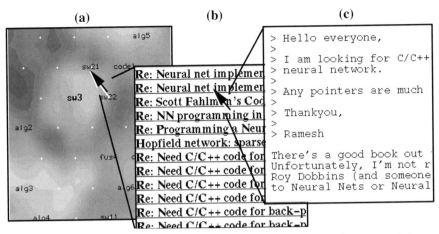

Fig. 7.16. Sample scenes of the WEBSOM interface. (**a**) Part of a zoomed document map display. The clustering tendency is visualised using the gray scale. (**b**) The map node contents. (**c**) An example of a newsgroup article picked up from the node

to the final state, the asymptotic convergence of the map can be made at least and order of magnitude faster.

Here is a recapitulation of the computing tricks used in the very large WEBSOM experiments, where a general-purpose computer (Silicon Graphics server) was used for all computations.

Estimating Initial Values for a Large SOM. Several suggestions for increasing the number of nodes of the SOM during its construction (cf., e.g. [3.43]) have been made. The new idea presented in Sect. 3.17.2 is to *estimate* good initial values for the model vectors of a very large map on the basis of asymptotic values of the model vectors of a much smaller map, which can be computed fast.

Fast Distance Computation. In word histograms, there are plenty of zeros, and if the pointer method of random projection is used, the number of zeros in the projected document vectors is still predominant.

The document vectors are mapped onto the SOM according to their inner products with the model vectors. Since the zero-valued components of the vectors do not contribute to inner products, it is possible to tabulate the indices of the non-zero components of each input vector, and thereafter consider only those components when computing distances.

A related method has been proposed for computing Euclidean distances between sparse vectors [7.50]. Our formulation is, however, simpler when only inner products are needed.

Addressing Old Winners. As explained in Sect. 3.17.1, the searching for the winner can be accelerated by orders of magnitude storing the pointers to the approximate (old) winner locations together with the training data. When a smaller map has been computed by this method, approximate pointers to the winners of the much larger map can be obtained by the interpolation-extrapolation procedure explained in Sect. 3.17.2.

Parallelized Batch Map Algorithm. Let us recall (Sect. 3.6) that in the asymptotic state of the SOM algorithm the model vectors must fulfill the condition

$$m_i = \frac{\sum_j n_j h_{ji} \bar{x}_j}{\sum_j n_j h_{ji}} , \tag{7.10}$$

where $h_{ji}\bar{x}_j = \sum_{k:c_k=j} x_k/n_j$ is the mean of the inputs that are closest to the model vector m_j, and n_j is the number of those inputs. The Batch Map algorithm consists of iterative application of equation (7.10).

Equation (7.10) allows for a very efficient parallel implementation, in which extra memory need not be reserved for the new values of m_i. At each iteration we first compute the pointer c_k to the best-matching unit for each input x_k. If the old value of the pointer can be assumed to be close to the final value, as is the case if the pointer has been initialized properly or obtained in the previous iteration of a relatively well-organized map, we need not perform an exhaustive winner search as discussed above. Moreover, since the model vectors do not change at this stage, the winner search can be easily implemented in parallel by dividing the data into the different processors in a shared-memory computer.

After the pointers have been computed, the previous values of the model vectors are not needed any longer. They can be replaced by the means $\bar{x}_j$ and therefore no extra memory is needed.

Finally, the new values of the model vectors can be computed based on (7.10). This computation can also be implemented in parallel and done within the memory reserved for the model vectors if a subset of the new values of the model vectors is held in a suitably defined buffer.

It should perhaps be noted that if the neighborhood function is very narrow or if there is only a small amount of data in relation to the map size, the sum $\sum_j n_j h_{ji}$ in (7.10) may become zero for some j. It has turned out in our experiments that there exists a simple remedy to this: the computation can be continued successfully by keeping the previous value of m_j in such situations.

Saving Memory by Reducing Representation Accuracy. The memory requirements can be reduced significantly by using a coarser quantization of the vectors, whereupon even a very large SOM can be kept in the main memory. We have used a common adaptive scale for all of the components of a model vector, representing each component with 8 bits only. If the dimensionality of the data vectors is large, the statistical accuracy of the distance computations is still sufficient.

It is advisable to choose the correct quantization level probabilistically when computing new values for the model vectors.

User Interface and Exploration of the Document Map. The document map has been presented as a series of HTML pages that enable exploration of the grid points: when clicking the latter with a mouse, links to the document data base enable reading the contents of the articles. If the grid is large, subsets of it can first be viewed by zooming.

Automatic Labeling. In order to help finding starting points for browsing, one needs descriptive "landmarks" or "signposts" to be assigned to map regions where a particular topic area is discussed. On the top level of the WEBSOM there can be fewer landmarks, but in "zooming" to magnified portion of the map, more landmarks are made to appear. These landmarks cannot be regarded as class labels, because the class regions are not delimited sharply; they are rather comparable to the main keywords that characterize an area.

The landmarks can be found automatically based on the statistical clustering properties of the words. The following method has been worked out by *K. Lagus* and *S. Kaski* [7.51], and it is quoted here in a slightly simplified way.

A good landmark should be a word that occurs often in the articles of that area and rarely elsewhere. These two criteria may be combined into an index that ranks the words w according to their goodness as landmarks for map unit or cluster j:

$$G_j(w) = F_j^{\text{clust}}(w) \times F_j^{\text{coll}}(w) , \tag{7.11}$$

where the first term, F_j^{clust}, describes word w in relation to other words within the cluster j to be described, whereas the second term, F_j^{coll}, relates the word to the whole collection.

Let $f_j(w)$ denote the number of times word w occurs in map unit j, i.e. the *frequency* of word w in j. Let $F_j(w)$ denote the *relative frequency* of word w, defined as

$$F_j(w) = \frac{f_j(w)}{\sum_v f_j(v)} . \tag{7.12}$$

Note that $0 \le F_j(w) \le 1$ and $\sum_w F_j(w) = 1$. The effect of this normalization is to disregard the sizes of the clusters, and instead to measure the relative importance of a word compared to the other words occurring in the cluster. The relative frequency $F_j(w)$ now seems a good candidate for F_j^{clust}.

Next, F_j^{coll} shall measure the ratio of the frequency of w in map unit j to the "background frequency" that describes how typical the word is in other parts of the collection. A straightforward measure for this comparison would be

$$F_j^{\text{coll}} = \frac{F_j(w)}{\sum_i F_i(w)} . \tag{7.13}$$

On a closer examination, however, a couple of heuristic improvements can be suggested. First, if a word occurs often in a map unit, it is probably relatively common also in the neighboring units. However, frequent appearance of a word in immediately neighboring units, say up to radius r_1 from unit j_1 should not reduce its goodness as keyword in the original map unit. Second, if we are looking for good descriptors for *larger map areas*, as in finding labels for large portions of the graphical map display, we would like to reward a word in unit j if it is a good descriptor of the neighboring units as well, up to another radius r_0 from unit j.

The goodness value may then be re-expressed in a way that explicitly rewards several map units within radius r_0 for forming a cluster in terms of word w, while not punishing words in unit j merely for being good descriptors also for the neighboring units up to radius r_1 from j:

$$G_j(w) = \frac{\left[\sum_{k \in A_0^j} F_k(w)\right]^2}{\sum_{i \notin A_1^j} F_i(w)} , \tag{7.14}$$

where

$d(j, i)$ is the distance on map grid between units i and j,

$k \in A_0^j$ if $d(j, k) < r_0$, and

$i \in A_1^j$ if $r_0 < d(j, i) < r_1$.

These goodness values are computed for all nodes j and the keywords are selected for an area in this rank order.

Some extra considerations are necessary for selecting keywords to the different "zooming levels" (*loc. cit.*).

Content-Addressable Search. The HTML page can be provided with a form field into which the user can type an own query in the form of a short "document," eventually consisting of one or a few keywords only. This query is preprocessed and a document vector (histogram) is formed in the same way as for the stored documents. This histogram is then compared with the models of all grid points, and a specified number of best-matching points are marked with a symbol: the better the match, the larger the symbol. These symbols provide good starting points for browsing.

For comparison, we have also provided an option for the keyword search, in which each word of the vocabulary is indexed by pointers to those map units where these words occur.

7.8.3 The WEBSOM of All Electronic Patent Abstracts

We have constructed a map of all the 6,840,568 patent abstracts that were available in English in electronic form. The average length of the abstracts was 132 words, and in total they contained 733,179 different words (base forms). The size of the SOM was 1,002,240 models (map units).

Preprocessing. From the raw patent abstracts we first extracted the titles and the texts for further processing. We then removed non-textual information. Mathematical symbols and numbers were converted into special symbols. All words were converted to their base form using a stemmer. The words occurring less than 50 times in the whole corpus, as well as a set of common words in a stopword list of 1,335 words were removed. The remaining vocabulary consisted of 43,222 words. Finally, we omitted the 122,524 abstracts in which less than 5 words remained.

Formation of Statistical Models. To reduce the dimensionality of the models we used the randomly projected word histograms. For the final dimensionality we selected 500, and 5 random pointers were used for each word (in the columns of the projection matrix R). The words were weighted using the Shannon entropy of their distribution which is related to their occurrence in the subsections of the patent classification system. There are 21 subsections in the patent classification system in total; examples of such subsections are agriculture, transportation, chemistry, building, engines, and electricity (cf. Fig. 7.17).

Formation of the Document Map. The final map was constructed in four stages. First, a 435-unit document map was computed very carefully. The small map was used to estimate a larger one which was then fine-tuned with the Batch Map algorithm. This process of estimation and fine-tuning was repeated in three steps with progressively larger maps.

With the newest versions of our programs the whole process of computation of the document map takes about six weeks on a six-processor SGI O2000 computer. We cannot yet provide exact figures of the real processing time since we have all the time developed the programs while carrying out the computations. We have computed the maps relatively carefully – reasonably well organized maps could have been obtained in a much shorter time.

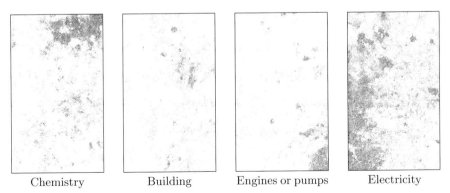

Chemistry   Building   Engines or pumps   Electricity

Fig. 7.17. Distribution of four sample subsections of the patent classification system on the document map. The gray level indicates the logarithm of the number of patents in each node

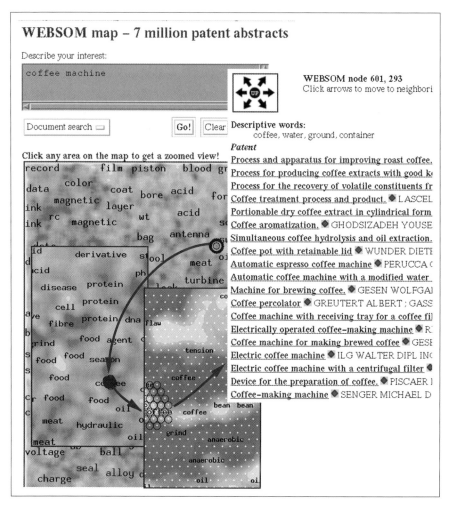

Fig. 7.18. Content-addressable search ("document search") was used to provide a starting point for exploration of the document map. The search for "coffee machine" directs the user to a map unit that contains several patents regarding various types of coffee machines. In the surrounding region related patents are found which concern coffee brewing, processing of coffee beans, etc. As suggested by the labels in the region, other food-related topics are located nearby. The automatically selected labels written on the display describe texts within the region. The shade of gray depicts the density of the documents in that region with light shade indicating a high density. Three zoom levels are used for such a large map

The maximal amount of memory required was about 800MB.

Forming the user interface took an additional week of computation. This time includes finding the keywords to label the map, forming the WWW-pages that are used in exploring the map, and indexing the map units for keyword searches.

Results. In order to get an idea of the quality of the organization of the final map we measured how the different subsections of the patent classification system were separated on the map. When each map node was labeled according to the majority of the subsections in the node and the abstracts belonging to the other subsections were considered as misclassifications, the resulting "accuracy" (actually, the "purity" of the nodes) was 64%. It should be noted that the subsections overlap partially – the same patent may have subclasses which belong to different subsections. The result corresponded well with the accuracies we have obtained with smaller maps computed on subsets of the document collection.

Distribution of patents on the final map has been visualized in Fig. 7.17. A case study of document search is depicted in Fig. 7.18.

7.9 Robot-Arm Control

7.9.1 Simultaneous Learning of Input and Output Parameters

Still another important application of the SOM is its use as a *look-up table*: an input pattern, by the "winner" matching, specifies a SOM location, where extra information can be made available. Such information may consist, e.g., of control variables or other output information. In this way, for instance, very specific and nonlinear control functions are definable. The table-look-up control has turned out to be particularly useful in *robotics*.

The control variables to be stored in the SOM locations may be defined heuristically or according to some control-theoretic considerations. They may also be *learned* adaptively, as is done in the following example [1.40].

Visuomotor Control of Robot Arm. Consider Fig. 7.19 that shows the positioning control system for a robot arm.

The two-dimensional coordinates u_1 and u_2 of a target point in the image planes of cameras 1 and 2 are combined into a four-dimensional (stereoscopic) input vector u, used as input to the SOM. A *three-dimensional SOM array* forms the *spatial representation* of the target point. The angular coordinates of the joints of the robot arm are defined by a *configuration vector* θ.

Let us restrict ourselves to the case in which there are no dynamic effects present, and the nonlinear input-output relation is the main problem. It is desirable to find the transformation $\theta(u)$ that brings the tip of the robot arm to the target point in space, from which the cameras can get the observation u. In vector quantization, the relation $\theta(u)$ can be linearized piecewise, whereby the origin of linearization is determined by the winner-take-all function, and the linearization parameters A_c and b_c are read from the "winner" location c:

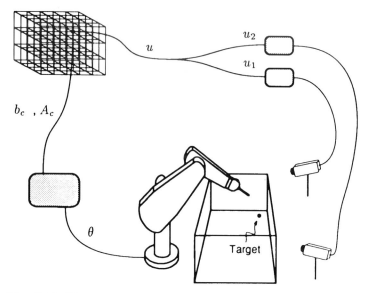

Fig. 7.19. Visuomotor coordination of robot arm. The image-plane coordinates u_1 and u_2 of the target are combined into a four-dimensional input vector u to a three-dimensional SOM, from which the configuration vector θ of the arm is obtained

$$\theta = A_c(u - m_c) + b_c \ . \tag{7.15}$$

Here m_c is the closest codebook vector to u. The factor A_c is called a *Jacobian matrix*, and (7.15) gives the first two terms of the Taylor expansion of $\theta(u)$. Linearization is carried out around m_c and is valid in the whole Voronoi set of u values around m_c.

The linearization approximation (7.15) is usually more or less erroneous, and so can A_c and b_c only be determined approximately. For this reason a small corrective control step can be carried out. The intermediate value for θ obtained in (7.15) is henceforth denoted by θ_i, i.e.,

$$\theta_i = A_c(u - m_c) + b_c \ , \tag{7.16}$$

whereby θ_i defines a new approximate tip position and a new camera vector u_i. The corrective step for the configuration vector, starting from θ_i, is

$$\theta_f = \theta_i + A_c(u - u_i) \ , \tag{7.17}$$

which again determines a new camera vector u_f.

Learning of Control Parameters. In real-world systems one might not want to calibrate the control parameters for all times. The positions of the cameras may change, there can be drifts in their imaging systems, and the mechanisms of the robot arm may change with aging, too. For these reasons the learning of $\theta(u)$ must be made adaptive.

In order to learn the relation $\theta(u)$, no topological constraints on the m_c are necessary in principle, and the classical vector quantization could be used to describe the space, whereby $\theta(u)$ could also be tabulated. There is even no stringent need to use VQ, since a regular three-dimensional lattice can define the codebook vectors over the space. Nonetheless it seems profitable to first learn and use the three-dimensional SOM array at least for the following reasons: 1. There can exist obstacles in the space; they must be learned in the mapping. 2. It is possible to allocate the codebook vectors to space in an optimal fashion, having higher density of lattice points where the control must be more accurate. 3. Topological constraints, such as the neighborhood function h_{ci}, significantly speed up learning (due to coherent corrections made at nearby nodes), and the mapping becomes smooth much faster than in a traditional VQ.

The idea used by *Ritter et al.* [1.40] in deriving the adaptation equations for A_c and b_c contain several approximations and may be delineated in the following way. Let us begin with the time-invariant codebook vector lattice, without considering any neighborhood effects first.

Let us first concentrate on the improvement of b_c, thereby considering A_c constant. If after the corrective step (7.17) the configuration vector θ_f defines a new tip position and the new camera vector u_f, (7.17) may be written

$$\theta_f = \theta_i + A_c^0(u_f - u_i) ,\qquad(7.18)$$

where the matrix A_c^0 is assumed to have a value that makes the linear relation as accurate as possible. In considering computation of b_c relative to A_c, we can assume A_c^0 equal to A_c. On the other hand, if θ_i is the real configuration vector for camera vector u_i, we can then linearly extrapolate on the basis of u_i and A_c for the value b_c^0 that would be the improved look-up table value at node c:

$$b_c^0 = \theta_i + A_c^0(m_c - u_i) .\qquad(7.19)$$

Substituting θ_i from (7.16), and assuming $A_c = A_c^0$ we finally obtain for the improved estimate:

$$b_c^0 = b_c + \delta_1 A_c(u - u_i) ,\qquad(7.20)$$

where δ_1 is the learning rate, a small number.

While it was possible to derive the adaptation equation for b_c as a linear compensation of the residual error, for the derivation of the A_c equation we have to minimize the following functional. Let next the summation index s only refer to terms in the sums for which the "winner" c is selected. Define the *mean-square error functional*

$$E = \frac{1}{2}\sum_s [(\theta_f(s) - \theta_i(s) - A_c(u_f(s) - u_i(s))]^2 .\qquad(7.21)$$

Using the stochastic-approximation philosophy, compute the gradient for every sample of E indexed by s; in forming the gradient A_s is regarded as a vector,

$$\nabla_{A_c} E = -\sum_s (\Delta\theta(s) - A_c \Delta u(s))(\Delta u(s)))^{\mathrm{T}} , \text{ with}$$
$$\Delta\theta(s) = \theta_f(s) - \theta_i(s) \text{ and}$$
$$\Delta u(s) = u_f(s) - u_i(s) . \tag{7.22}$$

Without more detailed consideration we now recapitulate the whole learning scheme of Ritter et al. [1.40], where the SOM is formed, and the control parameters are updated simultaneously:

1. Present a randomly chosen target point in the work space.
2. Let the cameras observe the corresponding input signal u.
3. Determine the map unit c corresponding to u.
4. Move the arm to an intermediate position by setting the joint angles to

$$\theta_i = A_c(u - m_c) + b_c ,$$

where A_c and b_c are found at location c, and record the corresponding coordinates u_i of the tip of the arm in the image planes of the cameras.
5. Execute a correction of the arm position according to

$$\theta_f = \theta_i + A_c(u - u_i) ,$$

and observe the corresponding camera coordinates u_f.
6. Execute a learning step of the SOM vectors according to

$$m_r^{new} = m_r^{old} + \epsilon h_{cr}(u - m_r^{old}) .$$

7. Determine improved values A^* and b^* using

$$A^* = A_r^{old} + \delta_2 \cdot A_r^{old}(u - u_f)(u_f - u_i)^{\mathrm{T}} ,$$
$$b^* = b_r^{old} + \delta_1 \cdot A_r^{old}(u - u_i) .$$

8. Execute a learning step of the output values stored at map unit c as well its neighbors r :

$$A_r^{new} = A_r^{old} + \epsilon' h'_{cr}(A^* - A_r^{old}) ,$$
$$b_r^{new} = b_r^{old} + \epsilon' h'_{cr}(b^* - b_r^{old}) ,$$

and continue with step 1.

Comment. It should be mentioned that the group of H. Ritter has later approached the robot-control problem quite differently. The coarse control conditions are learned in a somewhat similar fashion as estimates will be formed in Sect. 7.11.2. However, in order to define the control steps with high precision, they represent the map by a continuous manifold that can be constructed from a limited number of data points. Such a *parametrized SOM (PSOM)* has been described in [7.52–54].

7.9.2 Another Simple Robot-Arm Control

In the previous algorithm, the control parameters were determined by minimizing average control errors. In the method that this author conceived and successfully used in order to program a simple *robot-arm pole balancer* demonstration, the control parameters, stored in the "winner" locations, are *direct copies of the configuration vectors in equilibrium states* whereas the SOM is used in a special way to change the configuration vector θ during the control process in the proper direction. This may also give a hint of how biological control systems could operate, namely, by *fusing* different kind of information in the same SOM.

Consider a mechanism similar to the one shown in Fig. 3.15, except that only *one* arm is used, and it holds the lower end of the pole, while the eye is simultaneously made to look at the top of the pole. Training might mean moving a vertically oriented pole into random positions, whereby a map of these positions is formed into the SOM inputs. Let us recall that we have map inputs for both the eye and one arm.

"Learning" of control parameters in this model means that after the SOM inputs have been formed, the pole is moved horizontally into random positions once again. A copy of the arm configuration vector is always written to the "winner" location and eventually its neighbors, where the control parameters are held.

It was demonstrated in Sect. 3.4 that both the robot-arm position and the eye coordinates can be mapped onto the SOM in a *rectified scale*, although the configuration vector of the joint angles and the eye coordinates are highly nonlinear functions of the pole position.

What happens in a "learned" SOM of the above kind, if the pole is tilted, and the "eye" looks to a point that has the $r_1 \in \Re^2$ on the horizontal plane, but the tip of the arm has another location $r_2 \in \Re^2$ on the horizontal plane? The answer depends on the metric, but if the Euclidean metric is used and the codebook vectors of the SOM are distributed smoothly in the signal space, we may expect that the "winner" location would exist somewhere in the vicinity of $(r_1 + r_2)/2$. Its exact location is not important, as long as it is somewhere between r_1 and r_2. *The control parameters are picked up from this location.*

If the eye positioning and arm positioning do not agree, and the new configuration vector is picked from the region between r_1 and r_2, the control is automatically in the correct direction, however complex the configuration vector is.

7.10 Telecommunications

7.10.1 Adaptive Detector for Quantized Signals

Assume first that the transmission of signals from the transmitter to the receiver occurs through a single path as usual in wired telecommunications, not via multiple (reflected) paths like in radio traffic. Information, for instance of analog signal values, can then be converted into a sequence of discrete (pseudorandom) codes or symbols, each of which is transmitted as a corresponding quantized analog signal level or discretized amplitude, frequency, or phase modulation of a carrier wave. The main type of disturbance thereby consists of nonlinear and time-variant changes in modulation, caused by the transmitting and receiving electronics and variable attenuation, and white noise.

Consider here the *quadrature-amplitude modulation (QAM)*. In the QAM, the transmission channel is divided into two independent subchannels (defined as the "in-phase" (I) and "quadrature" (Q) components of the transmitted wave, respectively): two carrier waves that have the same frequency but a relative phase of 90° are transmitted through the same frequency channel and amplitude-modulated independently. In the 16QAM, the symbol is transmitted in parts, each consisting of four bits: two in the I part and two in the Q part, respectively. There are thus four discrete-valued modulation levels in each of the two subchannels. The ideal signal constellation in the two-dimensional I-Q-coordinate lattice is depicted in Figure 7.20(a). Thus only 16 symbols need to be encoded and decoded.

The square lattice formed of the quantized signals may be *deformed* during the operation in many ways, as depicted in Figure 7.20: e.g., a saturable nonlinearity of the frequency channel (b), and changing of the relative phase of the subchannels (c). Noise can be superimposed on each signal level. We shall show that a SOM, located at the receiving end, can be used as an effective adaptive detector for such a QAM system.

Consider that the I and Q signals at the receiving end, representing the amplitude modulation of the zero and 90° phase components, respectively, are regarded as the two scalar input signals to the SOM. The two-dimensional input vector x is then clustered, each cluster corresponding to one symbol in an orderly fashion. The input signal selects the closest unit in a SOM. For optimal resolution in reception, the m_i can now be updated for maximum separation of the signal states.

Each of the nodes or units of the 4 by 4 SOM is now supposed to follow up its own cluster under deformations of the types depicted in Figures 7.20(b) and (c), and their mixtures.

As the adaptive follow-up process continues indefinitely, the learning-rate factor α should be kept constant and at an optimal value. If the Batch Map process (Sect. 3.6) provided with the method for the elimination of border effects is used, the learning-rate factor need not be defined. Such a Batch Map is superior in this application.

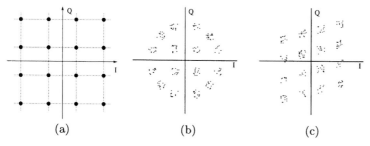

Fig. 7.20. Ideal signal constellation of quadrature amplitude modulation (16QAM) used in digital communication systems where the in-phase (I) and quadrature (Q) components occupy discrete lattice points in the I-Q -coordinate system as shown in Figure **(a)**. Typical nonlinear distortions occuring in practical systems using QAM modulation are the corner and lattice collapse situations, depicted in Figures **(b)** and **(c)**, respectively

7.10.2 Channel Equalization in the Adaptive QAM

Radio transmission introduces additional problems to telecommunications. Consider that the waves may be reflected through many paths; the different transmission paths have different delays. An earlier transmitted symbol can then be mixed with a subsequent symbol that has been transmitted through a longer path.

The so-called *linear transversal equalizers* or *decision-feedback equalizers (DFE)* [7.55] are standard means to compensate for dynamic distortions of the channel. However, they are generally not performing well under nonlinear distortions. To combine the advantages of the conventional methods and the SOM-based adaptive detection described in the previous section, we have developed nonlinear dynamic adaptation methods [7.38–43] . In these methods, called the *neural equalizers*, the SOM is used either in cascade or in parallel with the conventional equalizer.

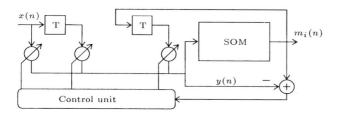

Fig. 7.21. Block diagram of the neural equalizer where the traditional Decision-Feedback-Equalizer (DFE) is used in cascade with the SOM. The DFE is first used to correct linear distortions and intersymbol interference due to the channel dynamics, while the SOM adapts to nonlinear distortions, changing the decision levels of the signal constellation

It is not possible to perform a complete analysis of the channel equalizer
here. Let is suffice to show the block diagram of the cascade combination
of the DFE structure with the SOM in Figure 7.21. The m_i values of the
SOM define the adaptive decision levels in the detector part. Let $y(n)$ be the
output from the channel equalizer (DFE) part for sample n. The output error,
$\epsilon(n) = y(n) - m_i(n)$, controls the adaptation of the DFE. The basic principle
of this neural equalizer is that the DFE corrects the linear distortions and
the channel dynamics, while the SOM is used to adaptively compensate for
nonlinearities.

7.10.3 Error-Tolerant Transmission of Images by a Pair of SOMs

In the previous example, the probability density function of the signal states
consisted of 16 clusters of equal probability mass, and the signal states formed
a regular array. Next we discuss another example in which the signal density is
continuous and has to be quantized optimally for error-tolerant transmission.
This problem is solved by having one SOM at the transmitting end as the
encoder, and an identical SOM at the receiving end as the decoder of signal
blocks, respectively [7.44, 45].

Consider *image compression* whereby information is transmitted in the
form of discrete symbols. If there are nonlinear and dynamic deformations in
the channel, they can be handled in the same way as before. In the following
discussion, however, we ignore the adaptive compensation and equalization
of the channel and concentrate on the problem of compressing and decom-
pressing the signal under presence of white noise. The coding scheme is also
different from the previous one.

Gray-level images were used in the example. They were divided into nonoverlapping blocks of 4 by 4 pixels, corresponding to 16-dimensional sample
vectors. For practical reasons the pixels were represented by 256 levels, but
in principle their signal domain could have been continuous. In the following
experiment, a set of 28 images was used for training, and an image, which
was not included in the training set, was used to test the system accuracy
and efficiency.

The first SOM, trained by representative sample vectors of the type of
images to be transmitted, will quantize the signal space almost optimally in
relation to the input density function. This gives the idea of using the *array
coordinates of the SOM as the codes*. Encoding thus means conversion of the
input vector, corresponding to the block of pixels to be transmitted, into
the discrete *array coordinates of the winner* m_c in the SOM, that are then
transmitted in the form of one or several codes. At the receiving end these
codes select the coordinates of an *identical SOM array*, and the weight vector
m_c' at that location is used as the replica of m_c, or the original block. The
latter step is the decoding process.

If now the transmitted code y is changed into another code y' due to
noise, ordering of the codebook vectors by the SOM guarantees that if y' is

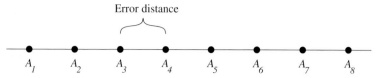

Fig. 7.22. Signal-space diagram for M-ary PAM signal (here M = 8). The distance between successive amplitude levels A_m defines the error probability. If noise amplitude on transmission channel exceeds half of the error distance, some of the codewords may be changed. In this work the noise amplitude probability was taken as normally distributed. At the noise levels discussed in this work, the probability for the noise causing errors corresponding to two or more amplitude levels is negligible

similar to y, the decoded sample vector m'_c will also be metrically similar to the sample vector m_c defined by the transmitted address.

To demonstrate the tolerance of a SOM-based image compression system to transmission channel errors, we used a simple *digital pulse amplitude modulation (PAM)* scheme [7.64, p. 164] with eight possible modulation amplitudes $\{A_m, m = 1, 2, \ldots, 8\}$ (see Fig. 7.22) for the transmission of the SOM coordinates. The errors in the PAM model, induced by noise in the transmission channel, are amplitude level changes. With the amount of noise discussed in this example, especially if the noise is normally distributed, the probability for the noise causing errors over two or more amplitude levels is negligible. Therefore, it will suffice to consider one-level errors (up and down) here. The noise is characterized by the probability for the one-level error being equal to p.

The SOM dimensions for codebooks had to be selected according to the accuracy of representation and principle of transmission. For a block of 16 pixels we decided to use 512 value combinations in all. For the eight-level PAM model, a three-dimensional SOM had then to be used, because if each dimension of the array has eight units, the total number of units in the SOM is $8^3 = 512$. The coordinate values in each of the SOM dimensions were used to select the amplitude levels in the PAM modulation. The codebook, corresponding to the array coordinates, was then composed of three 3-bit addresses of the array. Errors in the transmission channel mean that one or more of the coordinates are wrong by one unit, and a neighboring cell in the SOM array is selected at the receiving end.

For a comparative study, a similar encoding-decoding scheme with a *nonordered* codebook, i.e. a pair of identical vector quantizers was used. It is to be noted that due to a channel error and selection of a wrong codebook vector at the receiving end, a whole block of completely erroneous pixels will be replicated.

In Figure 7.23 two reconstructed images are shown. In these $p = 0.1$. (Because the transmission of three codes was required to transmit one nine-bit codeword and the error can happen either upward or downward in amplitude, the probability for an error in codeword $P(p)$ is considerably larger than the

Fig. 7.23. The encoded and decoded images after transmission through a very noisy ($p = 0.1$) channel. The image on the left has been vector quantized with a nonordered codebook, and the image on the right with an ordered codebook, respectively. The subjectively experienced qualities of the images differ significantly, although the same *number* of codeword errors was present in both images

probability for an error in a code. Here $P(0.1) = 1.0 - (1.0 - \frac{7}{4} * 0.1)^3 = 0.438$, i.e., more than 40 per cent of the codewords were erroneous.) In the image with random order in the codewords the errors are usually rather severe; e.g., in the middle of dark areas there are light blocks, and dark blocks are inserted in light areas, respectively. In the image with error-tolerant coding the errors are of different nature: for instance, in the dark area the erroneous blocks are never light, but "almost" dark. The subjectively experienced qualities of the images differ significantly, although the same *number* of codeword errors is present in both of the images.

7.11 The SOM as an Estimator

It already transpired in the robot-arm example that the SOM can be used as a *look-up table*. All input vectors that have a particular codebook vector as their nearest neighbor will first select this codebook vector for the "winner." At the winner location one can then store any kind of wanted output information. In this way one can also associate an output pattern with each input domain like in the so-called *associative mappings* and *estimators* [2.53]. The basic difference with respect to, say, the familiar *Gauss-Markov estimator* is that the mapping is no longer linear; the computational resources are also distributed into the input space optimally by the SOM.

7.11.1 Symmetric (Autoassociative) Mapping

The first principle for the implementation of an input-output mapping is somewhat similar to the *autoassociative encoding* discussed in [2.53]. Consider that the input vector x to the SOM is composed of *unconditioned input signals* $i_1, i_2, \ldots, i_p$ as well as a number of *wanted output signals* $o_1, o_2, \ldots, o_q$ given in conjunction with the i signals. Define the input vector to the SOM as

$$x = \begin{bmatrix} in \\ out \end{bmatrix}, \tag{7.23}$$

where $in = [i_1, i_2, \ldots, i_p]^\mathrm{T}$ and $out = [o_1, o_2, \ldots, o_q]^\mathrm{T}$. Apply a number of such x vectors as training inputs to a usual SOM. The weight vectors of this SOM have components corresponding to the unconditioned and wanted input signals, respectively:

$$m^{(i)} = \begin{bmatrix} m_i^{(in)} \\ m_i^{(out)} \end{bmatrix}. \tag{7.24}$$

After convergence, fix the m_i to their asymptotic values. If now an unknown unconditioned input vector is given, and the winner unit c is defined on the basis of the *in* part only, i.e. only the $m_i^{(in)}$ components are compared with the corresponding components of the new x vector, an estimate of the output in the sense of the SOM mapping is obtained as the vector $m_c^{(out)}$.

7.11.2 Asymmetric (Heteroassociative) Mapping

Construction of the symmetric mapping is straightforward, but it may be criticized on account of the partitioning of the input space depending on the complete x vector, i.e., both the unconditioned and the wanted signals. This may sometimes be justifiable (cf. the supervised SOM discussed in Sect. 5.10). In practical estimation applications, however, optimal partitioning of the input signal space and thus the resolution of the mapping should be determined by the unconditioned input signals solely. To this end, construction of the associative mapping can be accomplished in either of the two ways discussed next.

One-Phase Learning. If well-validated, labeled training data can be used for learning, the input vectors to be applied to the learning algorithm are defined as before:

$$x = \begin{bmatrix} in \\ out \end{bmatrix}. \tag{7.25}$$

However, this time, in order to quantize the input space according to the distribution of the unconditioned input signals only, *the winners shall now be determined on the basis of the input (in) parts of the x vectors only.* This kind of competitive learning will guarantee that the density of the *in* vectors will be approximated by the density of the codebook vectors.

Two-Phase Learning. Consider then that it is often much easier and less expensive to collect masses of *unlabeled* input data, in contrast to careful validation of a much smaller number of well-classified training samples. As the massively available information anyway describes the input density function that again defines the optimal resolution of the mapping, the unlabeled data should be utilized first for a preliminary unsupervised learning. As one of the most important benefits of the SOM is optimal vector quantization of the input space, the first phase in learning is indeed application of all the available unlabeled input data to a normal SOM algorithm, whereby the input vectors are of the form

$$x = \begin{bmatrix} in \\ \o \end{bmatrix} ; \tag{7.26}$$

here the symbol ø means the "dont't care" condition, i.e., when searching for the winner, only the in part of x is compared with the corresponding components of the weight vectors, and no output (out) part is thereby present in the learning algorithm.

After convergence of the SOM during this first phase, the well-validated, labeled calibration samples are applied to the SOM and training is continued. During this second phase the weight vectors corresponding to the in part of x are no longer changed; only the winners are located on the basis of the in parts. The SOM learning algorithm is now applied to the out parts of weight vectors only, The inputs at these two phases are thus:

$$
\begin{array}{cc}
\textit{During winner} & \textit{During learning} \\
\textit{search} & \textit{of the outputs}
\end{array}
\tag{7.27}
$$

$$x = \begin{bmatrix} in \\ \o \end{bmatrix} , \quad x = \begin{bmatrix} \o \\ out \end{bmatrix} .$$

Recall of the Estimate. When, after the learning phase and fixation of the weight vectors to constant values, an unknown input in is applied to a SOM of either of the above kinds, it always specifies the best-matching in part of the weight vectors, whereby this unit has the label c. The estimate of output, in analogy with Sect. 7.11.1, is then found as the value $m_c^{(out)}$.

Hybrid Estimators for Dynamical Problems. It is possible to combine different kinds of matching criteria in the different groups of inputs of the same neurons. For instance, one might add dynamical elements to some or all components of the in part of the x and m_i vectors as discussed in Sect. 5.6, or use the operator-map kind of matching as in Sect. 5.8. Alternatively, if the matching of the patterns in the in part were made according to the ASSOM principle described in Sect. 5.11, one might still associate any kind of responses to the out part. The reader may easily invent other functions that can be combined in the same SOM, for instance, for the estimation of $time$ $series$.

8. Software Tools for SOM

In order to obtain a "hands-on" experience of the methods, many people like to program neural network algorithms, such as the SOM, themselves. This is in general a good practice; however, many important factors should be taken into account. Neural-network methods in general, and the SOM process in particular do not produce such unique results as some deterministic mathematical algorithms, e.g., the Fast Fourier Transform. Adaptive processes may proceed in unexpected ways depending on the parameters and training sequences selected. At least in the beginning, the learning process may seem to fluctuate in an unpredictable fashion. Therefore special precautions must be taken.

The main reason for recommending the use of the readily available SOM software packages, at least in practical applications, is that there are many details in each phase of computation which are best appreciated by experience. Good software packages come with thoroughly tested recommendations for optimal use and many options for experimentation. Moreover, in good software packages there are also procedures for the monitoring of the self-organizing process and for the testing of the quality of the maps produced.

8.1 Necessary Requirements

The SOM Grid. As already discussed in Sect. 3.13, one should be able to select the SOM array to cope with the task and the data distribution. For most problems a 2D array is sufficient, but there may occur special tasks for which a higher-dimensional, e.g. 3D display is needed to span the data distribution better. The standard SOM software has only the 2D option, but if the source code is available, it may easily be modified for 3D arrays.

The hexagonal grid can usually represent the forms of the data clusters better than the rectangular one. The software package must allow for the definition of a hexagonal grid. The ratio of the sides of the array should be freely definable, because it should at least roughly correspond to the dimensions of the data distribution in the directions of its two principal axes. This could be defined automatically, but the standard packages do not have this option.

On the other hand, for some rare problems, e.g. those with cyclic data, a cyclic SOM array, for instance a toroid is better than a square array with open edges. Such special options for the grid topology are usually not available in standard SOM software.

Batch Map or Incremental SOM? A complete SOM software package should contain both the standard incremental SOM algorithm and the Batch Map version, at least for their comparison. The standard SOM has been studied mathematically in much greater depth than the Batch Map, and many industries are reluctant to accept software for which the mathematical foundations are not clear. Nonetheless the quality of the maps produced by either method can be tested easily, based on many different criteria, and especially in large problems the Batch Map has the advantage of being approximately an order of magnitude faster than the standard SOM.

One particular difference of these options seems to be that there is no explicit learning rate parameter in the Batch Map; however, in both versions one has anyway to define the time-dependent neighborhood function, a choice based on experience. If the initialization of the models is done carefully, the neighborhood function can be reasonably narrow all the time. Some software packages have a default definition of the neighborhood function, which helps an unexperienced user.

Initialization of the Models. Learning-Rate Factor and Neighborhood. It has been pointed out elsewhere in this book that random initialization was originally used to demonstrate the full organizing power of the SOM algorithm, which need not be proven every time the SOM is applied. With random initialization, the early ordering phases of the self-organizing process are critical and may take a long time, and for the global order to emerge one has to use a fairly wide neighborhood function during the first learning steps, which may also be computationally heavy. On the other hand, although it cannot be shown mathematically that there exists any ordered state in high-dimensional input spaces that is an absorbing state in the sense of Markov processes, nonetheless all practical experiences point to the following. If the model vectors of the array are initially ordered in any two-dimensional sequence, the convergence towards the asymptotic equilibrium proceeds at a rate which is orders of magnitude smoother, faster, and more reliable than with random initialization. Therefore we recommend that the initial values of the model vectors ought to be selectable as a regular two-dimensional sequence picked up from points on the 2D hyperplane that is oriented along the first two principal axes of the data distribution. This option is absolutely necessary in software packages. When this is done, one can continue the SOM process using a fairly narrow neighborhood function (that defines a suitable "stiffness" for the map), and if the standard SOM algorithm is used, fairly low learning-rate factors, of the order of a couple of per cent or less can also be used. If the Batch Map algorithm is used, no learning-rate factor need be defined.

So, the random-initialization option is not even necessary in standard software, and if the principal-axes initialization is made automatically, the supplier of the software ought to be able to specify the subsequent neighborhood width and learning-rate factor.

Missing Data. If the SOM software is used for data analysis, epecially for statistical applications, the possibility that a significant part of the descriptor values may be missing from the entries should be taken into account. The software must be able to handle the missing values, e.g., in the way described in Sect. 3.14.2.

Visualization. Since the SOM is mainly used to visualize clusters in the data, the quality of the software is also determined according to the ways in which the clusters are detected. For the time being, the so-called U-matrix method that visualizes the point density of the model vectors [3.38–39] is regarded as a necessary feature. Some recent studies (Kaski et al., 2000) have shown that the cluster borders can be detected more accurately by a gradient of the point density in regions where the sample density is low; this option will probably become a necessary feature in future standard software.

The component planes have also to be displayed in most applications. Moreover, visualization of the trajectory of the winner on the main map as well as on the component planes is considered as a necessary feature in most practical applications.

Monitoring. The average quantization error over all available input data is a sensitive measure of the mapping accuracy: if the configuration of the models has not yet reached the stable state in the learning process, or if there are unwanted "twists" in the map, the quantization error remains significantly higher than at the ordered optimum, and therefore a procedure for the computation of the quantization error is an unavoidable feature in the software. It is quite another question theoretically and also from the practical point of view whether the quantization error alone describes the topological order of the maps: some optional indicators of "topological errors" as suggested in Sect. 3.13 could be used but usually they are not included in the software packages, probably due to some ambiguities in their use.

8.2 Desirable Auxiliary Features

The features mentioned in Sect. 8.1 should not be lacking from any commercial SOM software, not even from programs written for special purposes. Moreover, the following features are desirable for general-purpose use.

Preprocessing. For a nonexpert user it is not always clear how important the preprocessing of raw data is, before feeding it to the SOM algorithm. The minimum requirement with stochastic measurements and statistical multivariate data is normalization of their scales so that the minimum and maximum values, respectively, of each normalized scale are the same, or that the variances of the scaled variables are the same. In some specific applications one may have to use different scales for the different variables or even to transform the primary data, but this can be carried out off-line and need not be included in standard software packages.

Line figures are seldom used as input to the SOM, but if this is done, some kind of blurring of the patterns (Sect. 7.1.1) is then necessary. For other optic patterns and acoustic signals the preprocessing is usually more complicated (Sects. 7.1 and 7.2).

Labeling. Often, especially in statistical applications, every entry has a unique set of descriptors and some identity (e.g., a name) that can be represented by a unique label. Labeling of the map units is then straightforward, it can be carried out after the models have converged to their stationary states, by inputting the entries again and assigning the labels to the winner units. This can be done automatically, since the labels are usually recorded in the data files with the entries. Some units in the map may thereupon remain unlabeled.

If the SOM is used for the classification of entries, for each of which there are several stochastic samples available, the problem is to divide the SOM area into nonoverlapping regions, each of which corresponds to a class. In other words, each map unit is assigned the most probable class label which is determined by first labeling the map units using *all* the available input samples, and then carrying out a majority voting over the labels at each map unit. In the case of a tie, a random choice between the winners is performed, or some secondary consideration applied. This labeling also proceeds automatically, and some map units may remain unlabeled. Sometimes one can use auxiliary clustering methods (cf. the Viscovery SOMine below) to make the class areas on the map connected and more uniform.

Sometimes it may be difficult to assign labels to the input samples a priori, for instance in process studies, where the SOM only performs an unsupervised clustering of the raw process measurements. For illustrative purposes, however, one may label the clusters manually, based on an expert evaluation of the process states. Such an evaluation and labeling of the SOM areas, e.g., by pseudocolors may be possible even if there are no clear cluster structures.

A different kind of labeling is needed in document maps (WEBSOM), where each entry contains a number of words, of which the most descriptive ones should be selected to describe a cluster or area on the SOM. This kind of labeling can be done automatically, and the procedure described in Sect. 7.8 is very powerful. Automatic labeling is the only possibility if the map is really big.

Tools for Preliminary Data Analysis. Before starting to use a new method such as the SOM for an unknown data set, it is necessary to have a conceptualization of the structures in the data distribution. It is always advisable to carry out a preliminary study, e.g., using some of the nonlinear projection methods mentioned in Sect. 1.3.2, to determine the topological relations between the data items. Therefore, it would be useful if some of these algorithms were already included in the software package, and they could use common data files along with the SOM procedures. For instance, Sammon's projection is a very useful tool in preliminary data analysis as well as for the monitoring of the model vectors.

Data Postprocessing and Visualization. In order to interpret the SOM mapping in terms of traditional statistical concepts, one may want to compute pairwise correlation, scatter plots, data histograms, cluster properties, principal components, and factor loadings on subsets of data segmented by the SOM. Many of these tasks require effective visualization methods, usually emphasized by pseudocolors.

User Interface. It is a common requirement in all contemporary software that it must have a user-friendly interface, with both effective command lines and sufficient graphics facilities. Some of the SOM programs run on standard software platform such as GUI, MatLab language, or graphics facilities provided by a major computer supplier. There is, of course, a big difference in the supports between public-domain and commercial program packages; a factor in favour of the former is the availability of their source codes.

8.3 SOM Program Packages

In this section we have descriptions of software that concentrates on the standard Self-Organizing Maps. The packages reviewed in this section have been developed by groups that have experience of some practical SOM applications. Only the main approaches are described, since the software is developing all the time. In Sect. 8.4 we shall discuss neural networks or other software that only includes the SOM as a special feature.

8.3.1 SOM_PAK

The first public-domain software package that was intended for a general-purpose SOM development tool is the SOM_PAK, released for the first time in 1990 by the Laboratory of Computer and Information Science of Helsinki University of Technology. Previously, in 1989 a similar package LVQ_PAK containing the Learning Vector Quantization algorithms had already been released. The original reason for the publication of these packages was that the SOM and LVQ methods contained plenty of details, such as the optimal number of models, their initialization, proper learning sequences etc. that could not be deduced from the general descriptions of the basic algorithms without extensive experience. Consequently, even known researchers, obviously starting with the algorithmic description of the SOM directly, published "benchmarkings" against other methods where these algorithms were not used correctly and produced inferior results. Our original idea was to give hands-on experience to newcomers, first by means of simple sessions using carefully selected, realistic exemplary data. The source code is completely available, and the only copyrighted restriction is that it or parts of it cannot be used in other software for commercial distribution.

One special remark has to be made. The SOM is an unsupervised classification method, comparable to clustering, whereas many of the "benchmarkings" were carried out with labeled data and compared with supervised algorithms such as Backpropagation. For a justified comparison, at least the LVQ algorithms (LVQ_PAK) should have been used.

Availability. The easiest access to the SOM_PAK and LVQ_PAK is from the Internet page http://www.cis.hut.fi/research/software.shtml. (The old ftp access is now down.) The programs, source code, and full documentation are available.

Platforms. These packages can be used under either UNIX or MS DOS, and they have been tested in many computers, ranging from supercomputers to pc's. They do not have special Windows interfaces.

SOM Features. The only SOM algorithm included in this package is the standard incremental-learning SOM, but the program code, which is written in ANSI C, contains many kinds of tricks developed since 1982 for its optimization. The main user interface is based on UNIX-type command lines on which the parameters of the algorithm are defined. The source code modules of the program have been given as C programming text files, which allow the users to make their own modifications. The SOM_PAK contains only rather simple graphics programs of its own; any more demanding visualization can be implemented with general-purpose graphics or visualization programs.

The SOM array topology can be selected as either rectangular or hexagonal, and the map size and the dimensionality of the vectors are unlimited, being only restricted by the computer's own resources. Very big data files can be processed in a buffered fashion.

Initialization of the models can be made at random or along the two principal axes of the data distribution. The initialization can be repeated automatically, and after short test runs the best map can be selected for continued learning.

The neighborhood function can have either the "bubble" or Gaussian form, and many different combinations can be defined for the learning sequences.

The algorithm automatically takes care of any missing data.

The standard visualization options, viz. the component planes with trajectories, and the U-matrix are provided.

For monitoring, evaluation of the average quantization error during the process is possible.

For the preliminary data analysis there is a procedure to compute the Sammon projection. If the projection is computed for the model vectors, the illustration is provided by the network of lines that links topologically neighboring model vectors.

The SOM_PAK has the important advantage of being compatible with the LVQ_PAK software: if the map shall be used for supervised classification, its fine-tuning with LVQ1, LVQ2, LVQ3 or OLVQ1 can be easily made.

8.3.2 SOM Toolbox

While the SOM_PAK and LVQ_PAK were compiled as a "missionary task" for a wide circle of potential users, the SOM Toolbox was created for very pragmatic reasons. The SOM algorithm had been used by us in our numerous cooperative projects, with industry as well as in financial applications, but the need for a better platform for experimentation, as well as more versatile visualization tools were felt necessary. On the other hand, in the industrial and finance applications the SOM dimensions were never very big, and a much lower computing capacity and speed than that provided by the SOM_PAK were sufficient. It has turned out that most practical applications are of this kind, and therefore, our next package was designed as a toolbox for the MatLab system: the versatile graphics facilities of the latter could then be exploited. The first version of the SOM Toolbox was released by the researchers of the Laboratory of Computer and Information Science of Helsinki University of Technology around 1996.

Availability and Requirements. In the same way as the SOM_PAK and LVQ_PAK, the SOM Toolbox can be downloaded from the Internet page http://www.cis.hut.fi/software.shtml. This is public-domain software with similar mild restrictions of use as the previous ones. Because the SOM Toolbox has very good graphics programs, it requires MatLab version 5 or higher. A GUI interface is also necessary.

SOM Features. Both the standard incremental SOM algorithm, as well as the Batch Map version are included. The map can be rectangular or hexagonal, and its size is unlimited (but the slower speed of the algorithm compared with LVQ_PAK has to be taken into account). Both random initialization, and initialization along the hyperplane spanned by the two principal axes is possible. The same neighborhoods and the same kinds of training sequences as in SOM_PAK can be used. The missing values are taken into account in similar ways to the SOM_PAK.

The input vectors can be scaled automatically in order to have the same variance in all components, or by some other methods. No automatic linking of file segments is possible.

For visualization, component planes with trajectories, the U-matrix, and a histogram of hits of the samples in the map units can be drawn. A range of other possibilities for the analysis using simple auxiliary programs can be found in [8.1].

For preliminary data analysis and monitoring, the Sammon projection and the average quantization error can be plotted.

8.3.3 Nenet (Neural Networks Tool)

This software was first released by graduate students, the Nenet Team of Helsinki University of Technology in 1997. It was intended to be more user-friendly than the ANN programs of that time. The SOM can be visualized in several ways.

The differences between SOM_PAK, SOM Toolbox and Nenet can be roughly characterized in the following way: SOM_PAK has been designed for very large and computationally heavy professional tasks. Nenet is easy to use and has good graphics programs, but is best suited for relatively small-scale problems, which illustrate the use of the SOM. The SOM Toolbox is a compromise, being versatile and easy to use and still being able to handle professional problems, although not as large as SOM_PAK.

Availability and Requirements. The programs can be downloaded from the page http://www.mbnet.fi/~phodju/nenet/Nenet/General.html. They use Windows 9x and Windows NT multitasking capabilities; for pc computers, the 32-bit Windows 95/NT is recommendable.

SOM Features. Data preprocessing can be made in the initialization phase, after which it is automatically taken into account in all successive phases. The scales can be normalized according to either the data maximum and minimum or variance. The preprocessing parameters are saved in a file for further phases.

The algorithm is the standard incremental SOM algorithm. The map topology can be rectangular or hexagonal, and initialization can be random or along the plane defined by the principal axes. The training sequences are similar to those in SOM_PAK and SOM Toolbox. Any missing data can be handled.

There are several visualization options: the component planes with trajectories, the U-matrix, 3D hit histrograms of the input samples and display of the active neuron coordinates. A label can be added by double clicking on a map unit. The initialization and training history is displayed in the map header.

The quantization and topographic errors are calculated and shown with the map.

8.3.4 Viscovery SOMine

This is a commercial SOM software package produced by Eudaptics GmbH in Austria. It is considered as user-friendly, flexible, and powerful, especially in statistical problems, such as financial, economics, and marketing applications, and is supplied as two versions: SOMine Pro and SOMine Lite. The Pro version has more extensive clustering options and can perform dependency analysis; it has the GUI, OLE, SQL and DB2 interfaces and some features for the processing of text files.

Availability and Requirements. The company's e-mail address is, office@eudaptics.co.at and their Internet page, http://www.eudaptics.co.at.

The programs run on Windows 95 and Windows NT 4.0.

SOM Features. The data preprocessing options are more versatile than in the previous packages: they include several scaling options, transformation of variables, prority setting etc.

In order to make the software user-friendly and minimize the prerequisites in its application, only the most central options of the SOM algorithm and map features have been selected for this package: the Batch Map algorithm, provided with some accelerated computing (growing map), has been used. Thus the computing speed is high. The map array is always hexagonal and no limitations to its size or input dimensionality are set. The initialization is always made along the plane spanned by the principal axes. The neighborhood function is always Gaussian. The training sequences follow predefined schedules. Any missing data are handled automatically. The reduced number of options makes the package easy to use and guarantees unique results.

A special feature in this software is that the SOM algorithm is combined with the *Ward* clustering algorithm, which segments the SOM area into more uniform subareas.

The visualization options include component planes with trajectories, the U-matrix, cluster windows and iso-contours of hit density.

Many different kinds of monitoring and data postprocessing facilities are provided.

8.4 Examples of the Use of SOM_PAK

Many neural-network program packages are used in the way illustrated below by the SOM_PAK commands, which are given as Unix-type command lines. They include the function symbol and the parameter values necessary for that function.

Before training, the data files for the input data and the model (reference) vectors must be specified.

8.4.1 File Formats

All data files (input vectors and maps) are stored as ASCII files for easy editing and checking. The files that contain training data and test data are formally similar, and can be used interchangeably.

Data File Format. The input data is stored in ASCII-form as a list of entries, one line being reserved for each vectorial sample.

The first line of the file is reserved for status knowledge of the entries; in the present version it is used to define the following items:

Compulsory:

– Dimensionality of the vectors (integer)

Optional:

– Topology type, either *hexa* or *rect* (string)
– Map dimension in x-direction (integer)
– Map dimension in y-direction (integer)
– Neighborhood type, either *bubble* or *gaussian* (string)

In data files the optional items are ignored during execution of the commands. As a matter of fact, the optional items are used when the file represents model vectors.

The subsequent lines consist of n floating-point numbers followed by an optional class label (that can be any string) and two optional qualifiers (see below) that determine the usage of the corresponding data entry in training programs. The data files can also contain an arbitrary number of comment lines that begin with '#', and are ignored. (One '#' for each comment line is needed.)

If some components of some data vectors are missing (due to data collection failures or any other reason) those components should be marked with 'x' (replacing the numerical value). For example, a part of a 5-dimensional data file might look like:

```
1.1   2.0   0.5   4.0   5.5
1.3   6.0    x    2.9    x
1.9   1.5   0.1   0.3    x
```

When the vectorial distances are calculated for winner detection and when the model vectors are modified, the components marked by x are ignored.

An example data file: Consider a hypothetical data file *exam.dat* that represents shades of colors in a three-component form. This file contains four color samples, each one comprising a three-dimensional data vector. (The dimensionality of the vectors is given on the first line.) The labels can be any strings; here 'yellow' and 'red' are the names of the classes.

exam.dat:

```
3
# First the yellow entries
181.0   196.0    17.0   yellow
251.0   217.0    49.0   yellow
# Then the red entries
248.0   119.0   110.0   red
213.0    64.0    87.0   red
```

Each data line may have two optional qualifiers that determine the usage of the data entry during training. The qualifiers are of the form *codeword=value*. The optional qualifiers are the following:

- Enhancement factor: e.g. *weight=3*.
  The training rate for the corresponding input pattern vector is multiplied by this parameter so that the model vectors are updated as if this input vector were repeated 3 times during training (i.e., as if the same vector had been stored 2 extra times in the data file).
- Fixed-point qualifier: e.g. *fixed=2,5*.
  In order to force a model vector into a given location on the map, the map unit defined by the fixed-point coordinates $(x = 2, y = 5)$ is selected instead of the best-matching unit for training. (See below for the definition of coordinates over the map.) If several inputs are forced to known locations, a wanted orientation results in the map.

Map File Format. The map files are produced by the SOM_PAK programs, and the user usually does not need to examine them by hand.

The reference vectors are stored in ASCII-form. The format of the entries is similar to that used in the input data files, except that the optional items on the first line of data files (topology type, x- and y-dimensions and neighborhood type) are now compulsory. In map files it is possible to include several labels for each entry.

An example: The map file *code.cod* contains a map of three-dimensional vectors, with three times two map units. This map corresponds to the training vectors in the *exam.dat* file.

code.cod:

```
3 hexa 3 2 bubble
191.105   199.014   21.6269
215.389   156.693   63.8977
242.999   111.141   106.704
241.07    214.011   44.4638
231.183   140.824   67.8754
217.914   71.7228   90.2189
```

The x-coordinates of the map (column numbers) may be thought to range from 0 to $n-1$, where n is the x-dimension of the map, and the y-coordinates (row numbers) from 0 to $m-1$, respectively, where m is the y-dimension of the map. The reference vectors of the map are stored in the map file in the following order:

In Fig. 8.1 the locations of the units in the two possible topological structures are shown. The distance between two units in the map is computed as the Euclidean distance between the grid points.

1	The unit with coordinates $(0,0)$.
2	The unit with coordinates $(1,0)$.
...	
n	The unit with coordinates $(n-1,0)$.
$n+1$	The unit with coordinates $(0,1)$.
...	
nm	The last unit is the one with coordinates $(n-1, m-1)$.

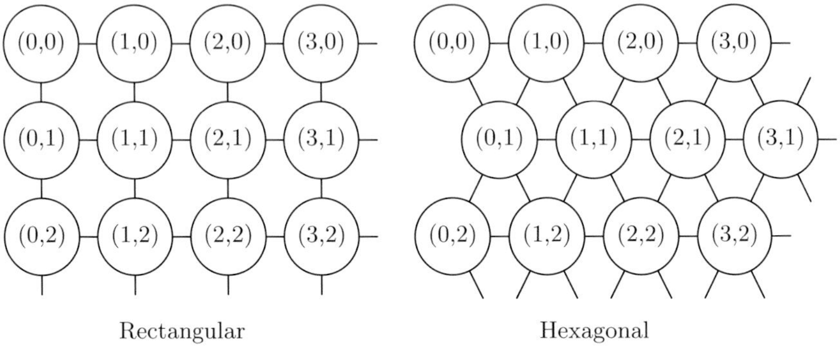

Rectangular   Hexagonal

Fig. 8.1. Map topologies

8.4.2 Description of the Programs in SOM_PAK

Program Parameters. Various programs need various parameters. All the parameters that are required by any program in this package are listed below. The meaning of the parameters is obvious in most cases. The parameters can be given in any order in the commands.

-din	Name of the input data file.
-dout	Name of the output data file.
-cin	Name of the file from which the reference vectors are read.
-cout	Name of the file to which the reference vectors are stored.
-rlen	Run length (number of steps) in training.
-alpha	Initial learning rate parameter. Decreases linearly to zero during training.
-radius	Initial radius of the training area in som-algorithm. Decreases linearly to one during training.
-xdim	Number of units in the x-direction.
-ydim	Number of units in the y-direction.
-topol	The topology type used in the map. Possible choices are the hexagonal lattice (*hexa*) and rectangular lattice (*rect*).
-neigh	The neighborhood function type used. Possible choices are the step function (*bubble*) and Gaussian (*gaussian*).
-plane	The component plane of the reference vectors that is displayed in the conversion routines.

-fixed   Defines whether the fixed-point qualifiers are used in the train-
                ing programs. A value one means that fixed-point qualifiers
                are taken into account. Default value is zero.

-weights   Defines whether the weighting qualifiers are used in the training
                programs. A value one means that qualifiers are taken into
                account. Default value is zero.

-alpha_type   The learning rate function type (in *vsom* and *vfind*). Possi-
                ble choices are the linear function (*linear*, the default) and
                inverse-time type function (*inverse_t*). The linear function
                is defined as $\alpha(t) = \alpha(0)(1.0 - t/rlen)$ and the inverse-time
                type function as $\alpha(t) = \alpha(0)C/(C + t)$ to compute $\alpha(t)$ for
                an iteration step t. In the package the constant C is defined
                $C = rlen/100.0$.

-qetype   The quantization error function type (in *qerror* and *vfind*). If a
                value greater than zero is given then a weighted quantization
                function is used.

-version   Gives the version number of SOM_PAK.

-rand   Parameter that defines whether a new seed for the random-
                number generator is defined; otherwise the seed is read from
                the system clock.

In addition to these, there are parameters for diagnostic output, and parameters for more advanced functions.

Initialization Programs. The initialization programs initialize the reference vectors.

– *randinit* - This program initializes the reference vectors to random values.
  The vector components are set to random values that are evenly distributed
  in the area of corresponding data vector components. The size of the map
  is given by defining the x-dimension (*-xdim*) and the y-dimension (*-ydim*)
  of the map. The topology of the map is defined with option (*-topol*) and is
  either hexagonal (*hexa*) or rectangular (*rect*). The neighborhood function
  is defined with option (*-neigh*) and is either a step function (*bubble*) or a
  Gaussian (*gaussian*).
  > *randinit -xdim 16 -ydim 12 -din file.dat -cout file.cod -neigh bubble -topol*
    hexa

– *lininit* - This program initializes the reference vectors in an orderly fashion
  along a two-dimensional subspace spanned by the two principal eigenvectors of the input data vectors.
  > *lininit -xdim 16 -ydim 12 -din file.dat -cout file.cod -neigh bubble -topol*
    hexa

Training Program. The *vsom* is the main program that constructs the
Self-Organizing Map.

– *vsom* - This program trains the reference vectors using the self-organizing
  map algorithm. The topology type and the neighborhood function defined

in the initialization phase are used throughout the training. The program finds the best-matching unit for each input sample vector and updates those units in its neighborhood according to the selected neighborhood function.

The initial value of the learning rate is defined and will decrease linearly to zero by the end of training. The initial value of the neighborhood radius is also defined and it will decrease linearly to one during training (in the end only the nearest neighbors are trained). If the qualifier parameters (-*fixed* and *-weight*) are given a value greater than zero, the corresponding definitions in the pattern vector file are used. The learning rate function α can be defined using the option *-alpha_type*. Possible choices are *linear* and *inverse_t*. The linear function is defined as $\alpha(t) = \alpha(0)(1.0-t/rlen)$ and the inverse-time type function as $\alpha(t) = \alpha(0)C/(C+t)$ to compute $\alpha(t)$ for an iteration step t. In the package the constant C is defined $C = rlen/100.0$.

> *vsom -din file.dat -cin file1.cod -cout file2.cod -rlen 10000 -alpha 0.03 -radius 10 [-fixed 1] [-weights 1] [-alpha_type linear]*

Notice that the degree of forcing data into specified map units can be controlled by alternating "fixed" and "nonfixed" training cycles.

Quantization Accuracy Program. The average quantization error of the final map is computed by this function.

– *qerror* - The average quantization error is evaluated. For each input sample vector the best-matching unit in the map is searched for and the average of the respective quantization errors is returned.

> *qerror -din file.dat -cin file.cod [-qetype 1] [radius 2]*

It is possible to compute a weighted quantization error $\sum h_{ci}||x - m_i||^2$ for each input sample and average these over the data files. If option *-qetype* is given a value greater than zero, then a weighted quantization error is used. Option *-radius* can be used to define the neighborhood radius for the weighting, default value for that is 1.0.

Monitoring Programs. The following functions display various results visually.

– *visual* - This program generates a list of coordinates corresponding to the best-matching unit in the map for each data sample in the data file. It also gives the individual quantization errors made and the class labels of the best matching units if the latter have been defined. The program will store the three-dimensional image points (coordinate values and the quantization error) in a way similar to that of the input data entries. If an input vector consists of missing components only, the program will skip the vector. If option *-noskip* is given the program will indicate the existence of such a line by saving line '*-1 -1 -1.0 EMPTY_LINE*' as a result.

> *visual -din file.dat -cin file.cod -dout file.vis [-noskip 1]*

– *sammon* - Generates the Sammon mapping from the n-dimensional input vectors to 2-dimensional points on a plane whereupon the distances between the image points tend to approximate the (general) distances of the input items. If option -*eps* is given an encapsulated postscript image of the result is produced. The name of the eps-file is generated by using the output file basename (up to the last dot in the name) and adding the ending _*sa.eps* to the output filename. If the option -*ps* is given a postscript image of the result is produced. The name of the ps-file is generated by using the output file basename (up to the last dot in the name) and adding the ending _*sa.ps* to the output filename.

In the following example, if the option -*eps 1* is given, an eps file named *file_sa.eps* is generated.

> *sammon -cin file.cod -cout file.sam -rlen 100 [-eps 1] [-ps 1]*

— *planes* - This program generates an encapsulated postscript (eps) code from one selected component plane (specified by the parameter -*plane*) of the map imaging the values of the components using gray levels. If the parameter given is zero, then all planes are converted. If the input data file is also given, the trajectory formed of the best-matching units is also converted to a separate file. The eps files are named using the map basename (up to the last dot in the name) and adding to it _*px.eps* (where *x* is replaced by the plane index, starting from one). The trajectory file is named accordingly, adding _*tr.eps* to the basename. If the -*ps* option is given a postscript code is generated instead and the produced files are named replacing .*eps* by .*ps*.

In the following example a file named *file_p1.eps* is generated containing the plane image. If the -*din* option is given, another file *file_tr.eps* is generated containing the trajectory. If the -*ps* option is given then the produced file is named *file_p1.ps*.

> *planes -cin file.cod [-plane 1] [-din file.dat] [-ps 1]*

— *umat* - This program generates an encapsulated postscript (eps) code of the U-matrix to visualize the distances between reference vectors of neighboring map units using gray levels. The eps file is named using the map basename (up to the last dot in the name) and adding .*eps* to it.

If the -*average* option is given the grey levels of the image are spatially filtered by averaging, and if the -*median* option is given median filtering is used. If the -*ps* option is given a postscript code is generated instead and a .*ps* ending is used in the filename.

In the following example a file named *file.eps* is generated containing the image.

> *umat -cin file.cod [-average 1] [-median 1] [-ps 1]*

— *vcal* - This program labels the map units according to the samples in the input data file. Each unit receives the labels of all the data vectors for which it is the best matching unit. The map units are then labeled according to the majority of labels 'hitting' a particular map unit. The units that get no 'hits' are left unlabeled. Giving the option -*numlabs* one can select the maximum number of labels saved for each codebook vector. The default value is one.

> *vcal -din file.dat -cin file.cod -cout file.cod [-numlabs 2]*

8.4.3 A Typical Training Sequence

First Step: Map Initialization. The reference (model) vectors of the map are first initialized to tentative values. As the initialization function precedes all training sequences, the main SOM features (map dimensions, grid topology, neighborhood type) are also given on its command line.

In the example the map is initialized using values picked up from the "principal plane." The lattice type is selected to be hexagonal (*hexa*) and the neighborhood function type is a step function (*bubble*). The map size here is 12 times 8 units.

> *lininit -xdim 16 -ydim 12 -din file.dat -cout file.cod -neigh bubble*

Second Step: Map Training. The map is trained by the self-organizing map algorithm using the program *vsom*.

With the *lininit* initialization only one phase of training is needed.

> *vsom -din ex.dat -cin ex.cod -cout ex.cod -rlen 10000 -alpha 0.02 -radius 3*

The reference vectors in each unit converge to their 'correct' values. After training the map is ready to be tested and to be used in monitoring applications.

Third Step: Map Visualization. The trained map can now be used for the visualization of the data samples. In SOM_PAK there are visualization programs that make an image of the map (actually one selected component plane of it) and plot the trajectory of the best-matching units vs. time on it.

Before visualization, the map units are calibrated using known input data samples. The sample file *ex_fts.dat* contains labeled samples from states of an overheating device.

> *vcal -din ex_fts.dat -cin ex.cod -cout ex.cod*

After calibration some of the units in the map have labels showing an area in the map which corresponds to fatal states.

The program *visual* generates a list of coordinates corresponding to all the best-matching units in the map for each data sample in the data file. It also returns the quantization errors and the class labels of the best-matching units, if the latter have been defined. The list of coordinates can then be processed for various graphical outputs.

The data file *ex_ndy.dat* contains samples collected over 24 hours from a device operating normally. The data file *ex_fdy.dat* contains samples collected during 24 hours from a device that has overheating problems during the day.

> *visual -din ex_ndy.dat -cin ex.cod -dout ex.nvs*
> *visual -din ex_fdy.dat -cin ex.cod -dout ex.fvs*

The program *visual* stores the three-dimensional image points (coordinate values of the responses and the quantization errors) in a similar fashion as the input data entries are stored.

The package also includes the program *planes* to convert the map planes to encapsulated postscript (eps) images and the program *umat* to compute the U-matrix visualization of the SOM reference vectors and to convert it to an encapsulated postscript (eps) image.

8.5 Neural-Networks Software with the SOM Option

It is very difficult to review the plethora of available software packages that include the SOM along with many other algorithms. We have studied some of these packages, not all of them, but it is hard to make comparisons, because many of the packages have only been made for a special purpose such as market segmentation, and the SOM algorithm may be combined with other methods. Normally these packages do not contain as many SOM features as mentioned in Sect. 8.3. The most serious criticism is that questionable training sequences, for which the asymptotic state is not guaranteed or may

Table 8.1. Neural Networks Software with SOM features

Product	Supplier	E-mail
SAS Neural Network Application	SAS Institute, Inc.	software@sas.sas.com
NeuralWorks	NeuralWare, Inc.	sales@neuralware.com
MatLab Neural Network Toolbox[1]	The MathWorks, Inc.	info@ mathworks.com
NeuroShell2/ NeuroWindows	Ward Systems Group, Inc.	WardSystems@ msn.com
NeuroSolutions v3.0	NeuroDimension, Inc.	info@nd.com
NeuroLab, A Neural Network Library	Mikuni Berkeley R&D Corporation	neurolab-info@ mikuni.com
havFmNet++	hav.Software	hav@neosoft.com
Neural Connection	Neural Connections	sales@spss.com
Trajan 2.0 Neural Network Simulator	Trajan Software Ltd.	andrew@tarjan-software.demon.co.uk

[1] not the same as SOM Toolbox described in Sect. 8.3.2

be ambiguous, have often been used, and usually no monitoring programs are included by means of which the quality of the resulting maps could be tested. The results obtained with these packages should be benchmarked against results obtained by the more extensive SOM packages.

Table 8.1 lists some of these products and their suppliers. It is based on the information in [10.3].

9. Hardware for SOM

If the dimensionality of an application problem is not very high (say, a few dozen inputs and outputs at most) and computing need not be performed in real time, most ANN algorithms can be implemented by pure software, especially if contemporary workstation computers are available. However, especially in real-time pattern recognition or robotics applications one might need special co-processor boards or even "neurocomputers." For really large problems, e.g. in the preprocessing stages for more complicated computer vision, special hardware networks may have to be developed. Such ANN circuits are sometimes necessary to miniaturize devices and make them cheap, for instance in medical applications or consumer electronics.

The principal methods for the implementation of ANNs are for the present: analog VLSI (very-large-scale integration), and parallel digital computer architectures, especially the SIMD (single instruction stream, multiple data stream) machines. Of ANN implementations that are still at the experimental stage one may mention optical computers.

Often the need for special hardware is overestimated. For instance, the rather sophisticated speech recognition system developed in our laboratory, being a hybrid of neural and more conventional solutions, is for the time being implemented by pure software written for a modern commercial workstation, without any auxiliary co-processor boards. It can recognize and type out arbitrary Finnish dictation in real time, with an overall accuracy of about 95 per cent, referring to the correctness of individual letters of Finnish orthography.

The material of this chapter will be presented in the order the ideas should interest the reader, not ranked according to practical importance. We shall start with some basic functions.

9.1 An Analog Classifier Circuit

Crossbar Architecture. A crossbar structure was first suggested for artificial neural networks by *Steinbuch* [9.1]. As a modernized version, which does not yet exist in the complete suggested form, we would have a *crossbar switch input structure*, and a *winner-take-all type output stage*. Whatever the implementation of the WTA, the architecture of a simple analog classifier

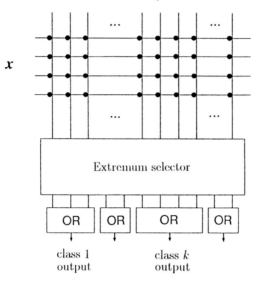

Each column corresponds
to a codebook vector m_i

x

Extremum selector

OR   OR   OR   OR

class 1   class k
output   output

Fig. 9.1. Architecture of a simple analog classifier circuit based on crossbar

circuit can be delineated like in Fig. 9.1, with the OR operation meaning the logical sum over respective output lines.

It may now be understandable why the nonlinear dynamic neuron equation in Chapt. 4 was written in the form (4.1) or (9.3) and not in any alternative way that also appears in literature. This "physiological" neuron model for analog computation of the ANN functions is very profitable, because the computationally most intensive part is the input activation, the expression for which we should keep as simple as possible, e.g.:

$$I_i = \sum_{j=1}^{n} \mu_{ij}\xi_j \ . \tag{9.1}$$

In the basic SOM algorithm and in some special hardware solutions I_i often occurs in the form

$$I_i = \sum_{j=1}^{n} (\xi_j - \mu_{ij})^2 \ , \tag{9.2}$$

which is also computable by rather simple analog means. In the latter case, the WTA circuit must be replaced by a *minimum selector* circuit.

The dot product expressed by (9.1) is implementable by the *Gilbert multiplier* [9.2], the principle of which is shown in Fig. 9.2.

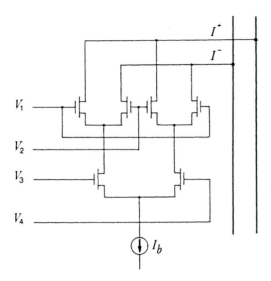

Fig. 9.2. Simplified principle of the Gilbert multiplier. The two lower MOS transistors act as a current divider for the constant-current source I_b, the division ratio depending on $V_3 - V_4$ (one of the multiplicands). Further division of these currents onto the output lines is controlled by $V_1 - V_2$. This circuit thus multiplies the values $V_1 - V_2$ and $V_3 - V_4$, where the latter difference signifies the input weight that may be stored in special analog memory cells. The product is represented by $I^+ - I^-$

Nonlinear Dynamic Output Circuit. To describe the neuron as a nonlinear dynamic element, the differential equation introduced in Chap. 4 may serve as a starting point:

$$d\eta_i/dt = I_i - \gamma(\eta_i) , \qquad (9.3)$$

where γ is a convex function of η_i, and $\eta_i \geq 0$. This equation can easily be implemented by elementary traditional analog computer (differential analyzer) circuits. Notice that (9.3) need only be computed once for every n input-line operations (9.1) or (9.2). A system of m nonlinear dynamic neurons of this type, with n inputs each, forming a WTA circuit, was discussed in Ch. 4.

Naturally the WTA function can be implemented much easier by nonlinear electronic circuits, as shown in Fig. 9.3.

A Simple Static WTA Circuit. If the dynamic operation defined by (9.3) is not essential to describe, say, dynamic systems, the winner-take-all circuit can be constructed much simpler as a static version, as shown in Fig. 9.3.

This circuit accepts signals as *input voltages*; so if the I_i are currents like at the outputs of an electronic crossbar, they must be converted into voltages by the so-called transconductance stages, familiar from elementary electronics.

In Fig. 9.3a, the usual OR (logic sum) gate of diode logic is used as maximum selector (MAX) for voltages. Due to the thermodynamic properties

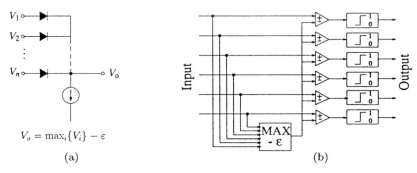

Fig. 9.3. Static WTA circuit. **(a)** Maximum selector (MAX), **(b)** The complete WTA

of semiconductor diodes, there is a small offset ε of the order of 50 millivolts at the output of the MAX stage.

The open triangular symbols followed by the rectangular ones in Fig. 9.3b represent so-called differential amplifiers with unsymmetric output stage: when the difference of input voltages exceeds a certain small limit, the output will attain a static high value, whereas if the difference of inputs is below another small limit, the output value is low. With standard electronic circuits the margin between these two limits can be made much smaller than the offset ε of the diode circuit. This WTA, however, has then an undefined state such that if two or more maximal inputs are approximately equal within the accuracy of ε, they may produce multiple output responses. In practice, however, such cases are rare, and it has even been shown in Sect. 3.5.2 that a multiple response in learning may not be harmful at all.

The complete classifier architecture described in this section is imagined to use fixed "synaptic" parameters, the values of which must be precomputed off-line and loaded, e.g. as electrical charges into the weight memories. For the time being, no solutions of practical importance exist for the implementation of on-line learning in the SOM or LVQ algorithms by analog circuits, although it may be imagined how this might be done.

9.2 Fast Digital Classifier Circuits

Before discussing genuine parallel computing architectures, it may be useful to demonstrate how fast certain simple special solutions can be. The architecture discussed in this section computes the classifier function serially, but with very high speed. It is possible that this same solution could find its way to parallel architectures, too.

Very Low Accuracy May Be Sufficient. Special hardware is only needed if we are working with a problem with high dimensionality. This may mean that we have over one hundred components in the input vectors and the SOM or LVQ may contain over 1000 codebook vectors. In classifier functions we have to sum up a large number of stochastic terms, whereby the individual noise contributions of the terms will be smoothed out statistically. Since inherent inaccuracy of the feature signals and overlap of class distributions constitute the biggest problems in statistical pattern recognition, it is obvious that an extra noise contribution due to round-off errors of the digital representations may not be harmful, especially since round-off errors can be regarded as white noise, and statistically independent errors usually add up quadratically. We will now show that the signal values can be rounded off radically, resulting in special computing solutions.

The round-off experiments reported in this section were carried out by the LVQ1 algorithm, using the LVQ_PAK software package (Sect. 8.3.1). In short, it has been shown that if the dimensionality of the input and codebook vectors is high ($>$ 100), their component values *during classification* (or determination of the "winner") can be approximated by as few as *three or four bits*, without noticeable reduction in recognition accuracy. Thus, if learning has been made off-line, preferably with a higher accuracy, the approximations of the vectorial values can be loaded onto very simple hardware devices. The quantized arguments also facilitate special computing solutions, such as replacement of arithmetic circuits by precomputed *tables* or memories, from which the values of the classifier functions can be *searched* rapidly.

One of the problems studied in our laboratory over the years has been speech recognition. The speech states are represented, e.g., every 10 ms by 20 cepstral coefficients [7.8]. To someone unfamiliar with this task it may not be quite clear how much statistical variance the same speech elements may have even for the same speaker. For this reason, increasing accuracy in feature detection is reasonable only up to a certain limit, whereas increasing the number of different feature variables is much more effective in improving statistical accuracy [7.11]. For instance, taking the cepstral coefficients as averages over seven subwindows the 220-ms "time window" shown in Fig. 9.4 and concatenating these 20-component subvectors into a 140-dimensional feature vector yields an about 14-fold accuracy in phonemic recognition compared with recognition of single 20-dimensional feature vectors.

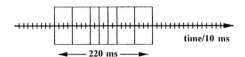

Fig. 9.4. The "time window" used in high-dimensional speech recognition experiments. Every 10 ms time division corresponds to 20 cepstral coefficients. An average of these coefficients in each of the seven subwindows was first formed, and then these 20-coefficient averages were concatenated into a 140-dimensional feature vector

It will now be demonstrated that (i) a very similar recognition accuracy is achievable, although the feature *amplitudes* are further *quantized* to a few, say, eight or 16 levels, (ii) with such a quantization, a very high computing power is achievable in decisions, using quite conventional circuit technology but unconventional architectures.

It was shown by simulation studies in [9.3] that the phonemic recognition accuracy is reduced only very little if the component values of the feature vectors, and those of the codebook vectors are represented in a quantized scale. The following percentages describe the *average* recognition accuracy over all the Finnish phonemes (for a single speaker): even /k/, /p/, and /t/ were separately included as individual phonemic classes. In one experiment the quantization levels were selected to be equidistant, whereby the level values were set experimentally to approximate to the dynamic range of the cepstral coefficients. In another experiment, the optimal quantization levels were determined for each vectorial component separately by the so-called *scalar quantization*, i.e., minimizing the average expected quantization errors of each component. The cases studied were: (i) 20-dimensional input feature vectors formed of the coefficients of single cepstra taken over 10 ms intervals and using 200 reference vectors, (ii) 140-dimensional input feature vectors taken from the "time window" (Fig. 9.4) and using 2000 reference vectors, respectively.

It is discernible from Table 9.1 that there is not much incentive for using a special quantized scale for the components. Equidistant quantization levels are as accurate in practice.

Table 9.1. Effect of quantization: recognition accuracy, per cent

No. of bits	No. of quantization levels	Dimensionality of input and size of codebook			
		20 times 200		140 times 2000	
		Equidistant	Scalar quant.	Equidistant	Scalar quant.
1	2	50.1	54.4	90.1	88.4
2	4	72.1	74.6	97.3	97.7
3	8	82.6	82.9	98.7	98.8
4	16	84.9	85.4	99.0	99.0
	Floating-point computing accuracy	85.9		99.0	

Evaluation of the Classifier Function by Tabular Search. The basic and most frequently computed expression in vector quantization methods (such as LVQ and SOM) is

$$c = \arg\min_i \left\{ \sum_{j=1}^n (\xi_j - \mu_{ij})^2 \right\} . \tag{9.4}$$

Here the ξ_j are the components of the pattern vector x, the μ_{ij} are the components of the codebook vector m_i, and c is the index of the codebook vector closest to x. It is to be noted that if the ξ_j and μ_{ij} are quantized, there exists only a finite and generally rather small number of their discrete-value combinations, for which the function $(\xi_j - \mu_{ij})^2$ can be *tabulated* completely. For instance, with 3-bit accuracy such a table contains 64 rows, and with 4 bits, 256 rows, respectively.

Some savings in computing time may be achieved by simultaneously searching for several elementary squares at a time. For instance, the address to the table might be formed of the pair of input vector components (ξ_j, ξ_{j+1}), concatenated with the pair of codebook vector components $(\mu_{ij}, \mu_{i(j+1)})$. The expression searched from the table is then $(\xi_j - \mu_{ij})^2 + (\xi_{j+1} - \mu_{i(j+1)})^2$. In this case, the required table size for 3-bit accuracy is 4096 rows, and for 4-bit accuracy 65536 rows, respectively; the computing time would be roughly half of that in the simpler case.

Another possibility is to compute one square term for *two different codebook vectors* μ_{ij} and $\mu_{(i+1)j}$ at the same time, i.e., searching from the table the distance elements for two codebook vectors simultaneously. The table would then be addressed by the concatenation of ξ_j, μ_{ij}, and $\mu_{(i+1)j}$. The table size for 3-bit accuracy would be 512 rows, and for 4-bit accuracy 4096 rows, respectively. Again the computing time would be roughly half of that in the simpler case.

A piece of typical i286 assembler program for software implementation of the first-type tabular search is given in Table 9.2. In the benchmarking reported below the innerloop instructions consumed at least 98 per cent of the computing time.

Table 9.2. Main parts of the i286-assembler classification program based on tabular search

---

...

```
@innerloop:
      mov   bx, dx                 ; beginning of function table is in dx
      add   bx, word ptr ss:[di]   ; add    32 * ξj (stored in word ptr ss:[di])
      add   bx, word ptr ss:[si]   ; add     2 * μij (stored in word ptr ss:[si])
      add   di, 2                  ; next   j
      add   si, 2                  ; next   ij
      add   ax, word ptr ss:[bx]   ; sum up  (ξj − μij)² to ax
      cmp   di, cx                 ; largest j + 2 is in cx
      jl    @innerloop
(outerloop:    save ax, reset ax, update ss, di, and si, test outerloop,
       jump to innerloop or minsearch)
(minsearch:    find the minimum distance)
```

System bus

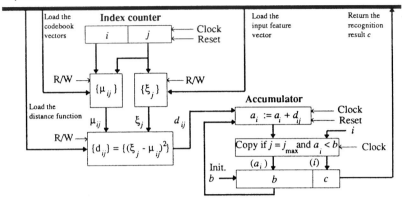

Fig. 9.5. Block scheme of a simple pattern-recognition architecture that makes use of tabular search in the evaluation of the classifier function. R/W = read/write control

Hardware Implementation. Expression (9.4) may be evaluated by a special co-processor depicted in Fig. 9.5. The ξ_j and μ_{ij} are loaded from a system bus into the argument memories. The expressions $(\xi_j - \mu_{ij})^2$ found from another memory are added up in a fast accumulator, and the minimum of distances is found by serial comparison. The fast accumulator can be used to test the sign of $a_i - b$ (using this subtraction followed by subsequent addition of b). This circuit can readily be paralleled.

Table 9.3. Comparison of recognition times of a 140-dimensional vector, in milliseconds; 2000 codebook vectors

Floating-point computations (LVQ_PAK)		
PC/486/66 MHz	Silicon Graphics Iris 4D/25	Silicon Graphics Iris Indigo R4000
540 [1]	300	73

Tabular search	
Software, PC/486/66 MHz	Co-processor, 66 MHz clock
70 [2]	4.2 [3]

[1] The allocated memory area was divided into eight codebooks, 250 vectors each.
[2] The i286 assembler refers to one 64 Kbyte block at a time. For larger codebooks the segment pointer in ss has to be changed upon need.
[3] This is still an estimated figure; a realistic implementation has been described in [9.4].

Benchmarking. We have performed comparisons of recognition times using different methods (Table 9.3). The tabular-search software algorithm was run on a PC, but the co-processor computing time was only estimated on the basis of the suggested architecture. Conventional LVQ programs from the software package LVQ_PAK (Sect. 8.3.1) were run on a PC and two workstation machines for reference.

9.3 SIMD Implementation of SOM

One of the most common parallel computer architectures is the SIMD (single instruction stream, multiple data stream) machine, which consists of a number of processors executing the same program for different data. The simplest distribution of tasks in SOM computation would be to allocate one processor for each m_i; the input vector x is then broadcast to all the processors, and most of the computations occur concurrently. The processors have to intercommunicate for two reasons only: 1. For the comparison of magnitude relations in the WTA computation. 2. In order to define the neighborhood function.

Most SIMD machines operate in the word-parallel, bit-serial fashion, whereby the variables are usually stored as binary numbers in shift registers and shifted bitwise through the arithmetic circuits. Fig. 9.6 delineates the organization.

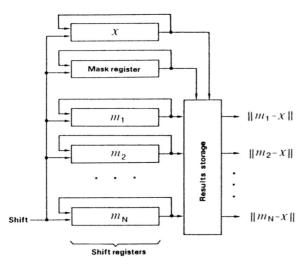

Fig. 9.6. The SIMD architecture for computing the winner-take-all function. The "mask register," which exists in most general-purpose SIMD machines, can be used to inactivate bits and fields in operands

One effective way to define the *neighborhood function* is to store the lattice coordinates of the SOM array in each processor. The lattice coordinates of the winner are also broadcast to all the processors, which, on the basis of the definition of the $h_{ci}(t)$ function, are able to make a *local* decision on the basis of distances relating to units c and i what value of h_{ci} they should use.

Distance Computations. Most SOM applications use the Euclidean distance $d(x, m_i)$ in defining the "winner" location:

$$d(x, m_i) = \sqrt{\sum_{j=1}^{n}(\xi_j - \mu_{ij})^2} . \tag{9.5}$$

Since comparisons can be based on the squares of distances, computation of the square root is superfluous.

Computationally much lighter would be the city-block distance

$$d_{CB}(x, m_i) = \sum_{j=1}^{n}|\xi_j - \mu_{ij}| , \tag{9.6}$$

but especially for visual representations that are the main applications of the SOM, the Euclidean distance is better, because a more isotropic display is obtained using it (i.e., the algorithm in hexagonal arrays does not so much prefer horizontal and vertical directions.)

Notice that

$$d^2(x, m_i) = ||x||^2 + ||m_i||^2 - 2m_i^T x ; \tag{9.7}$$

since $||x||^2$ is the same in all comparisons, it need not be computed. The $||m_i||^2$, on the other hand, which are scalars, can be buffered in the processors and they are only changed in updating, which in SOM algorithms is often made for a small subset of map nodes, especially during the long convergence phase. Therefore the main computing load consists of formation of the dot products $m_i^T x$. Many signal processors and neurocomputers have been designed for effective parallel computation of dot products. A drawback compared with direct computation of squares of differences, especially at high dimensionality of vectors and when using low numerical accuracy, is that the magnitudes of dot products are prone to overflow or saturate more easily.

Computation of products, or alternatively sums of squares of differences, can be made either serially or in parallel with respect to the components. The former mode is more common, since component-parallel circuitry may become intolerably expensive, taking into account the need of implementing very large SOMs (with thousands of nodes and even bigger).

Location of the "Winner." Of various solutions suggested for the search for extrema, it seems that the word-parallel, bit-serial comparison is most cost-effective (for a textbook account, see [1.17]).

A search for the *minimum* of binary numbers stored in parallel storage locations proceeds in the following way [9.5]. Consider Fig. 9.7.

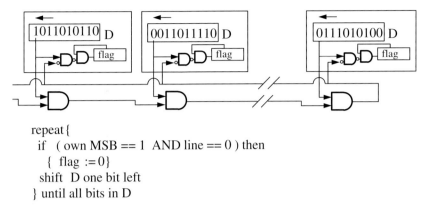

```
repeat {
    if  ( own MSB == 1  AND line == 0 ) then
        { flag := 0}
    shift  D one bit left
} until all bits in D
```

Fig. 9.7. Architecture for parallel search for minimum in a SIMD machine

Each bit memory called a *flag* is first set to 1. Comparison starts from the most significant bits (MSBs). They are connected to an AND circuit chain at the bottom of the picture. The output of the latter is 0 if any of the MSBs is 0. In the latter case all processors with MSB= 1 reset their flags, indicating that they are no longer candidates to be considered. The contents of each storage location D are shifted left by one bit position, whereby the next bits enter the MSB position, and if the AND line output is zero (there is at least one MSB= 0, the flags corresponding to MSB= 1 are reset etc. After all bits have been handled, the flag value 1 is only left for the minimum numbers; if there are several minima, only one of them can randomly be taken for "winner."

Updating. The learning equation

$$m_i(t+1) = m_i(t) + \alpha(t)[x(t) - m_i(t)] \text{ for all } i \in N_c \tag{9.8}$$

can be written in the following form with $\gamma(t) = 1 - \alpha(t)$:

$$m_i(t+1) = \gamma(t)m_i(t) + \alpha(t)x(t) , \; i \in N_c . \tag{9.9}$$

The right-hand side has the form in which it can be understood as a single dot product, thus saving time in usual parallel computers. The dot products can be computed for all processors in parallel, having the value 0 for $\gamma(t)$ and $\alpha(t)$ in elements outside N_c.

9.4 Transputer Implementation of SOM

The *Transputer* is a general-purpose commercial microprocessor chip meant mainly for MIMD (multiple instruction stream, multiple data stream) computers: a MIMD machine executes several concurrent programs, each controlling parallel computations on its own subset of data. The processors can

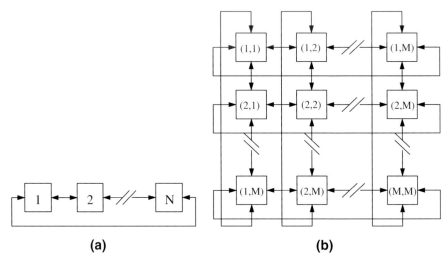

(a)   **(b)**

Fig. 9.8. Typical Transputer architectures: **(a)** One-dimensional, **(b)** Two-dimensional

be networked in many ways through their communication gates. Transputer systems and their software have been developed in joint projects within the European Community.

Several Transputer implementations of the SOM already exist [8.5, 6]. The processors can be arranged, e.g., into a linear ring or two-dimensional end-around (toroidal) array as shown in Fig. 9.8. Multi-dimensional networks may be even more effective in the searching for extrema, as will be pointed out below.

Selection of the "winner" is made by computing the distances locally and recycling them, say, horizontally, whereby at each step each processor selects the minimum of the distance stored in it and of the recirculated distance value, both of which are updated at each step. After a full revolution, every processor on each line will have recorded the minimum distance and its location. In the two-dimensional case, after a similar full revolution in the vertical direction, all processors of the array will have recorded the global minimum of distances and its location. In an N-dimensional ring architecture, with P processors, the number of steps in one revolution is $P^{1/N}$, whereas only N revolutions along the various dimensions are needed; thus maximum search is greatly accelerated.

Adaptation in the neighborhood of the "winner" is then made in parallel in a separate phase.

The main disadvantage of Transputer implementation of the SOM is that the MIMD function cannot be utilized fully effectively in this algorithm, as the SOM cannot easily be partitioned into separate and independent concurrent programs. The above-mentioned solutions only use partial MIMD properties.

9.5 Systolic-Array Implementation of SOM

The name *"systolic"* comes from the neuro-muscular activities that are propagated as "waves," say, in the heart muscles. In electronic, rectangular "systolic" arrays, which are available as commercial chips, data are transferred in one direction, say, downward, while the intermediate computing results move in the orthogonal direction, say, to the right. Computation is said to be spatially iterative, because the processor nodes receive their data and instructions from their adjacent processor nodes. No buslines are thus involved. No cyclic operation like that in the Transputer arrays does occur either.

There exist at least three systolic-array implementations of the SOM, e.g., [8.7–9]. Each row of the array contains one codebook vector, with one component in each processor. In computing the distances, the components of the input vector are propagated vertically, and accumulation of distance components occurs in the horizontal direction. The "winners" can be computed by the arrray, too. During adaptation, the components of the input vector and the learning-rate parameter are propagated vertically, and all components of the codebook vectors are adapted locally.

The systolic array is a two-dimensional pipeline (Fig. 9.9). For maximum utilization of its computing capacity, many consequent input vectors must be pipelined, and all distances computed for this "batch" must be buffered sequentially at the right end of the array before their comparison. Updating should then be made as a batch, too.

It is still unclear to what extent the systolic array can compete in learning with the more conventional SIMD arrays. According to some estimates we made, the SIMD machines can operate at the same speed compared with systolic arrays using only one third or one fourth of the hardware of the latter.

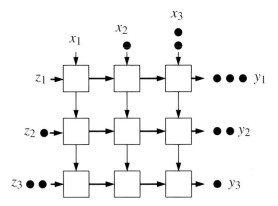

Fig. 9.9. Principle of the systolic-array architecture

9.6 The COKOS Chip

A SIMD processor chip that has been meant for a PC co-processor and is specifically designed for the SOM is the COKOS (COprocessor for KOhonen's Self-organizing map), due to *Speckmann et al.* [9.11–13]. Its prototype consists of eight parallel processing units called MAB (Memory Arithmetic Board). Each MAB contains a subtractor, a multiplier, and an adder in cascade, which accumulate their results in a RAM-memory. The weight vectors have been represented with 16-bit accuracy.

Each MAB computes the square of differences in one component of x and one m_i. These squares are summed up in a tree adder (Fig. 9.10).

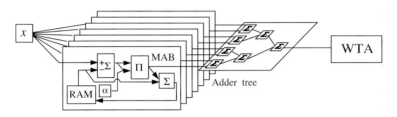

Fig. 9.10. The COKOS chip architecture

A separate WTA circuit that buffers and compares the differences in series is needed. The WTA sends the coordinates of the "winner" to the PC, which computes the coordinates of the neighborhood. The *previously used* input vector is sent a second time to the co-processors, which carry out a component-serial adaptation step on one of the neighborhood vectors, repeating this computation for codebook vectors of the whole neighborhood.

The computations are pipelined in the MABs, which saves computing time.

9.7 The TInMANN Chip

The VLSI chip of *Melton et al.* [9.14] called TInMANN has also been designed specifically for the SOM. In the prototype version there is one neuron per processor chip. Every processor contains arithmetic-logic circuits, an accumulator, RAM memory, and an *enable flag*. The processors are interconnected by a 10-bit common busline and one OR line. The architecture is presented in Fig. 9.11.

The codebook vectors and their coordinates on the SOM array are stored in RAM memories. The prototype version of TInMANN can accomodate three-component vectors with eight bits in each component. Computation of distances is made in the *city-block metric*.

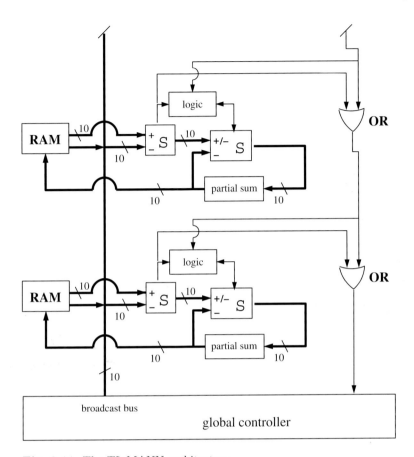

Fig. 9.11. The TInMANN architecture

The phases associated with learning are the following:

1. The controller sends the components of the input vector to the common busline in series. The processors compute the absolute values of the differences of the components of the input vector and the codebook vector, respectively, and sum up the terms in the accumulator. The city-block distance is then copied back to the RAMs.

2. The controller finds the "winner" in a series of successive binary approximations, by sending to all processors the values $2^{N-1}, 2^{N-2}, \ldots$, one at a time. If the difference of this value and the distance is negative, the value 1 is sent to the OR line. If there existed at least one smaller distance, the next binary power is sent to the common busline etc. A drawback of this method is that often when the search terminates, a rather low accuracy still remains, and an extra comparison of the remaining candidates must be made by some slower method.

3. The "winner" processor now sends its coordinates to all processors on the SOM array, on the basis of which the processors can decide whether they belong to the neighborhood or not.

4. The controller sends a random number that corresponds to the time-dependent learning-rate factor α to all processors. Those processors whose distance is greater than this number will be passivated (their *enable flag* will be turned off).

5. The controller again sends one component of the input vector to the common busline, and all processors compute the difference from their corresponding codebook vector component. If this difference is greater than another random number selected this time, the codebook vector component is changed by a fixed amount corresponding to the least significant bit in the direction toward the input vector.

6. All processors are re-activated (their *enable flags* are turned on). Steps 3 through 5 are repeated for all components, and steps 1 through 6 are iterated a sufficient number of times corresponding to the learning steps.

Above, the random-number process at steps 4 and 5 was meant to improve convergence in learning, by smoothing the effect of discrete steps otherwise caused by the 8-bit accuracy, especially when the learning-rate factor α is very small; notice that with small α the correction would otherwise remain smaller than the least-significant bit, and not be made at all. This author also experimented with a similar idea around 1982, but abandoned it, because practical problems are anyway usually very-high dimensional (over 100 components), whereby the quantization error of 8-bit numbers is not significant, and learning can be stopped when corrections reach the numerical accuracy.

The neighborhood set N_c on the SOM array in the TInMANN chip is rectangular, e.g. square, and therefore it can be defined by a city-block distance in the array. At step 3 another random number, from a support defining the neighborhood radius, is sent to the processors, and the latter compare their coordinates with this number to decide whether they belong to the neighborhood or not. Use of random numbers instead of fixed time-variable numbers smooths the self-organizing process, almost as if Gaussian neighborhood kernels were used.

The greatest merit of TInMANN is the very small chip area needed. This virtue is somewhat neutralized by the city-block distance used in computing.

9.8 NBISOM_25 Chip

The newest SOM hardware, which has already been manufactured and tested, is the high-performance system of *Rüping, Porrman* and *Rücket* [9.15]. It is a processor board with 16 customized chips, where each chip contains

25 processor elements. The surface area of the chip is 75 mm², including the pads. Each processor element has a calculation unit corresponding to one neuron, a local on-chip memory for 64 weights of 8-bit accuracy, and control registers with control circuitry. There is also a 14-bit buffer register for distance calculation is each processor element. Thus the maximum SOM array representable by the board consist of 400 map elements with 64 inputs each. As the system is very fast, its main applications are in adaptive on-line pattern matching and control problems.

In order to minimize the chip area and thus to maximize the number of processing elements per chip, the city-block distance metric was chosen for comparison operations. As the values of the learning-rate factor ("alpha") were also restricted to the set {1, 1/2, 1/4, 1/8, ...}, no multipliers were needed, neither in the distance calculation nor in adaptation; the multiplication by alpha could be replaced by shifting the other multiplicand. It has been shown in [9.16] that the simplified algorithm was still working properly in many pattern matching and control applications.

The frequency of the system clock is 16 MHz. The controlling unit of each processor element handles the synchronization and mode select of the other processor elements. There are connections to the 8-bit date bus, to the control bus, two clock lines, as well as the row and column control lines of the element.

The system performance can be characterized by the frequency of connections made in the matching mode, and the frequency of connection updates during learning. Table 9.4 gives these for the chip and the board, respectively.

The earlier phases of this system have been described in [9.17–9.20].

Table 9.4. Processing speed of the system parts, input dimensionality = 64

	Map units	Matching	Learning
Chip	5 * 5	256 MCPS	149 MCUPS
Board	20 * 20	4096 MCPS	2382 MCUPS

10. An Overview of SOM Literature

The number of scientific papers that analyze, extend, or apply the SOM algorithms or otherwise directly benefit from them is already on the order of 4000 at the moment of writing. In the second edition of IS30, this figure was still about 1500. One of the problems therefore was where to draw the line in referring to the literature. It would have been absurd to add to this edition some 150 pages of extra text relating to new references. As most of the SOM papers, with a complete documentation, are anyway listed and indexed in the Internet [2.80] at the address http://www.icsi.berkeley.edu/~jagota/NCS, only the newest books and some of the most central new papers have been added to this third edition. At the same time, this chapter has been thoroughly restructured.

As this chapter has been intended to provide an overview of the documented SOM and LVQ literature, the papers cited in the first and second edition have been retained as representative examples. Nonetheless, some parts of this chapter may still look a bit redundant with respect to the text in the other chapters. Works on other ANNs have been omitted from this chapter.

In addition to the papers that directly deal with the SOM algorithm, there also exist numerous papers on general aspects of competitive learning. A few of them have been included in this book. They may be much less familiar to the ANN community than, for instance, the more straightforward multilayer feedforward networks, or the feedback or state-transfer networks. The literature on competitive learning has also been somewhat esoteric all the time. Notwithstanding one should not underestimate its importance, especially since development of many biological neural functions is only explainable by means of competitive-learning models.

10.1 Books and Review Articles

This book is an outgrowth of certain parts of the author's previous book [2.53]. The monograph of *Ritter*, *Martinetz*, and *Schulten* [1.40] concentrates on the application of the SOM to robotics. The book of *Miikkulainen* [2.74] is another monograph that describes applications of the SOM to linguistics, i.e., subsymbolic computing. A Japanese monograph by *Tokutaka et al.* [10.1] deals with various applications of the SOM.

A quite new book on the information-theoretic aspects of the SOM and some lines for its generalization has been written by *Van Hulle* [10.2].

Financial applications of the SOM have been discussed in *Deboeck* and *Kohonen* [10.3].

Two special workshops, WSOM'97 and WSOM'99, have been dedicated to the SOM research. Their proceedings serve as up-to-date reports on some of the newest central results on the SOM [10.4–6].

The SOM and the LVQ have been described in numerous book chapters, review articles, and edited books found in the listings below. Their descriptions and details have often remained incomplete and many of them are already obsolete. For a reasonably up-to-date overview to a reader who is trying to find out the basic principles, this author would like to recommend his long 1990 paper in Proc. IEEE [10.7] and the introductory article in [10.5].

The state-of-the-art of the mathematical foundations of the SOM has been discussed rigorously by *Cottrell et al.* in two long papers [10.8, 9].

10.2 Early Works on Competitive Learning

The basic idea in competitive learning is to have several parallel "filters" for the same input, whereupon these filters must possess some pattern-normalizing properties. The filters are initially tuned differently, eventually randomly to the input patterns, and one or several filters then become "winners" in relation to a particular input, i.e., their outputs are highest and tend even higher. The "winners" tend to suppress the other cells by negative feedback, causing the output of one or a few best-matching filters remain active, while the other outputs are passivated. Only the "winner" filters, the selection of which depends on inputs, and eventually their neighbors in the network, will be updated in learning. In the long run, different filters will be automatically tuned to different domains of input signals, like in the classical Vector Quantization.

An early biological "winner-take-all" (WTA) model, without adaptive learning, was published (without mathematical proof) by Didday in 1970 [2.11]. The lateral interconnectivity typical to many competitive-learning networks was thereby assumed, and the work was related to frog's visual system. It is known that frog's eye contains many functions that the mammalia have in their visual cortex. A revised version of this work appeared later in [2.12].

The works of *Nass* and *Cooper* from 1975 [2.22] as well as of *Pérez, Glass,* and *Shlaer* from 1975 [2.23] are generally regarded as pioneering works in the area of adaptive formation of feature-sensitive cells. Another work from 1975 that captured the essentials of competitive learning is due to *Montalvo* [2.13]. Ideas somewhat similar to the previous ones appeared in *Grossberg's* works from 1976 [2.15, 16]; in the second one, his general ART1 architecture was also introduced. A very clear mathematical treatment of the basic WTA function was elaborated by *Amari* and *Arbib* in 1977 [2.14].

Competitive learning was in effect involved in the model of self-organized mapping of line orientations onto the visual cortex, published by *v.d. Malsburg* in 1973 [2.32], which is one of the pioneering works in this area. This network had the typical lateral interconnectivity structure but did not explicitly utilize the WTA function. Further studies along the same lines have been published by *Willshaw* and *v.d. Malsburg* [2.33] and *Swindale* [10.10].

The first mathematical analyses of the ordering of *neural connections* were published for the layered, continuous synaptic-field model of *Amari* [2.34], [10.11]. Although the self-organizing effect in it was very much related to that encountered in the SOM, nonetheless this theory does not apply as such to SOM processes. Neither is that model capable of ordering the maps globally. Therefore we shall restrict the discussion in this section only to theoretical works on the basic SOM or its minor modifications.

In the SOM publications of this author, starting with [2.35–37] the main purpose was to develop an effective computational algorithm for self-organization; this algorithm has resulted in a plethora of applications.

10.3 Status of the Mathematical Analyses

10.3.1 Zero-Order Topology (Classical VQ) Results

The vector quantization (VQ) methods have been known since the mid-1960s, and, e.g., *Gersho* [1.70] and *Zador* [1.74] derived the formula for the point density of the m_i vs. $p(x)$. It has to be pointed out that this result can be obtained *without computation of the optimal m_i*. This author has presented an alternative, shorter derivation of the VQ point density, based on the calculus of variations [10.12] (cf. also Sect. 1.5.3).

On the other hand, the proof presented by the present author in 1991 [1.75] showed rigorously that a gradient-descent formula for the "zero-order topology SOM" is derivable from the average expected quantization error functional (Sect. 1.5). Some complementary remarks to this study have been presented by *Pagès* in 1992 [10.13].

One might argue that if the neighborhood function were made very narrow at the last steps of the SOM process, the point density of the model vectors would approach that of VQ. In practice, however, the "elastic network" of the model vectors has to retain a certain degree of stiffness for good ordering, and with a neighborhood width that guarantees this, the model vectors no longer approximate $p(x)$: they are aligned along the two-dimensional surface defined by the local principal axes of $p(x)$. Accordingly, the point density of the model vectors can be quite different from that in the VQ case. As a matter of fact, the problem of the SOM point density in a higher-dimensional space has remained unsolved.

10.3.2 Alternative Topological Mappings

In a certain sense, the multidimensional scaling and the principal curves (Sect. 1.3.2) have originally been introduced for the same purpose as that of SOM. Contrary to them, the *Generative Topological Mapping (GTM)* [10.14, 15] was directly intended for an alternative computation of the SOM, in an attempt to define the metric relations between the models on the map grid with improved fidelity to those in the data space. The tradeoff thereby is a much higher computing load.

The SOM mapping derived by *Luttrell* using error-theoretic criteria [3.19], as well as the modification of *Heskes* and *Kappen* [3.20] where the winner is selected according to Euclidean distances weighted by the neighborhood function are also principally different from the basic SOM, since they can be shown to be derivable from a potential function, unlike the SOM. The tradeoff is again a much higher computing load in the identification of the winners.

Various other modifications or generalizations to improve the performance of SOM or to optimize a different objective function have been suggested: [2.74], [3.5–7], [5.3, 5], [10.16–90].

10.3.3 Alternative Architectures

Fixed Grid Topologies. The two basic grid alternatives are: a grid with open edges, which is the structure mainly discussed in this book, and a cyclic network (e.g., sphere or toroid). The toroidal structure, where the horizontal and vertical edges of a square grid are joined cyclically has been used 1. for problems where the data are supposed to be cyclic, e.g., in musical harmonies [10.91], or 2. when the SOM is used to represent process states. Notice that in a cyclic SOM there are no boundary effects, and all models are in a symmetric position with respect to their neighbors (neighboring states). While the open-edged network corresponds to a single-sheet "elastic network" that is fitted through the data distribution, the "elastic network" in the toroidal grid topology corresponds to a double sheet that is similarly fitted to data. A minor drawback is then that in a cyclic grid, some states which are adjacent in the input space may become mapped on opposite sides of the grid.

Of the two trivial alternatives for network topology, viz. the rectangular and the hexagonal grid, the latter is always to be preferred due to its better isotropy.

A quite different version of the grids is the hypercube topology, [5.11, 12], [10.92, 93] where the map units are located at the vertices of a hypercube, and the edges form the topological-neighborhood links. Such a topology is completely symmetric with respect to each map unit, and these networks have been used in signal coding.

Multistage, Multilevel, and Hierarchical SOMs. The three main reasons for suggesting *structured* SOMs and SOM organizations are: 1. Faster search for the "winner." 2. Faster convergence. 3. Better generalization and abstraction ability.

Such organizational modifications have been discussed in the augmented-output SOM [10.94], in the Hypermap-type SOM [10.95, 96], in the hierarchical SOM [10.97–102], in the multi-stage SOM [10.103], in competing maps [10.104], in the multilayer SOM [10.105–114], in multiresolution hierarchy [10.115], in combination with a counterpropagation network [10.116–119], in the tree-structured SOM [10.120, 121], in a combination of SOMs [5.9, 10], [10.123], in the modular SOM [10.124], in a multilayer semantic map [10.125, 126], in a twin SOM [10.127], in a two-phase SOM [10.128], and in a double two-stage classifier [10.129, 130]. One kind of hierarchy is encountered in an accelerated search structure [5.7], [10.131], which involves a multilayer SOM. Also the Hypermap discussed in Sect. 6.10 is a hierarchical structure.

Growing SOM Structures. The idea of making the topology of the SOM, or actually only the neighborhood function used during learning some function of ordering results, follows naturally from the basic idea of self-organization. Instead of using fixed one-level maps, or hierarchical structures made of fixed modules, it is possible to create arbitrarily structured SOM networks that match data distributions automatically. Such results have been reported in [3.40–54], [10.132–137].

Other Structural Variants. The following modifications of the basic SOM structure have also been suggested: splitting of neurons [10.139], multireference neighborhood search [10.140], growth in hypercubical output space [10.141], using small SOMs for the context-analysis of input vectors [10.142], multiple SOMs [10.143], added lateral interactions [10.144–10.146] , distance network [10.147], and parametrized SOMs [10.148–151].

10.3.4 Functional Variants

SOM for Sequential Inputs. Natural data are usually dynamic; the subsequent patterns cannot be assumed independent of each other. The dynamics of signal sequences can be taken into account in different ways (cf. Sects. 5.6, 5.8, and 5.11). The following works deal with dynamic SOMs: [5.4, 11–16], [10.152–157].

Fuzzy SOM and LVQ. Inclusion of fuzzy-set operations in the SOM has been suggested in [10.158–162].

For the LVQ, fuzzy-set operations have been used in [10.163–169].

A fuzzy associative memory related to the LVQ is discussed in [10.170–172].

Supervised SOM. As pointed out in Sect. 5.10, the SOM that is normally unsupervised can also be made supervised by adding class information to the input data. Various ways in which to do this are shown in [5.22], [10.113, 126, 173–179].

Other Functional Variants. Another nonlinear variant of the SOM has been suggested in [10.180]. Quantized weights have been used in [10.181]. Speeding up of learning has become possible by the use of a momentum term [10.182], topological interpolation [10.183, 184], and batch training [10.185]. Dynamic neighborhood widths have been suggested in [10.186] and [10.187]. Dynamic cell structures have been used in [10.188]. The engrams have been made to decay in [10.189]. A new principle in matching and updating, relating to several neurons, is the "mode map" [10.190]. Evolutionary optimization has been suggested in [10.191], and query-based learning in [10.192]. Avoidance of misleading information has been discussed in [10.193]. Cost-measure based addressed VQ has been used in [10.194]. Near-miss cases have been taken into account in [10.195]. Overlapping data have been removed in [10.196]. The equidistortion principle has been utilized to define the SOM [10.197]. Local weight updates have been made in [10.198]. Readability of the SOM can be improved as suggested in [10.199]. Time-frequency SOMs have been suggested by [10.200]. Yet another variant of SOM has been discussed by [10.201].

Matching of the dimensionality of the SOM with the fractal dimensionality of input data has been considered in [10.202].

Problems of dynamic systems handled by the SOM have been discussed in [10.203].

10.3.5 Theory of the Basic SOM

One might think that the structure of the basic regular-array SOM and its algorithm were so simple that the theory would already be established, after twenty years of intensive research. Therefore it may be proper to quote what the mathematical experts, *Marie Cottrell et al.* wrote in 1997 [10.9]: "The SOM algorithm is very astonishing. On the one hand, it is very simple to write down and to simulate, its practical properties are clear and easy to observe. But, on the other hand, its theoretical properties still remain without proof in the general case, despite the great efforts of several authors" and "... the Kohonen algorithm is surprisingly resistant to a complete mathematical study. As far as we know, the only case where a complete analysis has been achieved is the *one dimensional case* (the input space has dimension 1) for a *linear network* (the units are disposed along a one-dimensional array)."

The first complete proof of both ordering and converge properties in the one-dimensional case was presented by *Cottrell* and *Fort* in 1987 [3.11]. These results were generalized to a wide class of input distributions by *Bouton* and *Pagés* [10.204–210], and to a more general neighborhood by *Erwin et al.* [3.18]. Then, *Fort* and *Pagés* [10.211, 212] and *Benaïm* [10.213] were able to present a rigorous proof of the almost sure convergence towards a unique state, for a very general class of neighborhood functions.

Fort and *Pagés* [10.214], as well as *Flanagan* [10.215, 216] have discussed ordering of model vectors that are associated with a two-dimensional rectangular grid when the input distribution is uniform. There were some restrictions to the neighborhood function and input density, though. As a matter of fact *Fort* and *Pagés* [10.214] showed that there does not exist any absorbing state in the two-dimensional case.

Recently, *Flanagan* [10.219] has presented the proof in the one-dimensional case for a general (time-invariant) neighborhood function and input distribution that can be discontinuous. He has pointed out that the discontinuous distribution must fulfill certain properties similar to those encountered in the so-called self-organized criticality.

Ordering with Discrete Data. In one special case the average expected distortion measure (10.2) can be a potential function, even in a high-dimensional case, namely, when the input can only take on values from a *finite discrete set*. In this case there are no samples at the borders of the Voronoi tessellation, whereby the Voronoi sets of samples are not changed due to the shift of borders in differentiation with respect to the m_i. This fact made it possible for *Ritter et al.* [1.40] to resort to a gradient-descent method to prove the convergence in this special case, and then to apply the method to the Traveling-Salesman Problem, where the probability density function of input is discrete-valued. It was pointed out by *Ružica* in 1993 [10.220] that this potential function is not differentiable, but with certain conditions imposed on the learning rate factor and neighborhood function the convergence to a stationary state is true.

General Reviews. The following is a concise list of various *general* papers on partial aspects of the SOM, often discussed within the context of more traditional methods: [1.25], [2.35–37, 45, 53], [3.6, 7, 12, 13, 30], [6.1], [10.13, 10.221–401].

Derivations. Various theoretical papers differ significantly in their rigor in handling the fundamentals of the SOM. The most central works were reviewed in Sect. 10.3. Below, various derivations of the SOM and mathematical aspects of topological mappings are simply listed down: [3.29], [10.402–440]

On the Definition of Order. The most central question concerns the definition of *topological order* itself: In what general sense can, say, a two-dimensionally indexed array of samples taken from a higher-dimensional vector space be regarded as "ordered"? What exactly is "topological order"?

The concept of order is trivial in the one-dimensional case whereupon it is also possible to define an objective function J for it. Assume a set of scalar numbers $\mu_i \in \Re, i = 1, 2, ..., N$. Then

$$J = \sum_{i=2}^{N} |\mu_i - \mu_{i-1}| - |\mu_N - \mu_1| \qquad (10.1)$$

is minimum (zero) if and only if the m_i are numerically ordered in an ascending or descending sequence. Such an order can be achieved in a self-organizing process, in which the input x is one dimensional, and the μ_i correspond to scalar m_i, thereby being scalar network parameters associated with a linear array of nodes. The state with $J = 0$ is then an *absorbing state* in the sense that once reached, this condition is no longer changed in the learning process by any further choice of the external input. In general dimensions, however, especially if the dimensionality of the input vector is higher than that of the array of nodes with which the parameter vectors are associated, the existence of such an absorbing state seems unlikely. At least in painstaking work over the years, with many suggestions for the objective function J made, some counterexample has always been found that contradicts the existence of an absorbing state in the general case.

It must also be noted that the "order" resulting in the Self-Organizing Map always reflects properties of the probability density function $p(x)$. It thus seems, at least for the time being, that the "order" should be defined in terms of *the minimality condition of some properly defined average expected error function.* Then, in special cases, for instance when the dimensionalities of the signal space and the array are equal, *an order in vectorial parameter values that eventually complies with the geometric arrangement of the nodes is then only a necessary condition for the global minimum of the error functional in these special cases to exist.*

However, it would be desirable to relate the topological order of the arbitary-dimensional input space to the neighborhood relations on the low-dimensional grid. A very important result concerning the equivalence of topo-logical order between the Voronoi tessellation and neighborhood set has been derived in [3.14].

The definition of the quality of a SOM must then take into account both the average expected quantization error as well as the "topological errors," i.e., violation of the neighborhood relations on the grid [3.29–32].

Recapitulation of Attempts for Ordering and Convergence Proofs. In approaching these problems, the following mathematical tools are avail-able. First, one might try *constructive proofs.* It would mean that values generated by the SOM algorithm are listed, and possible transitions between them are analyzed. Constructive proofs yield determininistic results, but this method is only feasible for the lowest dimensionalities. Second, one might try to apply known results from the theory of *Markov processes*; and the SOM algorithm defines a special Markov process. The main problems are nonlin-earities (SOM is a decision process), structured boundary conditions at the edges of the array, and in the general case, very high dimensionality. Third, some computing techniques developed in *statistical physics*, especially sta-tistical mechanics might be used. Fourth, *the mathematical theory of errors* may form a foundation for some versions of the SOM. Fifth, results of *systems theory* might apply to some partial problems, such as convergence proofs.

It has been shown, e.g. in [3.18], that the basic SOM algorithm is not exactly derivable from an energy function. Notice that h_{ci} contains the index c of the "winner," which is a function of x and all the m_i. An energy function would be differentiable with respect to the m_i *only within the Voronoi polyhedrons*, not at the borders of the tessellation, which are changing when the m_i are changing.

Sixth, the *stochastic approximation* method of Robbins and Monro [1.43], as shown in Sect. 1.3.3, has been used for the description of a self-organizing process, and for a generic definition of a class of SOM algorithms. There exists a generalization of stochastic approximation called the *Dvoretzky algorithm* [10.441]. However, it was pointed out in Sect. 3.12.2 that the results thereby derived may not coincide accurately with those produced by the original SOM algorithm.

We shall now present a survey of mathematical SOM publications, without discussion of their details.

Attempts for Constructive Proofs. In [3.12] the changes of the ordering index expressed by (10.1) were studied in different configurations of reference values in a one-dimensional SOM with one-dimensional vectors, expressed in *continuous time*. It turned out that in solutions of the SOM differential equation, J decreased in 13 configurations of the SOM out of 16 possible ones. The most important result, however, was a proof of the existence of the *absorbing state* in the one-dimensional case, which means that *an ordered state cannot be disordered in further learning*. (This result may not hold for general dimensionalities.)

The above derivation has been criticized because in digital simulation implementations of the SOM the steps are always discrete, whereby the constructive proof would be somewhat changed. In principle this remark is right, but one has to realize that in biological and in general physical implementations, the SOM must be describable by time-continuous differential equations, and the simple proof was meant for that case.

Another constructive proof (cf. Sect. 3.5.2) was presented in [3.13], in which it has been proved that for a slightly modified one-dimensional system, ordering is almost sure (i.e., it occurs with probability one in an infinite learning time). The modification was a different definition of the neighborhood set: If $i = 1 \dots \ell$ are the array indices, then we define $N_1 = \{2\}, N_\ell = \{\ell - 1\}$, and $N_i = \{i - 1, i + 1\}$ for $2 \leq i \leq \ell - 1$. In case the "winner" c is unique, N_c is the neighborhood set. If c is not unique, the union of all N_c must be used as the neighborhood set, with all "winners" excluded from it.

Markov-Process Proofs. In spite of intensive theoretical research, the basic SOM algorithm, with general dimensionality of input and array dimensionality higher than one, has remained unproved; only partial results exist.

The proof presented in Sect. 3.5.1 was based on an idea communicated privately to the author by Professor *Ulf Grenander* of Brown University. It was published in [2.37]. The same idea, in a much more complete form, was elaborated by *Cottrell* and *Fort* [3.11].

As far as is known at this writing, the latter, with the complementary works mentioned in the beginning of Sect. 10.3.5, is the only fully rigorous analysis in the sense of Markov processes in which both self-ordering and convergence have been treated, but these proofs only relate to a *linear network* (i.e., one-dimensional, open-ended array indexed $0, 1, \ldots, M$) in which the neighborhood set of cell i is $N_i = \{\max(0, i-1), i, \min(M, i+1)\}$ and whereby one restricts the analysis to a uniform, singly connected distribution of input over a finite interval $x \in [a, b]$. Their discussion is far too long and detailed to be presented here. As mentioned earlier, *Bouton* and *Pagès* [10.204–210] generalized this problem for a fairly large class of input distributions, and a rigorous proof of convergence in these cases, after ordering, has been given by *Fort* and *Pagès* [10.211, 212]. The case with general (one-dimensional) input distributions has also been discussed by *Erwin et al.* in 1992 [3.18].

One may also mention the proof of *Thiran* and *Hasler* from 1992 [10.442] for a large class of neighborhood functions, when both the inputs and the weights are *quantized* to discrete values.

Energy-Function Formalisms. As the general ANN theory makes extensive use of system-theoretic tools such as Lyapunov functions and in particular energy functions, it is not surprising that in several contemporary attempts to create a theory for the SOM, energy functions are sought for objective functions.

However, as pointed out in [3.18], the original algorithm cannot be derived from energy (potential) functions. The energy functions written by *Tolat* [10.402] for every neuron separately are a different thing.

There exist many possibilities for the modification of the SOM equations to facilitate the use of energy-function formalisms. One of them, derived on the basis of error theory by *Luttrell* [3.24, 25] [9.14] is to redefine the matching of input x around neuron i on the basis of the *least distortion measure*, i.e.

$$d_i = \sum_j h_{ij} ||x - m_j||^2 ,$$ (10.2)

i.e., a weighted sum of distances. (In this derivation all weights must be normalized: $\sum_j h_{ij} = 1$.) A somewhat similar approach has been used in [3.20]. Mathematically this theory seems to yield a good convergence but a slightly worse approximation of $p(x)$ than the SOM does. Its further drawbacks are: 1. A computationally heavier and more complex winner-take-all operation (which might be alleviated by parallel computation). 2. The same neighborhood function h_{ij} must be used for both the WTA function and the learning law, for which no physical or physiological justification exists.

Further Analyses. Additional theoretical discussions of the following basic properties of the SOM have been published: ordering conditions [10.443], effect of concave vs. convex neighborhood function [10.444], ordering measures [10.445], topology-preserving properties [10.446], topology analysis [10.447], global ordering [10.448], metastable states [10.449], instability [10.450], level density [3.23], distribution and convergence of feature spaces [10.451], magnification factor [10.452], strong vs. weak self-organization [10.453], quantization vs. organization [10.454], comparison with elastic-net and energy function formalisms [10.455], matching the SOM dimensionality with fractal dimensions of data [10.456], comparison with Bayes classifier [10.457], comparison of supervised vs. unsupervised SOM [10.458], and characterization of the reference vectors [10.459].

Description of the SOM by differential equations has been made in [10.460]. An analysis using Sammon mapping can be found in [10.461], and an entropy approach in [10.462].

Convergence proofs usually require very complicated and stringent mathematical handling: [3.11, 18, 21], [10.204–208, 211, 212, 463–484].

Accelerated Convergence. By certain modifications of the basic algorithm, accelerated convergence has been obtained. These include modified learning resembling the Kalman filter [10.483, 485–487], use of generalized clustering methods [10.488–491], use of a momentum term in optimization of the objective function [10.418], taking the dependence of samples into account in optimization [10.152, 492, 493], probing in optimization [10.494, 495], using K-d trees for the accelerated search of the "winner" [10.496], using recurrent signal processing [10.497] and taking the changes of the Voronoi sets into account in optimization [1.75].

Point Density of the Model Vectors. We have discussed the relation of the point density of the m_i to the probability density function $p(x)$ in Sect. 1.5.3 and found a certain degree of proportionality between them. The inverse of this point density is called the "magnification factor," in the same sense as this term is used in neurophysiology to mean the scale factor of a particular brain map.

If the formalism of Luttrell [3.25] mentioned above is applied to a linear array with one-dimensional input, the point density of the m_i becomes proportional to $[p(x)]^{1/3}$, i.e., the same as in VQ. This expression is then independent of h_{ci}.

Benchmarkings. Several benchmarkings of SOM and LVQ algorithms in relation to and combination with other ANN algorithms have been performed. The following facts should thereby be noticed:

1. Only complete software packages, such as the LVQ_PAK and SOM_PAK published by us, and those mentioned in Chap. 8 should be benchmarked, because there are many details in them to be taken into account. If these details have not been taken into account, the benchmarkings have not much value.
2. The SOM was never intended for statistical (supervised) pattern recognition; all benchmarkings concerning classification accuracy should be based on the LVQ.

Most benchmarkings published, involving those that use LVQ algorithms, are not reliable, because they do not take essential conditions such as correct initialization, as discussed in Sect. 6.9, into account.

The following list of studies contains various comparative studies: [10.498–518].

One of the basic comparisons concerns the degree to which the point density of the SOM model vectors approximates the input probability density function. Such studies have been carried out in [10.519–523].

10.4 The Learning Vector Quantization

The basic LVQ algorithms can be modified in various ways: [10.21, 83, 137, 222, 524–562].

Initialization of LVQ has been discussed in detail in our software package LVQ_PAK, the first version of which was published in 1992, and in [10.563].

Fuzzy-set operations with the LVQ have been used in [10.158, 163–172].

Combination of LVQ with HMMs (Hidden Markov Models) has recently been a very important issue especially in speech recognition: [7.9, 10], [10.399, 433, 564–584]. A HMM/SOM combination is described in [10.585–587].

Various other analyses of the LVQ have appeared in [10.425, 588–595].

Related algorithms for LVQ have been suggested in [10.596], a dynamic LVQ has been introduced in [10.597], an LVQ with weighted objective function in [10.598], and a tree-structured LVQ in [10.599].

Theory. Initialization of the LVQ has been discussed by [10.600]. Optimal decision surfaces of the LVQ have been analyzed in [10.601], convergence in [10.602], and critical points in [10.603].

10.5 Diverse Applications of SOM

10.5.1 Machine Vision and Image Analysis

General. Major application areas of image processing in practice are: industrial machine vision (especially robot vision), printing, image transmission, medical imaging, and remote sensing. In addition to them, research is being pursued in basic problems such as target recognition. The SOM algorithm is thereby often used. Some general problems are discussed in [7.31], [10.168, 231, 362, 604–614].

Problems of vision have been discussed in [10.615]. Recognition of spatio-temporal patterns has been discussed in [10.616]. Binary visual patterns have been recognized in [10.617].

Image Coding and Compression. The first task in computer vision is digital encoding of the picture. *All the methods reported below are based on SOM or LVQ.* The following articles deal with the encoding of pictures: [10.618–624]. Special problems that occur in optimal coding of multiscale images have been described in [10.625], and image sequences in [10.626–628]. Adaptive vector quantization methods have thereby been used in [10.509, 629–639]. Image compression has been described in [10.111, 513, 638, 640–658].

Hough transforms are frequently used for image coding. Randomized Hough transforms have been based on SOM in [10.659–664].

Image Segmentation. Especially for pattern recognition purposes uniform areas in pictures must be segmented before their identification: [10.108, 299, 665–672]. Analysis of texture is a common task and differently textured areas can then be segmented: [2.71], [7.11–13, 17–20, 22, 24, 25], [10.673–687]. Texture generation has been discussed in [10.688].

Survey of Diverse Tasks in Machine Vision. Specific problems may occur in particular applications: image normalization [10.689], development support for visual inspection systems [10.690], detection of multiple "sparse" patterns in the image field [10.691, 692], contour detection in videophone images [10.693, 694], color classification [10.695], multispectral road classification [10.696], and error-tolerant encoding for transmission of images [10.697, 698]. It is further necessary to eliminate unnecessary information while preserving natural clustering [10.699].

Distortion-tolerant image recognition is an important problem [10.700], and homing based on images another [10.701]. Of other image analysis problems one may mention image clustering [10.702] and color quantization [10.703]. Visual analysis of air flow has been performed in [10.704]. Multivariable maps in microscopic analysis have been developed in [10.705].

Recognition of targets [10.706] may start with extraction of their features [10.507, 707–713], shape recognition [10.714], detection of curves [10.87, 715–719] or detection of objects from background [10.720, 721]. Sometimes special operations are necessary: redundancy reduction [10.722], detection of object orientation [10.723], tracking and perceiving moving objects [10.724–727], detecting anomalies in images [10.177], texture analysis [10.728, 729], finding similarity invariants [10.730], introducing attentive vision [10.731], feature linking [10.72], building structural image codebooks [10.732, 733], or reducing the search base for 3-D objects [10.169].

A frequently occurring task is recognition of faces: [10.734, 735]. An inverse operation to that is synthesis of facial expressions [10.736]. Related to face recognition is lipreading of videophone by deaf people [10.737]. Identification of body postures may be useful to control machines: [10.738–740].

The following particular application areas should also be mentioned: characterization of paper properties [7.29], remote sensing in general [5.9, 10], part-family classification [10.741], traffic sign recognition [10.742], recognition of underwater structures [10.743], radar classification of sea ice [10.744, 745], classification of clouds [7.29], [10.501, 682, 683, 746, 747], rainfall estimation [10.748], extraction of jet features [10.749], and analysis of neighbor-star kinematics [10.750].

Of diverse results obtained by SOM-based image processing one may mention thresholding of images [10.751], learning of minimal environment representation [10.752] and dimension reduction of working space [10.753]. Special devices constructed are 2-D color sensor [10.754] and visual displays [10.490].

Satellite Images and Data. General pattern recognition for satellite images has been developed in [10.755]. Antarctic satellite imagery has been studied in [10.756]. A Landsat thematic mapper is described in [10.757]. Cloud classification on the basis of textures is an important application [10.758].

Medical Imaging and Analysis. Medical image processing constitutes a major problem area of its own [10.759]. Positron-emission tomography data have been reconstructed in [10.760], and magnetic-resonance images have been segmented in [10.761]. Particular SOM-based applications in medical imaging [10.762] are detection of coronary-artery disease [10.763], cardio-angiographic sequence coding [10.764], classification of brain tumors [10.765], liver tissue classification [10.766], foreign-object detection [10.767], recognition of structural aberrations in chromosomes [10.768, 769], chromosome location and feature extraction [10.770], and identification of a new motif on nucleic acid sequences [10.771].

Qualitative morphological analyses of muscle biopsies can be found in [10.772].

10.5.2 Optical Character and Script Reading

The optical character readers (OCRs) represent one of the most straightforward applications of pattern recognition: SOM-based solutions to it have been given in [10.773–782]. Detection of the orientation of characters is one subtask [10.783], and invariant OCR another [10.784]. Character readers for Hangul [10.785], Chinese [10.52], and Korean and Chinese documents [10.786] are more demanding. Recognition of handprinted characters [10.422, 787], handwritten digits [10.788], handwritten alphabet and digits [10.789–796], and cursive handwriting [10.797–803] are progressively harder tasks solved by SOM principles. A system for digit recognition from pulp bales has been developed in [10.804].

10.5.3 Speech Analysis and Recognition

The main objectives in speech research are: isolated-word recognition, segmentation of speech, connected-word recognition, phoneme recognition, recognition of natural (continuous) speech, speaker identification, phonetic research, classification of speech disorders, speech transmission, and speech synthesis.

General. Problematics of speech recognition, and approaches based on the SOM are discussed in [7.5], [10.98, 399, 583, 584, 586, 587, 805–821]. Speaker normalization [10.822, 10.823], feature extraction from speech [10.824–826], compact speech coding [10.827–830], and other problems [10.831] are met in speech processing.

Isolated-Word Recognition. Most speech recognition devices have been designed to recognize carefully articulated isolated words, with a short (> 100 ms) pause held between them. Examples of SOM-based word recognizers are: [10.129, 832–837]. Spoken digit recognition is reported in [10.838].

Connected-Word and Continuous-Speech Recognition. Segmentation of speech is one of the first tasks in connected-word recognition [10.839]. After that, connected words can be distinguished: [10.424, 580, 581, 840].

Detection and classification of *phonemes* is often the first step toward continuous-speech recognition: [2.37], [5.16], [10.53, 585, 841–860].

Complete phoneme-based speech transcribers have been described in [5.17], [7.14–23], [10.574, 575, 861–868]. In these, fine-tuning of codebook vectors by LVQ is mostly used to improve recognition accuracy.

Speech coding and recognition has been successfully carried out in [10.869]. The Hypermap architecture has been used for speech recognition in [10.870].

Speaker Identification. Recognition of the speaker is needed, e.g., in safety systems, or to define personal speech parameters in a speech recognition system. Clustering of speakers into categories may be useful to alleviate the problem of speaker-independence in speech recognition, whereupon different SOMs can be used for different clusters. Speaker identification has been discussed in [10.871–877], speaker clustering in [10.878], and vector quantization of speech parameters in [10.879].

Phonetic Research. Speech research does not necessarily aim at speech recognition. Various approaches to speech research are mentioned below. Experimenting with ASSOM filters for speech has been reported in [10.880]. Other problems are determination of speech space dimensions [10.881, 882], classification of speech accents [10.883], multilingual speech database annotation [10.884], phoneme recognition by binaural cochlear models and stereausis representation [10.885], visualization and classification of voice quality [10.886–888], analysis of voice disorders [7.6, 7], [10.889–892], detection and assessing of /s/ misarticulation [10.893–895], of /r/ misarticulation [10.895], detection of fricative-vowel coarticulation [10.896, 897], visual feedback therapy of pronounciation using SOM [10.898, 899], representation of spoken words [10.900], synthetic speech [10.901], fuzzy systems in speech processing [10.902], HMM parameter estimation [10.903], and cochlear implants [10.904].

10.5.4 Acoustic and Musical Studies

Aside from speech research, the following types of acoustic problems have been solved by the SOM: recognition of acoustic signals [10.905], segmentation of acoustic features [10.906], representation of acoustic feature dynamics [10.566], analysis of acoustic-noise pollution in cities [10.907], classification of shallow-water acoustic sources [10.908], pitch classification of musical notes [10.909], analysis of musical patterns [10.910, 911], and timbre classification [10.912].

10.5.5 Signal Processing and Radar Measurements

General problems of signal processing by the SOM have been discussed in [10.913] and [10.914]. Unsupervised classification of signal sequences is described in [10.915]. The effect of signal phase is analyzed in [10.916]. Sonar signal analysis can be found in [10.917, 918], and ultrasonic signal analysis in [10.919].

Target identification from radar signals is presented in [10.920, 921], object detection in clutter in [10.922], and clutter classification in [10.923]. Analysis of radar behavior is discussed in [10.924].

10.5.6 Telecommunications

The SOM may be used to solve certain basic technical problems in telecommunications [10.925–927]: e.g., bandwidth compression [10.928] and reduction of transmission error effects [10.929]. It can be applied to the following special systems: enhancement of delta modulation and LPC tandem operation [10.930]; detection of discrete signals, e.g., by quadrature-amplitude modulation (QAM); [10.931–935]; and in combination with equalizers [10.936–939]. It can be used for clone detection in telecommunications software systems [10.940, 941].

Configuration and monitoring of telecommunications traffic networks is a particularly suitable application area for SOM: [10.942–947].

On the most general level of telecommunications, there occurs the problem of optimal traffic shaping [10.949]. In selecting robust communication techniques one can use vector-coding methods based on the SOM [10.950–10.955]. In the complexity analysis of telecommunications software [10.948], SOM methods can be used.

10.5.7 Industrial and Other Real-World Measurements

One of the basic problems in real-world applications is to combine or "fuse" signal data from different sources such as optical and infrared sensors: [10.956–959]. Sensor arrays [10.960] and special sensors, e.g., for surface classification [10.961–963] and odor classification [10.964] are proper applications for SOM.

Other practical tasks, not mentioned earlier, are: color matching of buttons [10.267, 965, 966, 1394–1397] , prediction of interference spectra [10.967], ship noise classification [10.968], acoustic leak detection [10.969], lithology classification [10.970], astrophysical source classification [10.971], environmental monitoring [10.972], vehicle routing [10.973, 974], and some other [10.975].

Undersea mine detection [10.976], and detection of fault conditions from an anaesthesia system [10.977, 978] are a couple of specific applications.

Particle-impact noise has been detected in [10.979].

Traffic routing is a special application of SOM [10.980].

10.5.8 Process Control

System-theoretic aspects and modeling of industrial processes form a category of problems of their own: [7.4], [10.94, 152, 170, 171, 981–986]. Identification of the process state is discussed in [7.4], [10.987–993]. Other general problems are process error detection [10.994, 995], fault diagnosis [10.996–999], diagnosis of machine vibrations [10.1000], and plant diagnostic symptoms [10.1001].

In power systems, the following types of operation conditions must be analyzed: electricity consumption [10.1002] consumption alarms [10.1003], load forecasting in electrical systems [10.1004, 1005], time of occurrence of electric utility peak loads [10.1006], power flow classification [10.1007], system stability [10.1008, 1009], power system static security assessment [10.1010–1014], impulse test fault diagnosis on power transformers [10.1015], partial discharge diagnosis [10.1016, 1017], and avoidance of the high-voltage mode of electromagnetic voltage transformers [10.1018].

A rather promising application area for the SOM is the visualization of system and engine conditions [7.5], [10.1019–1022], and large-scale diagnostic problems [10.1023].

The following listing contains diverse industrial applications: monitoring paper machine quality [10.1024], analysis of particle jets [10.1025], flow regime identification and flow rate measurement [10.1026, 1027], grading of beer quality [10.1028], velotopic maps for well-log inversion [10.1029], shear velocity estimation [10.1030], intrusion detection [10.1031], estimation of torque in switched reluctance motor [10.1032], composite damage assessment [10.1033], heating and cooling load prediction [10.1034], identification of car body steel [10.1035], operation guidance in a blast furnace [10.1036], controlling of 1000 amps [10.1037], calculation of energy losses in distributed systems [10.1038], evaluation of solid print quality [7.31], location of buried objects [10.1039], analysis of breaks in a paper machine [10.1040] and classification of wooden boards [10.130]. Other practical tasks are tracking of changes in chemical processes [10.1050], sensory diagnostic of steam turbines [10.1051], forecasting of electricity demand [10.1052], characterization of flotation froths [10.1053], classification of universal motors [10.1054], classification of the metal transfer mode [10.1055], and color bake inspection [10.1056].

A neuro-fuzzy controller has been described in [10.1041], and engineering pattern recognition has been described in [10.1049].

10.5.9 Robotics

General. Some researchers hold the view that autonomous robots constitute the ultimate goal of ANN research. Achievements in this area, ensuing from the SOM research, are discussed in this subsection. General papers are [10.114, 1042–1048].

Problems of machine vision were already surveyed in Sect. 10.5.1.

Robot learning strategies have been suggested in [10.1057–1059]. Classification of sensory-motor states [10.1060] and interpretation of tactile information [10.1061] are important detailed tasks. General control strategies of a mobile robot have been discussed in [10.1062], and monitoring tasks in a flexible manufacturing system in [10.1063].

Robot Arm. Control of the robot arm, coordination of multiple joints and visuomotor coordination is one of the basic tasks: cf. [10.60, 1064–1080].

Robot Navigation. Another basic problem is the collision-free navigation of the robot: [10.107, 1081–1106].

Further problems in robotics are intersensory coordination in goal-directed movements [10.1107], learning of motor shapes [10.1108], optimal information storage for robot control [10.1109 - 10.1111], representation of obstacles [10.1112], parametrization of motor degrees of freedom [10.1113], motion planning [10.1114], path planning [10.1115], and view-based cognitive map learning [10.1116].

10.5.10 Electronic-Circuit Design

Design of very-large-scale-integrated (VLSI) circuits involves optimization of the circuit. Its solutions with basic or modified SOM algorithms can be found in [10.31, 54, 55, 1117–1133]. Typical tasks are floorplanning of circuits [10.1135], component placement [10.1136] and minimal wiring [10.1137].

Other problems in VLSI production, solvable by SOM methods, are detection of partial discharge sources on the circuit [10.1016], detection and elimination of defective cells [10.1138, 1139], analysis of VLSI process data [10.1140], and solder joint classification [10.1141].

On the system design level of digital circuits, research has been done on digital coding of information [10.1142] and synthesis of synchronous digital systems using the SOM [10.1143–1146].

10.5.11 Physics

Mapping of infrared spectra [10.1147] onto the SOM can be used to visualize their structures. Separation of quark-gluon jets is another typical physical pattern-recognition task [10.1148, 1149]. Discrimination of pp to tt events has been performed in [10.1150].

Simulation of pulsed-laser material [10.78, 79, 82] and control of the grey-level capacity of bistable ferroelectric liquid crystal spatial light modulators [10.1151] are examples of the use of SOM in the development of special materials for physical research.

Geophysical inversion problems [10.1152] and classification of seismic events [10.1153–1155] are further examples of uses of the SOM in applied physics.

10.5.12 Chemistry

The complex atomic structures in chemistry, especially organic chemistry are suitable for SOM classification: [10.1156–1158]. Work on the classification of proteins has been reported in [10.1159–1168]. Evaluation of secondary structures of proteins from UV circular dichroism spectra with the aid of SOM has been studied in [10.1169] and protein homologies in [10.1170].

Chromosome feature extraction [10.1171] seems a very promising area.

A special application is automatic recognition of 2-D electrophoresis images [10.1172]. Characterization of microbial systems using pyrolysis mass spectrometry is another special application: [10.1173–1175].

10.5.13 Biomedical Applications Without Image Processing

A promising application area for the SOM in medicine is electro-encephalography (EEG); its signal analyses have been performed in [10.24, 1176–1082], and [10.1183–1187]. Closely related to it are electromyographic measurements [10.1188–1190], magneto-encephalography (MEG) studies [10.1191]. Automated sleep classification analysis is a characteristic application [10.1192] of the SOM.

Classification of electro-cardiogram (ECG) signals [10.1193–1195], blood pressure time series [10.1196], and lung sounds [10.1197] belong to medical signal analyses carried out by the SOM.

The article [10.1198] describes a diagnosis system, and [10.1199] a health monitoring system. Sub-classification of subjects at risk of catching Huntington's disease is reported in [10.1200].

Analysis of glaucomatous test data is described in [10.1201].

Analysis of saccades [10.1202], identification of gait patterns [10.1203], and analysis of motor cortical discharge pattern [10.1204] are other examples.

In vitro screening of chemoterapeutic agents [10.1205], nucleic acid analysis [10.1206], and drug discovery for cancer treatment [10.1207] are new important lines in SOM applications. Comparison of novice and expert evaluation [10.1208] seems to be a very interesting and promising application.

10.5.14 Neurophysiological Research

The SOM may also be used for the development of research methods for biology and especially neurophysiology. Signal classification [10.95, 1209] is a straightforward example of it.

A very effective and useful mode of use of the SOM is processing of observations, such as assessing the latency and peak pressure amplitude from auditory evoked potentials [10.1210], and neutralization of the temporal statistics of spontaneous movements in motor cortex functional characterization [10.1211]. The SOM can also be used in sleep classification [10.1209], to decode the side of hand movement from one-trial multichannel EEG data [10.1212], in primary cortical analysis of real-world scenes [10.1213], evaluation of sensori-motor relevance [10.1214], and analysis of functional topology in visuo-verbal judgment tasks [10.1215].

One should also mention the attempts to explain the brain maps [10.1216, 1217] . Numerous brain function models can be based on the SOM: physiological model of the cortex [10.1218], cortical representation of the external space [10.1219], temporal theory of auditory sensation [10.1220], modeling peripheral pre-attention and foveal fixation [10.712], ocular dominance bands [10.1221], sensory-controlled internal stimulation [10.1222], auditory brainstem responses [10.1223], structures in the cortex [2.53], [10.1224–1234], simulation of sensorimotor systems with cortical topology [10.1235], and simulation of Braitenberg vehicles [10.1236].

10.5.15 Data Processing and Analysis

One of the main applications of SOM is in the description of statistical data. General discussions of this area are found in [10.1237–1241].

A very promising application area is in the exploratory analysis of financial and economic data: [10.3, 1242–1249]. Profiling customers [10.1250] and appraisal of land value of shore parcels [10.1251] are similar problems.

In basic data-processing techniques one can find data-analysis problems such as data compression [10.1252], information retrieval [10.106, 270], information management [10.1253], knowledge extraction [10.1254, 1255], data analysis and interpretation methods [10.1256, 1257], spatio-temporal data coding [10.1258], sorting [10.1259], qualitative reasoning [10.1260], computer system performance tuning [10.1261], analysis of multiusers on a Prolog database [10.1262], user identification [10.1263], load-data analysis, load forecasting in information systems [10.1264], and reengineering software modularity [10.1265, 1266].

Data normalization is one of basic problems of data analysis [10.1267].

Another very promising application is classification of reusable software modules by SOM methods, in order to maintain a library for them [10.1268–1272].

10.5.16 Linguistic and AI Problems

It may sound surprising that vector-space methods such as SOM can handle highly structured, discrete, linguistic data. Therefore it has to be emphasized that philosophically, the logical concepts have always been preceded by and based on so-called "psychic concepts" that are some kind of clusters of perceptions. The latter can be formed in the SOM too on the basis of attributes and contextual patterns found in linguistic expressions.

A discussion relating to AI and creativity can be found in [10.1273].

Categories. The basis of language is the lexicon. It is usually divided into word categories. Association of meanings to lexical variables, e.g., using the SOM is thus the first task [2.74], [10.1274–1276]. When pieces of text are presented to the SOM (cf. Sect. 7.7), the representations of words or other linguistic items become clustered in categories: [2.73], [7.34], [10.1277–1283].

Concept development has been discussed in [10.1285] and [10.1286]. Semantic disambiguation [10.1287], and handling of attribute maps [10.1288] are fundamental tasks in data analysis that can be handled by the SOM.

Expressions and Sentences. The traditional approach to linguistic expressions is parsing. Data-oriented (automatic) parsing using the SOM is described in [10.1289–1294].

Formation and recognition of natural sentences is reported in [2.74], [10.45, 61–69, 1295–1297].

Full-Text Analysis. Coded text databases, such as books and journals, form the main source for automatic acquisition of languages. Their analysis by SOM methods has been done in [2.75], [10.1298–1309]. A specific task of text and data analyses handled by the SOM is the study of scientific journals [10.1310].

Knowledge Acquisition. For the acquisition of structured knowledge, many artificial-intelligence (AI) methods have been developed. To these one can now add ANN and in particular SOM methods: [10.1311–1317].

For natural discussion and communication, some additional abilities are necessary: [10.57, 1296]. Optimization of logical proofs is done in [10.1318].

Information Retrieval. One objective of linguistic research is flexible information retrieval. This aspect is discussed in [10.1319] and [10.1298, 1320, 1321].

Further Linguistic Studies. In text processing, a simple task is automatic hyphenation [10.1322].

Neurolinguistics in general bridges AI research with brain theory [10.1323, 1324]. Some of its results lead to enhanced human-computer interaction [10.1325]. Closely related to verbal communication is interpretation of sign languages [10.1326].

10.5.17 Mathematical and Other Theoretical Problems

Clustering is the operation that is naturally performed by the SOM. Such problems have been handled in [10.27, 104, 1327–1334].

Another natural application of the SOM, by definition, is nonparametric regression: [10.32–38, 1335–1338].

Approximation of functions by the SOM is done in [10.1339] and prediction of time series in [5.4] and [10.1340]. Other problems are smoothing [10.1341, 1342], Markov models [10.1343], reduced-kernel estimation [10.1344, 1345], nonlinear mapping [10.1346], nonlinear projection [10.1347], dimensionality reduction [10.1348, 1349], time-series and sequence models [10.1350–1352], association mapping of lookup tables [10.1353], decoding functions [10.1354], general vector quantization [10.1355], recursive vector quantization [10.1356], distance function estimation [10.1357], analysis of unlabeled data [10.1358, 1359], quadratic assignment [10.1360], and analysis of complex pattern spaces [10.1361].

In general, pattern recognition or classification involves several operations, such as pre-quantization [10.1362], harmonic-family classifier [10.1363], and feature extraction [10.1364, 1365], for which the SOM may be useful. See also [10.1366, 1367], and [10.1239].

Design of details of other ANNs using the SOM has been done in [10.1368–1374].

Diverse SOM applications are: adaptive state-space partitioning [10.1375], Hamming coding [10.1376], analyzing a contingence table [10.1377, 1378], canonical parametrization of fibre bundles [10.1379], computation of chaotic systems [10.1380], modeling of chaos [10.1381], numerical parametrization of manifolds [10.1382], generation of finite-element meshes [10.1383, 1384], nonlinear system identification [10.1385], finding multi-faculty structures [10.1386], graph matching [10.1387], statistical problems [10.1388–1391], multiple-correspondence analysis relating to a crosstabulation matrix [10.1392], action planning [10.1393] and selecting the color combination of buttons [10.268, 1394–1397].

Problem-specific applications are reviewed in [10.1398].

The Traveling-Salesman Problem. Finding the optimal path in a network, or routing, is one of the basic optimization problems [10.1399]. The traveling-salesman problem (TSP) is a stereotype of them, and it has become a benchmark for the ANNs as well as other "soft computing" methods. Therefore SOM problems also involve the TSP: [10.134, 1400–1415]. The multiple TSP has been handled by [10.1004, 1416, 1417].

Fuzzy Logic and SOM. Hybrids of fuzzy-set concepts and ANNs have great promises in practical applications. For instance, generation of membership functions and fuzzy rules by the SOM [10.1418], fuzzy-logic inference [10.1419], automatic fuzzy-system synthesis [10.1420], fuzzy-logic control in general [10.1421] and fuzzy adaptive control with reinforcement learning based on the SOM [10.1422] are good examples of it.

Hybridization of the SOM with Other Neural Networks. The following networks or algorithms have been combined with the SOM: multilayer Perceptron [10.1423], counterpropagation [10.1424], radial basis functions [10.1425], ART [10.1426, 1427], genetic algorithms [10.1428, 1429], evolutionary learning [10.1430], and fuzzy logic [10.1431].

10.6 Applications of LVQ

Although the SOM algorithm is more versatile in creating abstract representations of data, the accuracy of classification of stochastic samples by the LVQ can be several times greater compared with SOM. This is due to supervised learning, on account of which the LVQ algorithm defines the class borders optimally. Therefore statistical pattern recognition problems should be based on the LVQ, if supervised learning is at all possible. This is particularly true in speech recognition.

Image Analysis and OCR. The following image-processing applications are based on LVQ instead of SOM: color image quantization [10.540], image sequence coding for video conferences [10.1432–1435], image segmentation [10.167], reduced search base in 3-D object recognition [10.169], texture boundary detection [7.24], and in general, optimal classification of stochastic textures [2.71], [7.25–35].

Particular LVQ-based applications are construction of a stellar catalogue [10.1436], measuring paper quality [10.1437], classification of chest radiographs [10.1438] and detection of lesions from ultrasound images [10.1439].

In OCR research, the following problems have been handled by LVQ: shift-tolerant digit recognition [10.1440], recognition of handwritten characters [10.1441] and recognition of Japanese Kanji characters [10.1442].

Speech Analysis and Recognition. The stochastic nature of speech signals calls naturally for the LVQ. Problems thereby studied include: parametrization of the speech space [10.1443], phoneme classification [7.8], [10.576, 1444–1447], phoneme segmentation [10.1448], recognition of Japanese phonemes [10.1449], vowel recognition [10.1450], speaker-independent vowel classification [10.568], classification of unvoiced stops [10.1451], comparison of neural speech recognizers [10.1452], validation of ANN principles for speech recognition [10.1453], robust speech recognition [10.1454], digit recognition [10.1455], isolated-word recognition [10.1456], speech recognition in general [10.564, 567, 1457–1459], large-vocabulary speech recognition [10.1460], speaker-independent continuous Mandarine digit recognition [10.1461], speaker-independent large-vocabulary word recognition [10.565], speaker-independent recognition of isolated words [10.1462], and continuous-speech recognition [10.1463].

The Hypermap-LVQ principle for the classification of speech has been used in [10.96].

Combination of Hidden Markov Models and LVQ is a particular and very promising direction in continuous speech recognition [7.9–23], [10.399, 564–584, 1464].

Speaker adaptation [10.1465, 1466], speaker identification [10.1467, 1468], speech synthesis [10.1469], speech analysis [10.821], and speech processing [10.545] are further tasks in which LVQ has been applied.

Signal Processing and Radar. Classification of waveforms [10.1470] and EEG signals [10.1471–1473], ultrasonic signal analysis [10.919], radar signal classification [10.1474], and synthetic-aperture radar target recognition [10.1475] have been based on LVQ.

Industrial and Real-World Measurements and Robotics. Sensor fusion [10.1476], intelligent sensory architectures [10.172, 1477], expression of odor sensory quantity [10.1478], nuclear power plant transient diagnostics [10.1479], and gas recognition [10.1480] are good examples of LVQ in this category.

Path-planning and navigation of robots has been improved by the use of LVQ [10.1481].

Mathematical Problems. Optimization [10.561], monitoring of input signal subspace [10.1482], pattern discrimination [10.1483], design of radial basis functions [10.1484] and construction of a fuzzy reasoning system [10.1485] have been based on LVQ.

10.7 Survey of SOM and LVQ Implementations

The following listing contains both software and hardware results, but it does not express explicitly what architectural principles are still on the conceptual stage and which ones have already been implemented in VLSI. This situation is changing all the time.

Software Packages. Many software packages have been developed for neural computing. Of design tools that have been influenced strongly by SOM and LVQ ideas one may mention the definition and simulation language CARELIA [10.1486–1488], the LVQNET 1.10 program [10.1489], our LVQ_PAK and SOM_PAK program packages, of which have been described in [10.1490], some graphic programs [10.1491, 1292], and other packages mentioned in Chap. 8.

Programming SOM on Parallel Computers. Parallel computers for SOM calculations have been studied in [10.1493, 1494]. The article [10.1495] describes a conceptual framework for parallel-machine programming to describe ANNs, including the SOM. Other general reviews of parallel-computer implementations are in [10.1439, 1496–1504]. The dot-product implementation of the SOM architecture has been suggested in [10.1505].

Perhaps the most advanced parallel SOM software has been written for SIMD (single instruction stream, multiple data stream) computers such as the CNAPS [10.1506] and other [10.1507, 1508].

Transputer architectures for the SOM have been reported in [3.29], [9.5, 6], [10.911, 1238, 1509–1513].

Other parallel computing implementations of SOM exist for the Hypercube computer [10.1514], for the Connection Machine [10.1511–1513, 1515], for systolic computer architectures [9.7–9], and for a parallel associative processor [10.1516].

A special data structure with doubly linked lists has been described in [10.1517].

Analog SOM Architectures. The analog VLSI technology, although already feasible for more traditional ANNs, is still being studied for the SOM. The following designs may be mentioned: [10.1518–1525]. Optical-computing implementations of the SOM have deen discussed in [10.1526–1529].

A solution based on *active media* has been proposed in [10.1530].

An optical-holography computer that lends itself to the implementation of many neural-network algorithms including the SOM has been developed in [10.1531].

Digital SOM Architectures. Various architectural ideas for the SOM have been presented in [2.72], [9.3, 10–12], [10.790, 1532–1571]. A fast digital-signal-processor architecture for classification has been built [10.1572].

Analog-Digital SOM Architecture. A fully parallel mixed analog-digital architecture for the SOM has been suggested in [10.1573].

Digital Chips for SOM. Finally it may be interesting to mention detailed designs for electronic SOM chips, many of which have already been fabricated: [9.13–19], [10.1558, 1559, 1574–1583].

11. Glossary of "Neural" Terms

The basic neural-network literature may already be familiar to a wide circle of scientists and engineers. Nonetheless, the true meaning of several concepts, never explicitly defined, may have escaped the attention of many readers. This chapter is therefore devoted to all those who always wanted to know about these definitions but were afraid to ask. This glossary is by no means exhaustive: it only serves the reading of this book, the literature references given, and other collateral texts. The definitions have been meant to be more explanatory than distinctive. Among alternative explanations, only those that are relevant to neurophysiology or ANNs have been given. The partial vocabularies picked up from artificial-intelligence techniques, biophysics, bionics, image analysis, linguistics, optics, pattern-recognition techniques, phonetics, psychology, and medical science are extremely concise. Mathematical concepts have also been explained in more detail in Chap. 1.

absolute correction rule principle according to which every misclassification is immediately corrected by changing the classifier parameters by just a sufficient amount.

absorbing state state in a Markov process that, when reached, cannot be changed to any other state by any input signals.

acoustic processor preprocessor for acoustic signals, performing frequency filtering, enhancement of features, normalization of the representations, frequency analysis, etc.

action pulse neural impulse of fixed size and duration. Parts of the membranes of some neural cells, and their axons as a whole, constitute so-called active media. In a complicated dynamic process described, e.g., by the Hodgkin-Huxley equations [2.24], the membrane potentials resulting from ionic flows, and voltage-dependent permeability of the cell membrane to the ions create unstable cyclic electro-chemical phenomena. Such phenomena are most prominent in the axons, along which neural signals are propagated in the form of electrical impulses, like in active transmission lines that amplify and sustain signal energy without attenuation.

actuator mechanism by which something is moved or controlled.

Adaline adaptive element for multilayer feedforward networks introduced by Widrow [1.44].

adaptation automatic modification or control of parameters, usually to maximize some performance measure. In evolution: adjusting of species to environment by natural selection or by changing their behavior.

adaptive control control action in relation to a performance measure, which is increased due to adaptation, or when the system to be controlled is intermittently identified for the optimization of control.

adaptive resonance theory theory of a type of neural system introduced by Grossberg [2.16]. Such a system consists of two interacting networks, one of them at the input, and a number of control paths. This theory applies competitive learning and defines codebook vectors, the number of which is variable.

A/D conversion see *analog-to-digital conversion.*

afferent carrying inward; propagating neural signals from the periphery toward a nerve center.

AI see *artificial intelligence.*

algorithm open, cyclic, or arbitrarily structured sequence of exactly defined unconditional or conditional instructions.

analog computer usually an electrical network that solves differential or other equations by simulation (modeling). It seems that the brain operates according to analog computing principles.

analog representation quantitative description of one entity by another measurable entity.

analog-to-digital conversion encoding of continuous-valued variables by digital representations, whereby the latter approximate the analog representations by means of discrete values.

ANN see *artificial neural network.*

apical dendrite dendrite of a *pyramidal cell* in the most distal part of the neuron.

AR autoregressive.

ARAS see *arising reticular activation system.*

architecture overall organization for information transfer, storage, and handling.

arising reticular activation system network of small nuclei in the brain stem that receives sample information from the sensory systems and also from higher parts of the brain, and the main purpose of which is to control arousal.

ARMA process autoregressive moving-average process.

array processor set of identical processors having some means for communicating instructions and/or data.

ART see *adaptive resonance theory.*

artificial intelligence ability of an artificial system to perform tasks that are usually thought to require *intelligence.*

artificial neural network massively parallel interconnected network of simple (usually adaptive) elements and their hierarchical organizations, intended to interact with the objects of the real world in the same way as the biological nervous systems do. In a more general sense, artificial neural networks also encompass abstract schemata, such as mathematical estimators and systems of symbolic rules, constructed automatically from masses of examples, without heuristic design or other human intervention. Such schemata are supposed to describe the operation of biological or artificial neural networks in a highly idealized form and define certain performance limits.

artifical perception *pattern recognition* in a more general sense.

association cortex areas of the cerebral cortex outside and mainly between sensory and motor areas, and in general, areas of the cortex in which signals of different *modalities* are merged and which are not dedicated to definite sensory or motor functions.

associative memory generic name for any memory function that makes it possible to recall stored information by associations to other pieces of stored information; *associative recall* is triggered by some externally given cue information.

associative recall retrieval of pieces of stored information on the basis of related items, or parts or noisy versions of them.

attractor any of the stable states in a feedback system toward which the system state makes transitions, starting in the *basin of attraction.*

attribute component in a data vector having some descriptive property.

auditory cortex any of a set of areas in the temporal lobe of the cerebral cortex where functions associated with hearing are situated.

augmentation adding (concatenating) one or several elements, such as bias or class identifier to a pattern or state vector to simplify formal vector expressions and computations.

autoassociative memory content-addressable memory, from which information is recalled by its arbitrary, sufficiently large fragment or noisy version.

automaton device usually based on *sequential circuits* and being capable of producing responses and sequences of actions in relation to various sequences of inputs in a meaningful way, without human intervention.

axon output branch from a neuron. The axon has normally many collateral branches that contact different neurons.

axonal flow continuous active closed-loop hydrodynamic flow in the axon of a neural cell. The tubular organelles in the axon are actively transporting various particles between the synaptic terminal and the cell body. This is part of metabolism and has an important effect on the formation, strengthening, or decay of neural connections.

axon hillock place where the axon leaves the soma of neuron. The neuron has a prominence at this place. Triggering of the neuron usually starts at the axon hillock.

backpropagation weight-vector optimization method used in *multilayer feedforward networks*. The corrective steps are made starting at the output layer and proceeding toward the input layer.

basal dendrite dendrite of a *pyramidal cell* in the proximity of the soma of the neuron.

basilar membrane resonant membrane in the cochlea of inner ear, driven by acoustic waves and responding to their frequency.

basin of attraction set of initial states in a feedback system from which the system state uniquely converges to the corresponding *attractor*.

basis vector member in a set of elementary vectors used to express an arbitrary vector in terms of meaningful components.

batch set of tasks performed by a single computational procedure.

batch processing computation in batches, e.g., of the corrections relating to all training data before updating the parameters.

Bayes classifier statistical classification algorithm in which the class borders are determined decision-theoretically, on the basis of class distributions and misclassification costs.

bias constant value added to a set of variables.

binary attribute two-valued descriptor describing either presence or absence of a property; sometimes, especially in image processing, a property (such as length of street) that is a function of two reference points.

binary pattern pattern, the elements of which are two-valued.

blurring intentional blurring of a signal or image, usually by convolution, to increase translatorial or other invariance in its recognition.

Boltzmann machine neural-network model otherwise similar to a *Hopfield network* but having symmetric interconnects and stochastic processing elements. The input-output relation is optimized by adjusting the bistable values of its internal state variables one at a time, relating to a thermodynamically inspired rule, to reach a global optimum.

bootstrap learning mode in learning whereby an unlimited sequence of training samples is picked up at random from a finite data set. The purpose is to define an open data set, the elements of which are statistically independent, at least sequentially.

bottom-up relating to *inductive inference*, i.e. from particulars to generals.

BP see *backpropagation*.

brain area area in the cerebral cortex, distinguished by its neurophysiological function.

brain-state-in-a-box state-transition neural network suggested by J. Anderson [11.1], similar to a *Hopfield network*.

brain stem anatomical formation surrounding the upper end of the spinal cord (medulla oblongata) and containing numerous *nuclei* and other neural networks and mainly controlling arousal and certain states of emotion.

branching factor number of possible choices for the next word in relation to a recognized previous word (usually much smaller than the size of vocabulary); see *language model, perplexity.*

broad phonetic class combination of phonological units into categories that are believed to be detectable reliably, for instance, for the segmentation of the speech waveform, or quick lexical access. Types of such classes are, for instance, vocalic, voiced closure, noise, and silence.

Brodmann's area any of 52 specified brain areas traditionally identified on the basis of its neuroanatomical structure.

BSB see *brain-state-in-a-box.*

calibration determination of scaling or labeling for an analytical device by means of well-validated input data.

CAM see *content-addressable memory.*

category class of items defined on the basis of very general properties.

cellular array array processor, the processing elements of which communicate with their topological neighbors.

cellular automaton system consisting of identical automata configured as a cellular array.

central nervous system brain and spinal cord.

cepstrum result of a double integral transformation, whereby first the amplitude spectrum of a time function is computed, and then another amplitude spectrum is computed to make the cepstrum. The previous amplitude spectrum is often logarithmized.

cerebellum part of brain situated rear to and below the cerebrum; its most important task is coordination of movements.

cerebrum the dominant part of the brain especially in man, mainly consisting of two large hemispheres and extensive connections inside and between them.

channel system for the transmission of restricted type of information.

chemical receptor agglomeration of large organic molecules piercing the membrane of the receiving neuron, and forming gated channels for light ions. The chemical transmitter molecules control the transmittance of these gated channels. Influx and outflux of the light ions make the membrane potential change.

chemical transmitter organic liquid that is formed in and released from a synaptic terminal by neural signals. It controls the electric membrane potential in another neuron (postsynaptic neuron), and thus its triggering. There exist several dozens of different chemical transmitters.

closure closing of the vocal tract during the utterance of some phoneneme (e.g., /k,p,t,b,d,g/). See *glottal stop.*

cluster set of data points, such as pattern or codebook vectors, close to each other in the data space.

CNS see *central nervous system.*

cochlea part of the inner ear where mechanical acoustical resonances (in the *basilar membrane*) are converted into neural signals.

codebook vector parameter vector formally similar to a *weight vector*, except that it is used as a *reference vector* in vector quantization methods (and thus also in SOM and LVQ).

cognitive relating to human information processing especially on the conscious level, often in linguistic form.

column vertical array. In neurophysiology: organizational unit of the cortex. It contains all the different types of neuron occurring within that brain area. The neurophysiological column extends perpendicularly through the cortex, often having the form of an hourglass, whereas its lateral dimensions vary; typical diameter is on the order of 100 μm.

commissural type of tract or set of axons transfering information massively from one area to another in the cerebral cortex.

commuting property of operators being exchangeable in operator products.

component plane cross section of all model vectors of the SOM at a certain component position.

compression any kind of data reduction method that preserves the application-specific information.

connection link between neurons, coupling signals in proportion to its weight parameter.

connectionist attribute relating to artificial neural networks. This term is mainly used in artificial-intelligence techniques in contrast to symbolic.

content-addressable memory memory device from which pieces of stored information are recalled on the basis of arithmetic functions or fragments of stored items.

context information describing the situation or instance in which an item occurs.

continuous-speech recognition recognition of spoken expressions when words are uttered at varying speed and intonation, and eventually fused.

corpus callosum a very large bundle or fan of axons through which the left and right hemispheres of the *cerebral cortex* communicate.

correlation matrix memory model for distributed content-addressable memory, the internal state of which is the correlation matrix of input vectors.

cortex any of the layered areas in neural networks. The most important representative of cortices is the *cerebral cortex*, in man a 2–3 mm thick and about 2000 cm^2 wide sheet below the skull. The major part of the cerebral cortex consists of the *neocortex*, which is the newest part and apparently in a rapid state of evolutionary development. The cortices are nicknamed "grey matter", in contrast to *white matter*, which mainly consists of myelinated axons contacting various areas of the cortices.

crossbar switch network formed of two sets of crossing signal lines, with an individual connection strength defined at each crossing.

cups speed measure for neurocomputers: connection updates per second. The number of corrections made on weight values in a second. Often the same as the number of multiplication-addition pairs per second.

curse of dimensionality expression introduced by mathematician R. Bellman to describe the property of certain optimization problems in which memory demand and computing time increase very rapidly and even to intolerable amounts when the dimensionality of input vectors is increased.

daemon computational procedure automatically triggered by an event or need.

data matrix matrix-valued table in which columns signify data items and values on the rows their attribute values, respectively.

D/A conversion see *digital-to-analog conversion.*

DEC see *dynamically expanding context.*

decision border mathematically defined limit in the signal space between classes in pattern recognition or statistical decision-making.

decision-directed estimate an estimate made in sequential steps, whereby the result of the new step depends on decisions made on the basis of old results.

decision-directed learning learning on the basis of *decision-directed estimates.*

decision surface $(N-1)$-dimensional hypersurface or border between two classes in an N-dimensional signal space, for which the average decision error is supposed to be minimized.

decode to identify codes or patterns: to *segment* signals.

deductive inference deriving of conclusions or consequences from premises or general principles; reasoning from generals to particulars.

delta rule learning rule in which an additive correction to an old parameter value is a fraction of difference between the wanted and the true value.

dendrite branch of a neuron normally receiving input signals.

dendro-dendritic synapse synaptic connection between the dendrites of adjacent neurons.

depolarization reduction of voltage across a cell membrane due to excitatory synaptic transmission.

deterministic relating to inevitable consequence of antecedent sufficient conditions.

DFT discrete Fourier transform: computing scheme to evaluate the Fourier transform at periodic argument values by combining terms and avoiding multiplications.

differential analyzer computer, often analog, that solves differential equations by establishing physical analogies for them and into which various models of dynamic systems to be studied can be built.

digital image processing computerized processing of images, whereby the picture elements are represented numerically.

digital-to-analog conversion *analog representation* of digitally expressed numbers.

dimensionality number of elements in a vector or matrix.

dimensionality reduction preprocessing transformation that reduces the dimensionality of pattern vectors.

diphone transition from one phoneme to another; pair of phonemes.

discriminant function mathematical function defined for each class of patterns separately. Input data are determined to belong to that class for which the discrimination function attains the largest value.

distance specific measure of dissimilarity of sets of values.

distinctive feature member in a set of features, on the basis of which the observations can absolutely and error-free be divided into disjoint subsets.

distributed memory memory that stores spatially distributed transforms of items and superimposes them onto a common set of memory elements; the recollection is then formed as another transform.

domain set of argument values for which a function is defined.

dorsal situated on the back side.

dot product particular kind of *inner product* of vectors.

DTW see *dynamic time warping*.

dualism doctrine, according to which material and mental are separate kinds of existence or substance.

dynamic programming algorithmic optimization method that finds the optimal path between two points as a recursive expression of partial optima.

dynamic time warping nonlinear transformation of time scale in speech recognition for which signal values of two words or other phonetic expressions can be made to match optimally. The recognition results can be based on such optimal matching.

dynamically expanding context variable context used in input-output mappings. Its length is decided on the basis of conflicts that occur in the input-output relations based on shorter context.

early vision see *low-level vision*.

EEG see *electro-encephalogram*.

effector organ that is activated in response to stimulation.

efferent carrying outward; propagating neural signals from a nerve center to an effector.

eigenvalue scalar value λ for which an operator equation of the type $Fx = \lambda x$ can be satisfied, where F is an operator on x.

eigenvector solution for x in an operator equation $Fx = \lambda x$, where F is an operator and λ its *eigenvalue*.

electro-encephalogram measurement of bioelectrical sum potentials between pairs of electrodes mounted on the scalp; these potentials correlate with neurophysiological processes that take place within the brain.

electrotonic type of direct interaction between neurons through electric fields.

emulation computation of a task by imitating the way in which another computer operates.

energy function objective function to be minimized in some optimization problem. It is a *Lyapunov function*. *Error functions* are often energy functions.

engram memory trace.

enhancement amplification of certain features in signal or image preprocessing.

enrollment of speaker setting up a voice model for a particular speaker, or presenting his or her speech samples to tune up a speech recognition device or system.

epoch finite set of meaningful input patterns presented sequentially.

epoch learning learning by *epochs*.

EPSP excitatory *PSP* (postsynaptic potential), a voltage jump at the postsynaptic membrane in the direction of depolarization of the membrane, in relation to one action pulse at the presynaptic axon.

ergodic relating to a stochastic process in which every sizable subsequence is the same statistically, and every state will occur in the long run.

error function objective function or functional that expresses the average expected error in some optimization task.

error signal difference between the true and the wanted value of a signal.

estimator mathematical operator that defines optimal interpolation or extrapolation of a variable, or expected output from a system, based on a finite set of observed data.

Euclidean distance *distance* having certain specific geometric properties.

Euclidean metric *metric* having certain specific geometric properties.

Euclidean norm geometric length of vector in the *Euclidean space*.

Euclidean space mathematically defined set of vectorial elements, the relations of which are defined in terms of the *Euclidean distance*.

excitation synaptic control action that increases the activation of a neuron by depolarizing its membrane.

extracellular outside a cell.

fan-in number of input connections.

fan-out number of branching output connections.

fast Fourier transform any digital algorithm that computes the DFT effectively.

feature generic name for an elementary pattern of information that represents partial aspects or properties of an item.

feature distance distance measure relating to a set of feature values.

feature extraction preprocessing, in which a *feature vector* is formed.

feature vector vector, the elements of which represent feature values.

feedback set of signals connected from the output terminals to some input terminals.

feedback network network, in which signal paths can return to the same node.

feedforward network network, in which the signal paths can never return to the same node.

FFT see *fast Fourier transform.*

finite-state automaton sequential circuit or sequentially operating system having a finite number of internal states and deterministically defined transitions between them.

formal neuron neuron model in ANNs. Originally the same as *threshold-logic unit.*

formant acoustical resonance in the *vocal tract.*

formant tracker device or algorithm containing a number of frequency-controlled or phase-locked oscillators, each following up its own *formant* of speech in time.

frame set of neighboring phonemes surrounding a phoneme; e.g., if C means consonant and V vowel, then in a segment CVC the vowel V is in the frame of two consonants; see *phonotax, syntax.*

fricative see *voiceless.*

fundamental frequency see *pitch of speech.*

fuzzy logic continuous-valued logic consisting of maximum and minimum selection operators and making use of membership functions to define the graded affiliation to sets.

fuzzy reasoning act of applying *fuzzy logic* in reasoning.

fuzzy set theory set theory in which the operators obey rules of *fuzzy logic.*

Gabor filter two-dimensional kernel for image processing described by a spatial sinusoidal wave modulated by a Gaussian amplitude function. Gabor filters come in pairs, the wave in one of them being in zero phase, the other being 90 degrees out of phase with respect to the center of the kernel.

Gabor jet set of outputs from Gabor filters with different parameters and centered on the same location.

ganglion mass of neural cells that performs a particular type of processing such as sensory preprocessing. The brain of an insect mainly consists of ganglia.

generalization way of responding in the same way to a class of inputs, some of which do not belong to the training set of the same class.

genetic algorithm learning principle, in which learning results are found from generations of solutions by crossing and eliminating their members. An improved behavior usually ensues from selective stochastic replacements in subsets of system parameters.

glial cell cell supporting and providing for the nutrition of the neural cells. Glial cells do not take part in signal processing or transmission.

glottal stop complete closure of the glottis during articulation of voiceless stops, such as /tt/, or in the beginning of words starting with a vowel.

glottis space in the larynx and structures surrounding it where the voicing apparatus (such as the vocal cords) is situated.

gradient vector, the components of which are partial derivatives of a scalar function with respect to various dimensions in the data space. In physiology: difference of concentration across a membrane.

grammar in formal languages: a system of *rules* describing *syntax*.

Gram-Schmidt process algebraic method to find orthogonal basis vectors to a linear subspace.

gray matter brain tissue of the cortex containing neural cells.

gray scale set of available shades of gray for printing, representation, etc.

gyrus macroscopic ridge of the cortex.

halting see *stopping*.

Hamming distance number of dissimilar elements in two ordered sets.

hard-limited output output that is abruptly saturated when its value reaches its upper or lower bound.

hash coding storing data in computer memories, the addresses of which are certain arithmetic functions of the digital representations of data.

Heaviside function function $H(x)$ where $H(x) = 0$ for $x < 0$ and $H(x) = 1$ for $x \geq 1$.

Hebb's law the most frequently cited learning law in neural-network theory. The synaptic efficacy is thereby assumed to increase in proportion to the product of presynaptic and postsynaptic signal activity.

heteroassociative memory content-addressable memory, from which an item is recalled on the basis of another item memorized together with the former one.

heuristic relating to human-made designs and choices; based on intuition.

hierarchical relating to a multilevel organization, where lower levels are subordinates to higher levels.

hidden layer an intermediate layer of neurons in a multilayer feedforward network that has no direct signal connection to inputs or outputs.

hidden Markov model statistical model that describes input-output relations of sequentially occurring signals using internal, "hidden" states and transition probabilities between them. The probability functions of the hidden states are identified on the basis of training data, usually in learning processes.

hippocampus part of the brain consisting of two symmetric oblongated formations below the cortex, in close cooperation with the latter.

HMM see *hidden Markov model*.

Hopfield network state-transition neural network that has feedback connections between all neurons. Its energy-function formalism was worked out by Hopfield [2.29].

hypercolumn organizational unit of cortex consisting of several columns.

hypercube N-dimensional cube where $N > 3$.

Hypermap SOM or LVQ in which the winner is searched sequentially from subsets of candidates determined at previous searching phases.

hyperpolarization increase of voltage across a cell membrane due to inhibitory synaptic transmission.

idempotent property of operators when their integer powers are identical with themselves.

image analysis interpretation of the semantic content of an image; identification of items and their complexes from an image; thematic classification of areas in an image.

image understanding interpretation of an image relating to a knowledge base.

impulse response time-domain output signal from a system into which a short impulse is input.

inductive inference reasoning from particulars to generals; inference of the general validity of a law from its observed validity in particular cases.

inhibition synaptic control action that decreases the activation of a neuron by hyperpolarizing its membrane.

inner product two-argument, scalar-valued function describing similarity of ordered sets.

input activation implicit variable associated with a neuron. It is some function of input signals and parameters (synaptic weights). The output activity is determined as some function or by a differential equation involving the input activation.

input layer layer of neural cells, each of which directly receives input signals.

input-output relation mapping of the input data set onto the output data set.

input vector vector formed of input signal values.

intelligence ability to perform new tasks that are directly or indirectly vital to survival, and solve problems for which no precedent cases exist. Human intelligence is *measured* by standardized mental tests.

interconnect see *connection*.

interlocking property of information processes when they have to wait for results from other processes.

interneuron any of the smaller neurons that mediate signals between principal cells.

intracortical connection connection between neurons made within the cortex.

intracellular inside a cell.

invariance property associated with responses, when they are the same for some transformation group of input patterns.

ion channel pathway for ions through the cell membrane. There exist groups or agglomerations of large organic molecules attached to the cell membranes and forming pores or channels for light ions such as Na^+, K^+, and Ca^{++}. The molecules forming the channels often act as receptors for chemical transmitter molecules, and by this chemical control action, selectively control the amount of transflux of the light ions through the membrane. Ionic currents change the electric charge inside the cell and thus the voltage across the cell membrane, often causing dynamic electrical phenomena of the duration of some milliseconds to occur. Some of these phenomena carry information in the form of action pulses.

IPSP inhibitory *PSP* (postsynaptic potential), a voltage jump at the postsynaptic membrane in the direction of hyperpolarization of the membrane, in relation to one action pulse at the presynaptic axon.

isolated-word recognition recognition of words from a fixed vocabulary, when the words are uttered in isolation.

iteration repetition of a procedure (such as training) relating to one input or epoch.

Jacobian functional determinant factor used especially to express the true differential area or volume in a nonorthogonal or curvilinear coordinate system. The elements of the first column of the Jacobian functional determinant are the partial derivatives of the first dependent coordinate with respect to each of the independent coordinate variables, those of the second column the partial derivatives of the second dependent coordinate, and so on.

Karhunen-Loève expansion approximation method in the theory of stochastic processes, whereby the process is usually (in the simplest case) represented as an expansion in eigenvectors of the input covariance matrix. *Karhunen expansions* are even more general.

kernel mathematical operator in integral transforms, which often describes how an impulse or point is transformed.

K-means clustering vector quantization where new values for K codebook vectors are obtained as means over Voronoi sets defined by the old codebook vectors.

K-nearest-neighbor classification act, in which the majority of classes in K nearest samples to the unknown data vector in the data space determines its classification.

KNN-classification see *K-nearest-neighbor classification*.

knowledge base data base augmented by inference rules to describe the interdependence of stored items and their complexes.

Kronecker delta function δ_{ij} of two integer indices i and j such that $\delta_{ij} = 1$ when $i = j$ and $\delta_{ij} = 0$ when $i \neq j$.

label discrete symbol describing a class or meaningful cluster of representations.

language model set of linguistically based constraints on the possible joining of words, defined by a grammar; subset of possible next words in relation to a recognized word, statistically based on a Markov model of sequences of words.

Laplace filtering enhancement of edges or contours in a picture by a discrete approximation of the second spatial derivative of the intensity function (relating to the Laplace operator).

Laplace operator differential operator $\Delta = \partial^2/\partial x^2 + \partial^2/\partial y^2$, where x and y are the horizontal and vertical coordinate of the xy-plane, respectively.

lateral relating to the direction along the cortex or other layered network. Sometimes referring to direction toward the side of the body.

lattice regular, often two- or three-dimensional spatial configuration of nodes or neurons.

LBG algorithm see *Linde-Buzo-Gray algorithm.*

learning generic name for any behavioral changes that depend on experiences and improve the performance of a system. In a more restricted sense learning is identical with adaptation, especially selective modification of parameters of a system.

learning rate true learning rate, or rate of change in parameters especially relating to one learning step (or in time-continuous systems, to the time constants).

learning-rate factor factor usually multiplying a correction of system parameters and thus defining the learning rate.

learning subspace method supervised competitive-learning method in which each neuron is described by a set of basis vectors, defining a linear subspace of its input signals.

learning vector quantization supervised-learning vector quantization method in which the decision surfaces, relating to those of the Bayesian classifier, are defined by nearest-neighbor classification with respect to sets of codebook vectors assigned to each class and describing it.

leave-one-out testing principle to achieve high statistical accuracy with a limited number of exemplary data. The available data set is divided into N statistically independent subsets, of which one is set aside for testing, and the remaining $N - 1$ subsets are used for training, respectively. This division is iterated always setting a different subset aside and repeating the training and testing steps on the rest of the data. An average of test results obtained in these phases is formed.

Levenshtein distance minimum sum of replacements (substitutions), deletions, and additions to convert one symbol string into another.

lexical access algorithm or other method that searches words from a vocabulary on the basis of the symbolic output from the phonemic classifier.

limbic system organizational system consisting of the so-called parahippocampal gyrus (fringe) of the cerebral cortex, hippocampus, and a couple of nuclei. Signals are propagated through it in a circle. States of the limbic system are believed to correspond to emotional states and reactions, as well as to sleep states.

Linde-Buzo-Gray algorithm fast algorithm that computes vector quantization in batches, essentially the same as *K-means clustering*.

linear discriminant function discriminant function that is a linear function of the pattern vector.

linear predictive coding approximation of a new value in a time series by a linear recursive expression of the old values.

linear separability situation where all samples of two or more classes are separable by a linear discriminant function. Highly improbable in practical applications.

link see *connection*.

LMS least mean of squares (of errors).

lobe major division of the cerebral cortex. The latter is coarsely divided into frontal, temporal (lateral), parietal (upper), and occipital (rear) lobes.

local storage memory allocated to a processing element such as neuron.

locomotion controlled movement.

long-term memory permanent, high-capacity store for structured knowledge especially in man.

long-term potentiation neurophysiological phenomenon of the duration of about a day believed to represent an *engram*.

low-level vision enhancement and segmentation of, and extraction of two-dimensional features from images in machine vision.

LPC see *linear predictive coding*.

LSI large-scale integration (of electronic circuits).

LSM see *learning subspace method*.

LTP see *long-term potentiation*.

LVQ see *learning vector quantization*.

Lyapunov function nonnegative objective function that often can be defined for optimization problems: if there exists a Lyapunov function, every learning step decreases it, and a local minimum is then reached almost surely. Energy or potential functions belong to Lyapunov functions.

Madaline multilayer feedforward network made of *Adaline* elements.

magnification factor relative size of an area on the cortex or the SOM, occupied by the representation of an input item. This area is inversely proportional to the point density of the codebook vectors m_i, and a monotonic function of $p(x)$.

Mahalanobis distance vector-space distance weighted by the inverse of the input covariance matrix.

manifold topological space having certain specific properties.

Markov process stochastic process in which the new state of a system depends on the previous state only (or more generally, on a finite set of previous states).

matched filter linear filter, the kernel of which has the same shape as the pattern to be detected.

matching comparison for similarity or distance between two items.

membrane double layer formed of lipid molecules, forming the sheath or enclosure of biological cells.

mental relating to psychological functions.

mental image memory-based experience that has the same structure and behavior as a sensory occurrence.

messenger chemical molecule that makes cells communicate without electric signals, or mediates communication within a cell.

metric property defining *symmetric distance* between elements.

Mexican-hat function usually symmetric kernel function defined in space (usually along a plane) having positive values in the central region of its domain, surrounded by negative values.

midbrain middle part of the brain, especially in lower vertebrates and in the embryonic state, roughly corresponding to parts within and around the brain stem.

minimal spanning tree connected graph that links elements in a set and has the smallest sum of lengths of all links.

minimum description length principle principle suggested by Rissanen [1.78] in which the dimensionality of a pattern vector is decided on the basis of combined cost consisting of model dimensionality and misfit between the model and the data.

MLD coding see *minimum-length descriptive coding*.

MLF network see *multilayer feedforward network*.

MLP see *multilayer Perceptron*.

modality relating to a particular type of sensory (or sometimes motor) system.

model simplified and approximative description of a system or process, based on a finite set of essential variables and their analytically definable behavior.

momentum factor multiplier of an auxiliary term in learning, when the weight change does not only involve the gradient-step weight change but also an additional term that describes the previous weight change. The purpose of this term is to speed up and stabilize convergence.

monism doctrine, according to which material and mental belong to the same kind of existence or substance.

motoneuron large neuron that controls muscles.

motor cortex area in the cerebral cortex that controls motor functions.

MSB most significant bit (in a binary number).

MST see *minimal spanning tree.*

multivibrator oscillator producing nonsinusoidal waveforms, often using alternating switches.

multilayer feedforward network an architecture of ANNs consisting of successive layers of neurons that receive their inputs from a major part of the previous layer, except the input layer that receives input signals directly.

multilayer Perceptron multilayer feedforward network formed of *Perceptrons.*

multispectral relating to a multitude of pictures taken of the same scene at different wavelengths, especially in remote sensing.

myelin white fatlike substance covering certain nerve fibers.

neighborhood set of neurons (eventually including the neuron itself to which the neighborhood is related) located spatially in the neural network up to a defined radius from the neuron.

neighborhood function function describing the strength of control exercized by an active neuron onto the *synaptic plasticity* of neighboring neurons, as a function of distance and time, and relating to the network topology.

Neocognitron multilayer feedforward network introduced by Fukushima [11.2].

neural gas set of codebook vectors, the values of which depend on both input signals and mutual interactions defined in the input space.

neural model biophysical model for a biological neuron.

neurocomputer computer or computing device particularly suitable for solving ANN or neural-model equations. It can be an analog or digital computer and implemented by electronic, optical, or other means.

neurotransmitter see *chemical transmitter.*

neuron any of the numerous types of specialized cell in the brain or other nervous systems that transmit and process neural signals. The nodes of artificial neural networks are also called neurons.

node location of neuron or processing element in a network, especially in an ANN.

nondeterministic opposite of *deterministic.*

nonlinearly separable classes classes, the samples of which can be divided into disjoint sets using smooth, algebraically defined nonlinear discriminant functions.

nonparametric classifier classification method that is not based on any mathematical functional form for the description of class regions, but directly refers to the available exemplary data.

norm generalized length measure of a vector or matrix.

NP-complete problem problem of a certain complexity class. If an algorithm solves the problem in a time that is a deterministic polynomial function of the length of input, it is called a P-problem; NP means "nondeterministic" polynomial time. The most difficult problems are NP-complete problems, which usually require a computing time that is an exponential function of the length of input.

nucleus formation or mass of neural cells that samples and processes sensory information, takes part in preattentive information processing, and controls attention or states or functions of other parts of the brain. One of the main purposes of a nucleus is to control optimal operating conditions of the neural systems. It often has neurons with particular types of chemical transmitters. The main distinction between ganglia and nuclei is that the former are generally situated more peripherally, but there are also ganglia in the central nervous system.

observation matrix data matrix, the columns of which represent sets of signals or measurements.

OCR see *optical character reader*.

off-center relating to the negative part of the *Mexican-hat function*, or inhibitory part of receptive field, when the latter is surrounded by the excitatory part.

off-line relating to computing operations such as training performed prior to use of the device.

on-center relating to the positive part of the *Mexican-hat function* or excitatory part of a receptive field, when the latter is surrounded by the inhibitory part.

on-line relating to direct connection between machine and its environment and its operation in real time.

ontogenetic relating to the development of an individual.

operator mathematical concept that specifies a mapping of a certain class using algebraic expressions of symbolically defined basic operations.

optical character reader device for interpreting printed or handwritten letters and numerals.

optical flow field formed of velocity vectors of all points in a picture.

optimal associative mapping transformation of given vectorial input items into wanted output items such that, for instance, the crosstalk from other items is minimized.

organelle microscopic biological organ, usually located within a cell and associated with metabolism or energy supply of it.

orthogonality property of a pair of vectors when their inner product is zero.

orthonormality property of being orthogonal and normalized.

outlier statistical sample that does not belong to the theoretical distribution.

output layer layer of neural cells, each of which produces an output signal.

output vector vector formed of output signal values.

parametric classifier classification method in which the class regions are defined by specified mathematical functions involving free parameters.

parsing identification of the syntactic structure of an expression.

pattern ordered set of signal values used to represent an item.

pattern recognition in the most general sense the same as *artificial perception*.

PCA see *principal-component analyzer*.

Peano curve space-filling curve having a fractal form.

pel abbreviation for *pixel*.

perception identification and interpretation of an observable item or occurrence on the basis of sensory signals and memory contents.

Perceptron adaptive element for multilayer feedforward networks introduced by Rosenblatt [2.26].

performance index measure or objective function describing global, analytically expressible performance, such as average expected residual error, quantization error, classification accuracy, etc. On the other hand, speed of learning or classification, memory demand, number of structures in a network etc. are specific indicators that occasionally may be involved in the expressions of the performance index as extra restrictive conditions, but usually depend on the quality of hardware and are not mathematical concepts.

perplexity information-theoretic measure of *branching factor*.

PFM see *pulse frequency modulation*.

phoneme smallest meaningful unit of speech.

phoneme map SOM for the representation of phonemes.

phonemic classifier algorithm or circuit that converts the output of the acoustic processor into a string of phonemes.

phonemic labeling assignment of piece of the speech waveform into a phonemic class.

phonetic relating to speech.

phonetic knowledge representation modeling of expert judgement on the phonetic level (e.g., in spectrogram reading), utilization of linguistic information.

phonological unit any meaningful unit of speech (phoneme, syllable, etc.).

phonotax joining adjacent phonological units in natural speech; constraints in the possible ways of joining phonological units (see *syntax*).

phonotopic map two-dimensional array of processing units, in which every unit is sensitized and labeled to a phonemic token, the spatial order of the labels corresponding to the spectral similarity of the corresponding tokens.

phylogenetic relating to the development of species.

pipeline multiprocessor computer architecture, in which several programs relating to the same data are run in a sequence, and in which, at a given instant of time, the pipeline contains different data at successively more advanced phases of processing.

pitch of speech fundamental (exciting) frequency caused by vocal chords.

pixel numerically expressed picture element.

plasticity modifiability (of synapses), see *synaptic plasticity*.

point spread function kernel describing the transformation of a point in an image processing system.

pointer stored address reference to a memory or table.

polysynaptic connection cascade of synaptic connections made by several neurons in series.

postnatal after the birth.

postprocessing posterior operation performed on the outputs from, say, a neural network. Its purpose may be to segment sequences of output results into discrete elements, for instance to correct text on the basis of grammatical rules, or to improve labeling in image analysis.

postsynaptic relating to the synaptic junction on the side of the receiving neuron.

postsynaptic potential abrupt stepwise change in the electric membrane voltage of the postsynaptic neuron, caused by the depolarizing or hyper-polarizing effect of a single neural impulse at a presynaptic axon.

pragmatic relating to a consequence of an action.

preattentive at the level of automatic processing of sensory information, prior to voluntary or attentive selection or processing.

prenatal prior to birth.

preprocessing set of normalization, enhancement, feature-extraction, or other similar operations on signal patterns before they are presented to the central nervous system or other pattern recognition systems.

presynaptic relating to the synaptic junction on the side of the transmitting neuron.

primitive element, symbol, item, function, or any other similar elementary concept from which other concepts are derived.

principal cell the largest neural cell type in some major part of the brain. In the cerebral cortex and hippocampus, the pyramidal cells constitute the principal cells; in the cerebellum the Purkinje cells constitute the principal cells, respectively. They send axons to other parts of the brain and to the peripheral nervous system, whereas interneurons only connect locally within the cortical areas.

principal component coefficient in the expansion of a pattern vector in terms of the eigenvectors of the correlation matrix of a class of input patterns.

probabilistic reasoning decision-making based on probability calculus.

processing element local computing device, usually provided with local memory, in particular for the implementation of the function of a neuron in an ANN.

processor computing device, especially for the execution of a program in digital computing.

production rule see *rule*.

projection ordered mapping from a higher-dimensional to a lower-dimensional manifold. In neuroanatomy: Directed connectivity between brain areas or parts of the brain.

projector projection operator. Operator that defines a mathematical *projection*.

prosodics variation other than spectral in speech (such as pattern of intensity, etc.).

prototype typical sample from a class of items used for training.

pseudocolor sample of color values used to represent a classified zone in a picture or visual diagram and to enhance its visual perception.

pseudoinverse matrix Moore-Penrose generalized inverse matrix [1.1], [11.3] that always exists for any given matrix. Many estimators, projectors, and associative-memory models are based on it.

PSP see *postsynaptic potential*.

pulse frequency modulation representation of a signal value by a train of successive impulses, the inverse distances of which are proportional to the average signal value over the interpulse interval.

pyramidal cell the most important type of neuron in the *cerebral cortex* and *hippocampus*.

quantization error norm of difference of a signal vector from the closest *reference vector* in signal space.

quasiphoneme labeled segment of speech waveform identified at one of periodic sampling instants. The label describes similar acoustic quality as that of a *phoneme* in the middle of its stationarity region.

radial basis function particular basis function used in multilayer feedforward networks. Its value depends on the Euclidean distance of a pattern vector from a reference vector in the signal space of a layer.

ramification branching.

range image image in which the value of every pixel represents the distance of the corresponding target point from the camera.

rank number of linearly independent rows or columns in a matrix.

receptive field set of cells in a sensory organ from which a neuron or neural unit in the brain receives signals in an orderly fashion. Sometimes a domain in the signal space, such as a cone in space relative to retina from which a neuron receives stimuli.

receptive surface layered set of receptors, such as the retina, the basilar membrane, or the set of pressure receptors on the skin.

receptor any specific component in the nervous system, such as sensory cell or agglomeration of chemical macromolecules, that receives information and responds to it in some way. The receptor cell converts physical or chemical stimuli into electric signals.

recurrent network see *feedback network*.

recurrent signal see *feedback signal*.

recursion an iterative step in defining the new value of an expression as a function of its old values.

redundant hash addressing hash coding scheme for error-tolerant search of strings.

reference vector see *codebook vector*.

refractory period or **refractory phase** recovery period following a response or neural impulse before a second response or neural impulse is possible.

reinforcement learning learning mode in which adaptive changes of the parameters due to reward or punishment depend on the final outcome of a whole sequence of behavior. The results of learning are evaluated by some performance index.

relation ordered set of interrelated attributes.

relational database memory in which "relations" or *n*-tuples of values of related items are stored in the same memory location, accessible on the basis of any member of the relation.

relaxation iterative determination of values in an array or network by recursively averaging over neighboring elements or nodes; state transition that seeks an energetic minimum; iterative determination of best labels in a thematic image as a function of previous labeling.

remote sensing measurements made from a distant station, such as a satellite or an airplane.

representation encoding of information for computing.

response output signal from a network, relating to a meaningful input.

retina essentially two-dimensional neural network in the eye of higher animals onto which optical images are projected, and in which they are converted into neural signals.

retinotectal mapping ordered set of neural connections between *retina* and *tectum*.

retrograde messenger any messenger that transfers information from the postsynaptic neuron to the presynaptic neuron. Some kind of retrograde messenger action is necessary, if the synaptic plasticity of the neuron shall be describable by Hebb's law or any of its variants.

reward-punishment scheme supervised learning strategy in which corrections are either toward or away from the reference value, depending on correctness of classification result.

RHA see *redundant hash addressing.*

rms value see *root-mean-square value.*

root-mean-square value square root of the sum of squares of a sequence divided by its length; average effective value of signal describing its power.

row horizontal linear array.

rule prescribed elementary transformation.

Sammon's mapping clustering method [1.34] that visualizes mutual distances between high-dimensional data in a two-dimensional display.

scalar product see *dot product, inner product.*

scalar quantization special case of vector quantization when the data are one-dimensional.

scene analysis see *image understanding.*

schema large, complex unit of information used to organize data into knowledge. It may also contain nonspecified, unknown variables.

segmentation operation performed on signals or images to find meaningful intervals or areas in them; such intervals or areas can then be labeled uniformly.

self-organization in the original sense, simultaneous development of both structure and parameters in learning.

self-organizing feature map *SOM* used especially to extract features from input signals.

self-organizing map result of a nonparametric regression process that is mainly used to represent high-dimensional, nonlinearly related data items in an illustrative, often two-dimensional display, and to perform unsupervised classification and clustering.

semantic relating to the meaning or role of symbols.

semantic net abstract linguistic graph for verbal expressions where the nodes often represent verbs and the links other specifiers.

sensor detector of input information.

sensory cortex part of the cortex that receives sensory signals.

sensory register volatile, temporary store for primary sensory signals.

sequential circuit device or network that contains feedback paths and is multistable. Its output states can be made to occur in sequences, the structures of which are controlled by sequences of input signals. Memory elements and control circuits of digital computer technology belong to sequential circuits.

servomechanism power steering device based on closed-loop automatic control for the positioning, orientation, or navigation of a mechanical system, whereby very little energy is needed to set the command values.

shift register storage or buffer consisting of a number of elementary storage locations, each of which is capable of holding one bit, symbol, or real number and communicating it to the next location. Sequential signal values enter one end of the storage and at each clocking signals the contents of each storage location are transferred to the next location in the same direction. The contents of the last location overflow (are lost), unless the operation is end-around, in which case the overflowing contents are copied to the input location.

short-term memory working memory in man and higher animals cooperating with *sensory registers* and *long-term memory*. Its typical retention time is on the order of a few minutes and it can process simultaneously 5 to 9 symbolic items.

sigmoid function monotonically increasing, continuous and continuously differentiable scalar-valued function of a real scalar variable, the plot of which has a characteristic S-shape. It is bounded at low and high values. The so-called logistic function is often used to describe it mathematically. The sigmoid function is often used to describe the output nonlinearities of ANNs.

simulate to imitate the behavior of a system by its model.

simulated annealing thermodynamically motivated procedure for the adjusting of the internal state of an ANN such that the global error or energy function can either decrease or increase. The purpose is to find the global minimum of the energy function.

singular matrix matrix having zero *rank*.

SLM see *spatial light modulator*.

SOFM see *self-organizing feature map*.

SOM see *self-organizing map*.

soma cell body.

sonorant see *voiced*.

span to define a manifold.

spatial relating to space.

spatial light modulator controller of the local intensity of light when light is transmitted through a device. The control can be effected by electronic, optical, or electron-beam means.

spatiotemporal relating to or depending on both space and time.

speaker recognition verification of a speaker on the basis of his or her voice characteristics.

speaking mode the way of speaking under particular limitations on the utterances (e.g., isolated-word mode, connected-word mode).

spectrum set of eigenvalues of an operator; distribution of frequency components of a signal.

speech understanding interpretation of the meaning of an utterance; speech-induced actions or response; parsing of spoken utterances on a higher grammatical level.

spiking triggering *action pulses*.

spine bulge or bag on the postsynaptic membrane of neuron at the excitatory synapse. Its purpose is to enclose certain organelles necessary for the metabolism of the synapse, as well as to control the efficacy of the synapse by the change of its size and shape.

spurious state unwanted attractor that does not represent the correct outcome in a state-transfer neural network.

squashing nonlinear transformation of a signal scale in order to optimize the overall accuracy.

state often the same as output vector in a feedback system. In some models such as HMM it may also mean a set of signals internal to the system.

state transition transformation of one state into another in a dynamical system during some time interval.

steepest descent minimization of an objective function of the system by changing its parameters in the direction of its negative gradient with respect to the parameter vector.

step response output from a dynamic system in response to a threshold-function input with unit height.

stimulus something that temporarily influences the activity of an organism.

stochastic random, having a certain probability of occurrence.

stochastic approximation gradient-step optimization scheme in which the gradient is evaluated from stochastic samples.

stop consonant, in the articulation of which the breath passage is completely closed.

stopping halting an iterative algorithm, when training is considered complete or optimal.

stratification confinement in layers.

string sequence of symbols.

subcortical connection connection between neurons made through the white matter.

subsymbolic relating to representation or processing of information whereby the elements have no semantic content but constitute lower-level statistical descriptions of elementary features or properties.

subsynaptic membrane postsynaptic cell membrane directly below the *synapse*.

sulcus groove of the cortex between gyri.

superimpose to add linearly to previous values.

superposition linear addition to previous values.

superresolution separability of points in imaging exceeding the natural, physically limited resolution.

supervised learning learning with a teacher; learning scheme in which the average expected difference between wanted output for training samples, and the true output, respectively, is decreased.

supplementary motor cortex brain area frontal to the motor cortex, mediating mental commands (intentions) and converting them into motor and behavioral actions.

symbolic relating to representation or processing of information, whereby the elements of representation refer to complete items or their complexes.

synapse microscopic neurophysiological organ that creates the signal connection between two or more neurons; sometimes meaning the connection as a whole. There are two main types of synapses: *excitatory*, whereby the signal coupling is positive, and *inhibitory*, whereby the signal coupling is negative. An axon usually ends in a *synaptic terminal* (synaptic knob, synaptic bouton). Electrical neural signals make the synapse release chemical transmitter molecules that affect chemical receptor molecules of the receiving neuron. The latter control the transflux of light ions through the postsynaptic cell membrane and thus the membrane voltage on which the triggering of the neuron depends. There are also synapses that make a contact with another synapse; their gating operation is called *presynaptic control*.

synaptic cleft narrow spacing between the synaptic terminal and the postsynaptic (subsynaptic) membrane.

synaptic connection see *synapse*.

synaptic efficacy amount to which a presynaptic neuron controls a postsynaptic neuron through a single synapse.

synaptic junction see *synapse*.

synaptic plasticity the same as modifiability or adaptability of a synapse (whereby adaptation means changes in parameters, not evolution of forms), relating to changes in the efficacy of a synapse in neurophysiology.

synaptic terminal see *synapse*.

synaptic transmission complex act in propagation or processing of neural signals, whereby electric impulses first cause the release of chemical transmitter from the presynaptic terminal, which then affects the chemical receptor molecules at the postsynaptic membrane. The receptors control transflux of ions through the membrane, resulting in change of membrane voltage and eventually triggering of the postsynaptic neuron.

synaptic transmittance amount of depolarizing or hyperpolarizing effect in controlling the efficacy of a synapse, usually relating to an individual postsynaptic potential (PSP).

synaptic weight see *synaptic efficacy*.

syntactic relating to *syntax*.

syntax set of rules that describe how symbols can be joined to make meaningful strings; systematic joining of elements; especially combination of elements defined by a grammar (set of rules).

system identification determination of structure and/or parameters of a system from a sufficient number of observed input-output relations.

systolic array usually a two-dimensional array processor in which the computations are pipelined in both horizontal and vertical dimensions.

Tanimoto similarity specific similarity measure for unordered sets.

taxonomy orderly classification of items according to their similarity relationships.

TDNN see *time-delay neural network*.

tectum dorsal part of the midbrain.

template matching recognition of patterns using matched filters.

temporal relating to time; the temporal cortex means the lateral area of the cortex.

test set set of data items set aside for testing a learned system. This set must be statistically independent of the training set.

texture statistically and/or structurally defined repetitive set of elementary patterns on a surface.

thalamus part of brain situated below the cerebral cortex. It relays signals to the latter and also receives feedback projections from it.

thematic classification classification and labeling of groups of pixels (or other pattern elements) according to the attributes (theme) they represent, such as types of crop as seen from an aerial or satellite image of a field.

threshold-logic unit device that gives one of two possible responses to an input pattern depending on its classification.

threshold function the same as *Heaviside function* shifted to a threshold value and having an arbitrary step size.

threshold triggering abrupt transition of output from a low to a high value when the input activation exceeds a threshold value.

threshold value discriminative value for activation; triggering of a device occurs when activation exceeds the threshold value.

time constant characteristic time unit in dynamic processes; eigenvalue of dynamic operator (in the time domain); in the very simplest case (step response), the time to reach the value whereby the difference of the variable from its asymptotic value has reached a fraction of $1 - 1/e$ of the initial difference.

time-delay neural network usually a multilayer feedforward network, the layers of which perform static operations, but in which sequentially occurring signal values are first converted into parallel signal vectors using shift registers or equivalent delay lines.

TLU see *threshold-logic unit*.

top-down relating to *deductive inference*, i.e. from generals to particulars.

topology-preserving map see *SOM*.

tract system of parts acting in concert to implement some function; bundle or plexus of nerve fibers having a common origin and termination. See also *vocal tract*.

training forced learning, teaching.

training set set of data used as inputs in an adaptive process that teaches a neural network.

transfer function usually the Fourier transform of the impulse response or point spread function of a linear system.

transmission line linearly distributed medium along which signals are propagated.

transmitter vesicle elementary package of chemical transmitter. The chemical transmitter molecules in the synaptic terminal are packaged into small droplets, enclosed by a membrane consisting of similar molecules that form the cell membrane. In synaptic transmission the membranes of these vesicles are fused with the membrane of the presynaptic terminal, releasing their contents into the *synaptic cleft*.

traveling-salesman problem optimization problem in which a network of nodes has to be traveled along the shortest path.

tree structure branching graph structure in which lower nodes are subordinates to unique higher nodes.

trigger to set a course of action or to be released from a latched state; to elicit a signal when the input activation exceeds a certain threshold value.

TSP see *traveling-salesman problem*.

ultrametric relating to distance measure defined along a hierarchical-clustering tree structure.

unary attribute especially in image analysis: property of a single pixel.

unary number number code that defines the numerical value by the number of concatenated 1s; e.g., $111 = 3$.

unsupervised learning learning without *a priori* knowledge about the classification of samples; learning without a teacher. Often the same as formation of clusters, whereafter these clusters can be labeled. Also optimal allocation of computing resources when only unlabeled, unclassified data are input.

updating modification of parameters in learning; also bringing data up-to-time.

validation comparison of ANN performance with requirements set up in defining a task or objective, and evaluation of its usefulness. Validation of data: Selection of reliable data sets, especially for training an algorithm.

vector quantization representation of a distribution of vectorial data by a smaller number of reference or codebook data; optimal quantization of the signal space that minimizes the average expected quantization error.

ventral situated on the anterior or lower side.

verification of data determination of classification for data used in supervised training or calibration.

visual cortex central part of the occipital cortex where most of the neural functions of vision are situated.

VLSI very-large-scale integration (of electronic circuits).

vocal relating to non-turbulent sound in speech, especially in vowels and semivowels; "pseudo-vocal" relates to nasals (m, n, η) and liquids (l, r).

vocal cord either of the two pairs of folds in the glottis that produce the speech voice.

vocal tract acoustic transmission line encompassing the space from the vocal chords to lips, including the nasal cavity.

voiced excited or driven by pulsed airflow controlled by the vocal chords.

voiceless phonetic excitation driven by air friction only.

Voronoi polygon one partition in a (two-dimensional) *Voronoi tessellation*.

Voronoi polyhedron one partition in a *Voronoi tessellation*.

Voronoi set set of data vectors that are closest to one particular codebook vector in *Voronoi tessellation*.

Voronoi tessellation division of the data space by segments of hyperplanes that are midplanes between closest codebook vectors.

VQ see *vector quantization*.

wavelet basis function type for the representation of pieces of periodic signals. It has often a sinusoidal functional form, amplitude-modulated by a Gaussian function.

weight vector real-valued parametric vector that usually defines the input activation of a neuron as a dot product of the input vector and itself.

white matter brain tissue below the gray matter, mainly consisting of myelinated axons.

whitening filter operation or device that transforms a stochastic vector in such a way that its elements become statistically uncorrelated.

Widrow-Hoff rule learning law for Adaline elements where the LMS error criterion is applied, and where the optimization of its linear part is performed by *Robbins-Monro stochastic approximation* [1.43].

window interval in time from which elements of observations are picked. In the LVQ2 algorithms: a bounded zone between two neighboring codebook vectors of different classes.

winner neuron in competitive-learning neural networks that detects the maximum activation for, or minimum difference between input and weight vectors.

winner-take-all function maximum or minimum selector for input activation.

word spotting recognition of individual words from continuous speech.

WTA see *winner-take-all function*.

References

Chapter 1

[1.1]   A. Albert: *Regression and the Moore-Penrose Pseudoinverse* (Academic, New York, NY 1972)

[1.2]   T. Lewis, P. Odell: *Estimation in Linear Models* (Prentice-Hall, Englewood Cliffs, NJ 1971)

[1.3]   E. Oja: *Subspace Methods of Pattern Recognition* (Research Studies Press, Letchworth, UK 1983)

[1.4]   E. Oja, J. Karhunen: *T. Math. Analysis and Applications* **106**, 69 (1985)

[1.5]   D. Rogers, T. Tanimoto: Science **132**, 10 (1960)

[1.6]   L. Ornstein: J. M. Sinai Hosp. **32**, 437 (1965)

[1.7]   K. Sparck-Jones: *Automatic Keyword Classification and Information Retrieval* (Butterworth, London, UK 1971)

[1.8]   J. Minker, E. Peltola, G. A. Wilson: Tech. Report 201 (Univ. of Maryland, Computer Sci. Center, College Park, MD 1972)

[1.9]   J. Liénard, M. Młouka, J. Mariani, J. Sapaly: In *Preprints of the Speech Communication Seminar, Vol.3* (Almqvist & Wiksell, Uppsala, Sweden 1975) p. 183

[1.10]   T. Tanimoto: Undocumented internal report (IBM Corp. 1958)

[1.11]   J. Łukasiewicz: Ruch Filos. **5**, 169 (1920)

[1.12]   E. Post: Am. J. Math. **43**, 163 (1921)

[1.13]   L. Zadeh: IEEE Trans. Syst., Man and Cybern. SMC-**3**, 28 (1973)

[1.14]   R. Hamming: Bell Syst. Tech. J. **29**, 147 (1950)

[1.15]   V. Levenshtein: Sov. Phys. Dokl. **10**, 707 (1966)

[1.16]   T. Okuda, E. Tanaka, T. Kasai: IEEE Trans. C-**25**, 172 (1976)

[1.17]   T. Kohonen: *Content-Addressable Memories* (Springer, Berlin, Heidelberg 1980)

[1.18]   T. Kohonen, E. Reuhkala: In *Proc. 4IJCPR, Int. Joint Conf. on Pattern Recognition* (Pattern Recognition Society of Japan, Tokyo, Japan 1978) p. 807

[1.19]   O. Ventä, T. Kohonen: In *Proc. 8ICPR, Int. Conf. on Pattern Recognition* (IEEE Computer Soc. Press, Washington, D.C. 1986) p. 1214

[1.20]   T. Kohonen: *Pattern Recogn. letters* **3**, 309 (1985)

[1.21]   H. Hotelling: J. Educat. Psych. **24**, 498 (1933)

[1.22]   H. Kramer, M. Mathews: IRE Trans. Inform. Theory IT-**2**, 41 (1956)

[1.23]   E. Oja: Neural Networks **5**, 927 (1992)

[1.24]   J. Rubner, P. Tavan: Europhys. Lett. **10**, 693 (1989)

[1.25]   A. Cichocki, R. Unbehauen: Neural Networks for Optimization and Signal Processing (John Wiley, New York, NY 1993)

[1.26]   World Bank: *World Development Report 1992* (Oxford Univ. Press, New York, NY 1992)

[1.27]   S. Kaski: PhD Thesis, Acta Polytechnica Scandinavica Ma 82 (The Finnish Academy of Technology, Espoo 1997)

[1.28]   G. Young, A. S. Householder: Psychometrika **3**, 19 (1938)

[1.29]   W. S. Torgerson: Psychometrika **17**, 401 (1952)

[1.30]   J. B. Kruskal, M. Wish: *Multidimiensional Scaling* Sage University Paper Series on Quantitative Applications in the Social Sciences No.07-011. (Sage Publications, New bury Park, CA 1978)

[1.31]   J. de Leeuw, W. Heiser: In *Handbook of Statistics*, ed. by P. R. Krisnaiah, L. N. Kanal (North- Holland, Amsterdam 1982) p. II-317

[1.32]   M. Wish, J. D. Carroll: *ibid.*

[1.33]   F. W. Young: In *Encyclopedia of Statistical Sciences*, ed.by S. Koto, N. L. Johnson, C. B. Read (wiley, New York NY 1985) p. V-649

[1.34]   J. W. Sammon Jr.: IEEE Trans. Comp. C-**18**, 401 (1969)

[1.35]   J. B. Kruskal: Psychometrika **29**, 1 (1964)

[1.36]   R. N. Shepard: Psychometrika **27**, 125; **27**, 219 (1962)

[1.37]   J. K. Dixon: IEEE Trans. Syst., Man and Cybern. SMC-**9**, 617 (1979)

[1.38]   T. Hastie, W. Stuetzle: J. American Statistical Association **84**, 502 (1989)

[1.39]   F. Mulier, V. Cherkassky: Neural Computing **7**, 1165 (1995)

[1.40]   H. Ritter, T. Martinetz, K. Schulten: *Neural Computation and Self-Organizing Maps: An Introduction* (Addison-Wesley, Reading, MA 1992)

[1.41]   P. Demartines: PhD Thesis (Institut National Polytechnique de Grenoble, Grenoble, France 1994)

[1.42]   P. Demartines, J .Hérault: IEEE Trans. Neural Networks **8**, 148 (1997)

[1.43]   H. Robbins, S. Monro: Ann. Math. Statist. **22**, 400 (1951)

[1.44]   B. Widrow: In *Self-Organizing Systems 1962*, ed. by M. Yovits, G. Jacobi, G. Goldstein (Spartan Books, Washington, D.C 1962) p. 435

[1.45]   Y. Tsypkin: *Adaptation and Learning in Cybernetic Systems* (Nauka, Moscow, USSR 1968)

[1.46]   R. Tryon, D. Bailey: *Cluster Analysis* (McGraw-Hill, New York, NY 1973)

[1.47]   M. Anderberg: *Cluster Analysis for Applications* (Academic, New York, NY 1973)

[1.48]   E. Bijnen: *Cluster Analysis, Survey and Evaluation of Techniques* (Tilbury Univ. Press, Tilbury, Netherlands 1973)

[1.49]   H. Bock: *Automatische Klassifikation* (Vandenhoeck Ruprecht, Göttingen, Germany 1974)

[1.50]   E. Diday, J. Simon: In *Digital Pattern Recognition*, ed. by K. S. Fu (Springer, Berlin, Heidelberg 1976)

[1.51]   B. Duran, P. Odell: *Cluster Analysis, A Survey* (Springer, Berlin, Heidelberg 1974)

[1.52]   B. Everitt: *Cluster Analysis* (Heineman Educational Books, London, UK 1977)

[1.53]   J. Hartigan: *Clustering Algorithms* (Wiley, New York, NY 1975)

[1.54]   H. Späth: *Cluster Analysis Algorithms for Data Reduction and Classification of Objects* (Horwood, West Sussex, UK 1980)

[1.55]   D. Steinhauser, K. Langer: *Clusteranalyse, Einführung in Methoden und Verfahren der automatischen Klassifikation* (de Gruyter, Berlin, Germany 1977)

[1.56]   V. Yolkina, N. Zagoruyko: R.A.I.R.O. Informatique **12**, 37 (1978)

[1.57]   S. Watanabe, P. Lambert, C. Kulikowski, J. Buxton, R. Walker: In *Computer and Information Sciences*, ed. by J. Tou (Academic, New York, NY 1967) Vol. 2, p. 91

[1.58] E. Oja, J. Parkkinen: In *Pattern Recognition Theory and Applications* Ed. by P. A. Devijver, J. Kittler (Springer-Verlag, Berlin, Heidelberg 1987) p. 21 NATO ASI Series, Vol. F30

[1.59] E. Oja, T. Kohonen: In *Proc. of the Int. Conf. on Neural Networks* (IEEE 1988) p. I-277

[1.60] J. Laaksonen, E. Oja: to be published in the Proceedings of ICANN96 (Bochum, Germany, July 1996)

[1.61] K. Fukunaga, W.L. Koontz: IEEE Trans. Comp. C-**19**, 311 (1970)

[1.62] C.W. Therrien: IEEE Trans. Comp. C-**24**, 944 (1975)

[1.63] T. Kohonen, G. Nèmeth, K.-J. Bry, M. Jalanko, H. Riittinen: Report TKK-F-A348 (Helsinki University of Technology, Espoo, Finland 1978)

[1.64] T. Kohonen, G. Nèmeth, K.-J. Bry, M. Jalanko, E. Reuhkala, S. Haltsonen: In *Proc. ICASSP'79*, ed. by R. C. Olson (IEEE Service Center, Piscataway, NJ 1979) p. 97

[1.65] T. Kohonen, H. Riittinen, M. Jalanko, E. Reuhkala, S. Haltsonen: In *Proc. 5ICPR, Int. Conf. on Pattern Recognition*, ed. by R. Bajcsy (IEEE Computer Society Press, Los Alamitos, CA 1980) p. 158

[1.66] E. Oja: In *Electron. Electr. Eng. Res. Stud. Pattern Recognition and Image Processing Ser. 5* (Letchworth, UK 1984) p. 55

[1.67] C. Therrien: Tech. Note 1974-41 (Lincoln Lab., MIT, Lexington, MA 1974)

[1.68] S. Watanabe, N. Pakvasa: In *Proc. 1st Int. Joint Conf. on Pattern Recognition* (IEEE Computer Soc. Press, Washington, DC 1973) p. 25

[1.69] T. Kohonen, H. Riittinen, E. Reuhkala, S. Haltsonen: Inf. Sci. **33**, 3 (1984)

[1.70] A. Gersho: IEEE Trans. Inform. Theory IT-**25**, 373 (1979)

[1.71] Y. Linde, A. Buzo, R. Gray: IEEE Trans. Communication COM-**28**, 84 (1980)

[1.72] R. Gray: IEEE ASSP Magazine **1**, 4 (1984)

[1.73] J. Makhoul, S. Roucos, H. Gish: Proc. IEEE **73**, 1551 (1985)

[1.74] P. Zador: IEEE Trans. Inform. Theory IT-**28**, 139 (1982)

[1.75] T. Kohonen: In *Artificial Neural Networks*, ed. by T. Kohonen, K. Mäkisara, O. Simula, J. Kangas (North-Holland, Amsterdam, Netherlands 1991) p. II-981

[1.76] G. Voronoi: J. reine angew. Math. **134**, 198 (1908)

[1.77] T. Kohonen: In *Proc. 8ICPR, Int. Conf. on Pattern Recognition* (IEEE Computer Soc. Press, Washington, DC 1986) p. 1148

[1.78] J. Rissanen: Ann. Statist. **14**, 1080 (1986)

Chapter 2

[2.1] R. Sorabji: *Aristotle on Memory* (Brown Univ. Press, Providence, RI 1972)

[2.2] W. S. McCulloch, W. A. Pitts: Bull. Math. Biophys. **5**, 115 (1943)

[2.3] B. G. Farley, W. A. Clark: In Proc. 1954 Symp. on Information Theory (Inst. Radio Engrs., 1954) p. 76

[2.4] F. Rosenblatt: Psychoanal. Rev **65**, 386 (1958)

[2.5] B. Widrow, M. E. Hoff: In *Proc. 1960 WESCON Convention.* (Wescon Electronic Show and Convention, San Francisco, 1960) p. 60

[2.6] E. R. Caianiello: J. Theor. Biology **2**, 204 (1961)

[2.7] K. Steinbuch61 Kybernetik **1**, 36 (1961)

[2.8] P. Werbos: PhD Thesis (Harvard University, Cambridge, MA 1974)

[2.9]   S. P. Lloyd: Unpublished paper (1957); reprinted in IEEE Trans. Inf. Theory **IT-28**, No.2, 129 (1982)

[2.10]   E. W. Forgy: Biometrics **21**, 768 (1965)

[2.11]   R. Didday: PhD Thesis (Stanford University, Stanford, CA 1970)

[2.12]   R. Didday: Math. Biosci. **30**, 169 (1976)

[2.13]   F. Montalvo: Int. J. Man-Machine Stud. **7**, 333 (1975)

[2.14]   S. Amari, M. A. Arbib: In *Systems Neuroscience*, ed. by J. Metzler (Academic, New York, NY 1977) p. 119

[2.15]   S. Grossberg: Biol. Cyb. **23**, 121 (1976)

[2.16]   S. Grossberg: Biol. Cyb. **23**, 187 (1976)

[2.17]   G. Carpenter, S. Grossberg: Computer **21**, 77 (1988)

[2.18]   G. Carpenter, S. Grossberg: Neural Networks **3**, 129 (1990)

[2.19]   J. A. Anderson, E. Rosenfeld: *Neurocomputing* (MIT Press, Cambridge, MA 1988)

[2.20]   S. Grossberg: Neural Networks **1**, 17 (1988)

[2.21]   R. Hecht-Nielsen: *Neurocomputing* (Addison-Wesley, Reading, MA 1990)

[2.22]   M. Nass, L. Cooper: Biol. Cyb. **19**, 1 (1975)

[2.23]   R. Pérez, L. Glass, R. J. Shlaer: J. Math. Biol. **1**, 275 (1975)

[2.24]   A. Hodgkin: J. Physiol. **117**, 500 (1952)

[2.25]   G. Shepherd: *The Synaptic Organization of the Brain* (Oxford Univ. Press, New York, NY 1974)

[2.26]   F. Rosenblatt: *Principles of Neurodynamics: Perceptrons and the Theory of Brain Mechanisms* (Spartan Books, Washington, D.C. 1961)

[2.27]   D. E. Rumelhart, J. L. McClelland, and the PDP Research Group (eds.): *Parallel Distributed Processing: Explorations in the Microstructure of Cognition, Vol. 1: Foundations* (MIT Press, Cambridge, MA 1986)

[2.28]   T. Poggio, F. Girosi: Science **247**, 978 (1990)

[2.29]   J. Hopfield: Proc. Natl. Acad. Sci. USA **79**, 2554 (1982)

[2.30]   D. Ackley, G. Hinton, T. Sejnowski: Cognitive Science **9**, 147 (1985)

[2.31]   B. Kosko: IEEE Trans. on Systems, Man and Cybernetics **SMC 18**, 49 (1988)

[2.32]   C. v.d. Malsburg: Kybernetik **14**, 85 (1973)

[2.33]   D. Willshaw, C. v.d. Malsburg: Proc. R. Soc. London **B 194**, 431 (1976)

[2.34]   S.-i. Amari: Bull. Math. Biology **42**, 339 (1980)

[2.35]   T. Kohonen: In *Proc. 2SCIA, Scand. Conf. on Image Analysis*, ed. by E. Oja, O. Simula (Suomen Hahmontunnistustutkimuksen Seura r.y., Helsinki, Finland 1981) p. 214

[2.36]   T. Kohonen: Biol. Cyb. **43**, 59 (1982)

[2.37]   T. Kohonen: In *Proc. 6ICPR, Int. Conf. on Pattern Recognition* (IEEE Computer Soc. Press, Washington, D.C. 1982) p. 114

[2.38]   T. Kohonen: Report TKK-F-A601 (Helsinki University of Technology, Espoo, Finland 1986)

[2.39]   T. Kohonen, G. Barna, R. Chrisley: In *Proc. ICNN'88, Int. Conf. on Neural Networks* (IEEE Computer Soc. Press, Los Alamitos, CA 1988) p. I-61

[2.40]   T. Kohonen: Neural Networks **1**, 3 (1988)

[2.41]   T. Kohonen: Neural Networks **1**, 303 (1988)

[2.42]   E. Oja: Neurocomputing **17**, 25 (1997)

[2.43]   J. Karhunen, E. Oja, L. Wang, R. Vigario, J. Joutsensalo: IEEE Trans. On Neural Networks **8**, 486 (1997)

[2.44]   R. FitzHugh: Biophys. J. **1**, 17 (1988)

[2.45]   T. Kohonen: Neural Networks **6**, 895 (1993)

[2.46]   D. Hebb: *Organization of Behaviour* (Wiley, New York, NY 1949)

[2.47]   T. Kohonen: IEEE Trans. C-**21**, 353 (1972)
[2.48]   K. Nakano, J. Nagumo: In *Advance Papers, 2nd Int. Joint Conf. on Artificial Intelligence* (The British Computer Society, London, UK 1971) p. 101
[2.49]   J. Anderson: Math. Biosci. **14**, 197 (1972)
[2.50]   M. Fazeli: Trends in Neuroscience **15**, 115 (1992)
[2.51]   J. A. Gally, P. R. Montague, G. N. Reeke, Jr., G. M. Edelman: Proc. Natl. Acad. Sci. USA **87**, 3547 (1990)
[2.52]   E. Kandel: Scientific American **241**, 60 (1979)
[2.53]   T. Kohonen: *Self-Organization and Associative Memory* (Springer, Berlin, Heidelberg 1984). 3rd ed. 1989.
[2.54]   S. Geman: SIAM J. Appl. Math. **36**, 86 (1979)
[2.55]   E. Oja: J. Math. Biol. **15**, 267 (1982)
[2.56]   D. Gabor: J. IEE **93**, 429 (1946)
[2.57]   T. Kohonen: Neural Processing letters **9**, 153 (1999)
[2.58]   T. Kohonen: Int. J. of Neuroscience, **5**, 27 (1973)
[2.59]   R. Hunt, N. Berman: J. Comp. Neurol. **162**, 43 (1975)
[2.60]   G. E. Schneider: In *Neurosurgical Treatment in Psychiatry, Pain and Epilepsy*, ed. by W. H. Sweet, S. Abrador, J. G. Martin-Rodriquez (Univ. Park Press, Baltimore, MD 1977)
[2.61]   S. Sharma: Exp. Neurol. **34**, 171 (1972)
[2.62]   R. Sperry: Proc. Natl. Acad. Sci. USA **50**, 701 (1963)
[2.63]   R. Gaze, M. Keating: Nature **237**, 375 (1972)
[2.64]   R. Hope, B. Hammond, F. Gaze: Proc. Roy. Soc. London **194**, 447 (1976)
[2.65]   D. Olton: Scientific American **236**, 82 (1977)
[2.66]   E. I. Knudsen, M. Konishi: Science **200**, 795 (1978)
[2.67]   N. Suga, W. E. O'Neill: Science **206**, 351 (1971)
[2.68]   P. Gärdenfors: In *Logic, Methodology and Philosophy of Science IX*, ed. by D. Prawitz, B. Skyrns, D. Westerståhl, (Elsevier, Amsterdam, 1994)
[2.69]   S. Fahlman: In *Parallel Models of Associative Memory*, ed. by G. E. Hinton and J. A. Anderson (Lawrence Erlbaum, Hillsdale, N.J., 1981) p. 145
[2.70]   J. Saarinen, T. Kohonen: Perception **14**, 711 (1985)
[2.71]   A. Visa: In *Proc. 10ICPR, Int. Conf. on Pattern Recognition* (IEEE Service Center, Piscataway, NJ 1990) p. 518
[2.72]   K. Goser, U. Hilleringmann, U. Rueckert, K. Schumacher: IEEE Micro **9**, 28 (1989)
[2.73]   H. Ritter, T. Kohonen: Biol. Cyb. **61**, 241 (1989)
[2.74]   R. Miikkulainen: *Subsymbolic Natural Language Processing: An Integrated Model of Scripts, Lexicon, and Memory* (MIT Press, Cambridge, MA 1993)
[2.75]   J. C. Scholtes: PhD Thesis (Universiteit van Amsterdam, Amsterdam, Netherlands 1993)
[2.76]   G. Ojemann: Behav. Brain Sci. **2**, 189 (1983)
[2.77]   H. Goodglass, A. Wingfield, M. Hyde, J. Theurkauf: Cortex **22**, 87 (1986)
[2.78]   A. Caramazza: Ann. Rev. Neurosci. **11**, 395 (1988)
[2.79]   E. Alhoniemi, J. Hollmén, O. Simula, J. Vesanto: Integrated Computer-Aided Engineering **6**, 3 (1999)
[2.80]   S. Kaski, J. Kangas, T. Kohonen: Bibliography of self-organizing map (SOM) papers: 1981-1997. Neural Computing Surveys, 1(3&4):1-176, 1998.

Chapter 3

[3.1]   R. Durbin, D. Willshaw: Nature **326**, 689 (1987)
[3.2]   R. Durbin, G. Mitchison: Nature **343**, 644 (1990)
[3.3]   G. Goodhill, D. Willshaw: Network **1**, 41 (1990)
[3.4]   G. Goodhill: In *Advances in Neural Information Processing Systems 5*, ed. by L. Giles, S. Hanson, J. Cowan (Morgan Kaufmann, San Mateo, CA 1993) p. 985
[3.5]   J. A. Kangas, T. K. Kohonen, J. T. Laaksonen: IEEE Trans. Neural Networks **1**, 93 (1990)
[3.6]   T. Martinetz: PhD Thesis (Technische Universität München, München, Germany 1992)
[3.7]   T. Martinetz: In *Proc. ICANN'93, Int. Conf. on Artificial Neural Networks*, ed. by S. Gielen, B. Kappen (Springer, London, UK 1993) p. 427
[3.8]   Y. Cheng: Neural Computation **9(8)**, 1667 (1997)
[3.9]   U. Grenander. Private communication, 1981
[3.10]   S. Orey: *Limit Theorems for Markov Chain Transition Probabilities* (Van Nostrand, London, UK 1971)
[3.11]   M. Cottrell, J.-C. Fort: Annales de l'Institut Henri Poincaré **23**, 1 (1987)
[3.12]   T. Kohonen: Biol. Cyb. **44**, 135 (1982)
[3.13]   T. Kohonen, E. Oja: Report TKK-F-A474 (Helsinki University of Technology, Espoo, Finland 1982)
[3.14]   T. Martinetz, K. Schulten: Neural Networks **7** (1994)
[3.15]   T. Kohonen: In *Symp. on Neural Networks; Alliances and Perspectives in Senri* (Senri Int. Information Institute, Osaka, Japan 1992)
[3.16]   T. Kohonen: In *Proc. ICNN'93, Int. Conf. on Neural Networks* (IEEE Service Center, Piscataway, NJ 1993) p. 1147
[3.17]   F. Mulier, V. Cherkassky: In *Proc. 12 ICPR, Int. Conf. on Pattern Recognition* (IEEE Service Center, Piscataway, NJ 1994) p. II-224
[3.18]   E. Erwin, K. Obermayer, K. Schulten: Biol. Cyb. **67**, 35 (1992)
[3.19]   S. P. Luttrell: Technical Report 4669 (DRA, Malvern, UK 1992)
[3.20]   T. M. Heskes, B. Kappen: In *Proc. ICNN'93, Int. Conf. on Neural Networks* (IEEE Service Center, Piscataway, NJ 1993) p. III-1219
[3.21]   H. Ritter, K. Schulten: Biol. Cyb. **54**, 99 (1986)
[3.22]   H. Ritter: IEEE Trans. on Neural Networks **2**, 173 (1991)
[3.23]   D. R. Dersch, P. Tavan: IEEE Trans. on Neural Networks **6**, 230 (1995)
[3.24]   S. P. Luttrell: IEEE Trans. on Neural Networks **2**, 427 (1991)
[3.25]   S. P. Luttrell: *Memorandum 4669* (Defense Research Agency, Mahern, UK, 1992)
[3.26]   T. Kohonen, J. Hynninen, J. Kangas, J. Laaksonen: Technical Report A31 (Helsinki University of Technology, Laboratory of Computer and Information Science, Helsinki 1996)
[3.27]   T. Kohonen, J. Hynninen, J. Kangas, J. Laaksonen, K. Torkkola: Technical Report A30 (Helsinki University of Technology, Laboratory of Computer and Information Science, Helsinki 1996)
[3.28]   H.-U. Bauer, K. R. Pawelzik: IEEE Trans. on Neural Networks **3** 570 (1992)
[3.29]   S. Zrehen: In *Proc. ICANN'93, Int. Conf. on Artificial Neural Networks* (Springer-Verlag, London 1993) p. 609
[3.30]   T. Villmann, R. Der, T. Martinetz: In *Proc. ICNN'94, IEEE Int. Conf. on Neural Networks* (IEEE Service Center, Piscataway, NJ 1994), p. 645
[3.31]   K. Kiviluoto: In *Proc. ICNN'96, IEEE Int. Conf. on Neural Networks* (IEEE Service Center, Piscataway, NJ 1996), p. 294

[3.32]   S. Kaski, K. Lagus: In *Lecture Notes in Computer Science*, vol. 1112, ed. by C. v. d. Malsburg, W. von Seelen, J. C. Vorbrüggen, B. Sendhoff (Springer, Berlin 1996) p. 809

[3.33]   T. Samad, S. A. Harp: In *Proc. IJCNN'91, Int. Joint Conf. on Neural Networks* (IEEE Service Center, Piscataway, NJ 1991) p. II-949

[3.34]   T. Samad, S. A. Harp: Network: Computation in Neural Systems **3**, 205 (1992)

[3.35]   A. Ultsch, H. Siemon: Technical Report 329 (Univ. of Dortmund, Dortmund, Germany 1989)

[3.36]   M. A. Kraaijveld, J. Mao, A. K. Jain: In *Proc. 11ICPR, Int. Conf. on Pattern Recognition* (IEEE Comput. Soc. Press, Los Alamitos, CA 1992) p. 41

[3.37]   P. Koikkalainen: In *Proc. ECAI 94, 11th European Conf. on Artificial Intelligence*, ed.by A. Cohn (Wiley, New York, NY 1994) p. 211

[3.38]   P. Koikkalainen: In *Proc. ICANN, Int. Conf. on Artificial Neural Networks* (Paris, France 1995) p. II-63

[3.39]   T. Kohonen: Report A33 (Helsinki University of Technology, Laboratory of Computer and Information Science, Espoo 1996)

[3.40]   J. S. Rodrigues, L. B. Almeida: In *Proc. INNC'90, Int. Neural Networks Conference* (Kluwer, Dordrecht, Netherlands 1990) p. 813

[3.41]   J. S. Rodrigues, L. B. Almeida: In *Neural Networks: Advances and Applications*, ed. by E. Gelenbe (North-Holland, Amsterdam, Netherlands 1991) p. 63

[3.42]   B. Fritzke: In *Proc. IJCNN'91, Int. Joint Conf. on Neural Networks* (IEEE Service Center, Piscataway, NJ 1991) p. 531

[3.43]   B. Fritzke: In *Artificial Neural Networks*, ed. by T. Kohonen, K. Mäkisara, O. Simula, J. Kangas (North-Holland, Amsterdam, Netherlands 1991) p. I-403

[3.44]   B. Fritzke: Arbeitsbericht des IMMD, Universität Erlangen-Nürnberg **25**, 9 (1992)

[3.45]   B. Fritzke: In *Artificial Neural Networks, 2*, ed. by I. Aleksander, J. Taylor (North-Holland, Amsterdam, Netherlands 1992) p. II-1051

[3.46]   B. Fritzke: PhD Thesis (Technische Fakultät, Universität Erlangen-Nürnberg, Erlangen, Germany 1992)

[3.47]   B. Fritzke: In *Advances in Neural Information Processing Systems 5*, ed. by L. Giles, S. Hanson, J. Cowan (Morgan Kaufmann, San Mateo, CA 1993) p. 123

[3.48]   B. Fritzke: In *Proc. 1993 IEEE Workshop on Neural Networks for Signal Processing* (IEEE Service Center, Piscataway, NJ 1993)

[3.49]   B. Fritzke: Technical Report TR-93-026 (Int. Computer Science Institute, Berkeley, CA 1993)

[3.50]   B. Fritzke: In *Proc. ICANN'93, Int. Conf. on Artificial Neural Networks*, ed. by S. Gielen, B. Kappen (Springer, London, UK 1993) p. 580

[3.51]   J. Blackmore, R. Miikkulainen: Technical Report TR AI92-192 (University of Texas at Austin, Austin, TX 1992)

[3.52]   J. Blackmore, R. Miikkulainen: In *Proc. ICNN'93, Int. Conf. on Neural Networks* (IEEE Service Center, Piscataway, NJ 1993) p. I-450

[3.53]   C. Szepesvári, A. Lőrincz: In *Proc. ICANN-93, Int. Conf. on Artificial Neural Networks*, ed. by S. Gielen, B. Kappen (Springer, London, UK 1993) p. 678

[3.54]   C. Szepesvári, A. Lőrincz: In *Proc. WCNN'93, World Congress on Neural Networks* (INNS, Lawrence Erlbaum, Hillsdale, NJ 1993) p. II-497

Chapter 4

[4.1]   A. R. Tunturi: *Am. J. Physiol.* **168**, 712 (1952)
[4.2]   M. M. Merzenich, P. L. Knight, G. L. Roth: *J. Neurophysiol.* **38**, 231 (1975)
[4.3]   E. G. Jones, A. Peters, eds.: *Cerebral Cortex, vol. 2* (Plenum Press, New York, London 1984)
[4.4]   C. D. Gilbert: *Neuron* **9**, 1 (1992)
[4.5]   S. Kaski, T. Kohonen: *Neural Networks* **7**, 973 (1994)
[4.6]   W. C. Abraham, M. F. Bear: *Trends Neurosci.* **19**, 126 (1996)
[4.7]   D. V. Buonomano, M. M. Merzenich: *Annu. Rev. Neurosci.* **21**, 149 (1998)

Chapter 5

[5.1]   W. H. Press, B. P. Flannery, S. A. Teukolsky, W. T. Vetterling: *Numerical Recipes in C – The Art of Scientific Computing* (Press Syndicate of the University of Cambridge, Cambridge University Press 1988)
[5.2]   J. Kangas, T. Kohonen, J. Laaksonen, O. Simula, O. Ventä: In *Proc. IJCNN'89, Int. Joint Conf. on Neural Networks* (IEEE Service Center, Piscataway, NJ 1989) p. II-517
[5.3]   T. Martinetz, K. Schulten: In *Proc. Int. Conf. on Artificial Neural Networks* (Espoo, Finland), ed. by T. Kohonen, K. Mäkisara, O. Simula, J. Kangas (North-Holland, Amsterdam, Netherlands 1991) p. I-397
[5.4]   T. M. Martinetz, S. G. Berkovich, K. J. Schulten: IEEE Trans. on Neural Networks **4**, 558 (1993)
[5.5]   J. Lampinen, E. Oja: In *Proc. 6 SCIA, Scand. Conf. on Image Analysis*, ed. by M. Pietikäinen, J. Röning (Suomen Hahmontunnistustutkimuksen seura r.y., Helsinki, Finland 1989) p. 120
[5.6]   P. Koikkalainen, E. Oja: In *Proc. IJCNN-90, Int. Joint Conf. on Neural Networks* (IEEE Service Center, Piscataway, NJ 1990) p. 279
[5.7]   P. Koikkalainen: In *Proc. Symp. on Neural Networks in Finland*, ed. by A. Bulsari, B. Saxén (Finnish Artificial Intelligence Society, Helsinki, Finland 1993) p. 51
[5.8]   K. K. Truong: In *ICASSP'91, Int. Conf. on Acoustics, Speech and Signal Processing* (IEEE Service Center, Piscataway, NJ 1991) p. 2789
[5.9]   W. Wan, D. Fraser: In *Proc. IJCNN-93-Nagoya, Int. Joint Conf. on Neural Networks* (IEEE Service Center, Piscataway, NJ 1993) p. III-2464
[5.10] W. Wan, D. Fraser: In *Proc. of 5th Australian Conf. on Neural Networks*, ed. by A. C. Tsoi, T. Downs (University of Queensland, St Lucia, Australia 1994) p. 17
[5.11] N. M. Allinson, M. T. Brown, M. J. Johnson: In *IEE Int. Conf. on Artificial Neural Networks, Publication 313* (IEE, London, UK 1989) p. 261
[5.12] N. M. Allinson, M. J. Johnson: In *New Developments in Neural Computing*, ed. by J. G. Taylor, C. L. T. Mannion (Adam-Hilger, Bristol, UK 1989) p. 79
[5.13] R. Sedgewick: *Algorithms* (Addison-Wesley, Reading, MA 1983)
[5.14] J. Kangas: In *Proc. IJCNN-90-San Diego,Int. Joint Conf. on Neural Networks* (IEEE Comput. Soc. Press, Los Alamitos, CA 1990) p. II-331
[5.15] J. Kangas: In *Artificial Neural Networks*, ed. by T. Kohonen, K. Mäkisara, O. Simula, J. Kangas (North-Holland, Amsterdam, Netherlands 1991) p. II-1591

[5.16] J. Kangas: In *Proc. ICASSP'91, Int. Conf. on Acoustics, Speech and Signal Processing* (IEEE Service Center, Piscataway, NJ 1991) p. 101

[5.17] J. Kangas: In *Artificial Neural Networks, 2*, ed. by I. Aleksander, J. Taylor (North-Holland, Amsterdam, Netherlands 1992) p. I-117

[5.18] J. Kangas: PhD Thesis (Helsinki University of Technology, Espoo, Finland 1994)

[5.19] T. Kohonen: Report A42 (Helsinki University of Technology, Laboratory of Computer and Information Science, Espoo, Finalnd 1996)

[5.20] T. Kohonen, P. Somervuo: In *Proc. WSOM'97, Workshop on Self-Organizing Maps* (Helsinki University of Technology, Neural Networks Research Centre, Espoo, Finland 1997) p. 2

[5.21] R. Kalman, R. Bucy: J. Basic Engr. **83**, 95 (1961)

[5.22] T. Kohonen, K. Mäkisara, T. Saramäki: In *Proc. 7ICPR, Int. Conf. on Pattern Recognition* (IEEE Computer Soc. Press, Los Alamitos, CA 1984) p. 182

[5.23] T. Kohonen: Computer **21**, 11 (1988)

[5.24] T. Kohonen: In *Computational Intelligence, A Dynamic System Perspective*, ed. by M. Palaniswami, Y. Attikiouzel, R. J. Marks II, D. Fogel, T. Fukuda (IEEE Press, New York, NY 1995) p. 17

[5.25] T. Kohonen: Biol. Cybernetics **75(4)** 281 (1996)

[5.26] T. Kohonen, S. Kaski, H. Lappalainen: Neural Computation **9**, 1321 (1997)

[5.27] J. Daugman: IEEE Trans. Syst., Man, Cybern. **13**, 882 (1983)

[5.28] I. Daubechies: IEEE Trans. Inf. Theory **36**, 961 (1990)

[5.29] J. Daugman: Visual Research **20**, 847 (1980)

[5.30] J. Daugman: J. Opt. Soc. Am. **2**, 1160 (1985)

[5.31] S. Marcelja: J. Opt. Soc. Am. **70**, 1297 (1980)

[5.32] J. Jones, L. Palmer: J. Neurophysiol. **58**, 1233(1987)

Chapter 6

[6.1] T. Kohonen: In *Advanced Neural Networks*, ed. by R. Eckmiller (Elsevier, Amsterdam, Netherlands 1990) p. 137

[6.2] T. Kohonen: In *Proc. IJCNN-90-San Diego, Int. Joint Conf. on Neural Networks* (IEEE Service Center, Piscataway, NJ 1990) p. I-545

[6.3] T. Kohonen: In *Theory and Applications of Neural Networks, Proc. First British Neural Network Society Meeting* (BNNS, London, UK 1992) p. 235

[6.4] T. Kohonen: In *Artificial Neural Networks*, ed. by T. Kohonen, K. Mäkisara, O. Simula, J. Kangas (North-Holland, Amsterdam, Netherlands 1991) p. II-1357

Chapter 7

[7.1] J. Buhmann, J. Lange, C. von der Malsburg: In *Proc. IJCNN 89, Int. Joint Conf. on Neural Networks* (IEEE Service Center, Piscataway, NJ 1990) p. I-155

[7.2] H. Freeman: IRE Trans. EC-**20**, 260 (1961)

[7.3] R. Haralick: In *Proc. 4IJCPR, Int. Joint Conf. on Pattern Recognition* (Pattern Recognition Soc. of Japan, Tokyo, Japan 1978) p. 45

[7.4]   M. Kasslin, J. Kangas, O. Simula: In *Artificial Neural Networks, 2*, ed. by
        I. Aleksander, J. Taylor (North-Holland, Amsterdam, Netherlands 1992)
        p. II-1531
[7.5]   M. Vapola, O. Simula, T. Kohonen, P. Meriläinen: In *Proc. ICANN'94,
        Int. Conf. on Artificial Neural Networks* (Springer-Verlag, Beriln, Hei-
        delberg 1994) p. I-246
[7.6]   L. Leinonen, J. Kangas, K. Torkkola, A. Juvas: In *Artificial Neural
        Networks*, ed. by T. Kohonen, K. Mäkisara, O. Simula, J. Kangas (North-
        Holland, Amsterdam, Netherlands 1991) p. II-1385
[7.7]   L. Leinonen, J. Kangas, K. Torkkola, A. Juvas: J. Speech and Hearing
        Res. **35**, 287 (1992)
[7.8]   K. Torkkola, J. Kangas, P. Utela, S. Kaski, M. Kokkonen, M. Ku-
        rimo, T. Kohonen: In *Artificial Neural Networks*, ed. by T. Kohonen,
        K. Mäkisara, O. Simula, J. Kangas (North-Holland, Amsterdam, Nether-
        lands 1991) p. I-771
[7.9]   L. Rabiner: Proc. IEEE **77**, 257 (1989)
[7.10] K. Torkkola: In *Proc. COGNITIVA-90* (North-Holland, Amsterdam,
        Netherlands 1990) p. 637
[7.11] J. Mäntysalo, K. Torkkola, T. Kohonen: In *Proc. Int. Conf. on Spoken
        Language Processing* (University of Alberta, Edmonton, Alberta, Canada
        1992) p. 539
[7.12] J. Mäntysalo, K. Torkkola, T. Kohonen: In *Proc. 2nd Workshop on
        Neural Networks for Speech Processing*, ed. by M. Gori (Edizioni Lint
        Trieste, Trieste, Italy 1993) p. 39
[7.13] J. Mäntysalo, K. Torkkola, T. Kohonen: In *Proc. ICANN'93, Int. Conf.
        on Artificial Neural Networks*, ed. by S. Gielen, B. Kappen (Springer,
        London, UK 1993) p. 389
[7.14] M. Kurimo, K. Torkkola: In *ICSLP-92, Proc. Int. Conf. on Spoken Lan-
        guage Processing* (Univ. of Alberta, Edmonton, Alberta, Canada 1992)
        p. I-543
[7.15] M. Kurimo, K. Torkkola: In *Proc. SPIE's Conf. on Neural and Stochastic
        Methods in Image and Signal Processing* (SPIE, Bellingham, WA 1992)
        p. 726
[7.16] M. Kurimo, K. Torkkola: In *Proc. Workshop on Neural Networks for
        Signal Processing* (IEEE Service Center, Piscataway, NJ 1992) p. 174
[7.17] M. Kurimo: In *Proc. EUROSPEECH-93, 3rd European Conf. on Speech,
        Communication, and Technology* (ESCA, Berlin 1993) p. III-1731
[7.18] M. Kurimo: In *Proc. Int. Symp. on Speech, Image Processing and Neural
        Networks* (IEEE Hong Kong Chapter of Signal Processing, Hong Kong
        1994) p. II-718
[7.19] M. Kurimo: In *Proc. NNSP'94, IEEE Workshop on Neural Networks for
        Signal Processing* (IEEE Service Center, Piscataway, NJ 1994) p. 362
[7.20] M. Kurimo, P. Somervuo: In *Proc. ICSLP-96, Int. Conf. on Spoken Lan-
        guage Processing* (ICSLP, Philadelphia, PA 1996) p. 358
[7.21] M. Kurimo: In *Proc. WSOM'97, Workshop on Self-Organizing Maps*
        (Helsinki University of Technology, Neural Networks Research Centre,
        Espoo, Finland 1997) p. 8
[7.22] M. Kurimo: Computer Speech and Language **11**, 321 (1997)
[7.23] M. Kurimo: *Acta Polytechnica Scandinavica, Mathematics, Comput-
        ing, and Management in Engineering Series No. Ma 87* (The Finnish
        Academy of Technology, Espoo, Finland 1997)

[7.24]   A. Visa: In *Proc. 5th European Signal Processing Conf.*, ed. by L. Torres, E. Masgrau, M. A. Lagunes (Elsevier, Amsterdam, Netherlands 1990) p. 991

[7.25]   A. Visa: In *Proc. IJCNN-90-San Diego, Int. Joint Conf. on Neural Networks* (IEEE Service Center, Piscataway, NJ 1990) p. I-491

[7.26]   A. Visa: PhD Thesis (Helsinki University of Technology, Espoo, Finland 1990)

[7.27]   A. Visa: Report A13 (Helsinki University of Technology, Laboratory of Computer and Information Science, Espoo, Finland 1990)

[7.28]   A. Visa, A. Langinmaa, U. Lindquist: In *Proc. TAPPI, Int. Printing and Graphic Arts Conf.* (Canadian Pulp and Paper Assoc., Montreal, Canada 1990) p. 91

[7.29]   A. Visa: In *Proc. European Res. Symp. 'Image Analysis for Pulp and Paper Res. and Production'* (Center Technique du Papier, Grenoble, France 1991)

[7.30]   A. Visa: Graphic Arts in Finland **20**, 7 (1991)

[7.31]   A. Visa, A. Langinmaa: In *Proc. IARIGAI*, ed. by W. H. Banks (Pentech Press, London, UK 1992)

[7.32]   A. Visa, K. Valkealahti, O. Simula: In *Proc. IJCNN-91-Singapore, Int. Joint Conf. on Neural Networks* (IEEE Service Center, Piscataway, NJ 1991) p. 1001

[7.33]   A. Visa: In *Proc. Applications of Artificial Neural Networks II, SPIE Vol. 1469* (SPIE, Bellingham, WA 1991) p. 820

[7.34]   A. Visa: In *Proc. DECUS Finland ry. Spring Meeting* (DEC Users' Society, Helsinki, Finland 1992) p. 323

[7.35]   A. Visa: In *Proc. 11ICPR, Int. Conf. on Pattern Recognition* (IEEE Computer Society Press, Los Alamitos, CA 1992) p. 101

[7.36]   H. Ritter, T. Kohonen: Biol. Cyb. **61** 241 (1989)

[7.37]   J. C. Scholtes: PhD Thesis (University of Amsterdam, Amsterdam, Netherlands 1993)

[7.38]   T. Honkela, V. Pulkki, T. Kohonen: In *Proc. ICANN-95, Int. Conf. on Artificial Neural Networks* (EC2, Nanterre, France 1995) p. II-3

[7.39]   X. Lin, D. Soergel, G. Marchionini: In *Proc. 14th Ann. Int. ACM/SIGIR Conf. on R & D In Information Retrieval* (ACM, New York 1991) p. 262

[7.40]   J. C. Scholtes: In *Proc. IJCNN'91, Int. Joint Conf. on Neural Networks* (IEEE Service Center, Piscataway, NJ 1991) p. 18

[7.41]   D. Merkl, A. M. Tjoa, G. Kappel: In *Proc. ACNN'94, 5th Australian Conference on Neural Networks* (BrisBane, Australia 1994) p. 13

[7.42]   J. Zavrel: MA Thesis (University of Amsterdam, Amsterdam, Netherlands 1995)

[7.43]   D. Merkl: In *Proc. ICANN-95, Int. Conf on Artificial Neural Networks* (EC2, Nanterre, France 1995) p. II-239

[7.44]   S. Kaski, T. Honkela, K. Lagus, T. Kohonen: Neurocomputing **21**, 101 (1998)

[7.45]   G. Salton, M.J. McGill: *Introduction to Modern Information Retrieval* (McGraw-Hill, New York, NY 1983)

[7.46]   S. Deerwester, S. Dumais, G. Furnas, K. Landauer: J. Am. Soc. Inform. Sci. **41**, 391 (1990)

[7.47]   S. Kaski: *Acta Polytechnica Scandinavica, Mathematics, Computing and Management in Engineering Series No 82* (The Finnish Academy of Technology, Espoo, Finland 1997)

[7.48]   S. Kaski: In *Proc. IJCNN'98, Int. Joint Conf. on Neural Networks* (IEEE Press, Piscataway, NJ 1998) p. 413

[7.49]   T. Kohonen: In *Proc. ICANN'98, 8th Int. Conf. on Artificial Neural Networks*, ed. by L. Niklasson, M. Boden, T. Ziemke (Springer, London, UK 1998) p. 65

[7.50]   D. Roussinov, H. Chen: CC-AI–Communication, Cognition and Artificial Intelligence **15**, 81 (1998)

[7.51]   K. Lagus, S. Kaski: In *Proc. ICANN'99, Int. Conf. on Artificial Neural Networks* (IEE Press, London, UK 1999) p. 371

[7.52]   H. Ritter: In *Proc. ICANN'93, Int. Conf. on Artificial Neural Networks*, ed. by S. Gielen, B. Kappen (Springer, London, UK 1993) p. 568

[7.53]   H. Ritter: In *Proc. ICANN'94, Int. Conf. on Artificial Neural Networks*, ed. by M. Marinaro, P.G. Morasso (Springer, London, UK 1994) p. II-803

[7.54]   H. Ritter: In *Proc. ICANN'97, Int. Conf. on Artificial Neural Networks, vol. 1327 of Lecture Notes in Computer Science* (Springer, Berlin 1997) p. 675

[7.55]   J. G. Proakis: In *Advances in Communications Systems Theory and Applications*, Vol 4 (Academic Press, New York, NY 1975)

[7.56]   J. Henriksson: *Acta Polytechnica Scandinavica, Electrical Engineering Series No. 54* PhD Thesis (Helsinki, Finland 1984)

[7.57]   J. Henriksson, K. Raivio, T. Kohonen: Finnish Patent 85,548. U.S. Patent 5,233,635. Australian Patent 636,494.

[7.58]   T. Kohonen, K. Raivio, O. Simula, O. Ventä, J. Henriksson: In *Proc. IJCNN-90-San Diego, Int. Joint Conf. on Neural Networks* (IEEE Service Center, Piscataway, NJ 1990) p. I-223

[7.59]   T. Kohonen, K. Raivio, O. Simula, O. Ventä, J. Henriksson: In *Proc. IJCNN-90-WASH-DC, Int. Joint Conf. on Neural Networks* (Lawrence Erlbaum, Hillsdale, NJ 1990) p. II-249

[7.60]   T. Kohonen, K. Raivio, O. Simula, J. Henriksson: In *Proc. ICANN-91, Int. Conf. on Artificial Neural Networks* (North-Holland, Amsterdam, Netherlands) p. II-1677

[7.61]   T. Kohonen, K. Raivio, O. Simula, O. Ventä, J. Henriksson: In *Proc. IEEE Int. Conf. on Communications* (IEEE Service Center, Piscataway, NJ 1992) p. III-1523

[7.62]   J. Kangas, T. Kohonen: In *Proc. IMACS, Int. Symp. on Signal Processing, Robotics and Neural Networks* (IMACS, Lille, France 1994) p. 19

[7.63]   J. Kangas: In *Proc. ICANN'95, Int. Conf. on Artificial Neural Networks* (EC2, Nanterre, France 1995) p. I-287

[7.64]   J. G. Proakis: *Digital Communications* 2nd ed. (McGraw-Hill, Computer Science Series, New York, NY 1989)

Chapter 8

[8.1]   J.Vesanto, *Intelligent Data Analysis*, 3, p. 111-126 (1999)

Chapter 9

[9.1]   K. Steinbuch: *Automat und Mensch* (Springer, Berlin, Heidelberg 1963)

[9.2]   B. Gilbert: IEEE J. Solid-State Circuits, SC-**3** p. 365 (1968)

[9.3]   T. Kohonen: In *Proc. WCNN'93, World Congress on Neural Networks* (Lawrence Erlbaum, Hillsdale, NJ 1993) p. IV-1

[9.4]   J. Vuori, T. Kohonen: In Proc. ICNN'95, Int. Conf. on Neural Networks (IEEE Service Center, Piscataway, NJ 1995) p. IV-2019

[9.5]   V. Pulkki. M.Sc. Thesis (Helsinki University of Technology, Espoo, Finland 1994)

[9.6]   H. P. Siemon, A. Ultsch: In *Proc. INNC-90, Int. Neural Network Conf.* (Kluwer, Dordrecht, Netherlands 1990) p. 643

[9.7]   R. Togneri, Y. Attikiouzel: In *Proc. IJCNN-91-Singapore, Int. Joint Conf. on Neural Networks* (IEEE Comput. Soc. Press, Los Alamitos, CA 1991) p. II-1717

[9.8]   F. Blayo, C. Lehmann: In *Proc. INNC 90, Int. Neural Network Conf.* (Kluwer, Dordrecht, Netherlands 1990)

[9.9]   R. Mann, S. Haykin: In *Proc. IJCNN-90-WASH-DC, Int. Joint Conf. on Neural Networks* (Lawrence Erlbaum, Hillsdale, NJ 1990) p. II-84

[9.10]   C. Lehmann, F. Blayo: In *Proc. VLSI for Artificial Intelligence and Neural Networks*, ed. by J. G. Delgado-Frias, W. R. Moore (Plenum, New York, NY 1991) p. 325

[9.11]   H. Speckmann, P. Thole, W. Rosentiel: In *Artificial Neural Networks, 2*, ed. by I. Aleksander, J. Taylor (North-Holland, Amsterdam, Netherlands 1992) p. II-1451

[9.12]   H. Speckmann, P. Thole, W. Rosenstiel: In *Proc. ICANN-93, Int. Conf. on Artificial Neural Networks*, ed. by S. Gielen, B. Kappen (Springer, London, UK 1993) p. 1040

[9.13]   H. Speckmann, P. Thole, W. Rosenthal: In *Proc. IJCNN-93-Nagoya, Int. Joint Conf. on Neural Networks* (IEEE Service Center, Piscataway, NJ 1993) p. II-1951

[9.14]   M. S. Melton, T. Phan, D. S. Reeves, D. E. Van den Bout: IEEE Trans. on Neural Networks **3**, 375 (1992)

[9.15]   S. Rüping, M. Porrman, U. Rückert: In *Proc. WSOM'97, Workshop on Self-Organizing Maps* (Helsinki University of Technology, Neural Networks Research Centre, Espoo, Finland 1997) p. 136

[9.16]   S. Rüping: *VLSI-gerechte Umsetzung von Selbstorganisierenden Karten und ihre Einbindung in ein Neuro-Fuzzy Analysesystem;* Fortschritt-Berichte VDI, Reihe 9: Elektronik (VDI Verlag, Düsseldorf, Germany 1995)

[9.17]   S. Rüping, U. Rückert, K. Goser: New Trends in Neural Computation, Lecture Notes in Computer Science 686, ed. by J. Mira, J. Cabestony, A. Prieto (Springer, Berlin, Germany 1993) p. 488

[9.18]   S. Rüping, K. Goser, U. Rückert: IEEE Micro **15**, 57 (1995)

[9.19]   S. Rüping, U. Rückert: In Proc. MicroNeuro'96, Fifth Int. Conf. on Microelectronics for Neural Networks and Fuzzy Systems (Lausanne, Switzerland 1996) p. 285

[9.20]   U. Rückert: *IEE-Proc. Publication No. 395* (IEE, Norwich, UK 1994) p. 372

Chapter 10

[10.1]   H. Tokutaka, S. Kishida and K. Fujimura, *Applications of Self-Organizing Maps*, Kaibundo, Tokyo, Japan, 1999. (In Japanese)

[10.2]   M.M. Van Hulle: *Faithful Representations and Topographic Maps: From Distortion- to Information-Based Self-Organization* (Wiley, New York, NY 2000)

[10.3]   G. Deboeck and T. Kohonen (eds.): *Visual Explorations in Finance with Self-Organizing Maps* (Springer, London, UK 1998)

[10.4]   *Proc. WSOM'97, Workshop on Self-Organizing Maps* (Helsinki Univ. of Technology, Neural Networks Research Centre, Espoo, Finland 1997)

[10.5]   E. Oja (ed.): Neurocomputing, Special Volume on Self-Organizing Maps, vol. 21, Nos 1-3 (October 1998)

[10.6]   E. Oja, S. Kaski (eds.) *Kohonen Maps* (Elsevier, Amsterdam, Netherlands 1999)

[10.7]   T. Kohonen: Proc. IEEE **78**, 1464 (1990)

[10.8]   M. Cottrell, J. C. Fort, G. Pagés: In *Proc. ESANN'94, European Symposium on Neural Networks*, ed.by M. Verleysen (D Facto Conference Services, Brussels, Belgium 1994) p. 235

[10.9]   M. Cottrell, J. C. Fort, G. Pagés: In *Proc. WSOM'97, Workshop on Self-Organizing Maps* (Helsinki University of Technology, Neural Networks Research Centre, Espoo, Finland 1997) p. 246

[10.10] N. V. Swindale: Proc. Royal Society of London, **B 208**, 243 (1980)

[10.11] S.-i. Amari: In *Dynamic Interactions in Neural Networks: Models and Data*, ed. by M. A. Arbib, S.-i. Amari (Springer, Berlin, Germany 1989) p. 15

[10.12] T. Kohonen: Neural Computation **11**, 2081 (1999)

[10.13] G. Pagés: In *Proc. ESANN'93, European Symp. on Artificial Neural Networks* (D Facto Conference Services, Brussels, Belgium 1993) p. 221

[10.14] C. M. Bishop, M. Svensén, C. K. I. Williams: In *Advances in Neural Information Processing Systems 9* (MIT Press, Cambridge, MA 1997) p. 354

[10.15] C. M. Bishop, M. Svensén, C. K. I. Williams: Neural Computation **10**, 215 (1998)

[10.16] H. M. Jung, J. H. Lee, C. W. Lee: J. Korean Institute of Telematics and Electronics **26**, 228 (1989)

[10.17] B. Kappen, T. Heskes: In *Artificial Neural Networks, 2*, ed. by I. Aleksander, J. Taylor (North-Holland, Amsterdam, Netherlands 1992) p. I-71

[10.18] B. Kosko: In *Proc. IJCNN-90-Kyoto, Int. Joint Conf. on Neural Networks* (IEEE Service Center, Piscataway, NJ 1990) p. II-215

[10.19] R. Miikkulainen: Technical Report UCLA-AI-87-16 (Computer Science Department, University of California, Los Angeles, CA 1987)

[10.20] R. Miikkulainen, M. G. Dyer: In *Proc. 1988 Connectionist Models Summer School*, ed. by D. S. Touretzky, G. E. Hinton, T. J. Sejnowski (Morgan Kaufmann, San Mateo, CA 1989) p. 347

[10.21] G. Barna, K. Kaski: J. Physics A [Mathematical and General] **22**, 5174 (1989)

[10.22] M. Benaim: Neural Networks **6**, 655 (1993)

[10.23] J. C. Bezdek: Proc. SPIE – The Int. Society for Optical Engineering **1293**, 260 (1990)

[10.24] M. Blanchet, S. Yoshizawa, S.-I. Amari: In *Proc. IJCNN-93-Nagoya, Int. Joint Conf. on Neural Networks* (IEEE Service Center, Piscataway, NJ 1993) p. III-2476

[10.25] J. Buhmann, H. Kühnel: In *Proc. DCC'92, Data Compression Conf.*, ed. by J. A. Storer, M. Cohn (IEEE Comput. Soc. Press, Los Alamitos, CA 1992) p. 12

[10.26] J. Buhmann, H. Kühnel: IEEE Trans. Information Theory **39** (1993)

[10.27] J. Buhmann, H. Kühnel: Neural Computation **5**, 75 (1993)

[10.28] V. Chandrasekaran, M. Palaniswami, T. M. Caelli: In *Proc. WCNN'93, World Congress on Neural Networks* (Lawrence Erlbaum, Hillsdale, NJ 1993) p. IV-112

[10.29] V. Chandrasekaran, M. Palaniswami, T. M. Caelli: In *Proc. ICNN'93, Int. Conf. on Neural Networks* (IEEE Service Center, Piscataway, NJ 1993) p. III-1474

[10.30]   C.-C. Chang, C.-H. Chang, S.-Y. Hwang: In *Proc. 11ICPR, Int. Conf. on Pattern Recognition* (IEEE Comput. Soc. Press, Los Alamitos, CA 1992) p. III-522

[10.31]   R.-I. Chang, P.-Y. Hsiao: In *Proc. ICNN'93, Int. Conf. on Neural Networks* (IEEE Service Center, Piscataway, NJ 1993) p. I-103

[10.32]   V. Cherkassky, H. Lari-Najafi: In *Proc. INNC'90, Int. Neural Network Conf.* (Kluwer, Dordrecht, Netherlands 1990) p. I-370

[10.33]   V. Cherkassky, H. Lari-Najafi: In *Proc. INNC'90, Int. Neural Network Conf.* (Kluwer, Dordrecht, Netherlands 1990) p. II-838

[10.34]   V. Cherkassky, H. Lari-Najafi: Neural Networks **4**, 27 (1991)

[10.35]   V. Cherkassky, Y. Lee, H. Lari-Najafi: In *Proc. IJCNN 91, Int. Joint Conf. on Neural Networks* (IEEE Service Center, Piscataway, NJ 1991) p. I-79

[10.36]   V. Cherkassky: In *Proc. Workshop on Neural Networks for Signal Processing*, ed. by S. Y. Kung, F. Fallside, J. A. Sorensen, C. A. Kamm (IEEE Service Center, Piscataway, NJ 1992) p. 511

[10.37]   V. Cherkassky, H. Lari-Najafi: IEEE Expert **7**, 43 (1992)

[10.38]   V. Cherkassky, F. Mulier: In *Proc. SPIE Conf. on Appl. of Artificial Neural Networks* (SPIE, Bellingham, WA 1992)

[10.39]   A. C. C. Coolen, L. G. V. M. Lenders: J. Physics A [Mathematical and General] **25**, 2577 (1992)

[10.40]   A. C. C. Coolen, L. G. V. M. Lenders: J. Physics A [Mathematical and General] **25**, 2593 (1992)

[10.41]   G. R. D. Haan, O. Eğecioğlu: In *Proc. IJCNN'91, Int. Joint Conf. on Neural Networks* (IEEE Service Center, Piscataway, NJ 1991) p. 964

[10.42]   D. DeSieno: In *Proc. ICNN'88, Int. Conf. on Neural Networks* (IEEE Service Center, Piscataway, NJ 1988) p. 117

[10.43]   J. Göppert, W. Rosenstiel: In *Proc. ICANN'94, Int. Conf. on Artificial Neural Networks*, ed. by M. Marinaro, P. G. Morasso (Springer, London, UK 1994) p. I-330

[10.44]   Z. He, C. Wu, J. Wang, C. Zhu: In *Proc. of 1994 Int. Symp. on Speech, Image Processing and Neural Networks* (IEEE Hong Kong Chapt. of Signal Processing, Hong Kong 1994) p. II-654

[10.45]   A. Hoekstra, M. F. J. Drossaers: In *Proc. ICANN'93. Int. Conf. on Artificial Neural Networks*, ed. by S. Gielen, B. Kappen (Springer, London, UK 1993) p. 404

[10.46]   R. M. Holdaway: In *Proc. IJCNN'89, Int. Joint Conf. on Neural Networks* (IEEE Service Center. Piscataway, NJ 1989) p. II-523

[10.47]   R. M. Holdaway, M. W. White: Int. J. Neural Networks – Res. & Applications **1**, 227 (1990)

[10.48]   R. M. Holdaway, M. W. White: Int. J. Bio-Medical Computing **25**, 151 (1990)

[10.49]   L. Holmström, A. Hämäläinen: In *Proc. ICNN'93, Int. Conf. on Neural Networks* (IEEE Service Center, Piscataway, NJ 1993) p. I-417

[10.50]   A. Hämäläinen: In *Proc. ICNN'94, Int. Conf. on Neural Networks* (IEEE Service Center, Piscataway, NJ 1994) p. 659

[10.51]   A. Iwata, T. Tohma, H. Matsuo, N. Suzumura: Trans. of the Inst. of Electronics, Information and Communication Engineers **J73D-II**, 1261 (1990)

[10.52]   A. Iwata, T. Tohma, H. Matsuo, N. Suzumura: In *INNC'90, Int. Neural Network Conf.* (Kluwer, Dordrecht, Netherlands 1990) p. I-83

[10.53]   J.-X. Jiang, K.-C. Yi, Z. Hui: In *Proc. EUROSPEECH-91, 2nd European Conf. on Speech Communication and Technology* (Assoc. Belge Acoust.;

Assoc. Italiana di Acustica; CEC; et al, Istituto Int. Comunicazioni, Genova, Italy 1991) p. I-125

[10.54]   J. W. Jiang, M. Jabri: In *Proc. IJCNN'92, Int. Joint Conf. on Neural Networks* (IEEE Service Center, Piscataway, NJ 1992) p. II-510

[10.55]   J. W. Jiang, M. Jabri: In *Proc. ACNN'92, Third Australian Conf. on Neural Networks*, ed. by P. Leong, M. Jabri (Sydney Univ, Sydney, NSW, Australia 1992) p. 235

[10.56]   A. Kuh, G. Iseri, A. Mathur, Z. Huang: In *1990 IEEE Int. Symp. on Circuits and Systems* (IEEE Service Center, Piscataway, NJ 1990) p. IV-2512

[10.57]   N. Kunstmann, C. Hillermeier, P. Tavan: In *Proc. ICANN'93. Int. Conf. on Artificial Neural Networks*, ed. by S. Gielen, B. Kappen (Springer, London, UK 1993) p. 504

[10.58]   K.-P. Li: In *Artificial Neural Networks*, ed. by T. Kohonen, K. Mäkisara, O. Simula, J. Kangas (North-Holland, Amsterdam, Netherlands 1991) p. II-1353

[10.59]   P. Martín-Smith, F. J. Pelayo, A. Diaz, J. Ortega, A. Prieto: In *New Trends in Neural Computation, Lecture Notes in Computer Science No. 686*, ed. by J. Mira, J. Cabestany, A. Prieto (Springer, Berlin, Germany 1993)

[10.60]   T. Martinetz, K. Schulten: In *Proc. ICNN'93, Int. Conf. on Neural Networks* (IEEE Service Center, Piscataway, NJ 1993) p. II-820

[10.61]   R. Miikkulainen: Connection Science **2**, 83 (1990)

[10.62]   R. Miikkulainen: In *Proc. 12th Annual Conf. of the Cognitive Science Society* (Lawrence Erlbaum, Hillsdale, NJ 1990) p. 447

[10.63]   R. Miikkulainen: PhD Thesis (Computer Science Department, University of California, Los Angeles 1990). (Tech. Rep UCLA-AI-90-05)

[10.64]   R. Miikkulainen: In *Proc. Int. Workshop on Fundamental Res. for the Next Generation of Natural Language Processing* (ATR International, Kyoto, Japan 1991)

[10.65]   R. Miikkulainen, M. G. Dyer: Cognitive Science **15**, 343 (1991)

[10.66]   R. Miikkulainen: In *Artificial Neural Networks*, ed. by T. Kohonen, K. Mäkisara, O. Simula, J. Kangas (North-Holland, Amsterdam, Netherlands 1991) p. I-415

[10.67]   R. Miikkulainen: Biol. Cyb. **66**, 273 (1992)

[10.68]   R. Miikkulainen: In *Proc. Integrating Symbol Processors and Connectionist Networks for Artificial Intelligence and Cognitive Modeling*, ed. by V. Honavar, L. Uhr (Academic Press, New York, NY 1994) p. 483-508

[10.69]   R. Miikkulainen: In *Proc. Third Twente Workshop on Language Technology* (Computer Science Department, University of Twente, Twente, Netherlands 1993)

[10.70]   A. Ossen: In *Proc. ICANN-93, Int. Conf. on Artificial Neural Networks*, ed. by S. Gielen, B. Kappen (Springer, London, UK 1993) p. 586

[10.71]   K. Ozdemir, A. M. Erkmen: In *Proc. WCNN'93, World Congress on Neural Networks* (Lawrence Erlbaum, Hillsdale, NJ 1993) p. II-513

[10.72]   H. Ritter: In *Proc. INNC'90 Int. Neural Network Conf.* (Kluwer, Dordrecht, Netherlands 1990) p. 898

[10.73]   H. Ritter: In *Proc. ICANN-93 Int. Conf. on Artificial Neural Networks*, ed. by S. Gielen, B. Kappen (Springer, London, UK 1993) p. 568

[10.74]   J. Sirosh, R. Miikkulainen: Technical Report AI92-191 (The University of Texas at Austin, Austin, TX 1992)

[10.75]   J. Sirosh, R. Miikkulainen: Biol. Cyb. **71**, 65 (1994)

[10.76]   J. Sirosh, R. Miikkulainen: In *Proc. 1993 Connectionist Models Summer School* (Lawrence Erlbaum, Hillsdale, NJ 1994)
[10.77]   J. Sirosh, R. Miikkulainen: In *Proc. ICNN'93 Int. Conf. on Neural Networks* (IEEE Service Center, Piscataway, NJ 1993) p. III-1360
[10.78]   G. J. Tóth, T. Szakács, A. Lőrincz: Materials Science & Engineering B (Solid-State Materials for Advanced Technology) **B18**, 281 (1993)
[10.79]   G. J. Tóth, T. Szakács, A. Lőrincz: In *Proc. WCNN'93, World Congress on Neural Networks* (Lawrence Erlbaum, Hillsdale, NJ 1993) p. III-127
[10.80]   G. J. Tóth, A. Lőrincz: In *Proc. WCNN'93, World Congress on Neural Networks* (Lawrence Erlbaum, Hillsdale, NJ 1993) p. III-168
[10.81]   G. J. Tóth, A. Lőrincz: In *Proc. ICANN-93, Int. Conf. on Artificial Neural Networks*, ed. by S. Gielen, B. Kappen (Springer, London, UK 1993) p. 605
[10.82]   G. J. Tóth, T. Szakács, A. Lőrincz: In *Proc. ICANN-93 Int. Conf. on Artificial Neural Networks*, ed. by S. Gielen, B. Kappen (Springer, London, UK 1993) p. 861
[10.83]   N. Tschichold-Gürman, V. G. Dabija: In *Proc. ICNN'93, Int. Conf. on Neural Networks* (IEEE Service Center, Piscataway, NJ 1993) p. I-281
[10.84]   L. Z. Wang: In *Proc. IJCNN-93-Nagoya, Int. Joint Conf. on Neural Networks* (IEEE Service Center, Piscataway, NJ 1993) p. III-2452
[10.85]   A. Wichert: In *Proc. WCNN'93, World Congress on Neural Networks* (Lawrence Erlbaum, Hillsdale, NJ 1993) p. IV-59
[10.86]   P. Wu, K. Warwick: In *Proc. ICANN'91, Second Int. Conf. on Artificial Neural Networks (Conf. Publ. No.349)* (IEE, London, UK 1991) p. 350
[10.87]   L. Xu, E. Oja: Res. Report 16 (Department of Information Technology, Lappeenranta Univ. of Technology, Lappeenranta, Finland 1989)
[10.88]   L. Xu: Int. J. Neural Systems **1**, 269 (1990)
[10.89]   L. Xu, E. Oja: In *Proc. IJCNN-90-WASH-DC, Int. Joint Conf. on Neural Networks* (IEEE Service Center, Piscataway, NJ 1990) p. I-735
[10.90]   H. Yin, R. Lengelle, P. Gaillard: In *Proc. IJCNN'91, Int. Joint Conf. on Neural Networks* (IEEE Service Center, Piscataway, NJ 1991) p. I-839
[10.91]   M. Leman: *Music and Schema Theory: Cognitive Foundations of Systematic Musicology* (Springer, Berlin, Germany 1995)
[10.92]   M. Johnson, N. Allinson: In *Proc. SPIE – The Int. Society for Optical Engineering Vol. 1197* (SPIE, Bellingham, WA 1989) p. 109
[10.93]   M. J. Johnson, M. Brown, N. M. Allinson: In *Proc. Int. Workshop on Cellular Neural Networks and their Applications* (University of Budapest, Budapest, Hungary 1990) p. 254
[10.94]   N. R. Ball, K. Warwick: In *Proc. INNC'90, Int. Neural Network Conference* (Kluwer, Dordrecht, Netherlands 1990) p. I-242
[10.95]   B. Brückner, M. Franz, A. Richter: In *Artificial Neural Networks, 2*, ed. by I. Aleksander, J. Taylor (North-Holland, Amsterdam, Netherlands 1992) p. II-1167
[10.96]   B. Brückner, W. Zander: In *Proc. WCNN'93, World Congress on Neural Networks* (Lawrence Erlbaum, Hillsdale, NJ 1993) p. III-75
[10.97]   H.-L. Hung, W.-C. Lin: In *Proc. ICNN'94, Int. Conf. on Neural Networks* (IEEE Service Center, Piscataway, NJ 1994) p. 627
[10.98]   C. Kemke, A. Wichert: In *Proc. WCNN'93, World Congress on Neural Networks* (Lawrence Erlbaum, Hillsdale, NJ 1993) p. III-45
[10.99]   J. Lampinen, E. Oja: J. Mathematical Imaging and Vision **2**, 261 (1992)
[10.100] J. Lampinen: In *Artificial Neural Networks, 2*, ed. by I. Aleksander, J. Taylor (North-Holland, Amsterdam, Netherlands 1992) p. II-1219

[10.101] P. Weierich, M. v. Rosenberg: In *Proc. ICANN'94, Int. Conf. on Artificial Neural Networks*, ed. by M. Marinaro, P. G. Morasso (Springer, London, UK 1994) p. I-246

[10.102] P. Weierich, M. v. Rosenberg: In *Proc. ICNN'94, Int. Conf. on Neural Networks* (IEEE Service Center, Piscataway, NJ 1994) p. 612

[10.103] J. Li, C. N. Manikopoulos: Proc. SPIE – The Int. Society for Optical Engineering **1199**, 1046 (1989)

[10.104] Y. Cheng: In *Proc. IJCNN'92, Int. Joint Conf. on Neural Networks* (IEEE Service Center, Piscataway, NJ 1992) p. IV-785

[10.105] S. P. Luttrell: IEEE Trans. on Neural Networks **1**, 229 (1990)

[10.106] M. Gersho, R. Reiter: In *Proc. IJCNN-90-San Diego, Int. Joint Conf. on Neural Networks* (IEEE Service Center, Piscataway, NJ 1990) p. II-111

[10.107] O. G. Jakubowicz: In *Proc. IJCNN'89, Int. Joint Conf. on Neural Networks* (IEEE Service Center, Piscataway, NJ 1989) p. II-23

[10.108] J. Koh, M. Suk, S. M. Bhandarkar: In *Proc. ICNN'93, Int. Conf. on Neural Networks* (IEEE Service Center, Piscataway, NJ 1993) p. III-1270

[10.109] S. P. Luttrell: In *Proc. ICNN'88, Int. Conf. on Neural Networks* (IEEE Service Center, Piscataway, NJ 1988) p. I-93

[10.110] S. P. Luttrell: In *Proc. 1st IEE Conf. of Artificial Neural Networks* (British Neural Network Society, London, UK 1989) p. 2

[10.111] S. P. Luttrell: Pattern Recognition Letters **10**, 1 (1989)

[10.112] S. P. Luttrell: Proc. IEE Part I **136**, 405 (1989)

[10.113] S. P. Luttrell: In *Proc. 2nd IEE Conf. on Artificial Neural Networks* (British Neural Network Society, London, UK 1991) p. 5

[10.114] S. P. Luttrell: Technical Report 4467 (RSRE, Malvern, UK 1991)

[10.115] D. P. W. Graham, G. M. T. D'Eleuterio: In *Proc. IJCNN-91-Seattle, Int. Joint Conf. on Neural Networks* (IEEE Service Center, Piscataway, NJ 1991) p. II-1002

[10.116] R. Hecht-Nielsen: Appl. Opt. **26**, 4979 (1987)

[10.117] R. Hecht-Nielsen: In *Proc. ICNN'87, Int. Conf. on Neural Networks* (SOS Printing, San Diego, CA 1987) p. II-19

[10.118] R. Hecht-Nielsen: Neural Networks **1**, 131 (1988)

[10.119] Y. Lirov: In *IJCNN'91, Int. Joint Conf. on Neural Networks* (IEEE Service Center, Piscataway, NJ 1991) p. II-455

[10.120] T.-D. Chiueh, T.-T. Tang, L.-G. Chen: In *Proc. Int. Workshop on Application of Neural Networks to Telecommunications*, ed. by J. Alspector, R. Goodman, T. X. Brown (Lawrence Erlbaum, Hillsdale, NJ 1993) p. 259

[10.121] T. Li, L. Fang, A. Jennings: Technical Report CS-NN-91-5 (Concordia University, Department of Computer Science, Montreal, Quebec, Canada 1991)

[10.122] T.-D. Chiueh, T.-T. Tang, L.-G. Chen: IEEE Journal on Selected Areas in Communications **12**, 1594 (1994)

[10.123] H. J. Ritter: In *Proc. IJCNN'89, Int. Joint Conf. on Neural Networks, Washington DC* (IEEE Service Center, Piscataway, NJ 1989) p. II-499

[10.124] H. Ritter: In *Proc. COGNITIVA'90* (Elsevier, Amsterdam, Netherlands 1990) p. II-105

[10.125] H. Ichiki, M. Hagiwara, N. Nakagawa: In *Proc. IJCNN'91, Int. Conf. on Neural Networks* (IEEE Service Center, Piscataway, NJ 1991) p. I-357

[10.126] H. Ichiki, M. Hagiwara, M. Nakagawa: In *Proc. ICNN'93, Int. Conf. on Neural Networks* (IEEE Service Center, Piscataway, NJ 1993) p. III-1944

[10.127] L. Xu, E. Oja: In *Proc. IJCNN-90-WASH-DC, Int. Joint Conf. on Neural Networks* (Lawrence Erlbaum, Hillsdale, NJ 1990) p. II-531

[10.128]   D. S. Hwang, M. S. Han: In *Proc. ICNN'94, Int. Conf. on Neural Networks* (IEEE Service Center, Piscataway, NJ 1994) p. 742

[10.129]   A. S. Lazaro, L. Alonso, V. Cardenoso: In *Proc. Tenth IASTED Int. Conf. Applied Informatics*, ed. by M. H. Hamza (IASTED, Acta Press, Zurich, Switzerland 1992) p. 5

[10.130]   I. Yläkoski, A. Visa: In *Proc. 8SCIA, Scand. Conf. on Image Analysis* (NOBIM, Tromsø, Norway 1993) p. I-637

[10.131]   P. Koikkalainen, E. Oja: In *Proc. IJCNN-90-WASH-DC, Int. Joint Conf. on Neural Networks* (IEEE Service Center, Piscataway, NJ 1990) p. II-279

[10.132]   D.-I. Choi, S.-H. Park: Trans. Korean Inst. of Electrical Engineers **41**, 533 (1992)

[10.133]   G. Barna: Report A4 (Helsinki Univ. of Technology, Lab. of Computer and Information Science, Espoo, Finland 1987)

[10.134]   B. Fritzke, P. Wilke: In *Proc. IJCNN-90-Singapore, Int. Joint Conf. on Neural Networks* (IEEE Service Center, Piscataway, NJ 1991) p. 929

[10.135]   A. Sakar, R. J. Mammone: IEEE Trans. on Computers **42**, 291 (1993)

[10.136]   C. Szepesvári, L. Balázs, A. Lőrincz: Neural Computation **6**, 441 (1994)

[10.137]   W. X. Wen, V. Pang, A. Jennings: In *Proc. ICNN'93, Int. Conf. on Neural Networks* (IEEE Service Center, Piscataway, NJ 1993) p. III-1469

[10.138]   B. Fritzke: In *Proc. ESANN'96, European Symp. on Artificial Neural Networks*, ed. by M. Verleysen (D Facto Conference Services, Brussels, Belgium 1996) p. 61

[10.139]   L. L. H. Andrew: In *Proc. ANZIIS'94, Aust. New Zealand Intell. Info. Systems Conf.* (IEEE Service Center, Piscataway, NJ 1994) p. 10

[10.140]   K. W. Chan, K. L. Chan: In *Proc. ICNN'95, IEEE Int. Conf. on Neural Networks*, (IEEE Service Center, Piscataway, NJ 1995) p. IV-1898

[10.141]   H.-U. Bauer, T. Villmann: In *Proc. ICANN'95, Int. Conf. on Artificial Neural Networks*, ed. by F. Fogelman-Soulié, P. Gallinari, (EC2, Nanterre, France 1995) p. I-69

[10.142]   V. Pulkki: Report A27 (Helsinki Univ. of Technology, Laboratory of Computer and Information Science, Espoo, Finland 1995)

[10.143]   E. Cervera, A. P. del Pobil: In *Proc. CAEPIA'95, VI Conference of the Spanish Association for Artificial Intelligence* (1995) p. 471

[10.144]   T. Fujiwara, K. Fujimura, H. Tokutaka, S. Kishida: Technical Report NC94-49 (The Inst. of Electronics, Information and Communication Engineers, Tottori University, Koyama, Japan 1994)

[10.145]   T. Fujiwara, K. Fujimura, H. Tokutaka, S. Kishida: Technical Report NC94-100 (The Inst. of Electronics, Information and Communication Engineers, Tottori University, Koyama, Japan 1995)

[10.146]   K. Fujimura, T. Yamagishi, H. Tokutaka, T. Fujiwara, S. Kishida: In *Proc. 3rd Int. Conf. on Fuzzy Logic, Neural Nets and Soft Computing* (Fuzzy Logic Systems Institute, Iizuka, Japan 1994) p. 71

[10.147]   K. Yamane, H. Tokutaka, K. Fujimura, S. Kishida: Technical Report NC94-36 (The Inst. of Electronics, Information and Communication Engineers, Tottori University, Koyama, Japan 1994)

[10.148]   H. Ritter: In *Proc. ICANN'93, Int. Conf. on Artificial Neural Networks*, ed.by S. Gielen, B. Kappen (Springer, London, UK 1993) p. 568

[10.149]   H. Ritter: In *Proc. ICANN'94, Int. Conf. on Artificial Neural Networks*, ed.by M. Marinaro, P. G. Morasso (Springer, London, UK 1994) p. II-803

[10.150]   J. Walter, H. Ritter: In In *Proc. ICANN'95, Int. Conf. on Artificial Neural Networks*, ed. by F. Fogelman-Soulié, P. Gallinari, (EC2, Nanterre, France 1995) p. I-95

[10.151] H. Ritter: In *Proc. ICANN'97, Int. Conf. on Artificial Neural Networks*, vol.1327 of Lecture Notes in Computer Science, ed.by W. Gerstner, A. Germand, M. Haster, J. D. Nicoud (Springer, Berlin, Germany, 1997) p. 675

[10.152] H. Hyötyniemi: In *Proc. ICANN'93, Int. Conf. on Artificial Neural Networks*, ed. by S. Gielen, B. Kappen (Springer, London, UK 1993) p. 850

[10.153] G. J. Chappell, J. G. Taylor: Neural Networks **6**, 441 (1993)

[10.154] C. M. Privitera, P. Morasso: In *Proc. IJCNN-93-Nagoya, Int. Joint Conf. on Neural Networks* (IEEE Service Center, Piscataway, NJ 1993) p. III-2745

[10.155] J. K. Samarabandu, O. G. Jakubowicz: In *Proc. IJCNN-90-WASH-DC, Int. Joint Conf. on Neural Networks* (Lawrence Erlbaum, Hillsdale, NJ 1990) p. II-683

[10.156] V. V. Tolat, A. M. Peterson: In *Proc. IJCNN'89, Int. Joint Conf. on Neural Networks* (1989) p. II-561

[10.157] J. A. Zandhuis: Internal Report MPI-NL-TG-4/92 (Max-Planck-Institut für Psycholinguistik, Nijmegen, Netherlands 1992)

[10.158] J. C. Bezdek, E. C.-K. Tsao, N. R. Pal: In *Proc. IEEE Int. Conf. on Fuzzy Systems* (IEEE Service Center, Piscataway, NJ 1992) p. 1035

[10.159] Y. Lee, V. Cherkassky, J. R. Slagle: In *Proc. WCNN'94, World Congress on Neural Networks* (Lawrence Erlbaum, Hillsdale, NJ 1994) p. I-699

[10.160] H.-S. Rhee, K.-W. Oh: In *Proc. 3rd Int. Conf. on Fuzzy Logic, Neural Nets and Soft Computing* (Fuzzy Logic Systems Institute, Iizuka, Japan 1994) p. 335

[10.161] J. Sum, L.-W. Chan: In *Proc. ICNN'94, Int. Conf. on Neural Networks* (IEEE Service Center, Piscataway, NJ 1994) p. 1674

[10.162] J. Sum, L.-W. Chan: In *Proc. WCNN'94, World Congress on Neural Networks* (Lawrence Erlbaum, Hillsdale, NJ 1994) p. I-732

[10.163] F.-L. Chung, T. Lee: In *Proc. IJCNN-93-Nagoya, Int. Joint Conf. on Neural Networks* (IEEE Service Center, Piscataway, NJ 1993) p. III-2739

[10.164] F.-L. Chung, T. Lee: In *Proc. IJCNN-93-Nagoya, Int. Joint Conf. on Neural Networks* (IEEE Service Center, Piscataway, NJ 1993) p. III-2929

[10.165] Y. Sakuraba, T. Nakamoto, T. Moriizumi: Trans. Inst. of Electronics, Information and Communication Engineers **J73D-II**, 1863 (1990)

[10.166] Y. Sakuraba, T. Nakamoto, T. Moriizumi: Systems and Computers in Japan **22**, 93 (1991)

[10.167] E. C.-K. Tsao, J. C. Bezdek, N. R. Pal: In *NAFIPS'92, NASA Conf. Publication 10112* (North American Fuzzy Information Processing Society, 1992) p. I-98

[10.168] E. C.-K. Tsao, W.-C. Lin, C.-T. Chen: Pattern Recognition **26**, 553 (1993)

[10.169] E. C.-K. Tsao, H.-Y. Liao: In *Proc. ICANN'93, Int. Conf. on Artificial Neural Networks*, ed. by S. Gielen, B. Kappen (Springer, London, UK 1993) p. 249

[10.170] T. Yamaguchi, M. Tanabe, T. Takagi: In *Artificial Neural Networks*, ed. by T. Kohonen, K. Mäkisara, O. Simula, J. Kangas (North-Holland, Amsterdam, Netherlands 1991) p. II-1249

[10.171] T. Yamaguchi, M. Tanabe, K. Kuriyama, T. Mita: In *Int. Conf. on Control '91 (Conf. Publ. No.332)* (IEE, London, UK 1991) p. II-944

[10.172] T. Yamaguchi, T. Takagi, M. Tanabe: Electronics and Communications in Japan, Part 2 [Electronics] **75**, 52 (1992)

[10.173] H. Bayer: In *Proc. ICANN'93, Int. Conf. on Artificial Neural Networks*, ed. by S. Gielen, B. Kappen (Springer, London, UK 1993) p. 620

[10.174] M.-S. Chen, H.-C. Wang: Pattern Recognition Letters **13**, 315 (1992)

[10.175] T. Hrycej: Neurocomputing **4**, 17 (1992)

[10.176] K.-R. Hsieh, W.-T. Chen: IEEE Trans. Neural Networks **4**, 357 (1993)

[10.177] S. P. Luttrell: British Patent Application 9202752.3, 1992

[10.178] S. P. Luttrell: IEE Proc. F [Radar and Signal Processing] **139**, 371 (1992)

[10.179] S. Midenet, A. Grumbach: In *Proc. INNC'90 Int. Neural Network Conf.* (Kluwer, Dordrecht, Netherlands 1990) p. II-773

[10.180] J.-C. Fort, G. Pagès: In *Proc. ESANN'94, European Symp. on Artificial Neural Networks*, ed. by M. Verleysen (D Facto Conference Services, Brussels, Belgium 1994) p. 257

[10.181] P. Thiran: In *Proc. ESANN'95, European Symposium on Artificial Neural Networks*, ed. by M. Verleysen (D Facto Conference Services, Brussels, Belgium 1993) p. 203

[10.182] M. Hagiwara: Neurocomputing **10**, 71 (1996)

[10.183] J. Göppert, W. Rosenstiel: In *Proc. ESANN'95, European Symp. on Artificial Neural Networks*, ed. by M. Verleysen (D Facto Conference Services, Brussels, Belgium 1995) p. 15

[10.184] J. Göppert, W. Rosenstiel: In *Proc. ICANN'95, Int. Conf. on Artificial Neural Networks*, ed. by F. Fogelman-Soulié, P. Gallinari, (EC2, Nanterre, France 1995) p. II-69

[10.185] W.-P. Tai: In *Proc. ICANN'95, Int. Conf. on Artificial Neural Networks*, ed. by F. Fogelman-Soulié, P. Gallinari, (EC2, Nanterre, France 1995) p. II-33

[10.186] M. Herrmann: In *Proc. ICNN'95, IEEE Int. Conf. on Neural Networks*, (IEEE Service Center, Piscataway, NJ 1995) p.VI-2998

[10.187] K. Kiviluoto: In *Proc. ICNN'96, Int. Conf. on Neural Networks* (IEEE 1996 Service Center, Piscataway, NJ 1996) p. I-294

[10.188] I. Ahrns, J. Bruske, G. Sommer: In *Proc. ICANN'95, Int. Conf. on Artificial Neural Networks*, ed. by F. Fogelman-Soulié, P. Gallinari, (EC2, Nanterre, France 1995) p. II-141

[10.189] H. Copland, T. Hendtlass: In *Proc. ICNN'95, IEEE Int. Conf. on Neural Networks*, (IEEE Service Center, Piscataway, NJ 1995) p. I-669

[10.190] H. Hyötyniemi: In *Proc. EANN'95, Engineering Applications of Artificial Neural Networks* (Finnish Artificial Intelligence Society, Helsinki, Finland 1995) p. 147

[10.191] S. Jockusch, H. Ritter: Neural Networks **7**, 1229 (1994)

[10.192] R.-I. Chang, P.-Y. Hsiao: In *Proc. ICNN'95, IEEE Int. Conf. on Neural Networks*, (IEEE Service Center, Piscataway, NJ 1995) p.V-2610

[10.193] M. Cottrell, E. de Bodt: In *Proc. ESANN'96, European Symp. on Artificial Neural Networks*, ed. by M. Verleysen (D Facto Conference Services, Brussels, Belgium 1996) p. 103

[10.194] G. Poggi: IEEE Trans. on Image Processing **5**, 49 (1996)

[10.195] S. Rong, B. Bhanu: In *Proc. WCNN'95, World Congress on Neural Networks*, (INNS, Lawrence Erlbaum, Hillsdale, NJ 1995) p. I-552

[10.196] S.-I. Tanaka, K. Fujimura, H. Tokutaka, S. Kishida: Technical Report NC94-140 (The Inst. of Electronics, Information and Communication Engineers, Tottori University, Koyama, Japan 1995)

[10.197] N. Ueda, R. Nakano: Neural Networks **7**, 1211 (1994)

[10.198] M. M. Van Hulle: In *Proc. NNSP'95, IEEE Workshop on Neural Networks for Signal Processing* (IEEE Service Center, Piscataway, NJ 1995) p. 95

[10.199] A. Varfis: In *Proc. of NATO ASI Workshop on Statistics and Neural Networks* (1993)

[10.200] S. Puechmorel, E. Gaubet: In *Proc. WCNN'95, World Congress on Neural Networks*, (Lawrence Erlbaum, Hillsdale, NJ 1995) p. I-532

[10.201] M. Kobayashi, K. Tanahashi, K. Fujimura, H. Tokutaka, S. Kishida: Technical Report NC95-163 (The Inst. of Electronics, Information and Communication Engineers, Tottori University, Koyama, Japan 1996)

[10.202] H. Speckmann, G. Raddatz, W. Rosenstiel: In *Proc. ICANN'94, Int. Conf. on Artificial Neural Networks*, ed. by M. Marinaro, P. G. Morasso (Springer, London, UK 1994) p. I-342

[10.203] Z.-Z. Wang, D.-W. Hu, Q.-Y. Xiao: In *Proc. ICNN'94, Int. Conf. on Neural Networks* (IEEE Service Center, Piscataway, NJ 1994) p. 2793

[10.204] C. Bouton, G. Pagès: *Self-Organization and Convergence of the One-Dimensional Kohonen Algorithm with Non Uniformly Distributed Stimuli* (Laboratoire de Probabilités, Université Paris VI, Paris, France 1992)

[10.205] C. Bouton, G. Pagès: *Convergence in Distribution of the One-Dimensional Kohonen Algorithms when the Stimuli Are Not Uniform* (Laboratoire de Probabilités, Université Paris VI, Paris, France 1992)

[10.206] C. Bouton, G. Pagès: *Auto-organisation de l'algorithme de Kohonen en dimension 1*, in *Proc. Workshop 'Aspects Theoriques des Reseaux de Neurones'*, ed. by M. Cottrell, M. Chaleyat-Maurel (Université Paris I, Paris, France 1992)

[10.207] C. Bouton, G. Pagès: *Convergence p.s. et en loi de l'algorithme de Kohonen en dimension 1*, in *Proc. Workshop 'Aspects Theoriques des Reseaux de Neurones'*, ed. by M. Cottrell, M. Chaleyat-Maurel (Université Paris I, Paris, France 1992)

[10.208] J.-C. Fort, G. Pagès: C. R. Acad. Sci. Paris p. 389 (1993)

[10.209] C. Bouton, G. Pagès: Stochastic Processes and Their Applications **47**, 249 (1993)

[10.210] C. Bouton, G. Pagès: Adv. Appl. Prob. **26**, 80 (1994)

[10.211] J.-C. Fort, G. Pagès: In *Proc. ICANN'94, International Conference on Artificial Neural Networks*, ed.by M. Marinaro, P. G. Morasso (Springer, London 1994) p. 318

[10.212] J.-C. Fort, G. Pagès: Ann. Appl. Prob. **5(4)**, 1177 (1995)

[10.213] M. Benaïm, J.-C. Fort, G. Pagès: In *Proc. ESANN'97, European Symposium on Neural Networks*, ed.by M. Verkysen (D Facto Conference Services, Brussels, Belgium 1997) p. 193

[10.214] J.-C. Fort, G. Pagès: Neural Networks **9(5)**, 773 (1995)

[10.215] J. A. Flanagan: PhD Thesis (Ecole Polytechnique Fédérale de Lausanne, Lausanne, Switzerland 1994)

[10.216] J. A. Flanagan: Neural Networks **6(7)**, 1185 (1996)

[10.217] J, A. Flanagan: To be published in *Proc. ESANN'2000, European Symposium on Artifical Neural Networks*

[10.218] J. A. Flanagan: To be published in *Proc. IJCNN'2000, Int. Joint Conf. on Neural Networks*

[10.219] P. Bak, C. Tang, K. Wiesenfeld: Phys. Rev A**38**, 364 (1987)

[10.220] P. Ružička: Neural Network World **4**, 413 (1993)

[10.221] T. Ae: Joho Shori **32**, 1301 (1991)

[10.222] S. C. Ahalt, A. K. Krishnamurty, P. Chen, D. E. Melton: Neural Networks **3**, 277 (1990)

[10.223] P. Ajjimarangsee, T. L. Huntsberger: In *Proc. 4th Conf. on Hypercubes, Concurrent Computers and Applications* (Golden Gate Enterprises, Los Altos, CA 1989) p. II-1093

[10.224] K. Akingbehin, K. Khorasani, A. Shaout: In *Proc. 2nd IASTED Int. Symp. Expert Systems and Neural Networks*, ed. by M. H. Hamza (Acta Press, Anaheim, CA 1990) p. 66

[10.225] N. M. Allinson: In *Theory and Applications of Neural Networks*, ed. by J. G. Taylor, C. L. T. Mannion (Springer, London, UK 1992) p. 101

[10.226] M. Andres, O. Schlüter, F. Spengler, H. R. Dinse: In *Proc. ICANN'94, Int. Conf. on Artificial Neural Networks*, ed. by M. Marinaro, P. G. Morasso (Springer, London, UK 1994) p. 306

[10.227] L. L. H. Andrew, M. Palaniswami: In *Proc. ICNN'94 IEEE Int. Conf. on Neural Networks* (IEEE Service Center, Piscataway, NJ 1994) p. 4159

[10.228] N. R. Ball: In *Artificial Neural Networks, 2*, ed. by I. Aleksander, J. Taylor (North-Holland, Amsterdam, Netherlands 1992) p. I-703

[10.229] N. R. Ball: In *Proc. ICANN'94, Int. Conf. on Artificial Neural Networks*, ed. by M. Marinaro, P. G. Morasso (Springer, London, UK 1994) p. I-663

[10.230] W. Banzhaf, H. Haken: Neural Networks **3**, 423 (1990)

[10.231] G. N. Bebis, G. M. Papadourakis: In *Artificial Neural Networks*, ed. by T. Kohonen, K. Mäkisara, O. Simula, J. Kangas (North-Holland, Amsterdam, Netherlands 1991) p. II-1111

[10.232] J. C. Bezdek, N. R. Pal: In *Proc. 5th IFSA World Congress '93 – Seoul, Fifth Int. Fuzzy Systems Association World Congress* (Korea Fuzzy Mathematics and Systems Society, Seoul, Korea 1993) p. I-36

[10.233] J. M. Bishop, R. J. Mitchell: In *Proc. IEE Colloquium on 'Neural Networks for Systems: Principles and Applications' (Digest No.019)* (IEE, London, UK 1991) p. 1

[10.234] M. de Bollivier, P. Gallinari, S. Thiria: In *Proc. IJCNN'90, Int. Joint Conf. on Neural Networks* (IEEE Service Center, Piscataway, NJ 1990) p. I-113

[10.235] M. de Bollivier, P. Gallinari, S. Thiria: In *Proc. INNC 90, Int. Neural Network Conf.* (Kluwer, Dordrecht, Netherlands 1990) p. II-777

[10.236] R. W. Brause: In *Proc. ICANN'94, Int. Conf. on Artificial Neural Networks*, ed. by M. Marinaro, P. G. Morasso (Springer, London, UK 1994) p. I-701

[10.237] M. Budinich, J. G. Taylor: In *Proc. ICANN'94, Int. Conf. on Artificial Neural Networks*, ed. by M. Marinaro, P. G. Morasso (Springer, London, UK 1994) p. I-347

[10.238] S. Cammarata: Sistemi et Impresa **35**, 688 (1989)

[10.239] M. Caudill: In *Proc. Fourth Annual Artificial Intelligence and Advanced Computer Technology Conference* (Tower Conf. Management, Glen Ellyn, IL 1988) p. 298

[10.240] M. Caudill: AI Expert **8**, 16 (1993)

[10.241] R.-I. Chang, P.-Y. Hsiao: In *Proc. 1994 Int. Symp. on Speech, Image Processing and Neural Networks* (IEEE Hong Kong Chapt. of Signal Processing, Hong Kong 1994) p. I-85

[10.242] J. Chao, K. Minowa, S. Tsujii: In *Proc. ICANN'94, Int. Conf. on Artificial Neural Networks*, ed. by M. Marinaro, P. G. Morasso (Springer, London, UK 1994) p. II-1460

[10.243] S. Clippingdale, R. Wilson: In *Proc. IJCNN-93-Nagoya, Int. Joint Conf. on Neural Networks* (IEEE Service Center, Piscataway, NJ 1993) p. III-2504

[10.244] A. M. Colla, N. Longo, G. Morgavi, S. Ridella: In *Proc. ICANN'94, Int. Conf. on Artificial Neural Networks*, ed. by M. Marinaro, P. G. Morasso (Springer, London, UK 1994) p. I-230

[10.245] M. Cottrell, J.-C. Fort: Biol. Cyb. **53**, 405 (1986)

[10.246] M. Cottrell, J.-C. Fort, G. Pagès: Technical Report 31 (Université Paris 1, Paris, France 1994)

[10.247] D. D'Amore, V. Piuri: In *Proc. IMACS Int. Symp. on Signal Processing, Robotics and Neural Networks* (IMACS, Lille, France 1994) p. 534

[10.248] G. R. D. Haan, Ö. Eğecioğlu: In *Proc. IJCNN'91, Int. Joint Conf. on Neural Networks* (IEEE Service Center, Piscataway, NJ 1991) p. I-887

[10.249] P. Demartines, F. Blayo: Complex Systems **6**, 105 (1992)

[10.250] R. Der, M. Herrmann: In *Proc. ICANN'93, Int. Conf. on Artificial Neural Networks*, ed. by S. Gielen, B. Kappen (Springer, London, UK 1993) p. 597

[10.251] R. Der, M. Herrmann: In *Proc. ICANN'94, Int. Conf. on Artificial Neural Networks*, ed. by M. Marinaro, P. G. Morasso (Springer, London, UK 1994) p. I-322

[10.252] I. Dumitrache, C. Buiu: In *Proc. IMACS Int. Symp. on Signal Processing, Robotics and Neural Networks* (IMACS, Lille, France 1994) p. 530

[10.253] S. J. Eglen, G. Hill, F. J. Lazare, N. P. Walker: GEC Review **7**, 146 (1992)

[10.254] M. Eldracher, H. Geiger: In *Proc. ICANN'94, Int. Conf. on Artificial Neural Networks*, ed. by M. Marinaro, P. G. Morasso (Springer, London, UK 1994) p. I-771

[10.255] H. Elsherif, M. Hambaba: In *Proc. ICNN'94, Int. Conf. on Neural Networks* (IEEE Service Center, Piscataway, NJ 1994) p. 535

[10.256] P. Érdi, G. Barna: Biol. Cyb. **51**, 93 (1984)

[10.257] E. Erwin, K. Obermayer, K. Schulten: In *Proc. Fourth Conf. on Neural Networks*, ed. by S. I. Sayegh (Indiana University at Fort Wayne, Fort Wayne, IN 1992) p. 115

[10.258] W. Fakhr, M. Kamel, M. I. Elmasry: In *Proc. ICNN'94, Int. Conf. on Neural Networks* (IEEE Service Center, Piscataway, NJ 1994) p. 401

[10.259] W. Fakhr, M. Kamel, M. I. Elmasry: In *Proc. WCNN'94, World Congress on Neural Networks* (Lawrence Erlbaum, Hillsdale, NJ 1994) p. III-123

[10.260] E. Fiesler: In *Proc. ICANN'94, Int. Conf. on Artificial Neural Networks*, ed. by M. Marinaro, P. G. Morasso (Springer, London, UK 1994) p. I-793

[10.261] J. A. Flanagan, M. Hasler: In *Proc. Conf. on Artificial Intelligence Res. in Finland*, ed. by C. Carlsson, T. Järvi, T. Reponen, Number 12 in Conf. Proc. of Finnish Artificial Intelligence Society (Finnish Artificial Intelligence Society, Helsinki, Finland 1994) p. 13

[10.262] D. Flotzinger, M. Pregenzer, G. Pfurtscheller: In *Proc. ICNN'94, Int. Conf. on Neural Networks* (IEEE Service Center, Piscataway, NJ 1994) p. 3448

[10.263] F. Fogelman-Soulié, P. Gallinari: Bull. de liaison de la recherche en informatique et en automatique, p. 19 (1989)

[10.264] T. Fomin, C. Szepesvári, A. Lőrincz: In *Proc. ICNN'94, Int. Conf. on Neural Networks* (IEEE Service Center, Piscataway, NJ 1994) p. 2777

[10.265] J.-C. Fort, G. Pagès: In *Proc. ICANN'94, Int. Conf. on Artificial Neural Networks*, ed. by M. Marinaro, P. G. Morasso (Springer, London, UK 1994) p. 318

[10.266] J.-C. Fort, G. Pagès: Technical Report 29 (Université Paris 1, Paris, France 1994)

[10.267] B. Fritzke: In *Artificial Neural Networks 2*, ed. by I. Aleksander, J. Taylor (North-Holland, Amsterdam, Netherlands 1992) p. 1273

[10.268] K. Fujimura, T. Yamagishi, H. Tokutaka, T. Fujiwara, S. Kishida: In *Proc. 3rd Int. Conf. on Fuzzy Logic, Neural Nets and Soft Computing* (Fuzzy Logic Systems Institute, Iizuka, Japan 1994) p. 71

[10.269]   S. Garavaglia: In *Proc. WCNN'94, World Congress on Neural Networks* (Lawrence Erlbaum, Hillsdale, NJ 1994) p. I-502

[10.270]   L. Garrido (ed.). *Statistical Mechanics of Neural Networks. Proc. XI Sitges Conference* (Springer, Berlin, Germany 1990)

[10.271]   M. Gersho, R. Reiter: In *Proc. INNC'90, Int. Neural Network Conf.* (Kluwer, Dordrecht, Netherlands 1990) p. 1-361

[10.272]   T. Geszti: *Physical Models of Neural Networks* (World Scientific, Singapore 1990)

[10.273]   J. Heikkonen: PhD Thesis (Lappeenranta University of Technology, Lappeenranta, Finland 1994)

[10.274]   J. Henseler: PhD Thesis (University of Limburg, Maastricht, Netherlands 1993)

[10.275]   J. A. Hertz, A. Krogh, R. G. Palmer: *Introduction to the Theory of Neural Computation* (Addison-Wesley, Redwood City, CA 1991)

[10.276]   T. Heskes, S. Gielen: Phys. Rev. A **44**, 2718 (1991)

[10.277]   T. Heskes, B. Kappen, S. Gielen: In *Artificial Neural Networks*, ed. by T. Kohonen, K. Mäkisara, O. Simula, J. Kangas (North-Holland, Amsterdam, Netherlands 1991) p. I-15

[10.278]   T. Heskes, B. Kappen: Physical Review A **45**, 8885 (1992)

[10.279]   T. Heskes, E. Slijpen, B. Kappen: Physical Review A **46**, 5221 (1992)

[10.280]   T. Heskes, E. Slijpen: In *Artificial Neural Networks, 2*, ed. by I. Aleksander, J. Taylor (North-Holland, Amsterdam, Netherlands 1992) p. I-101

[10.281]   T. Heskes, B. Kappen: In *Mathematical Foundations of Neural Networks*, ed. by J. Taylor (Elsevier, Amsterdam, Netherlands 1993)

[10.282]   T. Heskes: PhD Thesis (Katholieke Universiteit Nijmegen, Nijmegen, Netherlands 1993)

[10.283]   A. Hiotis: AI Expert **8**, 38 (1993)

[10.284]   R. E. Hodges, C.-H. Wu: In *Proc. ISCAS'90, Int. Symp. on Circuits and Systems* (IEEE Service Center, Piscataway, NJ 1990) p. I-204

[10.285]   R. Hodges, C.-H. Wu: In *Proc. IJCNN'90-Wash, Int. Joint Conf. on Neural Networks* (Lawrence Erlbaum, Hillsdale, NJ 1990) p. I-517

[10.286]   L. Holmström, T. Kohonen: In *Tekoälyn ensyklopedia*, ed. by E. Hyvönen, I. Karanta, M. Syrjänen (Gaudeamus, Helsinki, Finland 1993) p. 85

[10.287]   T. Hrycej: In *Proc. IJCNN'90-San Diego, Int. joint Conf. on Neural Networks* (IEEE Service Center, Piscataway, NJ 1990) p. 2-307

[10.288]   T. L. Huntsberger, P. Ajjimarangsee: Int. J. General Systems **16**, 357 (1990)

[10.289]   J. Iivarinen, T. Kohonen, J. Kangas, S. Kaski: In *Proc. Conf. on Artificial Intelligence Res. in Finland*, ed. by C. Carlsson, T. Järvi, T. Reponen, Number 12 in Conf. Proc. of Finnish Artificial Intelligence Society (Finnish Artificial Intelligence Society, Helsinki, Finland 1994) p. 122

[10.290]   S. Jockusch: In *Parallel Processing in Neural Systems and Computers*, ed. by R. Eckmiller, G. Hartmann, G.Hauske (Elsevier, Amsterdam, Netherlands 1990) p. 169

[10.291]   T.-P. Jung, A. K. Krishnamurthy, S. C. Ahalt: In *Proc. IJCNN-90-San Diego, Int. Joint Conf. on Neural Networks* (IEEE Service Center, Piscataway, NJ 1990) p. III-251

[10.292]   C. Jutten, A. Guerin, H. L. N. Thi: In *Proc. IWANN'91, Int. Workshop on Artificial Neural Networks*, ed. by A. Prieto (Springer, Berlin, Heidelberg 1991) p. 54

[10.293]   I. T. Kalnay, Y. Cheng: In *IJCNN-91-Seattle, Int. Joint Conf. on Neural Networks* (IEEE Service Center, Piscataway, NJ 1991) p. II-981

[10.294] B.-H. Kang, D.-S. Hwang, J.-H. Yoo: In *Proc. 3rd Int. Conf. on Fuzzy Logic, Neural Nets and Soft Computing* (Fuzzy Logic Systems Institute, Iizuka, Japan 1994) p. 333

[10.295] J. Kangas, T. Kohonen: In *Proc. IMACS Int. Symp. on Signal Processing, Robotics and Neural Networks* (IMACS, Lille, France 1994) p. 19

[10.296] D. Keymeulen, J. Decuyper: In *Toward a Practice of Autonomous Systems. Proc. First European Conf. on Artificial Life*, ed. by F. J. Varela, P. Bourgine (MIT Press, Cambridge, MA 1992) p. 64

[10.297] C. Khunasaraphan, T. Tanprasert, C. Lursinsap: In *Proc. WCNN'94, World Congress on Neural Networks* (Lawrence Erlbaum, Hillsdale, NJ 1994) p. IV-234

[10.298] D. S. Kim, T. L. Huntsberger: In *Tenth Annual Int. Phoenix Conf. on Computers and Communications* (IEEE Comput. Soc. Press, Los Alamitos, CA 1991) p. 39

[10.299] K. Y. Kim, J. B. Ra: In *Proc. IJCNN-93-Nagoya, Int. Joint Conf. on Neural Networks* (IEEE Service Center, Piscataway, NJ 1993) p. II-1219

[10.300] J. Kim, J. Ahn, C. S. Kim, H. Hwang, S. Cho: In *Proc. ICNN'94, Int. Conf. on Neural Networks* (IEEE Service Center, Piscataway, NJ 1994) p. 692

[10.301] T. Kohonen: Report TKK-F-A450 (Helsinki University of Technology, Espoo, Finland 1981)

[10.302] T. Kohonen: Report TKK-F-A461 (Helsinki University of Technology, Espoo, Finland 1981)

[10.303] T. Kohonen: In *Competition and Cooperation in Neural Nets, Lecture Notes in Biomathematics, Vol. 45*, ed. by S.-i. Amari, M. A. Arbib (Springer, Berlin, Heidelberg 1982) p. 248

[10.304] T. Kohonen: In *Proc. Seminar on Frames, Pattern Recognition Processes, and Natural Language* (The Linguistic Society of Finland, Helsinki, Finland 1982)

[10.305] T. Kohonen: In *Topics in Technical Physics, Acta Polytechnica Scandinavica, Applied Physics Series No. 138*, ed. by V. Kelhä, M. Luukkala, T. Tuomi (Finnish Academy of Engineering Sciences, Helsinki, Finland 1983) p. 80

[10.306] T. Kohonen: In *Synergetics of the Brain*, ed. by E. Başar, H. Flohr, H. Haken, A. J. Mandell (Springer, Berlin, Heidelberg 1983) p. 264

[10.307] T. Kohonen: In *Proc. 3SCIA, Scand. Conf. on Image Analysis* (Studentlitteratur, Lund, Sweden 1983) p. 35

[10.308] T. Kohonen, P. Lehtiö: Tiede 2000 (Finland), No. 2, p. 19 (1983)

[10.309] T. Kohonen: In *Cybernetic Systems: Recognition, Learning, Self-Organization*, ed. by E. R. Caianiello, G. Musso (Res. Studies Press, Letchworth, UK 1984) p. 3

[10.310] T. Kohonen: Sähkö (Finland) **57**, 48 (1984)

[10.311] T. Kohonen: In *Proc. COGNITIVA 85* (North-Holland, Amsterdam, Netherlands 1985) p. 585

[10.312] T. Kohonen: In *Proc. of the XIV Int. Conf. on Medical Physics, Espoo, Finland, August 11-16* (Finnish Soc. Med. Phys. and Med. Engineering, Helsinki, Finland 1985) p. 1489

[10.313] T. Kohonen: In *Proc. 4SCIA, Scand. Conf. on Image Analysis* (Tapir Publishers, Trondheim, Norway 1985) p. 97

[10.314] T. Kohonen, K. Mäkisara: In *AIP Conf. Proc. 151, Neural Networks for Computing*, ed. by J. Denker (Amer. Inst. of Phys., New York, NY 1986) p. 271

[10.315]   T. Kohonen: In *Physics of Cognitive Processes*, ed. by E. R. Caianiello (World Scientific, Singapore 1987) p. 258

[10.316]   T. Kohonen: In *Second World Congr. of Neuroscience, Book of Abstracts. Neuroscience, Supplement to Volume 22* (1987) p. S100

[10.317]   T. Kohonen: In *Proc. ICNN'87, Int. Conf. on Neural Networks* (IEEE Service Center, Piscataway, NJ 1987) p. I-79

[10.318]   T. Kohonen: In *Optical and Hybrid Computing, SPIE Vol. 634* (SPIE, Bellingham, WA 1987) p. 248

[10.319]   T. Kohonen: In *Computer Simulation in Brain Science*, ed. by R. M. J. Cotterill (Cambridge University Press, Cambridge, UK 1988) p. 12

[10.320]   T. Kohonen: In *Kognitiotiede*, ed. by A. Hautamäki (Gaudeamus, Helsinki, Finland 1988) p. 100

[10.321]   T. Kohonen: In *European Seminar on Neural Computing* (British Neural Network Society, London, UK 1988)

[10.322]   T. Kohonen: Neural Networks **1**, 29 (1988)

[10.323]   T. Kohonen: In *First IEE Int. Conf. on Artificial Neural Networks* (IEE, London, UK 1989) p. 1

[10.324]   T. Kohonen, K. Mäkisara: Physica Scripta **39**, 168 (1989)

[10.325]   T. Kohonen: In *Parallel Processing in Neural Systems and Computers*, ed. by R. Eckmiller, G. Hartman, G. Hauske (Elsevier, Amsterdam, Netherlands 1990) p. 177

[10.326]   T. Kohonen: In *Neural Networks: Biological Computers or Electronic Brains, Proc. Int. Conf. Les Entrétiens de Lyon* (Springer, Paris, France 1990) p. 29

[10.327]   T. Kohonen: In *Brain Organization and Memory: Cells, Systems, and Circuits*, ed. by J. L. McGaugh, N. M. Weinberger, G. Lynch (Oxford University Press, New York, NY 1990) p. 323

[10.328]   T. Kohonen: In *New Concepts in Computer Science: Proc. Symp. in Honour of Jean-Claude Simon* (AFCET, Paris, France 1990) p. 181

[10.329]   T. Kohonen: In *Proc. Third Italian Workshop on Parallel Architectures and Neural Networks* (SIREN, Vietri sul Mare, Italy 1990) p. 13

[10.330]   T. Kohonen: In *Proc. IJCNN-90-WASH-DC, Int. Joint Conf. on Neural Networks Washington* (Lawrence Erlbaum, Hillsdale, NJ 1990) p. II-253

[10.331]   T. Kohonen, S. Kaski: In *Abstracts of the 15th Annual Meeting of the European Neuroscience Association* (Oxford University Press, Oxford, UK 1992) p. 280

[10.332]   T. Kohonen: In *Proc. ICANN'94, Int. Conf. on Artificial Neural Networks*, ed. by M. Marinaro, P. G. Morasso (Springer, London, UK 1994) p. I-292

[10.333]   P. Koikkalainen: In *Proc. ICANN'94, Int. Conf. on Artificial Neural Networks*, ed. by M. Marinaro, P. G. Morasso (Springer, London, UK 1994) p. II-1137

[10.334]   P. Koistinen, L. Holmström: Research Reports A10 (Rolf Nevanlinna Institute, Helsinki, Finland 1993)

[10.335]   G. A. Korn: IEEE Trans. on Syst., Man and Cybern. **20**, 1146 (1990)

[10.336]   A. Kumar, V. E. McGee: In *Proc. WCNN'94, World Congress on Neural Networks* (Lawrence Erlbaum, Hillsdale, NJ 1994) p. II-278

[10.337]   M. Lalonde, J.-J. Brault: In *Proc. WCNN'94, World Congress on Neural Networks* (Lawrence Erlbaum, Hillsdale, NJ 1994) p. III-110

[10.338]   J. Lampinen: PhD Thesis (Lappenranta University of Technology, Lappeenranta, Finland 1992)

[10.339]   R. Lancini, F. Perego, S. Tubaro: In *Proc. GLOBECOM'91, Global Telecommunications Conf. Countdown to the New Millennium. Featur-*

ing a Mini-Theme on: Personal Communications Services (PCS). (IEEE Service Center, Piscataway, NJ 1991) p. I-135

[10.340]   A. B. Larkin, E. L. Hines, S. M. Thomas: Neural Computing & Applications **2**, 53 (1994)

[10.341]   E. Lebert, R. H. Phaf: In *Proc. ICANN'93, Int. Conf. on Artificial Neural Networks*, ed. by S. Gielen, B. Kappen (Springer, London, UK 1993) p. 59

[10.342]   X. Li, J. Gasteiger, J. Zupan: Biol. Cyb. **70**, 189 (1993)

[10.343]   S. Z. Li: In *Proc. IJCNN-93-Nagoya, Int. Joint Conf. on Neural Networks* (IEEE Service Center, Piscataway, NJ 1993) p. II-1173

[10.344]   R. Linsker: In *Neural Information Processing Systems*, ed. by D. Z. Anderson (Amer. Inst. Phys., New York, NY 1987) p. 485

[10.345]   R. P. Lippmann: IEEE Acoustics, Speech and Signal Processing Magazine p. 4 (1987)

[10.346]   R. P. Lippmann: In *Proc. ICASSP'88, Int. Conf. on Acoustics, Speech and Signal Processing* (IEEE Service Center, Piscataway, NJ 1988) p. 1

[10.347]   R. P. Lippmann: In *Proc. ICS'88, Third Int. Conf. on Supercomputing*, ed. by L. P. Kartashev, S. I. Kartashev (Int. Supercomputing Inst., St.Petersburg, FL 1988) p. I-35

[10.348]   R. P. Lippmann: IEEE Communications Magazine **27**, 47 (1989)

[10.349]   H.-D. Litke: NET **44**, 330 (1990)

[10.350]   Z.-P. Lo, B. Bavarian: Pattern Recognition Letters **12**, 549 (1991)

[10.351]   Z.-P. Lo, M. Fujita, B. Bavarian: In *Proc. Conf. IEEE Int. Conf. on Syst., Man, and Cybern. 'Decision Aiding for Complex Systems'* (IEEE Service Center, Piscataway, NJ 1991) p. III-1599

[10.352]   Z.-P. Lo, B. Bavarian: In *IJCNN-91-Seattle: Int. Joint Conf. on Neural Networks* (IEEE Service Center, Piscataway, NJ 1991) p. I-263

[10.353]   Z.-P. Lo, M. Fujita, B. Bavarian: In *Proc. Fifth Int. Parallel Processing Symp.* (IEEE Comput. Soc. Press, Los Alamitos, CA 1991) p. 246

[10.354]   S. Maekawa, H. Kita, Y. Nishikawa: In *Proc. ICNN'94, Int. Conf. on Neural Networks* (IEEE Service Center, Piscataway, NJ 1994) p. 2813

[10.355]   E. Maillard, J. Gresser: In *Proc. ICNN'94, Int. Conf. on Neural Networks* (IEEE Service Center, Piscataway, NJ 1994) p. 704

[10.356]   E. Masson, Y.-J. Wang: European J. Operational Res. **47**, 1 (1990)

[10.357]   M. McInerney, A. Dhawan: In *Proc. ICNN'94, Int. Conf. on Neural Networks* (IEEE Service Center, Piscataway, NJ 1994) p. 641

[10.358]   A. Meyering, H. Ritter: In *Maschinelles Lernen – Modellierung von Lernen mit Maschinen*, ed. by K. Reiss, M. Reiss, H. Spandl (Springer, Berlin, Germany 1992)

[10.359]   J. Monnerjahn: In *Proc. ICANN'94, Int. Conf. on Artificial Neural Networks*, ed. by M. Marinaro, P. G. Morasso (Springer, London, UK 1994) p. I-326

[10.360]   P. Morasso, A. Pareto, V. Sanguineti: In *Proc. WCNN'93, World Congress on Neural Networks* (Lawrence Erlbaum, Hillsdale, NJ 1993) p. III-372

[10.361]   K.-L. Mou, D.-Y. Yeung: In *Proc. Int. Symp. on Speech, Image Processing and Neural Networks* (IEEE Hong Kong Chapter of Signal Processing, Hong Kong 1994) p. II-658

[10.362]   E. Oja: In *Proc. Symp. on Image Sensing and Processing in Industry* (Pattern Recognition Society of Japan, Tokyo, Japan 1991) p. 143

[10.363]   E. Oja: In *Proc. NORDDATA* (Tietojenkäsittelyliitto, Helsinki, Finland 1992) p. 306

[10.364] E. Oja: In *Proc. Conf. on Artificial Intelligence Res. in Finland*, ed. by C. Carlsson, T. Järvi, T. Reponen, Number 12 in Conf. Proc. of Finnish Artificial Intelligence Society (Finnish Artificial Intelligence Society, Helsinki, Finland 1994) p. 2

[10.365] Y.-H. Pao: *Adaptive Pattern Recognition and Neural Networks* (Addison-Wesley, Reading, MA 1989)

[10.366] S.-T. Park, S.-Y. Bang: Korea Information Science Society Rev. **10**, 5 (1992)

[10.367] H. Ritter: PhD Thesis (Technical University of Munich, Munich, Germany 1988)

[10.368] H. Ritter, K. Schulten: In *Proc. ICNN'88 Int. Conf. on Neural Networks* (IEEE Service Center, Piscataway, NJ 1988) p. I-109

[10.369] H. J. Ritter: Psych. Res. **52**, 128 (1990)

[10.370] H. Ritter, K. Obermayer, K. Schulten, J. Rubner: In *Models of Neural Networks*, ed. by J. L. von Hemmen, E. Domany, K. Schulten (Springer, New York, NY 1991) p. 281

[10.371] H. Ritter: In *Artificial Neural Networks.*, ed. by T. Kohonen, K. Mäkisara, O. Simula, J. Kangas (Elsevier, Amsterdam, Netherlands 1991) p. 379

[10.372] H. Ritter: In *Proc. NOLTA, 2nd Symp. on Nonlinear Theory and its Applications* (Fukuoka, Japan 1991) p. 5

[10.373] H. Ritter: In *Proc. ICANN'94, Int. Conf. on Artificial Neural Networks*, ed. by M. Marinaro, P. G. Morasso (Springer, London, UK 1994) p. II-803

[10.374] S. K. Rogers, M. Kabrisky: In *Proc. IEEE National Aerospace and Electronics Conf.* (IEEE Service Center, Piscataway, NJ 1989) p. 688

[10.375] T. Röfer: In *Proc. ICANN'94, Int. Conf. on Artificial Neural Networks*, ed. by M. Marinaro, P. G. Morasso (Springer, London, UK 1994) p. II-1311

[10.376] H. Sako: In *Proc. ICNN'94, Int. Conf. on Neural Networks* (IEEE Service Center, Piscataway, NJ 1994) p. 3072

[10.377] H. Sano, Y. Iwahori, N. Ishii: In *Proc. ICNN'94, Int. Conf. on Neural Networks* (IEEE Service Center, Piscataway, NJ 1994) p. 1537

[10.378] O. Scherf, K. Pawelzik, F. Wolf, T. Geisel: In *Proc. ICANN'94, Int. Conf. on Artificial Neural Networks*, ed. by M. Marinaro, P. G. Morasso (Springer, London, UK 1994) p. I-338

[10.379] J. C. Scholtes: In *Proc. IJCNN'91, Int. Joint Conf. on Neural Networks* (IEEE Service Center, Piscataway, NJ 1991) p. I-95

[10.380] H. P. Siemon: In *Artificial Neural Networks, 2*, ed. by I. Aleksander, J. Taylor (North-Holland, Amsterdam, Netherlands 1992) p. II-1573

[10.381] Y. T. So, K. P. Chan: In *Proc. ICNN'94, Int. Conf. on Neural Networks* (IEEE Service Center, Piscataway, NJ 1994) p. 681

[10.382] C. Szepesvári, T. Fomin, A. Lőrincz: In *Proc. ICANN'94, Int. Conf. on Artificial Neural Networks*, ed. by M. Marinaro, P. G. Morasso (Springer, London, UK 1994) p. II-1261

[10.383] J. G. Taylor, C. L. T. Mannion (eds.): *New Developments in Neural Computing. Proc. Meeting on Neural Computing* (Adam Hilger, Bristol, UK 1989)

[10.384] P. Thiran, M. Hasler: In *Proc. Workshop 'Aspects Theoriques des Reseaux de Neurones'*, ed. by M. Cottrell, M. Chaleyat-Maurel (Université Paris I, Paris, France 1992)

[10.385] R. Togneri, D. Farrokhi, Y. Zhang, Y. Attikiouzel: In *Proc. Fourth Australian Int. Conf. on Speech Science and Technology* (Brisbane, Australia 1992) p. 173

[10.386] C. Touzet: *Reseaux de neurones artificiels: introduction au connexionnisme (Artificial neural nets: introduction to connectionism)* (EC2, Nanterre, France 1992)

[10.387] P. C. Treleaven: Int. J. Neurocomputing **1**, 4 (1989)

[10.388] V. Tryba, K. M. Marks, U. Ruckert, K. Goser: ITG-Fachberichte **102**, 407 (1988)

[10.389] V. Tryba, K. M. Marks, U. Rückert, K. Goser: In *Tagungsband der ITG-Fachtagung "Digitale Speicher"* (ITG, Darmstadt, Germany 1988) p. 409

[10.390] R. W. M. Van Riet, P. C. Duives: Informatie **33**, 368 (1991)

[10.391] G. A. v. Velzen: Technical Report UBI-T-92.MF-077 (Utrecht Biophysics Res. Institute, Utrecht, Netherlands 1992)

[10.392] T. Villmann, R. Der, T. Martinetz: In *Proc. ICANN'94, Int. Conf. on Artificial Neural Networks*, ed. by M. Marinaro, P. G. Morasso (Springer, London, UK 1994) p. I-298

[10.393] T. Villmann, R. Der, T. Martinetz: In *Proc. ICNN'94, Int. Conf. on Neural Networks* (IEEE Service Center, Piscataway, NJ 1994) p. 645

[10.394] A. Visa: In *Artificial Neural Networks 2*, ed. by I. Aleksander, J. Taylor (Elsevier, Amsterdam, Netherlands 1992) p. 803

[10.395] L. Xu: In *Proc. ICNN'94, Int. Conf. on Neural Networks* (IEEE Service Center, Piscataway, NJ 1994) p. 315

[10.396] M. M. Yen, M. R. Blackburn, H. G. Nguyen: In *Proc. IJCNN'90-WASH-DC, Int. Joint Conf. on Neural Networks* (IEEE Service Center, Piscataway, NJ 1990) p. II-149

[10.397] J.-H. Yoo, B.-H. Kang, J.-W. Kim: In *Proc. 3rd Int. Conf. on Fuzzy Logic, Neural Nets and Soft Computing* (Fuzzy Logic Systems Institute, Iizuka, Japan 1994) p. 79

[10.398] A. Zell, H. Bayer, H. Bauknecht: In *Proc. WCNN'94, World Congress on Neural Networks* (Lawrence Erlbaum, Hillsdale, NJ 1994) p. IV-269

[10.399] Z. Zhao: In *Artificial Neural Networks, 2*, ed. by I. Aleksander, J. Taylor (North-Holland, Amsterdam, Netherlands 1992) p. I-779

[10.400] M. Cottrell, J.-C. Fort, G. Pagès: In *Proc. ESANN'94, European Symp. on Artificial Neural Networks*, ed. by M. Verleysen (D Facto Conference Services, Brussels, Belgium 1994) p. 235

[10.401] E. Oja: In *Neural Networks for Chemical Engineers, Computer-Aided Chemical Engineering, 6*, Unsupervised neural learning. (Elsevier, Amsterdam 1995) p. 21

[10.402] V. V. Tolat: Biol. Cyb. **64**, 155 (1990)

[10.403] S. P. Luttrell: Neural Computation **6**, 767 (1994)

[10.404] H.-U. Bauer, K. R. Pawelzik: IEEE Trans. on Neural Networks **3**, 570 (1992)

[10.405] H.-U. Bauer, K. Pawelzik, T. Geisel: In *Advances in Neural Information Processing Systems 4*, ed. by J. E. Moody, S. J. Hanson, R. P. Lippmann (Morgan Kaufmann, San Mateo, CA 1992) p. 1141

[10.406] J. C. Bezdek, N. R. Pal: In *Proc. IJCNN-93-Nagoya, Int. Joint Conf. on Neural Networks* (IEEE Service Center, Piscataway, NJ 1993) p. III-2435

[10.407] P. Brauer, P. Knagenhjelm: In *Proc. ICASSP'89, Int. Conf. on Acoustics, Speech and Signal Processing* (IEEE Service Center, Piscataway, NJ 1989) p. 647

[10.408] P. Brauer, P. Hedelin, D. Huber, P. Knagenhjelm: In *Proc. ICASSP'91, Int. Conf. on Acoustics, Speech and Signal Processing* (IEEE Service Center, Piscataway, NJ 1991) p. I-133

[10.409] G. Burel: Bull. d'information des Laboratoires Centraux de Thomson CSF (1992) p. 3

[10.410] G. Burel: Traitement du Signal **10**, 41 (1993)

[10.411] P. Conti, L. D. Giovanni: In *Artificial Neural Networks*, ed. by T. Kohonen, K. Mäkisara, O. Simula, J. Kangas (North-Holland, Amsterdam, Netherlands 1991) p. II-1809

[10.412] R. Der, M. Herrmann, T. Villmann: In *Proc. WCNN'93, World Congress on Neural Networks* (Lawrence Erlbaum, Hillsdale, NJ 1993) p. II-461

[10.413] E. A. Ferrán: In *Artificial Neural Networks, 2*, ed. by I. Aleksander, J. Taylor (North-Holland, Amsterdam, Netherlands 1992) p. I-165

[10.414] E. A. Ferrán: Network **4**, 337 (1993)

[10.415] D. Fox, V. Heinze, K. Möller, S. Thrun, G. Veenker: In *Artificial Neural Networks*, ed. by T. Kohonen, K. Mäkisara, O. Simula, J. Kangas (North-Holland, Amsterdam, Netherlands 1991) p. I-207

[10.416] T. Geszti, I. Csabai, F. Czakó, T. Szakács, R. Serneels, G. Vattay: In *Statistical Mechanics of Neural Networks: Sitges, Barcelona, Spain* (Springer, Berlin, Germany 1990) p. 341

[10.417] I. Grabec: Biol. Cyb. **63**, 403 (1990)

[10.418] M. Hagiwara: In *Proc. IJCNN'93-Nagoya, Int. Joint Conf. on Neural Networks* (IEEE Service Center, Piscataway, NJ 1993) p. I-267

[10.419] S. A. Harp, T. Samad: In *Proc. IJCNN-91-Seattle, Int. Joint Conf. on Neural Networks* (IEEE Service Center, Piscataway, NJ 1991) p. I-341

[10.420] P. Koistinen: In *Proc. Symp. on Neural Networks in Finland*, ed. by A. Bulsari, B. Saxén (Finnish Artificial Intelligence Society, Helsinki, Finland 1993) p. 1

[10.421] P. Koistinen: In *Proc. ICANN'93, Int. Conf. on Artificial Neural Networks, Amsterdam*, ed. by S. Gielen, B. Kappen (Springer, London, UK 1993) p. 219

[10.422] C.-Y. Liou, H.-C. Yang: In *Proc. ICANN'93 Int. Conf. on Artificial Neural Networks*, ed. by S. Gielen, B. Kappen (Springer, London, UK 1993) p. 918

[10.423] C.-Y. Liou, W.-P. Tai: In *Proc. IJCNN-93-Nagoya, Int. Joint Conf. on Neural Networks* (IEEE Service Center, Piscataway, NJ 1993) p. II-1618

[10.424] Z.-P. Lo, Y. Qu, B. Bavarian: In *Proc. IJCNN'92, Int. Joint Conf. on Neural Networks* (IEEE Service Center, Piscataway, NJ 1992) p. I-589

[10.425] Z.-P. Lo, Y. Yu, B. Bavarian: In *Proc. IJCNN'92, Int. Joint Conference on Neural Networks* (IEEE Service Center, Piscataway, NJ 1992) p. IV-755

[10.426] S. P. Luttrell: In *Proc. IJCNN'89, Int Joint Conf. on Neural Networks* (IEEE Service Center, Piscataway, NJ 1989) p. II-495

[10.427] S. P. Luttrell: Technical Report 4742 (DRA, Malvern, UK 1993)

[10.428] Y. Matsuyama, M. Tan: In *Proc. IJCNN-93-Nagoya, Int. Joint Conf. on Neural Networks* (IEEE Service Center, Piscataway, NJ 1993) p. III-2061

[10.429] L. Owsley, L. Atlas: In *Proc. Neural Networks for Signal Processing* (IEEE Service Center, Piscataway, NJ 1993) p. 141

[10.430] F. Peper, M. N. Shirazi, H. Noda: IEEE Trans. on Neural Networks **4**, 151 (1993)

[10.431] J. C. Platt, A. H. Barr: In *Neural Information Processing Systems*, ed. by D. Z. Anderson (American Inst. of Physics, New York, NY 1987) p. 612

[10.432] G. Poggi, E. Sasso: In *Proc. ICASSP'93, Int. Conf. on Acoustics, Speech and Signal Processing* (IEEE Service Center, Piscataway, NJ 1993) p. V-587

[10.433] G. Rigoll: In *Proc. IJCNN'90-San Diego, Int. Joint Conf. on Neural Networks* (IEEE Service Center, Piscataway, NJ 1990) p. I-569

[10.434]   V. T. Ruoppila, T. Sorsa, H. N. Koivo: In *Proc. ICNN'93, Int. Conf. on Neural Networks* (IEEE Service Center, Piscataway, NJ 1993) p. III-1480

[10.435]   G. Tambouratzis, T. J. Stonham: In *Proc. ICANN'93, Int. Conf. on Artificial Neural Networks*, ed. by S. Gielen, B. Kappen (Springer, London, UK 1993) p. 76

[10.436]   V. Tryba, K. Goser.: In *Proc. of the 2nd Int. Conf. on Microelectronics for Neural Networks*, ed. by U. Ramacher, U. Ruckert, J. A. Nossek (Kyrill & Method Verlag, Munich, Germany 1991) p. 83

[10.437]   V. Tryba, K. Goser: In *Proc. IWANN, Int. Workshop on Artificial Neural Networks*, ed. by A. Prieto (Springer, Berlin, Germany 1991) p. 33

[10.438]   V. Tryba, K. Goser: In *Digest of ESANN'93* (D Facto Conference Services, Brussels, Belgium 1993)

[10.439]   M. M. Van Hulle, D. Martinez: Neural Computation **5**, 939 (1993)

[10.440]   W. Y. Yan, H. Kan, Z. Liangzhu, W. J. Wei: In *Proc. NAECON 1992, National Aerospace and Electronics Conference* (IEEE Service Center, Piscataway, NJ 1992) p. I-108

[10.441]   A. Dvoretzky: In *Proc. 3rd Berkeley Symp. on Mathematical Statistics and Probability* (Univ. of California Press, Berkeley, CA 1956) p. 39

[10.442]   P. Thiran, M. Hasler: In *Proc. Workshop 'Aspects Theoriques des Reseaux de Neurones'*, ed. by M. Cottrell, M. Chaleyat-Maurel (Université Paris I, Paris, France 1992)

[10.443]   M. Budinich, J. G. Taylor: Neural Computation **7**, 284 (1995)

[10.444]   Y. Zheng, J. F. Greenleaf: IEEE Trans. on Neural Networks **7**, 87 (1996)

[10.445]   W. Duch: Open Systems & Information Dynamics **2**, 295 (1994)

[10.446]   G. Tambouratzis, T. J. Stonham: Pattern Recognition Letters **14**, 927 (1993)

[10.447]   K. Minamimoto, K. Ikeda, K. Nakayama: In *Proc. ICNN'95, IEEE Int. Conf. on Neural Networks*, (IEEE Service Center, Piscataway, NJ 1995) p. II-789

[10.448]   W. Duch, A. Naud: In *Proc. ESANN'96, European Symp. on Artificial Neural Networks*, ed. by M. Verleysen (D Facto Conference Services, Brussels, Belgium 1996) p. 91

[10.449]   J. A. Flanagan, M. Hasler: In *Proc. ESANN'95, European Symp. on Artificial Neural Networks*, ed. by M. Verleysen (D Facto Conference Services, Brussels, Belgium 1995) p. 1

[10.450]   R. Der, M. Herrmann: In *Proc. ESANN'94, European Symp. on Artificial Neural Networks*, ed. by M. Verleysen (D Facto Conference Services, Brussels, Belgium 1994) p. 271

[10.451]   H. Yin, N. M. Allinson: Neural Computation **7**, 1178 (1995)

[10.452]   M. Herrmann, H.-U. Bauer, R. Der: In *Proc. ICANN'95, Int. Conf. on Artificial Neural Networks*, ed. by F. Fogelman-Soulié, P. Gallinari, (EC2, Nanterre, France 1995) p. I-75

[10.453]   J.-C. Fort, G. Pagès: In *Proc. ESANN'95, European Symp. on Artificial Neural Networks*, ed. by M. Verleysen (D Facto Conference Services, Brussels, Belgium 1995) p. 9

[10.454]   J.-C. Fort, G. Pagès: In *Proc. ESANN'96, European Symp. on Artificial Neural Networks*, ed. by M. Verleysen (D Facto Conference Services, Brussels, Belgium 1996) p. 85

[10.455]   O. Scherf, K. Pawelzik, F. Wolf, T. Geisel: In *Proc. ICANN'94, Int. Conf. on Artificial Neural Networks*, ed. by M. Marinaro, P. G. Morasso, (Springer, London, UK 1994) p. I-338

[10.456] H. Speckmann, G. Raddatz, W. Rosenstiel: In *Proc. ESANN'94, European Symp. on Artificial Neural Networks*, ed. by M. Verleysen (D Facto Conference Services, Brussels, Belgium 1994) p. 251

[10.457] H. Yin, N. M. Allinson: In *Proc. ICANN'95, Int. Conf. on Artificial Neural Networks*, ed. by F. Fogelman-Soulié, P. Gallinari, (EC2, Nanterre, France 1995) p. II-45

[10.458] G. Tambouratzis: Electronics Letters **30**, 248 (1993)

[10.459] A. Varfis, C. Versino: In *Proc. ESANN'93, European Symposium on Artificial Neural Networks*, ed. by M. Verleysen (D Facto Conference Services, Brussels, Belgium 1993) p. 229

[10.460] A. Kanstein, K. Goser: In *Proc. ESANN'94, European Symp. on Artificial Neural Networks*, ed. by M. Verleysen (D Facto Conference Services, Brussels, Belgium 1994) p. 263

[10.461] S. Garavaglia: In *Proc. WCNN'95, World Congress on Neural Networks*, (INNS, Lawrence Erlbaum, Hillsdale, NJ 1995) p. I-203

[10.462] D. Miller, A. Rao, K. Rose, A. Gersho: In *Proc. NNSP'95, IEEE Workshop on Neural Networks for Signal Processing* (IEEE Service Center, Piscataway, NJ 1995) p. 58

[10.463] C. Bouton, M. Cottrell, J.-C. Fort, G. Pagès: In *Probabilités Numériques*, ed. by N. Bouleau, D. Talay (INRIA, Paris, France 1991) Chap. V.2, p. 163

[10.464] C. Bouton, G. Pagès: Stochastic Processes and Their Applications **47**, 249 (1993)

[10.465] M. Cottrell: PhD Thesis (Université Paris Sud, Centre d'Orsay, Orsay, France 1988)

[10.466] M. Cottrell, J.-C. Fort, G. Pagès: Technical Report 32 (Université Paris 1, Paris, France 1994)

[10.467] D. A. Critchley: In *Artificial Neural Networks, 2*, ed. by I. Aleksander, J. Taylor (North-Holland, Amsterdam, Netherlands 1992) p. I-281

[10.468] I. Csabai, T. Geszti, G. Vattay: Phys. Rev. A [Statistical Physics, Plasmas, Fluids, and Related Interdisciplinary Topics] **46**, R6181 (1992)

[10.469] R. Der, T. Villmann: In *Proc. WCNN'93, World Congress on Neural Networks* (Lawrence Erlbaum, Hillsdale, NJ 1993) p. II-457

[10.470] E. Erwin, K. Obermeyer, K. Schulten: In *Artificial Neural Networks*, ed. by T. Kohonen, K. Mäkisara, O. Simula, J. Kangas (Elsevier, Amsterdam, Netherlands 1991) p. 409

[10.471] J.-C. Fort, G. Pagès: note aux C. R. Acad. Sci. Paris **Série I**, 389 (1993)

[10.472] O. Francois, J. Demongeot, T. Herve: Neural Networks **5**, 277 (1992)

[10.473] T. M. Heskes, E. T. P. Slijpen, B. Kappen: Physical Review E **47**, 4457 (1993)

[10.474] T. Heskes: In *Proc. ICANN'93, Int. Conf. on Artificial Neural Networks* (Springer, London, UK 1993) p. 533

[10.475] K. Hornik, C.-M. Kuan: Neural Networks **5**, 229 (1992)

[10.476] C.-M. Kuan, K. Hornik: IEEE Trans. on Neural Networks **2** (1991)

[10.477] Z.-P. Lo, B. Bavarian: Biol. Cyb. **65**, 55 (1991)

[10.478] Z.-P. Lo, B. Bavarian: In *Proc. IJCNN'91, Int. Joint Conf. on Neural Networks* (IEEE Service Center, Piscataway, NJ 1991) p. II-201

[10.479] Z.-P. Lo, Y. Yu, B. Bavarian: IEEE Trans. on Neural Networks **4**, 207 (1993)

[10.480] J. Sum, L.-W. Chan: In *Proc. Int. Symp. on Speech, Image Processing and Neural Networks* (IEEE Hong Kong Chapt. of Signal Processing, Hong Kong 1994) p. I-81

[10.481] H. Yang, T. S. Dillon: Neural Networks **5**, 485 (1992)

[10.482] H. Yin, N. M. Allinson: In *New Trends in Neural Computation*, ed. by
          J. Mira, J. Cabestany, A. Prieto (Springer, Berlin, Germany 1993) p. 291
[10.483] J. Zhang: Neural Computation **3**, 54 (1991)
[10.484] M. Cottrell, J.-C. Fort, G. Pagès: IEEE Trans. on Neural Networks **6**,
          797 (1995)
[10.485] N. M. Allinson, H. Yin: In *Proc. ICANN'93, Int. Conf. on Artificial
          Neural Networks*, ed. by S. Gielen, B. Kappen (Springer, London, UK
          1993)
[10.486] P. D. Picton: In *IEE Colloquium on 'Adaptive Filtering, Non-Linear Dy-
          namics and Neural Networks' (Digest No.176)* (IEE, London, UK 1991)
          p. 7/1
[10.487] K. Watanabe, S. G. Tzafestas: J. Intelligent and Robotic Systems: The-
          ory and Applications **3**, 305 (1990)
[10.488] N. R. Pal, J. C. Bezdek, E. C. Tsao: In *SPIE Vol. 1710, Science of
          Artificial Neural Networks* (SPIE, Bellingham, WA 1992) p. 500
[10.489] N. R. Pal, J. C. Bezdek, E. C.-K. Tsao: IEEE Trans. on Neural Networks
          4, 549 (1993)
[10.490] N. R. Pal, J. C. Bezdek: In *Proc. IJCNN-93-Nagoya, Int. Joint Conf. on
          Neural Networks* (IEEE Service Center, Piscataway, NJ 1993) p. III-2441
[10.491] N. R. Pal, J. C. Bezdek, E. C.-K. Tsao: IEEE Trans. on Neural Networks
          4, 549 (1993)
[10.492] K. Ishida, Y. Matsumoto, N. Okino: In *Artificial Neural Networks, 2*,
          ed. by I. Aleksander, J. Taylor (North-Holland, Amsterdam, Netherlands
          1992) p. I-353
[10.493] K. Ishida, Y. Matsumoto, N. Okino: In *Proc. IJCNN-93-Nagoya, Int.
          Joint Conf. on Neural Networks* (IEEE Service Center, Piscataway, NJ
          1993) p. III-2460
[10.494] J. Lampinen, E. Oja: In *Proc. IJCNN'89, Int. Joint Conf. on Neural
          Networks* (IEEE Service Center, Piscataway, NJ 1989) p. II-503
[10.495] J. Lampinen, E. Oja: In *Neurocomputing: Algorithms, Architectures, and
          Applications, NATO ASI Series F: Computer and Systems Sciences, vol.
          68*, ed. by F. Fogelman-Soulié, J. Herault (Springer, Berlin, Germany
          1990) p. 65
[10.496] M. Kelly: In *Artificial Neural Networks*, ed. by T. Kohonen, K. Mäkisara,
          O. Simula, J. Kangas (North-Holland, Amsterdam, Netherlands 1991)
          p. II-1041
[10.497] P. Tavan, H. Grubmüller, H. Kühnel: Biol. Cyb. **64**, 95 (1990)
[10.498] G. Barna, R. Chrisley, T. Kohonen: Neural Networks **1**, 7 (1988)
[10.499] G. Barna, K. Kaski: Physica Scripta **T33**, 110 (1990)
[10.500] F. Blayo, R. Demartines: In *Proc. IWANN'91, Int. Workshop on Artifi-
          cial Neural Networks*, ed. by A. Prieto (Springer, Berlin, Germany 1991)
          p. 469
[10.501] P. Blonda, G. Pasquariello, J. Smith: In *Proc. IJCNN-93-Nagoya, Int.
          Joint Conf. on Neural Networks* (IEEE Service Center, Piscataway, NJ
          1993) p. II-1231
[10.502] X. Driancourt, L. Bottou, P. Gallinari: In *Proc. IJCNN 91, Int. Joint
          Conf. on Neural Networks* (IEEE Service Center, Piscataway, NJ 1991)
          p. II-815
[10.503] X. Driancourt, L. Bottou, P. Gallinari: In *Artificial Neural Networks*,
          ed. by T. Kohonen, K. Mäkisara, O. Simula, J. Kangas (North-Holland,
          Amsterdam, Netherlands 1991) p. II-1649
[10.504] J. S. Gruner: M.Sc. Thesis, (Air Force Inst. of Tech., Wright-Patterson
          AFB, OH 1992)

[10.505]   D. R. Hush, J. M. Salas: In *Proc. ISE '89, Eleventh Annual Ideas in Science and Electronics Exposition and Symposium*, ed. by C. Christmann (Ideas in Sci. & Electron.; IEEE, Ideas in Sci. & Electron, Albuquerque, NM 1989) p. 107

[10.506]   D. R. Hush, B. Horne: Informatica y Automatica **25**, 19 (1992)

[10.507]   R. A. Hutchinson, W. J. Welsh: In *Proc. First IEE Int. Conf. on Artificial Neural Networks* (IEE, London, UK 1989) p. 201

[10.508]   A. E. Lucas, J. Kittler: In *Proc. First IEE Int. Conf. on Artificial Neural Networks* (IEE, London, UK 1989) p. 7

[10.509]   J. D. McAuliffe, L. E. Atlas, C. Rivera: In *Proc. ICASSP'90, Int. Conf. on Acoustics, Speech and Signal Processing* (IEEE Service Center, Piscataway, NJ 1990) p. IV-2293

[10.510]   F. Peper, B. Zhang, H. Noda: In *Proc. IJCNN'93-Nagoya, Int. Joint Conf. on Neural Networks* (IEEE Service Center, Piscataway, NJ 1993) p. II-1425

[10.511]   W. Poechmueller, M. Glesner, H. Juergs: In *Proc. ICNN'93, Int. Conf. on Neural Networks* (IEEE Service Center, Piscataway, NJ 1993) p. III-1207

[10.512]   C. Pope, L. Atlas, C. Nelson: In *Proc. IEEE Pacific Rim Conf. on Communications, Computers and Signal Processing.* (IEEE Service Center, Piscataway, NJ 1989) p. 521

[10.513]   S. Lin: In *Proc. China 1991 Int. Conf. on Circuits and Systems* (IEEE Service Center, Piscataway, NJ 1991) p. II-808

[10.514]   T. Tanaka, M. Saito: Trans. Inst. of Electronics, Information and Communication Engineers **J75D-II**, 1085 (1992)

[10.515]   C.-D. Wann, S. C. A. Thomopoulos: In *Proc. of the World Congress on Neural Networks* (INNS, Lawrence Erlbaum, Hillsdale, NJ 1993) p. II-549

[10.516]   P. C. Woodland, S. G. Smyth: Speech Communication **9**, 73 (1990)

[10.517]   F. H. Wu, K. Ganesan: In *Proc. ICASSP'89 Int. Conf. on Acoustics, Speech and Signal Processing, Glasgow, Scotland* (IEEE Service Center, Piscataway, NJ 1989) p. 751

[10.518]   F. H. Wu, K. Ganesan: In *Proc. Ninth Annual Int. Phoenix Conf. on Computers and Communications* (IEEE Comput. Soc. Press, Los Alamitos, CA 1990) p. 263

[10.519]   A. Hämäläinen: Licentiate's Thesis (University of Jyväskylä, Jyväskylä, Finland 1992)

[10.520]   S. P. Luttrell: Technical Report 4392 (RSRE, Malvern, UK 1990)

[10.521]   S. P. Luttrell: IEEE Trans. on Neural Networks **2**, 427 (1991)

[10.522]   H. Ritter: Report A9 (Helsinki University of Technology, Laboratory of Computer and Information Science, Espoo, Finland 1989)

[10.523]   H. Ritter: IEEE Trans. on Neural Networks **2**, 173 (1991)

[10.524]   A. Ando, K. Ozeki: In *Artificial Neural Networks*, ed. by T. Kohonen, K. Mäkisara, O. Simula, J. Kangas (North-Holland, Amsterdam, Netherlands 1991) p. 421

[10.525]   J. C. Bezdek: In *SPIE Vol. 1826, Intelligent Robots and Computer Vision XI: Biological, Neural Net, and 3-D Methods* (SPIE, Bellingham, WA 1992) p. 280

[10.526]   J. C. Bezdek, N. R. Pal, E. C. Tsao: In *Proc. Third Int. Workshop on Neural Networks and Fuzzy Logic, Houston, Texas, NASA Conf. Publication 10111*, ed. by C. J. Culbert (NASA 1993) p. II-199

[10.527]   H. Bi, G. Bi, Y. Mao: In *Proc. ICNN'94, Int. Conf. on Neural Networks* (IEEE Service Center, Piscataway, NJ 1994) p. 622

[10.528] H. Bi, G. Bi, Y. Mao: In *Proc. Int. Symp. on Speech, Image Processing and Neural Networks* (IEEE Hong Kong Chapter of Signal Processing, Hong Kong 1994) p. II-650

[10.529] P. Burrascano: IEEE Trans. on Neural Networks **2**, 458 (1991)

[10.530] S. Cagnoni, G. Valli: In *Proc. ICNN'94, Int. Conf. on Neural Networks* (IEEE Service Center, Piscataway, NJ 1994) p. 762

[10.531] C. Diamantini, A. Spalvieri: In *Proc. ICANN'94, Int. Conf. on Artificial Neural Networks*, ed. by M. Marinaro, P. G. Morasso (Springer, London, UK 1994) p. II-1091

[10.532] T. Geszti, I. Csabai: Complex Systems **6**, 179 (1992)

[10.533] T. Geszti: In *From Phase Transitions to Chaos*, ed. by G. Györgyi, I. Kondor, L. Sasvari, T. Tel (World Scientific, Singapore 1992)

[10.534] S. Geva, J. Sitte: IEEE Trans. on Neural Networks **2**, 318 (1991)

[10.535] S. Geva, J. Sitte: In *Proc. ACNN'91, Second Australian Conf. on Neural Networks*, ed. by M. Jabri (Sydney Univ. Electr. Eng, Sydney, Australia 1991) p. 13

[10.536] S. Geva, J. Sitte: Neural Computation **3**, 623 (1991)

[10.537] P. Israel, F. R. Parris: In *Proc. WCNN'93, World Congress on Neural Networks* (Lawrence Erlbaum, Hillsdale, NJ 1993) p. III-445

[10.538] P. M. Kelly, D. R. Hush, J. M. White: In *Proc. IJCNN'92, Int. Joint Conf. on Neural Networks* (IEEE Service Center, Piscataway, NJ 1992) p. IV-196

[10.539] D. Knoll, J. T.-H. Lo: In *Proc. IJCNN'92, Int. Joint Conf. on Neural Networks* (IEEE Service Center, Piscataway, NJ 1992) p. III-573

[10.540] C. Kotropoulos, E. Augé, I. Pitas: In *Proc. EUSIPCO-92, Sixth European Signal Processing Conf.*, ed. by J. Vandewalle, R. Boite, M. Moonen, A. Oosterlinck (Elsevier, Amsterdam, Netherlands 1992) p. II-1177

[10.541] Y.-C. Lai, S.-S. Yu, S.-L. Chou: In *Proc. IJCNN-93-Nagoya, Int. Joint Conf. on Neural Networks* (IEEE Service Center, Piscataway, NJ 1993) p. III-2587

[10.542] E. Maillard, B. Solaiman: In *Proc. ICNN'94, Int. Conf. on Neural Networks* (IEEE Service Center, Piscataway, NJ 1994) p. 766

[10.543] A. LaVigna: PhD Thesis (University of Maryland, College Park, MD 1989)

[10.544] S. Lee, S. Shimoji: In *Proc. ICNN'93, Int. Conf. on Neural Networks* (IEEE Service Center, Piscataway, NJ 1993) p. III-1354

[10.545] F. Poirier: In *Proc. ICASSP'91, Int. Conf. on Acoustics, Speech and Signal Processing* (IEEE Service Center, Piscataway, NJ 1991) p. I-649

[10.546] F. Poirier: In *Proc. EUROSPEECH-91, 2nd European Conf. on Speech Communication and Technology* (Assoc. Belge Acoust.; Assoc. Italiana di Acustica; CEC; et al, Istituto Int. Comunicazioni, Genova, Italy 1991) p. II-1003

[10.547] F. Poirier, A. Ferrieux: In *Artificial Neural Networks*, ed. by T. Kohonen, K. Mäkisara, O. Simula, J. Kangas (North-Holland, Amsterdam, Netherlands 1991) p. II-1333

[10.548] M. Pregenzer, D. Flotzinger, G. Pfurtscheller: In *Proc. ICANN'94, Int. Conf. on Artificial Neural Networks*, ed. by M. Marinaro, P. G. Morasso (Springer, London, UK 1994) p. II-1075

[10.549] M. Pregenzer, D. Flotzinger, G. Pfurtscheller: In *Proc. ICNN'94, Int. Conf. on Neural Networks* (IEEE Service Center, Piscataway, NJ 1994) p. 2890

[10.550] V. R. de Sa, D. H. Ballard: In *Advances in Neural Information Processing Systems 5*, ed. by L. Giles, S. Hanson, J. Cowan (Morgan Kaufmann, San Mateo, CA 1993) p. 220

[10.551] A. Sato, K. Yamada, J. Tsukumo: In *Proc. ICNN'93, Int. Conf. on Neural Networks* (IEEE Service Center, Piscataway, NJ 1993) p. II-632

[10.552] A. Sato, J. Tsukumo: In *Proc. ICNN'94, Int. Conf. on Neural Networks* (IEEE Service Center, Piscataway, NJ 1994) p. 161

[10.553] B. Solaiman, M. C. Mouchot, E. Maillard: In *Proc. ICNN'94, Int. Conf. on Neural Networks* (IEEE Service Center, Piscataway, NJ 1994) p. 1772

[10.554] C. Tadj, F. Poirier: In *Proc. EUROSPEECH-93, 3rd European Conf. on Speech, Communication and Technology* (ESCA, Berlin, Germany 1993) p. II-1009

[10.555] K. S. Thyagarajan, A. Eghbalmoghadam: Archiv für Elektronik und Übertragungstechnik **44**, 439 (1990)

[10.556] N. Ueda, R. Nakano: In *Proc. of IEEE Int. Conf. on Neural Networks, San Francisco* (IEEE Service Center, Piscataway, NJ 1993) p. III-1444

[10.557] M. Verleysen, P. Thissen, J.-D. Legat: In *New Trends in Neural Computation, Lecture Notes in Computer Science No. 686*, ed. by J. Mira, J. Cabestany, A. Prieto (Springer, Berlin, Germany 1993) p. 340

[10.558] W. Xinwen, Z. Lihe, H. Zhenya: In *Proc. China 1991 Int. Conf. on Circuits and Systems* (IEEE Service Center, Piscataway, NJ 1991) p. II-523

[10.559] Z. Wang, J. V. Hanson: In *Proc. WCNN'93, World Congress on Neural Networks* (INNS, Lawrence Erlbaum, Hillsdale, NJ 1993) p. IV-605

[10.560] L. Wu, F. Fallside: In *Proc. Int. Conf. on Spoken Language Processing* (The Acoustical Society of Japan, Tokyo, Japan 1990) p. 1029

[10.561] T. Yoshihara, T. Wada: In *Proc. IJCNN'91, Int. Joint Conf. on Neural Networks* (IEEE Service Center, Piscataway, NJ 1991)

[10.562] A. Zell, M. Schmalzl: In *Proc. ICANN'94, Int. Conf. on Artificial Neural Networks*, ed. by M. Marinaro, P. G. Morasso (Springer, London, UK 1994) p. II-1095

[10.563] B.-s. Kim, S. H. Lee, D. K. Kim: In *Proc. IJCNN-93-Nagoya, Int. Joint Conf. on Neural Networks* (IEEE Service Center, Piscataway, NJ 1993) p. III-2456

[10.564] H. Iwamida, S. Katagiri, E. McDermott, Y. Tohkura: In *Proc. ICASSP'90, Int. Conf. on Acoustics, Speech and Signal Processing* (IEEE Service Center, Piscataway, NJ 1990) p. 1-489

[10.565] H. Iwamida, et al.: In *Proc. ICASSP'91, Int. Conf. on Acoustics, Speech and Signal Processing* (IEEE Service Center, Piscataway, NJ 1991) p. I-553

[10.566] S. Katagiri, E. McDermott, M. Yokota: In *Proc. ICASSP'89, Int. Conf. on Acoustics, Speech and Signal Processing* (IEEE Service Center, Piscataway, NJ 1989) p. I-322

[10.567] S. Katagiri, C. H. Lee: In *Proc. GLOBECOM'90, IEEE Global Telecommunications Conf. and Exhibition. 'Communications: Connecting the Future'* (IEEE Service Center, Piscataway, NJ 1990) p. II-1032

[10.568] D. G. Kimber, M. A. Bush, G. N. Tajchman: In *Proc. ICASSP'90, Int. Conf. on Acoustics, Speech and Signal Processing* (IEEE Service Center, Piscataway, NJ 1990) p. I-497

[10.569] M. Kurimo. M.Sc. Thesis (Helsinki University of Technology, Espoo, Finland 1992)

[10.570] M. Kurimo, K. Torkkola: In *Proc. Int. Conf. on Spoken Language Processing* (University of Alberta, Edmonton, Alberta, Canada 1992) p. 1-543

[10.571] M. Kurimo, K. Torkkola: In *Proc. Workshop on Neural Networks for Signal Processing* (IEEE Service Center, Piscataway, NJ 1992) p. 174

[10.572] M. Kurimo, K. Torkkola: In *Proc. SPIE's Conf. on Neural and Stochastic Methods in Image and Signal Processing* (SPIE, Bellingham, WA 1992) p. 726

[10.573] M. Kurimo: In *Proc. EUROSPEECH-93, 3rd European Conf. on Speech, Communication and Technology* (ESCA, Berlin 1993) p. III-1731

[10.574] S. Makino, A. Ito, M. Endo, K. Kido: IEICE Trans. **E74**, 1773 (1991)

[10.575] S. Makino, A. Ito, M. Endo, K. Kido: In *Proc. ICASSP'91, Int. Conf. on Acoustics, Speech and Signal Processing* (IEEE Service Center, Piscataway, NJ 1991) p. I-273

[10.576] S. Makino, M. Endo, T. Sone, K. Kido: J. Acoustical Society of Japan [E] **13**, 351 (1992)

[10.577] S. Mizuta, K. Nakajima: In *Proc. ICSLP, Int. Conf. on Spoken Language Processing* (University of Alberta, Edmonton, Alberta, Canada 1990) p. I-245

[10.578] S. Nakagawa, Y. Hirata: In *Proc. ICASSP'90, Int. Conf. on Acoustics Speech and Signal Processing* (IEEE Service Center, Piscataway, NJ 1990) p. I-509

[10.579] P. Ramesh, S. KATAGIRI, C. -H. Lee: In *Proc. ICASSP'91, Int. Conf. on Acoustics, Speech and Signal Processing* (IEEE Service Center, Piscataway, NJ 1991) p. I-113

[10.580] G. Rigoll: In *Neural Networks. EURASIP Workshop 1990 Proceedings*, ed. by L. B. Almeida, C. J. Wellekens (Springer, Berlin, Germany 1990) p. 205

[10.581] G. Rigoll: In *Proc. ICASSP'91, Int. Conf. on Acoustics, Speech and Signal Processing* (IEEE Service Center, Piscataway, NJ 1991) p. I-65

[10.582] P. Utela, S. Kaski, K. Torkkola: In *Proc. ICSLP'92 Int. Conf. on Spoken Language Processing (ICSLP 92)* (Personal Publishing Ltd., Edmonton, Canada 1992) p. 551

[10.583] Z. Zhao, C. Rowden: In *Second Int. Conf. on Artificial Neural Networks (Conf. Publ. No.349)* (IEE, London, UK 1991) p. 175

[10.584] Z. Zhao, C. G. Rowden: IEE Proc. F [Radar and Signal Processing] **139**, 385 (1992)

[10.585] T. Koizumi, J. Urata, S. Taniguchi: In *Artificial Neural Networks*, ed. by T. Kohonen, K. Mäkisara, O. Simula, J. Kangas (North-Holland, Amsterdam, Netherlands 1991) p. I-777

[10.586] E. Monte, J. B. Mariño: In *Proc. IWANN'91, Int. Workshop on Artificial Neural Networks*, ed. by A. Prieto (Springer, Berlin, Germany 1991) p. 370

[10.587] E. Monte, J. B. Mariño, E. L. Leida: In *Proc. ICSLP'92, Int. Conf. on Spoken Language Processing* (University of Alberta, Edmonton, Alberta, Canada 1992) p. I-551

[10.588] J. S. Baras, A. LaVigna: In *Proc. IJCNN-90-San Diego, Int. Joint Conf. on Neural Networks* (IEEE Service Center, Piscataway, NJ 1990) p. III-17

[10.589] J. S. Baras, A. LaVigna: In *Proc. INNC 90, Int. Neural Network Conf.* (Kluwer, Dordrecht, Netherlands 1990) p. II-1028

[10.590] J. S. Baras, A. L. Vigna: In *Proc. 29th IEEE Conf. on Decision and Control* (IEEE Service Center, Piscataway, NJ 1990) p. III-1735

[10.591] Z.-P. Lo, Y. Yu, B. Bavarian: In *Proc. IJCNN'92, Int. Joint Conf. on Neural Networks* (IEEE Service Center, Piscataway, NJ 1992) p. III-561

[10.592] N. Mohsenian, N. M. Nasrabadi: In *Proc. ICASSP'93, Int. Conf. on Acoustics, Speech and Signal Processing* (IEEE Service Center, Piscataway, NJ 1993) p. V

[10.593] W. Suewatanakul, D. M. Himmelblau: In *Proc. Third Workshop on Neural Networks: Academic/Industrial/NASA/Defense WNN92* (Auburn Univ.; Center Commersial Dev. Space Power and Adv. Electron.; NASA, Soc. Comput. Simulation, San Diego, CA 1993) p. 275

[10.594] R. Togneri, Y. Zhang, C. J. S. deSilva, Y. Attikiouzel: In *Proc. Third Int. Symp. on Signal Processing and its Applications* (1992) p. II-384

[10.595] A. Visa: In *Proc. INNC'90, Int. Neural Network Conf.* (Kluwer, Dordrecht, Netherlands 1990) p. 729

[10.596] J. C. Bezdek, N. R. Pal: Neural Networks **8**, 729 (1995)

[10.597] M. Cappelli, R. Zunino: In *Proc. WCNN'95, World Congress on Neural Networks*, (Lawrence Erlbaum, Hillsdale, NJ 1995) p. I-652

[10.598] S.-J. You, C.-H. Choi: In *Proc. ICNN'95, IEEE Int. Conf. on Neural Networks*, (IEEE Service Center, Piscataway, NJ 1995) p. V-2763

[10.599] J. Xuan, T. Adali: In *Proc. WCNN'95, World Congress on Neural Networks*, (Lawrence Erlbaum, Hillsdale, NJ 1995) p. I-756

[10.600] N. Kitajima: In *Proc. ICNN'95, IEEE Int. Conf. on Neural Networks*, (IEEE Service Center, Piscataway, NJ 1995) p. V-2775

[10.601] M. Verleysen, P. Thissen, J.-D. Legat: In *Proc. ESANN'93, European Symposium on Artificial Neural Networks*, ed. by M. Verleysen (D Facto Conference Services, Brussels, Belgium 1993) p. 209

[10.602] E. B. Kosmatopoulos, M. A. Christodoulou: IEEE Trans. on Image Processing **5**, 361 (1996)

[10.603] D. Lamberton, G. Pagès: In *Proc. ESANN'96, European Symp. on Artificial Neural Networks*, ed. by M. Verleysen (D Facto Conference Services, Brussels, Belgium 1996) p. 97

[10.604] H. K. Aghajan, C. D. Schaper, T. Kailath: Opt. Eng. **32**, 828 (1993)

[10.605] R. A. Beard, K. S. Rattan: In *Proc. NAECON 1989, IEEE 1989 National Aerospace and Electronics Conf.* (IEEE Service Center, Piscataway, NJ 1989) p. IV-1920

[10.606] G. N. Bebis, G. M. Papadourakis: Pattern Recognition **25**, 25 (1992)

[10.607] J. Kopecz: In *Proc. ICNN'93, Int. Conf. on Neural Networks* (IEEE Service Center, Piscataway, NJ 1993) p. I-138

[10.608] C. Nightingale, R. A. Hutchinson: British Telecom Technology J. **8**, 81 (1990)

[10.609] E. Oja: In *Neural Networks for Perception, vol. 1: Human and Machine Perception*, ed. by H. Wechsler (Academic Press, New York, NY 1992) p. 368

[10.610] C. Teng: In *Proc. Second IASTED International Symposium. Expert Systems and Neural Networks*, ed. by M. H. Hamza (IASTED, Acta Press, Anaheim, CA 1990) p. 35

[10.611] C. Teng, P. A. Ligomenides: Proc. SPIE – The Int. Society for Optical Engineering **1382**, 74 (1991)

[10.612] W. Wan, D. Fraser: In *Proc. 7th Australasian Remote Sensing Conference, Melborne, Australia* (Remote Sensing and Photogrammetry Association Australia, Ltd 1994) p. 423

[10.613] W. Wan, D. Fraser: In *Proc. 7th Australasian Remote Sensing Conference, Melborne, Australia* (Remote Sensing and Photogrammetry Association Australia, Ltd 1994) p. 145

[10.614]   W. Wan, D. Fraser: In *Proc. 7th Australasian Remote Sensing Confer-ence, Melborne, Australia* (Remote Sensing and Photogrammetry Asso-ciation Australia, Ltd 1994) p. 151

[10.615]   J. C. Guerrero: In *Proc. TTIA'95, Transferencia Tecnológica de In-teligencia Artificial a Industria, Medicina y Aplicaciones Sociales*, ed. by R. R. Aldeguer, J. M. G. Chamizo (Spain 1995) p. 189

[10.616]   Y. Chen, Y. Cao: In *Proc. ICNN'95, IEEE Int. Conf. on Neural Net-works*, (IEEE Service Center, Piscataway, NJ 1995) p. III-1414

[10.617]   J. Tanomaru, A. Inubushi: In *Proc. ICNN'95, IEEE Int. Conf. on Neural Networks*, (IEEE Service Center, Piscataway, NJ 1995) p.V-2432

[10.618]   M. Antonini, M. Barlaud, P. Mathieu: In *Signal Processing V. Theories and Applications. Proc. EUSIPCO-90, Fifth European Signal Process-ing Conference*, ed. by L. Torres, E. Masgrau, M. A. Lagunas (Elsevier, Amsterdam, Netherlands 1990) p. II-1091

[10.619]   E. Carlson: In *Proc. ICANN'93, Int. Conf. on Artificial Neural Networks*, ed. by S. Gielen, B. Kappen (Springer, London, UK 1993) p. 1018

[10.620]   S. Carrato, G. L. Siguranza, L. Manzo: In *Neural Networks for Signal Processing. Proc. 1993 IEEE Workshop* (IEEE Service Center, Piscat-away, NJ 1993) p. 291

[10.621]   D. D. Giusto, G. Vernazza: In *Proc. ICASSP'90, Int. Conf. on Acoustics, Speech and Signal Processing* (IEEE Service Center, Piscataway, NJ 1990) p. III-2265

[10.622]   J. Kennedy, F. Lavagetto, P. Morasso: In *Proc. INNC'90 Int. Neural Network Conf.* (Kluwer, Dordrecht, Netherlands 1990) p. I-54

[10.623]   H. Liu, D. Y. Y. Yun: In *Proc. Int. Workshop on Application of Neural Networks to Telecommunications*, ed. by J. Alspector, R. Goodman, T. X. Brown (Lawrence Erlbaum, Hillsdale, NJ 1993) p. 176

[10.624]   J. F. Nunes, J. S. Marques: European Trans. on Telecommunications and Related Technologies **3**, 599 (1992)

[10.625]   M. Antonini, M. Barlaud, P. Mathieu, J. C. Feauveau: Proc. SPIE – The Int. Society for Optical Engineering **1360**, 14 (1990)

[10.626]   S. S. Jumpertz, E. J. Garcia: In *Proc. ICANN'93, Int. Conf. on Artificial Neural Networks*, ed. by S. Gielen, B. Kappen (Springer, London, UK 1993) p. 1020

[10.627]   L. L. H. Andrew, M. Palaniswami: In *Proc. ICNN'95, IEEE Int. Conf. on Neural Networks*, (IEEE Service Center, Piscataway, NJ 1995) p. IV-2071

[10.628]   L. L. H. Andrew: In *Proc. IPCS'6 Int. Picture Coding Symposium* (1996) p. 569

[10.629]   G. Burel, I. Pottier: Revue Technique Thomson-CSF **23**, 137 (1991)

[10.630]   G. Burel, J.-Y. Catros: In *Proc. ICNN'93, Int. Conf. on Neural Networks* (IEEE Service Center, Piscataway, NJ 1993) p. II-727

[10.631]   A. H. Dekker: Technical Report TR10 (Department of Information and Computer Science, National University of Singapore, Singapore 1993)

[10.632]   E. L. Bail, A. Mitiche: Traitement du Signal **6**, 529 (1989)

[10.633]   M. Lech, Y. Hua: In *Neural Networks for Signal Processing. Proc. of the 1991 IEEE Workshop*, ed. by B. H. Juang, S. Y. Kung, C. A. Kamm (IEEE Service Center, Piscataway, NJ 1991) p. 552

[10.634]   M. Lech, Y. Hua: In *Int. Conf. on Image Processing and its Applications* (IEE, London, UK, 1992)

[10.635]   N. M. Nasrabadi, Y. Feng: Proc. SPIE – The Int. Society for Optical Engineering **1001**, 207 (1988)

[10.636] N. M. Nasrabadi, Y. Feng: In *Proc. ICNN'88, Int. Conf. on Neural Networks* (IEEE Service Center, Piscataway, NJ 1988) p. I-101

[10.637] N. M. Nasrabadi, Y. Feng: Neural Networks **1**, 518 (1988)

[10.638] Y. H. Shin, C.-C. Lu: In *Conf. Proc. 1991 IEEE Int. Conf. on Systems, Man, and Cybern. 'Decision Aiding for Complex Systems'* (IEEE Service Center, Piscataway, NJ 1991) p. III-1487

[10.639] R. Li, E. Sherrod, H. Si: In *Proc. WCNN'95, World Congress on Neural Networks*, (INNS, Lawrence Erlbaum, Hillsdale, NJ 1995) p. I-548

[10.640] W. Chang, H. S. Soliman, A. H. Sung: In *Proc. ICNN'94, Int. Conf. on Neural Networks* (IEEE Service Center, Piscataway, NJ 1994) p. 4163

[10.641] O. T.-C. Chen, B. J. Sheu, W.-C. Fang: In *Proc. ICASSP'92, Int. Conf. on Acoustics, Speech and Signal processing* (IEEE Service Center, Piscataway, NJ 1992) p. II-385

[10.642] X. Chen, R. Kothari, P. Klinkhachorn: In *Proc. WCNN'93, World Congress on Neural Networks* (Lawrence Erlbaum, Hillsdale 1993) p. I-555

[10.643] L.-Y. Chiou, J. Limqueco, J. Tian, C. Lirsinsap, H. Chu: In *Proc. WCNN'94, World Congress on Neural Networks* (Lawrence Erlbaum, Hillsdale, NJ 1994) p. I-342

[10.644] K. B. Cho, C. H. Park, S.-Y. Lee: In *Proc. WCNN'94, World Congress on Neural Networks* (Lawrence Erlbaum, Hillsdale, NJ 1994) p. III-26

[10.645] J. A. Corral, M. Guerrero, P. J. Zufiria: In *Proc. ICNN'94, Int. Conf. on Neural Networks* (IEEE Service Center, Piscataway, NJ 1994) p. 4113

[10.646] K. R. L. Godfrey: In *Proc. ICNN'93, Int. Conf. on Neural Networks* (IEEE Service Center, Piscataway, NJ 1993) p. III-1622

[10.647] N.-C. Kim, W.-H. Hong, M. Suk, J. Koh: In *Proc. IJCNN-93-Nagoya, Int. Joint Conf. on Neural Networks* (IEEE Service Center, Piscataway, NJ 1993) p. III-2203

[10.648] A. Koenig, M. Glesner: In *Artificial Neural Networks*, ed. by T. Kohonen, K. Mäkisara, O. Simula, J. Kangas (North-Holland, Amsterdam, Netherlands 1991) p. II-1345

[10.649] R. Krovi, W. E. Pracht: In *Proc. IEEE/ACM Int. Conference on Developing and Managing Expert System Programs*, ed. by J. Feinstein, E. Awad, L. Medsker, E. Turban (IEEE Comput. Soc. Press, Los Alamitos, CA 1991) p. 210

[10.650] C.-C. Lu, Y. H. Shin: IEEE Trans. on Consumer Electronics **38**, 25 (1992)

[10.651] L. Schweizer, G. Parladori, G. L. Sicuranza, S. Marsi: In *Artificial Neural Networks*, ed. by T. Kohonen, K. Mäkisara, O. Simula, J. Kangas (North-Holland, Amsterdam 1991) p. I-815

[10.652] W. Wang, G. Zhang, D. Cai, F. Wan: In *Proc. Second IASTED International Conference. Computer Applications in Industry*, ed. by H. T. Dorrah (IASTED, ACTA Press, Zurich, Switzerland 1992) p. I-197

[10.653] D. Anguita, F. Passaggio, R. Zunino: In *Proc. WCNN'95, World Congress on Neural Networks*, (INNS, Lawrence Erlbaum, Hillsdale, NJ 1995) p. I-739

[10.654] R. D. Dony, S. Haykin: Proc. of the IEEE **83**, 288 (1995)

[10.655] J. Kangas: In *Proc. NNSP'95, IEEE Workshop on Neural Networks for Signal Processing* (IEEE Service Center, Piscataway, NJ 1995) p. 343

[10.656] J. Kangas: In *Proc. ICANN'95, Int. Conf. on Artificial Neural Networks*, ed. by F. Fogelman-Soulié, P. Gallinari, (EC2, Nanterre, France 1995) p. I-287

[10.657]   J. Kangas: In *Proc. ICNN'95, IEEE Int. Conf. on Neural Networks*, (IEEE Service Center, Piscataway, NJ 1995) p. IV-2081

[10.658]   K. O. Perlmutter, S. M. Perlmutter, R. M. Gray, R. A. Olshen, K. L. Oehler: IEEE Trans. on Image Processing **5**, 347 (1996)

[10.659]   P. Kultanen, E. Oja, L. Xu: In *Proc. IAPR Workshop on Machine Vision Applications* (International Association for Pattern Recognition, New York, NY 1990) p. 173

[10.660]   P. Kultanen, L. Xu, E. Oja: In *Proc. 10ICPR, Int. Conf. on Pattern Recognition* (IEEE Comput. Soc. Press, Los Alamitos, CA 1990) p. 631

[10.661]   L. Xu, E. Oja, P. Kultanen: Res. Report 18 (Lappeenranta University of Technology, Department of Information Technology, Lappeenranta, Finland 1990)

[10.662]   L. Xu, E. Oja: In *Proc. 11ICPR, Int. Conf. on Pattern Recognition* (IEEE Comput. Soc. Press, Los Alamitos, CA 1992) p. 125

[10.663]   L. Xu, A. Krzyzak, E. Oja: In *Proc. 11ICPR, Int. Conf. on Pattern Recognition* (IEEE Comput. Soc. Press, Los Alamitos, CA 1992) p. 496

[10.664]   L. Xu, E. Oja: Computer Vision, Graphics, and Image Processing: Image Understanding **57**, 131 (1993)

[10.665]   E. Ardizzone, A. Chella, R. Rizzo: In *Proc. ICANN'94, Int. Conf. on Artificial Neural Networks*, ed. by M. Marinaro, P. G. Morasso (Springer, London, UK 1994) p. II-1161

[10.666]   K. A. Baraghimian: In *Proc. 13th Annual Int. Computer Software and Applications Conf.* (IEEE Computer Soc. Press, Los Alamitos, CA 1989) p. 680

[10.667]   A. P. Dhawan, L. Arata: In *Proc. ICNN'93, Int. Conf. on Neural Networks* (IEEE Service Center, Piscataway, NJ 1993) p. III-1277

[10.668]   S. Haring, M. A. Viergever, J. N. Kok: In *Proc. IJCNN-93-Nagoya, Int. Joint Conf. on Neural Networks* (IEEE Service Center, Piscataway, NJ 1993) p. I-193

[10.669]   A. Finch, J. Austin: In *Proc. ICANN'94, Int. Conf. on Artificial Neural Networks*, ed. by M. Marinaro, P. G. Morasso (Springer, London, UK 1994) p. II-1141

[10.670]   W. C. Lin, E. C. K. Tsao, C. T. Chen: Pattern Recognition **25**, 679 (1992)

[10.671]   R. Natowicz, F. Bosio, S. Sean: In *Proc. ICANN'93, Int. Conf. on Artificial Neural Networks (ICANN-93)*, ed. by S. Gielen, B. Kappen (Springer, London, UK 1993) p. 1002

[10.672]   G. Tambouratzis, D. Patel, T. J. Stonham: In *Proc. ICANN'93, Int. Conf. on Artificial Neural Networks*, ed. by S. Gielen, B. Kappen (Springer, London, UK 1993) p. 903

[10.673]   N. R. Dupaguntla, V. Vemuri: In *Proc. IJCNN'89, Int. Joint Conf. on Neural Networks* (IEEE Service Center, Piscataway, NJ 1989) p. I-127

[10.674]   S. Ghosal, R. Mehrotra: In *Proc. IJCNN'92, Int. Joint Conference on Neural Networks* (IEEE Service Center, Piscataway, NJ 1992) p. III-297

[10.675]   S. Ghosal, R. Mehrotra: In *Proc. ICNN'93, Int. Conf. on Neural Networks* (IEEE Service Center, Piscataway, NJ 1993) p. II-721

[10.676]   H. Greenspan, R. Goodman, R. Chellappa: In *Proc. IJCNN-91-Seattle, Int. Joint Conf. on Neural Networks* (IEEE Service Center, Piscataway, NJ 1991) p. I-639

[10.677]   S.-Y. Lu, J. E. Hernandez, G. A. Clark: In *Proc. IJCNN'91, Int. Joint Conf. on Neural Networks* (IEEE Service Center, Piscataway, NJ 1991) p. I-683

[10.678] E. Maillard, B. Zerr, J. Merckle: In *Proc. EUSIPCO-92, Sixth European Signal Processing Conference*, ed. by J. Vandewalle, R. Boite, M. Moonen, A. Oosterlinck (Elsevier, Amsterdam, Netherlands 1992) p. II-1173

[10.679] S. Oe, M. Hashida, Y. Shinohara: In *Proc. IJCNN-93-Nagoya, Int. Joint Conf. on Neural Networks* (IEEE Service Center, Piscataway, NJ 1993) p. I-189

[10.680] S. Oe, M. Hashida, M. Enokihara, Y. Shinohara: In *Proc. ICNN'94, Int. Conf. on Neural Networks* (IEEE Service Center, Piscataway, NJ 1994) p. 2415

[10.681] P. P. Raghu, R. Poongodi, B. Yegnanarayana: In *Proc. IJCNN-93-Nagoya, Int. Joint Conf. on Neural Networks* (IEEE Service Center, Piscataway, NJ 1993) p. III-2195

[10.682] O. Simula, A. Visa: In *Artificial Neural Networks, 2*, ed. by I. Aleksander, J. Taylor (North-Holland, Amsterdam, Netherlands 1992) p. II-1621

[10.683] O. Simula, A. Visa, K. Valkealahti: In *Proc. ICANN'93, Int. Conf. on Artificial Neural Networks*, ed. by S. Gielen, B. Kappen (Springer, London, UK 1993) p. 899

[10.684] K. Valkealahti, A. Visa, O. Simula: In *Proc. Fifth Finnish Artificial Intelligence Conf. (SteP-92): New Directions in Artificial Intelligence* (Finnish Artificial Intelligence Society, Helsinki, Finland 1992) p. II-189

[10.685] A. Visa: In *Proc. 3rd Int. Conf. on Fuzzy Logic, Neural Nets and Soft Computing* (Fuzzy Logic Systems Institute, Iizuka, Japan 1994) p. 145

[10.686] H. Yin, N. M. Allinson: In *Proc. ICANN'94, Int. Conf. on Artificial Neural Networks*, ed. by M. Marinaro, P. G. Morasso (Springer, London, UK 1994) p. II-1149

[10.687] E. Oja, K. Valkealahti: In *Proc. ICNN'95, IEEE Int. Conf. on Neural Networks*, (IEEE Service Center, Piscataway, NJ 1995) p. II-1160

[10.688] R.-I. Chang, P.-Y. Hsiao: In *Proc. ICNN'94, Int. Conf. on Neural Networks* (IEEE Service Center, Piscataway, NJ 1994) p. 4123

[10.689] S. Jockusch, H. Ritter: In *Proc. ICANN'94, Int. Conf. on Artificial Neural Networks*, ed. by M. Marinaro, P. G. Morasso (Springer, London, UK 1994) p. II-1105

[10.690] H. Kauniskangas, O. Silvén: In *Proc. Conf. on Artificial Intelligence Res. in Finland*, ed. by C. Carlsson, T. Järvi, T. Reponen, Number 12 in Conf. Proc. of Finnish Artificial Intelligence Society (Finnish Artificial Intelligence Society, Helsinki, Finland 1994) p. 149

[10.691] R. Anand, K. Mehrotra, C. K. Mohan, S. Ranka: In *Proc. Int. Joint Conf. on Artificial Intelligence (IJCAI)* (University of Sydney, Sydney, Australia 1991)

[10.692] R. Anand, K. Mehrotra, C. K. Mohan, S. Ranka: Pattern Recognition **26**, 1717 (1993)

[10.693] R. Gemello, C. Lettera, F. Mana, L. Masera: In *Artificial Neural Networks*, ed. by T. Kohonen, K. Mäkisara, O. Simula, J. Kangas (North-Holland, Amsterdam, Netherlands 1991) p. II-1305

[10.694] R. Gemello, C. Lettera, F. Mana, L. Masera: CSELT Technical Reports **20**, 143 (1992)

[10.695] A. Verikas, K. Malmqvist, M. Bachauskene, L. Bergman, K. Nilsson: In *Proc. ICNN'94, Int. Conf. on Neural Networks* (IEEE Service Center, Piscataway, NJ 1994) p. 2938

[10.696] J. Wolfer, J. Robergé, T. Grace: In *Proc. WCNN'94, World Congress on Neural Networks* (Lawrence Erlbaum, Hillsdale, NJ 1994) p. I-260

[10.697]   J. Kangas: Technical Report A21 (Helsinki University of Technology, Laboratory of Computer and Information Science, SF-02150 Espoo, Finland 1993)

[10.698]   W. Gong, K. R. Rao, M. T. Manry: In *Conf. Record of the Twenty-Fifth Asilomar Conf. on Signals, Systems and Computers* (IEEE Comput. Soc. Press, Los Alamitos, CA 1991) p. I-477

[10.699]   W. Cheng, H. S. Soliman, A. H. Sung: In *Proc. ICNN'93, Int. Conf. on Neural Networks* (IEEE Service Center, Piscataway, NJ 1993) p. II-661

[10.700]   J. Lampinen, E. Oja: IEEE Trans. on Neural Networks **6**, 539 (1995)

[10.701]   T. Röfer: In *Proc. ICANN'95, Int. Conf. on Artificial Neural Networks*, ed. by F. Fogelman-Soulié, P. Gallinari, (EC2, Nanterre, France 1995) p. I-475

[10.702]   K. A. Han, J. C. Lee, C. J. Hwang: In *Proc. ICNN'95, IEEE Int. Conf. on Neural Networks*, (IEEE Service Center, Piscataway, NJ 1995) p. I-465

[10.703]   A. H. Dekker: Network: Computation in Neural Systems **5**, 351 (1994)

[10.704]   J. Heikkonen: In *Proc. EANN'95, Engineering Applications of Artificial Neural Networks* (Finnish Artificial Intelligence Society, Helsinki, Finland 1995) p. 33

[10.705]   N. Bonnet: Ultramicroscopy **57**, 17 (1995)

[10.706]   M. M. Moya, M. W. Koch, L. D. Hostetler: In *Proc. WCNN'93, World Congress on Neural Networks* (Lawrence Erlbaum, Hillsdale, NJ 1993) p. III-797

[10.707]   D. Hrycej: Neurocomputing **3**, 287 (1991)

[10.708]   J. Iivarinen, K. Valkealahti, A. Visa, O. Simula: In *Proc. ICANN'94, Int. Conf. on Artificial Neural Networks*, ed. by M. Marinaro, P. G. Morasso (Springer, London, UK 1994) p. I-334

[10.709]   J. Lampinen, E. Oja: In *Proc. INNC'90, Int. Neural Network Conf.* (Kluwer, Dordrecht, Netherlands 1990) p. I-301

[10.710]   J. Lampinen: Proc. SPIE – The Int. Society for Optical Engineering **1469**, 832 (1991)

[10.711]   J. Lampinen: In *Artificial Neural Networks*, ed. by T. Kohonen, K. Mäkisara, O. Simula, J. Kangas (North-Holland, Amsterdam, Netherlands 1991) p. II-99

[10.712]   A. Luckman, N. Allinson: Proc. Society of Photo-optical Instrumentation Engineers **1197**, 98 (1990)

[10.713]   A. J. Luckman, N. M. Allinson: In *Visual Search*, ed. by D. Brogner (Taylor & Francis, London, UK 1992) p. 169

[10.714]   A. Järvi, J. Järvi: In *Proc. Conf. on Artificial Intelligence Res. in Finland*, ed. by C. Carlsson, T. Järvi, T. Reponen, Number 12 in Conf. Proc. of Finnish Artificial Intelligence Society (Finnish Artificial Intelligence Society, Helsinki, Finland 1994) p. 104

[10.715]   E. Oja, L. Xu, P. Kultanen: In *Proc. INNC'90, Int. Neural Network Conference* (Kluwer, Dordrecht, Netherlands 1990) p. I-27

[10.716]   A. Visa: In *Proc. of SPIE Aerospace Sensing, Vol. 1709 Science of Neural Networks* (SPIE, Bellingham, USA 1992) p. 642

[10.717]   L. Xu, E. Oja: Res. Report 18 (Lappeenranta University of Technology, Department of Information Technology, Lappeenranta, Finland 1989)

[10.718]   L. Xu, E. Oja, P. Kultanen: Pattern Recognition Letters **11**, 331 (1990)

[10.719]   L. Xu, A. Krzyżak, E. Oja: IEEE Trans. on Neural Networks **4**, 636 (1993)

[10.720]   A. Ghosh, S. K. Pal: Pattern Recognition Letters **13**, 387 (1992)

[10.721]   A. Ghosh, N. R. Pal, S. R. Pal: IEEE Trans. on Fuzzy Systems **1**, 54 (1993)

[10.722] A. N. Redlich: Proc. SPIE – The Int. Society for Optical Engineering **1710**, 201 (1992)
[10.723] R. J. T. Morris, L. D. Rubin, H. Tirri: IEEE Trans. on Pattern Analysis and Machine Intelligence **12**, 1107 (1990)
[10.724] J. Hogden, E. Saltzman, P. Rubin: In *Proc. WCNN'93, World Congress on Neural Networks* (Lawrence Erlbaum, Hillsdale, NJ 1993) p. II-409
[10.725] J. Heikkonen, P. Koikkalainen: In *Proc. 4th Int. Workshop: Time-Varying Image Processing and Moving Object Recognition*, ed. by V. Cappellini (Elsevier, Amsterdam, Netherlands 1993) p. 327
[10.726] J. Marshall: In *Proc. IJCNN'89, Int. Joint Conf. on Neural Networks* (IEEE Service Center, Piscataway, NJ 1989) p. II-227
[10.727] J. Marshall: Neural Networks **3**, 45 (1990)
[10.728] M. Köppen: In *Proc. 3rd Int. Conf. on Fuzzy Logic, Neural Nets and Soft Computing* (Fuzzy Logic Systems Institute, Iizuka, Japan 1994) p. 149
[10.729] S. P. Luttrell: Technical Report 4437 (RSRE, Malvern, UK 1990)
[10.730] A. P. Kartashov: In *Artificial Neural Networks*, ed. by T. Kohonen, K. Mäkisara, O. Simula, J. Kangas (North-Holland, Amsterdam, Netherlands 1991) p. II-1103
[10.731] N. M. Allinson, M. J. Johnson: In *Proc. Fourth Int. IEE Conf. on Image Processing and its Applications*, Maastricht, Netherlands (1992) p. 193
[10.732] K. K. Truong, R. M. Mersereau: In *Proc. ICASSP'90, Int. Conf. on Acoustics, Speech and Signal Processing* (IEEE Service Center, Piscataway, NJ 1990) p. IV-2289
[10.733] K. K. Truong: In *Proc. ICASSP'91, Int. Conf. on Acoustics, Speech and Signal Processing* (IEEE Service Center, Piscataway, NJ 1991) p. IV-2789
[10.734] N. M. Allinson, A. W. Ellis: IEE Electronics and Communication J. **4**, 291 (1992)
[10.735] M. F. Augusteijn, T. L. Skufca: In *Proc. ICNN'93, Int. Conf. on Neural Networks* (IEEE Service Center, Piscataway, NJ 1993) p. I-392
[10.736] S. Jockusch, H. Ritter: In *Proc. IJCNN-93-Nagoya, Int. Joint Conf. on Neural Networks* (IEEE Service Center, Piscataway, NJ 1993) p. III-2077
[10.737] A. Carraro, E. Chilton, H. McGurk: In *IEE Colloquium on 'Biomedical Applications of Digital Signal Processing' (Digest No.144)* (IEE, London, UK 1989)
[10.738] C. Maggioni, B. Wirtz: In *Artificial Neural Networks*, ed. by T. Kohonen, K. Mäkisara, O. Simula, J. Kangas (North-Holland, Amsterdam, Netherlands 1991) p. I-75
[10.739] A. Meyering, H. Ritter: In *Artificial Neural Networks, 2*, ed. by I. Aleksander, J. Taylor (North-Holland, Amsterdam, Netherlands 1992) p. I-821
[10.740] A. Meyering, H. Ritter: In *Proc. IJCNN'92, Int. Joint Conf. on Neural Networks* (IEEE Service Center, Piscataway, NJ 1992) p. IV-432
[10.741] K. Chakraborty, U. Roy: Computers & Industrial Engineering **24**, 189 (1993)
[10.742] R. C. Luo, H. Potlapalli, D. Hislop: In *Proc. IJCNN-93-Nagoya, Int. Joint Conf. on Neural Networks* (IEEE Service Center, Piscataway, NJ 1993) p. II-1306
[10.743] S. Taraglio, S. Moronesi, A. Sargeni, G. B. Meo: In *Fourth Italian Workshop. Parallel Architectures and Neural Networks*, ed. by E. R. Caianiello (World Scientific, Singapore 1991) p. 378
[10.744] J. Orlando, R. Mann, S. Haykin: In *Proc. IJCNN-90-WASH-DC, Int. Joint Conf. of Neural Networks* (Lawrence Erlbaum, Hillsdale, NJ 1990) p. 263

[10.745] J. R. Orlando, R. Mann, S. Haykin: IEEE J. Oceanic Engineering **15**, 228 (1990)

[10.746] R.-J. Liou, M. R. Azimi-Sadjadi, D. L. Reinke: In *Proc. ICNN'94, Int. Conf. on Neural Networks* (IEEE Service Center, Piscataway, NJ 1994) p. 4327

[10.747] A. Visa, K. Valkealahti, J. Iivarinen, O. Simula: In *Proc. SPIE – The Int. Society for Optical Engineering, Applications of Artificial Neural Networks V*, ed. by S. K. R. a Dennis W. Ruck (SPIE, Bellingham, WA 1994) p. 2243-484

[10.748] H. Murao, I. Nishikawa, S. Kitamura: In *Proc. IJCNN-93-Nagoya, Int. Joint Conf. on Neural Networks* (IEEE Service Center, Piscataway, NJ 1993) p. II-1211

[10.749] L. Lönnblad, C. Peterson, H. Pi, T. Rögnvaldsson: Computer Physics Communications **67**, 193 (1991)

[10.750] M. Hernandez-Pajares, R. Cubarsi, E. Monte: Neural Network World **3**, 311 (1993)

[10.751] Z.-P. Lo, B. Bavarian: In *Proc. Fifth Int. Parallel Processing Symp.* (IEEE Comput. Soc. Press, Los Alamitos, CA 1991) p. 228

[10.752] S. Najand, Z.-P. Lo, B. Bavarian: In *Proc. IJCNN'92, Int. Joint Conference on Neural Networks* (IEEE Service Center, Piscataway, NJ 1992) p. II-87

[10.753] M. Sase, T. Hirano, T. Beppu, Y. Kosugi: Robot p. 106 (1992)

[10.754] M. Takahashi, H. Hashimukai, H. Ando: In *Proc. IJCNN'91, Int. Joint Conf. on Neural Networks* (IEEE Service Center, Piscataway, NJ 1991) p. II-932

[10.755] J. Waldemark, P.-O. Dovner, J. Karlsson: In *Proc. ICNN'95, IEEE Int. Conf. on Neural Networks*, (IEEE Service Center, Piscataway, NJ 1995) p. I-195

[10.756] D. Kilpatrick, R. Williams: In *Proc. ICNN'95, IEEE Int. Conf. on Neural Networks*, (IEEE Service Center, Piscataway, NJ 1995) p. I-32

[10.757] J. Wolfer, J. Robergé, T. Grace: In *Proc. WCNN'95, World Congress on Neural Networks*, (INNS, Lawrence Erlbaum, Hillsdale, NJ 1995) p. I-157

[10.758] E. Kwiatkowska, I. S. Torsun: In *Proc. ICNN'95, IEEE Int. Conf. on Neural Networks*, (IEEE Service Center, Piscataway, NJ 1995) p. IV-1907

[10.759] H. Bertsch, J. Dengler: In *9. DAGM-Symp. Mustererkennung*, ed. by E. Paulus (Deutche Arbeitsgruppe für Mustererkennung, 1987) p. 166

[10.760] C. Comtat, C. Morel: IEEE Trans. on Neural Networks **6**, 783 (1995)

[10.761] R. Deaton, J. Sun, W. E. Reddick: In *Proc. WCNN'95, World Congress on Neural Networks*, (INNS, Lawrence Erlbaum, Hillsdale, NJ 1995) p. II-815

[10.762] A. Manduca: In *Proc. ICNN'94, Int. Conf. on Neural Networks* (IEEE Service Center, Piscataway, NJ 1994) p. 3990

[10.763] K. J. Cios, L. S. Goodenday, M. Merhi, R. A. Langenderfer: In *Proc. Computers in Cardiology* (IEEE Comput. Soc. Press, Los Alamitos, CA 1990) p. 33

[10.764] C.-H. Joo, J.-S. Choi: Trans. Korean Inst. of Electrical Engineers **40**, 374 (1991)

[10.765] L. Vercauteren, G. Sieben, M. Praet: In *Proc. INNC'90, Int. Neural Network Conference* (Kluwer, Dordrecht, Netherlands 1990) p. 387

[10.766] H.-L. Pan, Y.-C. Chen: Pattern Recognition Letters **13**, 355 (1992)

[10.767] D. Patel, I. Hannah, E. R. Davies: In *Proc. WCNN'94, World Congress on Neural Networks* (Lawrence Erlbaum, Hillsdale, NJ 1994) p. I-631

[10.768] M. Turner, J. Austin, N. Allinson, P. Thomson: In *Artificial Neural Networks, 2*, ed. by I. Aleksander, J. Taylor (North-Holland 1992) p. I-799

[10.769] M. Turner, J. Austin, N. Allinson, P. Thompson: In *Proc. British Machine Vision Association Conf.* (1992) p. 257

[10.770] M. Turner, J. Austin, N. M. Allinson, P. Thompson: Image and Vision Computing **11**, 235 (1993)

[10.771] P. Arrigo, F. Giuliano, F. Scalia, A. Rapallo, G. Damiani: Comput. Appl. Biosci. **7**, 353 (1991)

[10.772] G. Gabriel, C. N. Schizas, C. S. Pattichis, R. Constantinou, A. Hadjianastasiou, A. Schizas: In *Proc. IJCNN-93-Nagoya, Int. Joint Conf. on Neural Networks* (IEEE Service Center, Piscataway, NJ 1993) p. I-943

[10.773] J.-M. Auger, Y. Idan, R. Chevallier, B. Dorizzi: In *Proc. IJCNN'92, Int. Joint Conf. on Neural Networks* (IEEE Service Center, Piscataway, NJ 1992) p. IV-444

[10.774] N. Baykal, N. Yalabik, A. H. Goktogan: In *Computer and Information Sciences VI. Proc. 1991 Int. Symposium*, ed. by M. Baray, B. Ozguc (Elsevier, Amsterdam, Netherlands 1991) p. II-923

[10.775] J. Henseler, J. C. Scholtes, C. R. J. Verhoest. M.Sc. Thesis, Delft University, Delft, Netherlands, 1987

[10.776] J. Henseler, H. J. van der Herik, E. J. H. Kerchhoffs, H. Koppelaar, J. C. Scholtes, C. R. J. Verhoest: In *Proc. Summer Comp. Simulation Conf., Seattle* (1988) p. 14

[10.777] H. J. van der Herik, J. C. Scholtes, C. R. J. Verhoest: In *Proc. European Simulation Multiconference* (1988) p. 350

[10.778] S. W. Khobragade, A. K. Ray: In *Proc. ICNN'93, Int. Conf. on Neural Networks* (IEEE Service Center, Piscataway, NJ 1993) p. III-1606

[10.779] I.-B. Lee, K.-Y. Lee: Korea Information Science Soc. Review **10**, 27 (1992)

[10.780] J. Loncelle, N. Derycke, F. Fogelman-Soulié: In *Proc. IJCNN'92, Int. Joint Conf. on Neural Networks* (IEEE Service Center, Piscataway, NJ 1992) p. III-694

[10.781] G. M. Papadourakis, G. N. Bebis, M. Georgiopoulos: In *Proc. INNC'90, Int. Neural Network Conf.* (Kluwer, Dordrecht, Netherlands 1990) p. I-392

[10.782] S. Taraglio: In *Proc. INNC'90, Int. Neural Network Conf.* (Kluwer, Dordrecht, Netherlands 1990) p. I-103

[10.783] R. J. T. Morris, L. D. Rubin, H. Tirri: In *Proc. IJCNN'89 Int. Joint Conf. on Neural Networks* (IEEE Service Center, Piscataway, NJ 1989) p. II-291

[10.784] A. P. Azcarraga, B. Amy: In *Artificial Intelligence IV. Methodology, Systems, Applications. Proc. of the Fourth International Conf. (AIMSA '90)*, ed. by P. Jorrand, V. Sgurev (North-Holland, Amsterdam, Netherlands 1990) p. 209

[10.785] D. H. Choi, S. W. Ryu, H. C. Kang, K. T. Park: J. Korean Inst. of Telematics and Electronics **28B**, 1 (1991)

[10.786] W. S. Kim, S. Y. Bang: J. Korean Inst. of Telematics and Electronics **29B**, 50 (1992)

[10.787] A. M. Colla, P. Pedrazzi: In *Proc. ICANN'94, Int. Conf. on Artificial Neural Networks*, ed. by M. Marinaro, P. G. Morasso (Springer, London, UK 1994) p. II-969

[10.788] Y. Idan, R. C. Chevallier: In *Proc. IJCNN-91-Singapore, Int. Joint Conf. on Neural Networks* (IEEE Service Center, Piscataway, NJ 1991) p. III-2576

[10.789] K. Nakayama, Y. Chigawa, O. Hasegawa: In *Proc. IJCNN'92, of the Int. Joint Conf. on Neural Networks* (IEEE Service Center, Piscataway, NJ 1992) p. IV-235

[10.790] M. Gioiello, G. Vassallo, F. Sorbello: In *The V Italian Workshop on Parallel Architectures and Neural Networks* (World Scientific, Singapore 1992) p. 293

[10.791] P. J. G. Lisboa: Int. J. Neural Networks – Res. & Applications **3**, 17 (1992)

[10.792] K. Yamane, K. Fuzimura, H. Tokimatu, H. Tokutaka, S. Kisida: Technical Report NC93-25 (The Inst. of Electronics, Information and Communication Engineers, Tottori University, Koyama, Japan 1993)

[10.793] K. Yamane, K. Fujimura, H. Tokutaka, S. Kishida: Technical Report NC93-86 (The Inst. of Electronics, Information and Communication Engineers, Tottori University, Koyama, Japan 1994)

[10.794] Z. Chi, H. Yan: Neural Networks **8**, 821 (1995)

[10.795] N. Natori, K. Nishimura: In *Proc. ICNN'95, IEEE Int. Conf. on Neural Networks*, (IEEE Service Center, Piscataway, NJ 1995) p. VI-3089

[10.796] J. Wu, H. Yan: In *Proc. ICNN'95, IEEE Int. Conf. on Neural Networks*, (IEEE Service Center, Piscataway, NJ 1995) p. VI-3074

[10.797] P. Morasso: In *Proc. IJCNN'89, Int. Joint Conf. on Neural Networks* (IEEE Service Center, Piscataway, NJ 1989) p. II-539

[10.798] P. Morasso, S. Pagliano: In *Fourth Italian Workshop. Parallel Architectures and Neural Networks*, ed. by E. R. Caianiello (World Scientific, Singapore 1991) p. 250

[10.799] P. Morasso: In *Artificial Neural Networks*, ed. by T. Kohonen, K. Mäkisara, O. Simula, J. Kangas (North-Holland, Amsterdam, Netherlands 1991) p. II-1323

[10.800] P. Morasso, L. Barberis, S. Pagliano, D. Vergano: Pattern Recognition **26**, 451 (1993)

[10.801] P. Morasso, L. Gismondi, E. Musante, A. Pareto: In *Proc. WCNN'93, World Congress on Neural Networks* (Lawrence Erlbaum, Hillsdale, NJ 1993) p. III-71

[10.802] L. Schomaker: Pattern Recognition **26**, 443 (1993)

[10.803] F. Andianasy, M. Milgram: In *Proc. EANN'95, Engineering Applications of Artificial Neural Networks* (Finnish Artificial Intelligence Society, Helsinki, Finland 1995) p. 61

[10.804] J. Heikkonen, M. Mäntynen: In *Proc. EANN'95, Engineering Applications of Artificial Neural Networks* (Finnish Artificial Intelligence Society, Helsinki, Finland 1995) p. 75

[10.805] G. D. Barmore: M.Sc. Thesis (Air Force Inst. of Tech., Wright-Patterson AFB, OH 1988)

[10.806] H. Behme, W. D. Brandt, H. W. Strube: In *Proc. ICANN'93, Int. Conf. on Artificial Neural Networks*, ed. by S. Gielen, B. Kappen (Springer, London, UK 1993) p. 416

[10.807] H. Behme, W. D. Brandt, H. W. Strube: In *Proc. IJCNN-93-Nagoya, Int. Joint Conf. on Neural Networks* (IEEE Service Center, Piscataway, NJ 1993) p. I-279

[10.808] L. D. Giovanni, S. Montesi: In *Proc. 1st Workshop on Neural Networks and Speech Processing, November 89, Roma*, ed. by A. Paoloni (Roma, Italy 1990) p. 75

[10.809]   C. Guan, C. Zhu, Y. Chen, Z. He: In *Proc. Int. Symp. on Speech, Image Processing and Neural Networks* (IEEE Hong Kong Chapter of Signal Processing, Hong Kong 1994) p. II-710

[10.810]   J. He, H. Leich: In *Proc. Int. Symp. on Speech, Image Processing and Neural Networks* (IEEE Hong Kong Chapt. of Signal Processing, Hong Kong 1994) p. I-109

[10.811]   H.-P. Hutter: Mitteilungen AGEN p. 9 (1992)

[10.812]   O. B. Jensen, M. Olsen, T. Rohde: Technical Report DAIMI IR-101 (Computer Science Department, Aarhus University, Aarhus, Denmark 1991)

[10.813]   C.-Y. Liou, C.-Y. Shiah: In *Proc. IJCNN-93-Nagoya, Int. Joint Conf. on Neural Networks* (IEEE Service Center, Piscataway, NJ 1993) p. I-251

[10.814]   E. López-Gonzalo, L. A. Hernández-Gómez: In *Proc. EUROSPEECH-93, 3rd European Conf. on Speech, Communication and Technology* (ESCA, Berlin, Germany 1993) p. I-55

[10.815]   A. Paoloni: In *Proc. 1st Workshop on Neural Networks and Speech Processing, November 89, Roma.*, ed. by A. Paoloni (1990) p. 5

[10.816]   G. N. di Pietro: Bull. des Schweizerischen Elektrotechnischen Vereins & des Verbandes Schweizerischer Elektrizitätswerke **82**, 17 (1991)

[10.817]   W. F. Recla: M.Sc. Thesis (Air Force Inst. of Tech., Wright-Patterson AFB, OH 1989)

[10.818]   F. S. Stowe: M.Sc. Thesis (Air Force Inst. of Tech., School of Engineering, Wright-Patterson AFB, OH 1990)

[10.819]   K. Torkkola, M. Kokkonen: In *Proc. ICASSP'91, Int. Conf. on Acoustics, Speech and Signal Processing* (IEEE Service Center, Piscataway, NJ 1991) p. I-261

[10.820]   K. Torkkola: PhD Thesis (Helsinki University of Technology, Espoo, Finland 1991)

[10.821]   P. Utela, K. Torkkola, L. Leinonen, J. Kangas, S. Kaski, T. Kohonen: In *Proc. SteP'92, Fifth Finnish Artificial Intelligence Conf., New Directions in Artificial Intelligence* (Finnish Artificial Intelligence Society, Helsinki, Finland 1992) p. II-178

[10.822]   L. Knohl, A. Rinscheid: In *Proc. EUROSPEECH-93, 3rd European Conf. on Speech, Communication and Technology* (ESCA, Berlin 1993) p. I-367

[10.823]   L. Knohl, A. Rinscheid: In *Proc. IJCNN-93-Nagoya, Int. Joint Conf. on Neural Networks* (IEEE Service Center, Piscataway, NJ 1993) p. I-243

[10.824]   P. Dalsgaard, O. Andersen, W. Barry: In *Proc. EUROSPEECH-91, 2nd European Conf. on Speech Communication and Technology Proceedings* (Istituto Int. Comunicazioni, Genova, Italy 1991) p. II-685

[10.825]   J. Kangas, K. Torkkola, M. Kokkonen: In *Proc. ICASSP'92, Int. Conf. on Acoustics, Speech and Signal Processing* (IEEE Service Center, Piscataway, NJ 1992)

[10.826]   J.-S. Kim, C.-M. Kyung: In *International Symp. on Circuits and Systems* (IEEE Service Center, Piscataway, NJ 1989) p. III-1879

[10.827]   T. M. English, L. C. Boggess: In *Proc. Cooperation, ACM Eighteenth Annual Computer Science Conf.* (ACM, New York, NY 1990) p. 444

[10.828]   L. A. Hernandez-Gomez, E. Lopez-Gonzalo: In *Proc. ICASSP'93, Int. Conf. on Acoustics, Speech and Signal Processing* (IEEE Service Center, Piscataway, NJ 1993) p. II-628

[10.829]   J. Thyssen, S. D. Hansen: In *Proc. ICASSP'93, Int. Conf. on Acoustics, Speech and Signal Processing* (IEEE Service Center, Piscataway, NJ 1993) p. II-431

[10.830] H. C. Card, S. Kamarsu: In *Proc. WCNN'95, World Congress on Neural Networks*, (Lawrence Erlbaum, Hillsdale, NJ 1995) p. I-128

[10.831] R. Togneri, M. Alder, Y. Attikiouzel: In *Proc. Third Australian Int. Conf. on Speech Science and Technology* (Melbourne, Australia 1990) p. 304

[10.832] A. Canas, J. Ortega, F. J. Fernandez, A. Prieto, F. J. Pelayo: In *Proc. IWANN'91, Int. Workshop on Artificial Neural Networks*, ed. by A. Prieto (Springer, Berlin, Germany 1991) p. 340

[10.833] V. H. Chin: In *C-CORE Publication no. 91-15* (C-CORE 1991)

[10.834] Z. Huang, A. Kuh: IEEE Trans. Signal Processing **40**, 2651 (1992)

[10.835] V. Z. Kepuska, J. N. Gowdy: In *Proc. Annual Southeastern Symp. on System Theory 1988* (IEEE Service Center, Piscataway, NJ 1988) p. 388

[10.836] V. Z. Kepuska, J. N. Gowdy: In *SOUTHEASTCON '90* (IEEE Service Center, Piscataway, NJ 1990) p. I-64

[10.837] L. S. Javier Tuya, E. A, J. A. Corrales: In *Proc. IWANN'93, Int. Workshop on Neural Networks, Sitges, Spain*, ed. by A. P. J. Mira, J. C (Springer, Berlin, Germany 1993) p. 550

[10.838] T. Matsuoka, Y. Ishida: In *Proc. ICNN'95, IEEE Int. Conf. on Neural Networks*, (IEEE Service Center, Piscataway, NJ 1995) p. V-2900

[10.839] O. Anderson, P. Cosi, P. Dalsgaard: In *Proc. 1st Workshop on Neural Networks and Speech Processing, November 89, Roma*, ed. by A. Paoloni (Roma, Italy 1990) p. 18

[10.840] U. Dagitan, N. Yalabik: In *Neurocomputing, Algorithms, Architectures and Applications. Proc. NATO Advanced Res. Workshop*, ed. by F. Fogelman-Soulié, J. Herault (Springer, Berlin, Germany 1990) p. 297

[10.841] M. A. Al-Sulaiman, S. I. Ahson, M. I. Al-Kanhal: In *Proc. WCNN'93, World Congress on Neural Networks* (Lawrence Erlbaum, Hillsdale, NJ 1993) p. IV-84

[10.842] T. R. Anderson.: In *Proc. ICASSP'91, Int. Conf. on Acoustics, Speech and Signal Processing* (IEEE Service Center, Piscataway, NJ 1991) p. I-149

[10.843] T. Anderson: In *Proc. ICNN'94, Int. Conf. on Neural Networks* (IEEE Service Center, Piscataway, NJ 1994) p. 4466

[10.844] D. Chen, Y. Gao: In *Proc. INNC'90, Int. Neural Network Conference* (Kluwer, Dordrecht, Netherlands 1990) p. I-195

[10.845] P. Dalsgaard: Computer Speech and Language **6**, 303 (1992)

[10.846] S. Danielson: In *Proc. IJCNN-90-WASH-DC, Int. Joint Conf. on Neural Networks* (IEEE Service Center, Piscataway, NJ 1990) p. III-677

[10.847] M. Hanawa, T. Hasega-Wa: Trans. Inst. of Electronics, Information and Communication Engineers D-II **J75D-II**, 426 (1992)

[10.848] J. Kangas, O. Naukkarinen, T. Kohonen, K. Mäkisara, O. Ventä: Report TKK-F-A585 (Helsinki University of Technology, Espoo, Finland 1985)

[10.849] J. Kangas: M.Sc. Thesis (Helsinki University of Technology, Espoo, Finland 1988)

[10.850] J. Kangas, T. Kohonen: In *Proc. First Expert Systems Applications World Conference* (IITT International, France 1989) p. 321

[10.851] J. Kangas, T. Kohonen: In *Proc. EUROSPEECH-89, European Conf. on Speech Communication and Technology* (ESCA, Berlin, Germany 1989) p. 345

[10.852] N. Kasabov, E. Peev: In *Proc. ICANN'94, Int. Conf. on Artificial Neural Networks*, ed. by M. Marinaro, P. G. Morasso (Springer, London, UK 1994) p. I-201

[10.853] V. Z. Kepuska, J. N. Gowdy: In *Proc. IEEE SOUTHEASTCON* (IEEE Service Center, Piscataway, NJ 1989) p. II-770

[10.854] V. Z. Kepuska, J. N. Gowdy: In *Proc. ICASSP'89, Int. Conf. on Acoustics, Speech and Signal Processing* (IEEE Service Center, Piscataway, NJ 1989) p. I-504

[10.855] K. Kiseok, K. I. Kim, H. Heeyeung: In *Proc. 5th Jerusalem Conf. on Information Technology (JCIT). Next Decade in Information Technology* (IEEE Comput. Soc. Press, Los Alamitos, CA 1990) p. 364

[10.856] D.-K. Kim, C.-G. Jeong, H. Jeong: Trans. Korean Inst. of Electrical Engineers **40**, 360 (1991)

[10.857] P. Knagenhjelm, P. Brauer: Speech Communication **9**, 31 (1990)

[10.858] T. Kohonen: Report TKK-F-A463 (Helsinki University of Technology, Espoo, Finland 1981)

[10.859] F. Mihelic, I. Ipsic, S. Dobrisek, N. Pavesic: Pattern Recognition Letters **13**, 879 (1992)

[10.860] P. Wu, K. Warwick, M. Koska: Neurocomputing **4**, 109 (1992)

[10.861] T. Kohonen, K. Torkkola, J. Kangas, O. Ventä: In *Papers from the 15th Meeting of Finnish Phoneticians, Publication 31, Helsinki University of Technology, Acoustics Laboratory* (Helsinki University of Technology, Espoo, Finland 1988) p. 97

[10.862] T. Kohonen: In *Neural Computing Architectures*, ed. by I. Aleksander (North Oxford Academic Publishers/Kogan Page, Oxford, UK 1989) p. 26

[10.863] T. Kohonen: In *The Second European Seminar on Neural Networks, London, UK, February 16-17* (British Neural Networks Society, London, UK 1989)

[10.864] T. Kohonen: In *Proc. IEEE Workshop on Neural Networks for Signal Processing* (IEEE Service Center, Piscataway, NJ 1991) p. 279

[10.865] T. Kohonen: In *Applications of Neural Networks* (VCH, Weinheim, Germany 1992) p. 25

[10.866] M. Kokkonen, K. Torkkola: In *Proc. EUROSPEECH-89, European Conf. on Speech Communication and Technology*, ed. by J. P. Tubach, J. J. Mariani (Assoc. Belge des Acousticiens; Assoc. Recherche Cognitive; Comm. Eur. Communities; et al, CEP Consultants, Edinburgh, UK 1989) p. II-561

[10.867] M. Kokkonen, K. Torkkola: Speech Communication **9**, 541 (1990)

[10.868] M. Kokkonen: M.Sc. Thesis (Helsinki University of Technology, Espoo, Finland 1991)

[10.869] H. Skinnemoen. *New Advances and Trends in Speech Recognition and Coding*, MOR-VQ for Speech Coding over Noisy Channels. NATO ASI Series F. (Springer, Berlin, Germany 1993)

[10.870] B. Brückner, T. Wesarg, C. Blumenstein: In *Proc. ICNN'95, IEEE Int. Conf. on Neural Networks*, (IEEE Service Center, Piscataway, NJ 1995) p.V-2891

[10.871] J. M. Colombi: M.Sc. Thesis (Air Force Inst. of Tech., School of Engineering, Wright-Patterson AFB, OH 1992)

[10.872] J. M. Colombi, S. K. Rogers, D. W. Ruck: In *Proc. ICASSP'93, Int. Conf. on Acoustics, Speech and Signal Processing* (IEEE Service Center, Piscataway, NJ 1993) p. II-700

[10.873] S. Hadjitodorov, B. Boyanov, T. Ivanov, N. Dalakchieva: Electronics Letters **30**, 838 (1994)

[10.874] X. Jiang, Z. Gong, F. Sun, h Chi: In *Proc. WCNN'94, World Congress on Neural Networks* (Lawrence Erlbaum, Hillsdale, NJ 1994) p. IV-595

[10.875] J. Naylor, A. Higgins, K. P. Li, D. Schmoldt: Neural Networks **1**, 311 (1988)

[10.876] D. Çetin, F. Yildirim, D. Demirekler, B. Nakiboğlu, B. Tüzün: In *Proc. EANN'95, Engineering Applications of Artificial Neural Networks* (Finnish Artificial Intelligence Society, Helsinki, Finland 1995) p. 267

[10.877] J. He, L. Liu, G. Palm: In *Proc. ICNN'95, IEEE Int. Conf. on Neural Networks*, (IEEE Service Center, Piscataway, NJ 1995) p. IV-2052

[10.878] S. Nakamura, T. Akabane: In *ICASSP'91. 1991 Int. Conf. on Acoustics, Speech and Signal Processing* (IEEE Service Center, Piscataway, NJ 1991) p. II-853

[10.879] J. Naylor, K. P. Li: Neural Networks **1**, 310 (1988)

[10.880] H. Hase, H. Matsuyama, H. Tokutaka, S. Kishida: Technical Report NC95-140 (The Inst. of Electronics, Information and Communication Engineers, Tottori University, Koyama, Japan 1996)

[10.881] M. Alder, R. Togneri, E. Lai, Y. Attikiouzel: Pattern Recognition Letters **11**, 313 (1990)

[10.882] R. Togneri, M. Alder, Y. Attikiouzel: IEE Proceedings-I **139**, 123 (1992)

[10.883] M. V. Chan, X. Feng, J. A. Heinen, R. J. Niederjohn: In *Proc. ICNN'94, Int. Conf. on Neural Networks* (IEEE Service Center, Piscataway, NJ 1994) p. 4483

[10.884] W. Barry, P. Dalsgaard: In *Proc. EUROSPEECH'93, 3rd European Conf. on Speech, Communication and Technology* (1993) p. I-13

[10.885] M. P. DeSimio, T. R. Anderson: In *Proc. ICASSP'93, Int. Conf. on Acoustics, Speech and Signal Processing* (IEEE Service Center, Piscataway, NJ 1993) p. I-521

[10.886] T. Hiltunen, L. Leinonen, J. Kangas: In *Proc. ICANN'93, Int. Conf. on Artificial Neural Networks*, ed. by S. Gielen, B. Kappen (Springer, London, UK 1993) p. 420

[10.887] J. Kangas, P. Utela: Tekniikka logopediassa ja foniatriassa p. 36 (1992)

[10.888] P. Utela, J. Kangas, L. Leinonen: In *Artificial Neural Networks, 2*, ed. by I. Aleksander, J. Taylor (North-Holland, Amsterdam, Netherlands 1992) p. I-791

[10.889] J. Kangas, L. Leinonen, A. Juvas: University of Oulu, Publications of the Department of Logopedics and Phonetics p. 23 (1991)

[10.890] L. Leinonen, J. Kangas, K. Torkkola, A. Juvas, H. Rihkanen, R. Mujunen: Suomen Logopedis-Foniatrinen Aikakauslehti **10**, 4 (1991)

[10.891] L. Leinonen, J. Kangas, K. Torkkola: Tekniikka logopediassa ja foniatriassa p. 41 (1992)

[10.892] L. Leinonen, T. Hiltunen, J. Kangas, A. Juvas, H. Rihkanen: Scand. J. Log. Phon. **18**, 159 (1993)

[10.893] M. Beveridge: M.Sc. Thesis (University of Edinburgh, Department of Linguistics, Edinburgh, UK 1993)

[10.894] R. Mujunen, L. Leinonen, J. Kangas, K. Torkkola: Folia Phoniatrica **45**, 135 (1993)

[10.895] T. Räsänen, S. K. Hakumäki, E. Oja, M. O. K. Hakumäki: Folia Phoniatrica **42**, 135 (1990)

[10.896] L. Leinonen, T. Hiltunen, K. Torkkola, J. Kangas: J. Acoust. Soc. of America **93**, 3468 (1993)

[10.897] L. Leinonen, R. Mujunen, J. Kangas, K. Torkkola: Folia Phoniatrica **45**, 173 (1993)

[10.898] J. Reynolds: Report OUEL 1914/92 (Univ. of Oxford, Oxford, UK 1992)

[10.899] J. Reynolds, L. Tarassenko: Neural Computing & Application **1**, 169 (1993)

[10.900] L. P. J. Veelenturf: In *Twente Workshop on Language Technology 3: Connectionism and Natural Language Processing*, ed. by A. N. Marc F.

[10.985]   T. Yamaguchi, M. Tanabe, J. Murakami, K. Goto: Trans. Inst. of Elec-
           trical Engineers of Japan, Part C **111-C**, 40 (1991)
[10.986]   C. P. Matthews, K. Warwick: In *Proc. EANN'95, Engineering Applica-
           tions of Artificial Neural Networks* (Finnish Artificial Intelligence Society,
           Helsinki, Finland 1995) p. 449
[10.987]   K. Goser, K. M. Marks, U. Rueckert, V. Tryba: In *3. Internationaler
           GI-Kongress über Wissensbasierte Systeme, München, October 16-17*
           (Springer, Berlin, Heidelberg 1989) p. 225
[10.988]   E. Govekar, E. Susič, P. Mužič, I. Grabec: In *Artificial Neural Networks,
           2*, ed. by I. Aleksander, J. Taylor (North-Holland, Amsterdam, Nether-
           lands 1992) p. I-579
[10.989]   J. O'Brien, C. Reeves: In *Proc. 5th Int. Congress on Condition Monitor-
           ing and Diagnostic Engineering Management*, ed. by R. B. K. N. Rao,
           G. J. Trmal (University of the West of England, Bristol. UK 1993) p. 395
[10.990]   P. J. C. Skitt, M. A. Javed, S. A. Sanders, A. M. Higginson: J. Intelligent
           Manufacturing **4**, 79 (1993)
[10.991]   V. Tryba, K. Goser: In *Artificial Neural Networks*, ed. by T. Kohonen,
           K. Mäkisara, O. Simula, J. Kangas (North-Holland, Amsterdam, Nether-
           lands 1991) p. 847
[10.992]   A. Ultsch: In *Proc. ICANN'93, Int. Conf. on Artificial Neural Networks*,
           ed. by S. Gielen, B. Kappen (Springer, London, UK 1993) p. 864
[10.993]   O. Simula, J. Kangas.   *Neural Networks for Chemical Engineers,
           Computer-Aided Chemical Engineering, 6*, Process monitoring and visu-
           alization using self-organizing maps. (Elsevier, Amsterdam 1995) p. 371
[10.994]   J. T. Alander, M. Frisk, L. Holmström, A. Hämäläinen, J. Tuominen: In
           Artificial Neural Networks, ed. by T. Kohonen, K. Mäkisara, O. Simula,
           J. Kangas (North-Holland, Amsterdam, Netherlands 1991) p. II-1229
[10.995]   J. T. Alander, M. Frisk, L. Holmström, A. Hämäläinen, J. Tuominen:
           Res. Reports A5 (Rolf Nevanlinna Institute, Helsinki, Finland 1991)
[10.996]   F. Firenze, L. Ricciardiello, S. Pagliano: In   *Proc. ICANN'94, Int.
           Conf. on Artificial Neural Networks*, ed. by M. Marinaro, P. G. Morasso
           (Springer, London, UK 1994) p. II-1239
[10.997]   T. Sorsa, H. N. Koivo, H. Koivisto: IEEE Trans. on Syst., Man, and
           Cyb. **21**, 815 (1991)
[10.998]   T. Sorsa, H. N. Koivo: Automatica **29**, 843 (1993)
[10.999]   P. Tse, D. D. Wang, D. Atherton: In *Proc. ICNN'95, IEEE Int. Conf. on
           Neural Networks*, (IEEE Service Center, Piscataway, NJ 1995) p. II-927
[10.1000] J.-M. Wu, J.-Y. Lee, Y.-C. Tu, C.-Y. Liou: In *Proc. IECON '91, Int.
           Conf. on Industrial Electronics, Control and Instrumentation* (IEEE Ser-
           vice Center, Piscataway, NJ 1991) p. II-1506
[10.1001] H. Furukawa, T. Ueda, M. Kitamura: In *Proc.3rd Int. Conf. on Fuzzy
           Logic, Neural Nets and Soft Computing* (Fuzzy Logic Systems Institute,
           Iizuka, Japan 1994) p. 555
[10.1002] C. Muller, M. Cottrell, B. Girard, Y. Girard, M. Mangeas: In   *Proc.
           WCNN'94, World Congress on Neural Networks* (Lawrence Erlbaum,
           Hillsdale, NJ 1994) p. I-360
[10.1003] N. Ball, L. Kierman, K. Warwick, E. Cahill, D. Esp, J. Macqueen: Neu-
           rocomputing **4**, 5 (1992)
[10.1004] Y.-Y. Hsu, C.-C. Yang: IEE Proc. C [Generation, Transmission and
           Distribution] **138**, 407 (1991)
[10.1005] N. Macabrey, T. Baumann, A. J. Germond: Bulletin des Schweiz-
           erischen Elektrotechnischen Vereins & des Verbandes Schweizerischer
           Elektrizitätswerke **83**, 13 (1992)

[10.1006] D. J. Nelson, S.-J. Chang, M. Chen: In *Proc. 1992 Summer Computer Simulation Conference. Twenty-Fourth Annual Computer Simulation Conference*, ed. by P. Luker (SCS, San Diego, CA 1992) p. 217

[10.1007] D. Niebur, A. J. Germond: In *Proc. Third Symp. on Expert Systems Application to Power Systems* (Tokyo & Kobe 1991)

[10.1008] H. Mori, Y. Tamaru, S. Tsuzuki: In *Conf. Papers. 1991 Power Industry Computer Application Conf. Seventeenth PICA Conf.* (IEEE Service Center, Piscataway, NJ 1991) p. 293

[10.1009] H. Mori, Y. Tamaru, S. Tsuzuki: IEEE Trans. Power Systems **7**, 856 (1992)

[10.1010] R. Fischl: In *Proc. ICNN'94, Int. Conf. on Neural Networks* (IEEE Service Center, Piscataway, NJ 1994) p. 3719

[10.1011] D. Niebur, A. J. Germond: In *Conf. Papers. 1991 Power Industry Computer Application Conference. Seventeenth PICA Conference.* (IEEE Service Center, Piscataway, NJ 1991) p. 270

[10.1012] D. Niebur, A. J. Germond: In *Proc. First Int. Forum on Applications of Neural Networks to Power Systems*, ed. by M. A. El-Sharkawi, R. J. M. II (IEEE Service Center, Piscataway, NJ 1991) p. 83

[10.1013] D. Niebur, A. J. Germond: Int. J. Electrical Power & Energy Systems **14**, 233 (1992)

[10.1014] D. Niebur, A. J. Germond: IEEE Trans. Power Systems **7**, 865 (1992)

[10.1015] T. Baumann, A. Germond, D. Tschudi: In *Proc. Third Symp. on Expert Systems Application to Power Systems* (Tokyo & Kobe 1991)

[10.1016] A. Schnettler, V. Tryba: Archiv für Elektrotechnik **76**, 149 (1993)

[10.1017] A. Schnettler, M. Kurrat: In *Proc. 8th Int. Symp. on High Voltage Engineering, Yokohama* (1993) p. 57

[10.1018] J. Yu, Z. Gue, Z. Liu: In *Proc. First Int. Forum on Applications of Neural Networks to Power Systems*, ed. by M. A. El-Sharkawi, R. J. M. II (IEEE Service Center, Piscataway, NJ 1991) p. 293

[10.1019] S. Cumming: Neural Computing & Applications **1**, 96 (1993)

[10.1020] J. T. Gengo. M.Sc. Thesis (Naval Postgraduate School, Monterey, CA 1989)

[10.1021] H. Ogi, Y. Izui, S. Kobayashi: Mitsubishi Denki Giho **66**, 63 (1992)

[10.1022] S. Zhang, T. S. Sankar: In *Proc. IMACS, Int. Symp. on Signal Processing, Robotics and Neural Networks* (IMACS, Lille, France 1994) p. 183

[10.1023] L. Monostori, A. Bothe: In *Industrial and Engineering Applications of Artificial Intelligence and Expert Systems. 5th Int. Conf., IEA/AIE-92*, ed. by F. Belli, F. J. Radermacher (Springer, Berlin, Heidelberg 1992) p. 113

[10.1024] J. Lampinen, O. Taipale: In *Proc. ICNN'94, Int. Conf. on Neural Networks* (IEEE Service Center, Piscataway, NJ 1994) p. 3812

[10.1025] K.-H. Becks, J. Dahm, F. Seidel: In *Industrial and Engineering Applications of Artificial Intelligence and Expert Systems. 5th International Conference, IEA/AIE-92*, ed. by F. Belli, F. J. Radermacher (Springer, Berlin, Heidelberg 1992) p. 109

[10.1026] S. Cai, H. Toral, J. Qiu: In *Proc. ICANN-93, Int. Conf. on Artificial Neural Networks*, ed. by S. Gielen, B. Kappen (Springer, London, UK 1993) p. 868

[10.1027] S. Cai, H. Toral: In *Proc. IJCNN-93-Nagoya, Int. Joint Conf. on Neural Networks* (IEEE Service Center, Piscataway, NJ 1993) p. II-2013

[10.1028] Y. Cai: In *Proc. WCNN'94, World Congress on Neural Networks* (Lawrence Erlbaum, Hillsdale, NJ 1994) p. I-516

[10.1029] P. Burrascano, P. Lucci, G. Martinelli, R. Perfetti: In *Proc. IJCNN'90-WASH-DC, Int. Joint Conf. on Neural Networks* (IEEE Service Center, Piscataway, NJ 1990) p. I-311

[10.1030] P. Burrascano, P. Lucci, G. Martinelli, R. Perfetti: In *Proc. ICASSP'90, Int. Conf. on Acoustics, Speech and Signal Processing* (IEEE Service Center, Piscataway, NJ 1990) p. IV-1921

[10.1031] K. L. Fox, R. R. Henning, J. H. Reed, R. P. Simonian: In *Proc. 13th National Computer Security Conference. Information Systems Security. Standards – the Key to the Future* (NIST, Gaithersburg, MD 1990) p. I-124

[10.1032] J. J. Garside, R. H. Brown, T. L. Ruchti, X. Feng: In *Proc. IJCNN'92, Int. Joint Conf. on Neural Networks* (IEEE Service Center, Piscataway, NJ 1992) p. II-811

[10.1033] B. Grossman, X. Gao, M. Thursby: Proc. SPIE – The Int. Soc. for Opt. Eng. **1588**, 64 (1991)

[10.1034] N. Kashiwagi, T. Tobi: In *Proc. IJCNN-93-Nagoya, Int. Joint Conf. on Neural Networks* (IEEE Service Center, Piscataway, NJ 1993) p. I-939

[10.1035] W. Kessler, D. Ende, R. W. Kessler, W. Rosenstiel: In *Proc. ICANN-93, Int. Conf. on Artificial Neural Networks*, ed. by S. Gielen, B. Kappen (Springer, London, UK 1993) p. 860

[10.1036] M. Konishi, Y. Otsuka, K. Matsuda, N. Tamura, A. Fuki, K. Kadoguchi: In *Third European Seminar on Neural Computing: The Marketplace* (IBC Tech. Services, London, UK 1990) p. 13

[10.1037] R. R. Stroud, S. Swallow, J. R. McCardle, K. T. Burge: In *Proc. IJCNN-93-Nagoya, Int. Joint Conf. on Neural Networks* (IEEE Service Center, Piscataway, NJ 1993) p. II-1857

[10.1038] W. Fushuan, H. Zhenxiang: In *Third Biennial Symp. on Industrial Electric Power Applications* (Louisiana Tech. Univ, Ruston, LA, USA 1992) p. 268

[10.1039] G. A. Clark, J. E. Hernandez, N. K. DelGrande, R. J. Sherwood, S.-Y. Lu, P. C. Schaich, P. F. Durbin: In *Conf. Record of the Twenty-Fifth Asilomar Conf. on Signals, Systems and Computers* (IEEE Comput. Soc. Press, Los Alamitos, CA 1991) p. II-1235

[10.1040] T. Sorsa, H. N. Koivo, R. Korhonen: In *Preprints of the IFAC Symp. on On-Line Fault Detection and Supervision in the Chemical Process Industries, Newark, Delaware, April 1992* (1992) p. 162

[10.1041] P. Vuorimaa: In *Proc. Conf. on Artificial Intelligence Res. in Finland*, ed. by C. Carlsson, T. Järvi, T. Reponen, Number 12 in Conf. Proc. of Finnish Artificial Intelligence Society (Finnish Artificial Intelligence Society, Helsinki, Finland 1994) p. 177

[10.1042] P. Franchi, P. Morasso, G. Vercelli: In *Proc. ICANN'94, Int. Conf. on Artificial Neural Networks*, ed. by M. Marinaro, P. G. Morasso (Springer, London, UK 1994) p. II-1287

[10.1043] E. Littman, A. Meyering, J. Walter, T. Wengerek, H. Ritter: In *Applications of Neural Networks*, ed. by K. Schuster (VCH, Weinheim, Germany 1992) p. 79

[10.1044] T. Martinetz, K. Schulten: Computers & Electrical Engineering **19**, 315 (1993)

[10.1045] B. W. Mel: In *Proc. First IEEE Conf. on Neural Information Processing Systems*, ed. by D. Z. Anderson (IEEE Service Center, Piscataway, NJ 1988) p. 544

[10.1046] H. Ritter, T. Martinetz, K. Schulten: MC-Computermagazin **2**, 48 (1989)

[10.1047] P. van der Smagt, F. Groen, F. van het Groenewoud: In *Proc. ICNN'94, Int. Conf. on Neural Networks* (IEEE Service Center, Piscataway, NJ 1994) p. 2787

[10.1048] F. B. Verona, F. E. Lauria, M. Sette, S. Visco: In *Proc. IJCNN-93-Nagoya, Int. Joint Conf. on Neural Networks* (IEEE Service Center, Piscataway, NJ 1993) p. II-1861

[10.1049] D. A. C. Barone, A. R. M. Ramos: In *Proc. EANN'95, Engineering Applications of Artificial Neural Networks* (Finnish Artificial Intelligence Society, Helsinki, Finland 1995) p. 95

[10.1050] J. S. J. v. Deventer: In *Proc. ICNN'95, IEEE Int. Conf. on Neural Networks*, (IEEE Service Center, Piscataway, NJ 1995) p. VI-3068

[10.1051] T. Harris, L. Gamlyn, P. Smith, J. MacIntyre, A. Brason, R. Palmer, H. Smith, A. Slater: In *Proc. ICNN'95, IEEE Int. Conf. on Neural Networks*, (IEEE Service Center, Piscataway, NJ 1995) p. II-686

[10.1052] M. Mangeas, A. S. Weigend, C. Muller: In *Proc. WCNN'95, World Congress on Neural Networks*, (Lawrence Erlbaum, Hillsdale, NJ 1995) p. II-48

[10.1053] D. W. M. a. C. Aldrict, J. S. J. v. Deventer. *Neural Networks for Chemical Engineers, Computer-Aided Chemical Engineering*, The videographic characterization of flotation froths using neural networks. (Elsevier, Amsterdam, Netherlands 1995) p. 535

[10.1054] K. Röpke, D. Filbert: In *Proc. SAFEPROCESS'94, IFAC Symp. on Fault Detection, Supervision and Technical Processes*, (IFAL 1994) p. II-720

[10.1055] D. Vincent, J. McCardle, R. Stroud: In *Proc. ICNN'95, IEEE Int. Conf. on Neural Networks*, (IEEE Service Center, Piscataway, NJ 1995) p. I-522

[10.1056] J. C. H. Yeh, L. G. C. Hamey, T. Westcott, S. K. Y. Sung: In *Proc. ICNN'95, IEEE Int. Conf. on Neural Networks*, (IEEE Service Center, Piscataway, NJ 1995) p. I-37

[10.1057] J. L. Buessler, D. Kuhn, J. P. Urban: In *Proc. WCNN'95, World Congress on Neural Networks*, (INNS, Lawrence Erlbaum, Hillsdale, NJ 1995) p. II-384

[10.1058] A. H. Dekker, P. K. Piggott: In *Proc. of Robots for Australian Industries, National Conference of the Australian Robot Association* (Australian Robot Association, 1995) p. 369

[10.1059] J. Heikkonen, J. del R. Millán, E. Cuesta: In *Proc. EANN'95, Engineering Applications of Artificial Neural Networks* (Finnish Artificial Intelligence Society, Helsinki, Finland 1995) p. 119

[10.1060] D. Lambrinos, C. Scheier, R. Pfeifer: In *Proc. ICANN'95, Int. Conf. on Artificial Neural Networks*, ed. by F. Fogelman-Soulié, P. Gallinari, (EC2, Nanterre, France 1995) p. II-467

[10.1061] E. Cervera, A. P. del Pobil, E. Marta, M. A. Serna: In *Proc. CAEPIA'95, VI Conference of the Spanish Association for Artificial Intelligence* (1995) p. 415

[10.1062] J. Heikkonen, M. Surakka, J. Riekki: In *Proc. EANN'95, Engineering Applications of Artificial Neural Networks* (Finnish Artificial Intelligence Society, Helsinki, Finland 1995) p. 53

[10.1063] E. Cervera, A. P. del Pobil, E. Marta, M. A. Serna: In *Proc. TTIA'95, Transferencia Tecnológica de Inteligencia Artificial a Industria, Medicina y Aplicaciones Sociales*, ed. by R. R. Aldeguer, J. M. G. Chamizo (1995) p. 3

[10.1064] D. Graf, W. LaLonde: In *Proc. IJCNN'89, Int. Joint Conf. on Neural Networks* (IEEE Service Center, Piscataway, NJ 1989) p. II-543

[10.1065] T. Hesselroth, K. Sarkar, P. P. v. d. Smagt, K. Schulten: IEEE Trans. on Syst., Man and Cyb. **24**, 28 (1993)

[10.1066] T. Hirano, M. Sase, Y. Kosugi: Trans. Inst. Electronics, Information and Communication Engineers **J76D-II**, 881 (1993)

[10.1067] M. Jones, D. Vernon: Neural Computing & Applications **2**, 2 (1994)

[10.1068] S. Kieffer, V. Morellas, M. Donath: In *Proc. Int. Conf. on Robotics and Automation* (IEEE Comput. Soc. Press, Los Alamitos, CA 1991) p. III-2418

[10.1069] T. Martinetz, H. Ritter, K. Shulten: In *Proc. IJCNN'89, Int. Joint Conf. on Neural Networks* (IEEE Service Center, Piscataway, NJ 1989) p. II-351

[10.1070] T. Martinetz, H. Ritter, K. Schulten: In *Proc. Int. Conf. on Parallel Processing in Neural Systems and Computers (ICNC), Düsseldorf* (Elsevier, Amsterdam, Netherlands 1990) p. 431

[10.1071] T. Martinetz, H. Ritter, K. Schulten: In *Proc. ISRAM-90, Third Int. Symp. on Robotics and Manufacturing* (Vancouver, Canada 1990) p. 521

[10.1072] T. M. Martinetz, K. J. Schulten: In *Proc. IJCNN-90-WASH-DC, Int. Joint Conf. on Neural Networks* (IEEE Service Center, Piscataway, NJ 1990) p. II-747

[10.1073] T. M. Martinetz, H. J. Ritter, K. J. Schulten: IEEE Trans. on Neural Networks **1**, 131 (1990)

[10.1074] H. Ritter, K. Schulten: In *Neural Networks for Computing, AIP Conference Proc. 151, Snowbird, Utah*, ed. by J. S. Denker (American Inst. of Phys., New York, NY 1986) p. 376

[10.1075] H. Ritter, K. Schulten: In *Neural Computers*, ed. by R. Eckmiller, C. v. d. Malsburg (Springer, Berlin, Heidelberg 1988) p. 393.

[10.1076] H. Ritter, T. M. Martinetz, K. J. Schulten: Neural Networks **2**, 159 (1989)

[10.1077] H. Ritter, T. Martinetz, K. Schulten: In *Neural Networks, from Models to Applications*, ed. by L. Personnaz, G. Dreyfus (EZIDET, Paris, France 1989) p. 579

[10.1078] J. A. Walter, T. M. Martinetz, K. J. Schulten: In *Artificial Neural Networks*, ed. by T. Kohonen, K. Mäkisara, O. Simula, J. Kangas (North-Holland, Amsterdam, Netherlands 1991) p. I-357

[10.1079] J. A. Walter, K. Schulten: IEEE Trans. on Neural Networks **4**, 86 (1993)

[10.1080] P. Morasso, V. Sanguineti: In *Proc. Conf. on Prerational Intelligence – Phenomenology of Complexity Emerging in Systems of Agents Interagtion Using Simple Rules*, (, Center for Interdisciplinary Research, University of Bielefeld 1993) p. II-71

[10.1081] N. Ball, K. Warwick: In *Proc. American Control Conf.* (American Automatic Control Council, Green Valley, AZ 1992) p. 3062

[10.1082] N. R. Ball, K. Warwick: IEE Proc. D (Control Theory and Applications) **140**, 176 (1993)

[10.1083] N. R. Ball: In *Proc. IMACS Int. Symp. on Signal Processing, Robotics and Neural Networks* (IMACS, Lille, France 1994) p. 294

[10.1084] D. H. Graf, W. R. LaLonde: In *Proc. ICNN'88, Int. Conf. on Neural Networks* (IEEE Service Center, Piscataway, NJ 1988) p. I-77

[10.1085] J. Heikkonen, P. Koikkalainen, E. Oja, J. Mononen: In *Proc. Symp. on Neural Networks in Finland, Åbo Akademi, Turku, January 21.*, ed. by A. Bulsari, B. Saxén (Finnish Artificial Intelligence Society, Helsinki, Finland 1993) p. 63

[10.1086] J. Heikkonen, P. Koikkalainen, E. Oja: In *Proc. ICANN'93, Int. Conf. on Artificial Neural Networks*, ed. by S. Gielen, B. Kappen (Springer, London, UK 1993) p. 262

[10.1087] J. Heikkonen, E. Oja: In *Proc. IJCNN-93-Nagoya, Int. Joint Conf. on Neural Networks* (IEEE Service Center, Piscataway, NJ 1993) p. I-669

[10.1088] J. Heikkonen, P. Koikkalainen, E. Oja: In *Proc. WCNN'93, World Congress on Neural Networks* (Lawrence Erlbaum, Hillsdale, NJ 1993) p. III-141

[10.1089] O. G. Jakubowicz: Proc. SPIE – The Int. Society for Optical Engineering **1192**, 528 (1990)

[10.1090] B. J. A. Kröse, M. Eecen: In *Proc. ICANN'94, Int. Conf. on Artificial Neural Networks*, ed. by M. Marinaro, P. G. Morasso (Springer, London, UK 1994) p. II-1303

[10.1091] R. C. Luo, H. Potlapalli: In *Proc. ICNN'94, Int. Conf. on Neural Networks* (IEEE Service Center, Piscataway, NJ 1994) p. 2703

[10.1092] P. Morasso, G. Vercelli, R. Zaccaria: In *Proc. IJCNN-93-Nagoya, Int. Joint Conf. on Neural Networks* (IEEE Service Center, Piscataway, NJ 1993) p. II-1875

[10.1093] U. Nehmzow, T. Smithers: Technical Report DAI-489 (Department of Artificial Intelligence, University of Edinburgh, Edinburgh, Scotland 1990)

[10.1094] U. Nehmzow, T. Smithers, J. Hallam: In *Information Processing in Autonomous Mobile Robots. Proc. of the Int. Workshop*, ed. by G. Schmidt (Springer, Berlin, Germany 1991) p. 267

[10.1095] U. Nehmzow, T. Smithers: In *Toward a Practice of Autonomous Systems. Proc. First European Conf. on Artificial Life*, ed. by F. J. Varela, P. Bourgine (MIT Press, Cambridge, MA, USA 1992) p. 96

[10.1096] U. Nehmzow: PhD Thesis (University of Edinburgh, Department of Artificial Intelligence, Edinburgh, UK 1992)

[10.1097] H.-G. Park, S.-Y. Oh: In *Proc. ICNN'94, Int. Conf. on Neural Networks* (IEEE Service Center, Piscataway, NJ 1994) p. 2754

[10.1098] H. Ritter: In *Neural Networks for Sensory and Motor Systems*, ed. by R. Eckmiller (Elsevier, Amsterdam, Netherlands 1990)

[10.1099] H. Ritter: In *Advanced Neural Computers*, ed. by R. Eckmiller (Elsevier, Amsterdam, Netherlands 1990) p. 381

[10.1100] W. D. Smart, J. Hallam: In *Proc. IMACS Int. Symp. on Signal Processing, Robotics and Neural Networks* (IMACS, Lille, France 1994) p. 449

[10.1101] J. Tani, N. Fukumura: In *Proc. IJCNN-93-Nagoya, Int. Joint Conf. on Neural Networks* (IEEE Service Center, Piscataway, NJ 1993) p. II-1747

[10.1102] N. W. Townsend, M. J. Brownlow, L. Tarassenko: In *Proc. WCNN'94, World Congress on Neural Networks* (Lawrence Erlbaum, Hillsdale, NJ 1994) p. II-9

[10.1103] G. Vercelli: In *Proc. ICANN'94, Int. Conf. on Artificial Neural Networks*, ed. by M. Marinaro, P. G. Morasso (Springer, London, UK 1994) p. II-1307

[10.1104] J. M. Vleugels, J. N. Kok, M. H. Overmars: In *Proc. ICANN-93, Int. Conf. on Artificial Neural Networks*, ed. by S. Gielen, B. Kappen (Springer, London, UK 1993) p. 281

[10.1105] A. Walker, J. Hallam, D. Willshaw: In *Proc. ICNN'93, Int. Conf. on Neural Networks* (IEEE Service Center, Piscataway, NJ 1993) p. III-1451

[10.1106] U. R. Zimmer, C. Fischer, E. von Puttkamer: In *Proc. 3rd Int. Conf. on Fuzzy Logic, Neural Nets and Soft Computing* (Fuzzy Logic Systems Institute, Iizuka, Japan 1994) p. 131

[10.1107] Y. Coiton, J. C. Gilhodes, J. L. Velay, J. P. Roll: Biol. Cyb. **66**, 167 (1991)

[10.1108] J. L. Velay, J. C. Gilhodes, B. Ans, Y. Coiton: In *Proc. ICANN-93, Int. Conf. on Artificial Neural Networks*, ed. by S. Gielen, B. Kappen (Springer, London, UK 1993) p. 51

[10.1109] R. Brause: In *Proc. INNC'90, Int. Neural Network Conference* (Kluwer, Dordrecht, Netherlands 1990) p. I-221

[10.1110] R. Brause: In *Proc.2nd Int. IEEE Conference on Tools for Artificial Intelligence* (IEEE Comput. Soc. Press, Los Alamitos, CA 1990) p. 451

[10.1111] R. Brause: Int. J. Computers and Artificial Intelligence **11**, 173 (1992)

[10.1112] N. R. Ball: In *Proc. ESANN'96, European Symp. on Artificial Neural Networks*, ed. by M. Verleysen (D Facto Conference Services, Brussels, Belgium 1996) p. 155

[10.1113] D. DeMers, K. Kreutz-Delgado: IEEE Trans. on Neural Networks **7**, 43 (1996)

[10.1114] A. J. Knobbe, J. N. Kok, M. H. Overmars: In *Proc. ICANN'95, Int. Conf. on Artificial Neural Networks*, ed. by F. Fogelman-Soulié, P. Gallinari, (EC2, Nanterre, France 1995) p. II-375

[10.1115] S. Sehad, C. Touzet: In *Proc. WCNN'95, World Congress on Neural Networks*, (INNS, Lawrence Erlbaum, Hillsdale, NJ 1995) p. II-350

[10.1116] H. A. Mallot, H. H. Bülthoff, P. Georg, B. Schölkopf, K. Yasuhara: In *Proc. ICANN'95, Int. Conf. on Artificial Neural Networks*, ed. by F. Fogelman-Soulié, P. Gallinari, (EC2, Nanterre, France 1995) p. II-381

[10.1117] D. D. Caviglia, G. M. Bisio, F. Curatelli, L. Giovannacci, L. Raffo: In *Proc. EDAC, European Design Automation Conf., Glasgow, Scotland* (IEEE Comput. Soc. Press, Washington, DC 1990) p. 650

[10.1118] R.-I. Chang, P.-Y. Hsiao: In *Proc. ICNN'94, Int. Conf. on Neural Networks* (IEEE Service Center, Piscataway, NJ 1994) p. 3381

[10.1119] A. Hemani, A. Postula: Neural Networks **3**, 337 (1990)

[10.1120] S.-S. Kim, C.-M. Kyung: In *Proc. 1991 IEEE Int. Symp. on Circuits and Systems* (IEEE Service Center, Piscataway, NJ 1991) p. V-3122

[10.1121] B. Kiziloglu, V. Tryba, W. Daehn: In *Proc. IJCNN-93-Nagoya, Int. Joint Conf. on Neural Networks* (IEEE Service Center, Piscataway, NJ 1993) p. III-2413

[10.1122] L. Raffo, D. D. Caviglia, G. M. Bisio: In *Proc. COMPEURO'92, The Hague, Netherlands, May 4-8* (IEEE Service Center, Piscataway, NJ 1992) p. 556

[10.1123] R. Sadananda, A. Shestra: In *Proc. IJCNN-93-Nagoya, Int. Joint Conf. on Neural Networks* (IEEE Service Center, Piscataway, NJ 1993) p. II-1955

[10.1124] T. Shen, J. Gan, L. Yao: In *Proc. IJCNN'92, Int. Joint Conf. on Neural Networks* (IEEE Service Center, Piscataway, NJ 1992) p. IV-761

[10.1125] T. Shen, J. Gan, L. Yao: Chinese J. Computers **15**, 641 (1992)

[10.1126] T. Shen, J. Gan, L. Yao: Chinese J. Computers **15**, 648 (1992)

[10.1127] T. Shen, J. Gan, L. Yao: Acta Electronica Sinica **20**, 100 (1992)

[10.1128] M. Takahashi, K. Kyuma, E. Funada: In *Proc. IJCNN-93-Nagoya, Int. Joint Conf. on Neural Networks* (IEEE Service Center, Piscataway, NJ 1993) p. III-2417

[10.1129] V. Tryba, S. Metzen, K. Goser: In *Neuro-Nîmes '89. Int. Workshop on Neural Networks and their Applications* (EC2, Nanterre, France 1989) p. 225

[10.1130] C.-X. Zhang, D. A. Mlynski: In *Proc. Int. Symp. on Circuits and Systems, New Orleans, Luisiana, May* (IEEE Service Center, Piscataway, NJ 1990) p. 475

[10.1131] C. Zhang, D. Mlynski: GME Fachbericht **8**, 297 (1991)

[10.1132] C. Zhang, A. Vogt, D. Mlynski: Elektronik **15**, 68 (1991)

[10.1133] C.-X. Zhang, A. Vogt, D. A. Mlynski: In *Proc. Int. Symp. on Circuits and Systems, Singapore* (IEEE Service Center, Piscataway, NJ 1991) p. 2060

[10.1134] C.-X. Zhang, D. A. Mlynski: In *Proc. IJCNN-91-Singapore, Int. Joint Conf. on Neural Networks* (IEEE Service Center, Piscataway, NJ 1991) p. 863

[10.1135] M. S. Zamani, G. R. Hellestrand: In *Proc. EANN'95, Engineering Applications of Artificial Neural Networks* (Finnish Artificial Intelligence Society, Helsinki, Finland 1995) p. 279

[10.1136] M. Z. Zamani, G. R. Hellestrand: In *Proc. ICNN'95, IEEE Int. Conf. on Neural Networks*, (IEEE Service Center, Piscataway, NJ 1995) p. V-2185

[10.1137] G. Mitchison: Neural Computation **7**, 25 (1995)

[10.1138] M. Yasunaga, M. Asai, K. Shibata, M. Yamada: Trans. of the Inst. of Electronics, Information and Communication Engineers **J75D-I**, 1099 (1992)

[10.1139] M. Collobert, D. Collobert: In *Proc. Int. Workshop on Applications of Neural Networks to Telecommunications 2*, ed. by J. Alspector, R. Goodman, T. X. Brown (Lawrence Erlbaum, Hillsdale, NJ 1995) p. 334

[10.1140] K. M. Marks, K. F. Goser: In *Proc. of Neuro-Nîmes, Int. Workshop on Neural Networks and their Applications* (EC2, Nanterre, France 1988) p. 337

[10.1141] V. Sankaran, M. J. Embrechts, L.-E. Harsson, R. P. Kraft: In *Proc. WCNN'95, World Congress on Neural Networks*, (INNS, Lawrence Erlbaum, Hillsdale, NJ 1995) p. II-642

[10.1142] A. C. Izquierdo, J. C. Sueiro, J. A. H. Mendez: In *Proc. IWANN'91, Int. Workshop on Artificial Neural Networks.*, ed. by A. Prieto (Springer, Berlin, Heidelberg 1991) p. 401

[10.1143] A. Hemani, A. Postula: In *Proc. EDAC, European Design Automation Conference* (IEEE Comput. Soc. Press, Washington, DC 1990) p. 136

[10.1144] A. Hemani, A. Postula: In *Proc. IJCNN-90-WASH-DC, Int. Joint Conf. on Neural Networks* (IEEE Service Center, Piscataway, NJ 1990) p. II-543

[10.1145] A. Hemani: PhD Thesis (The Royal Inst. of Technology, Stockholm, Sweden 1992)

[10.1146] A. Hemani: In *Proc. 6th Int. Conf. on VLSI Design, Bombay* (IEEE Service Center, Piscataway, NJ 1993)

[10.1147] W. J. Melssen, J. R. M. Smits, G. H. Rolf, G. Kateman: Chemometrics and Intelligent Laboratory Systems **18**, 195 (1993)

[10.1148] I. Csabai, F. Czako, Z. Fodor: Phys. Rev. D **44**, R1905 (1991)

[10.1149] I. Scabai, F. Czakó, Z. Fodor: Nuclear Physics **B374**, 288 (1992)

[10.1150] A. Cherubini, R. Odorico: Z. Physik C [Particles and Fields] **53**, 139 (1992)

[10.1151] M. Killinger, J. L. D. B. D. L. Tocnaye, P. Cambon: Ferroelectrics **122**, 89 (1991)

[10.1152] A. Raiche: Geophysical J. International **105**, 629 (1991)

[10.1153] G.-S. Jang: J. Franklin Inst. **330**, 505 (1993)

[10.1154] W. J. Maurer, F. U. Dowla, S. P. Jarpe: In *Australian Conf. on Neural Networks* (Department of Energy, Washington, DC, 1991)

[10.1155] W. J. Maurer, F. U. Dowla, S. P. Jarpe: In *Proc. Third Australian Conf. on Neural Networks (ACNN '92)*, ed. by P. Leong, M. Jabri (Sydney Univ, Sydney, Australia 1992) p. 162

[10.1156] B. Bienfait: J. Chemical Information and Computer Sciences **34**, 890 (1994)

[10.1157] J. Gasteiger, J. Zupan: Angewandte Chemie, Intrenational Edition in English **32**, 503 (1993)

[10.1158] A. Zell, H. Bayer, H. Bauknecht: In *Proc. ICNN'94, Int. Conf. on Neural Networks* (IEEE Service Center, Piscataway, NJ 1994) p. 719

[10.1159] E. A. Ferrán, P. Ferrara: Biol. Cyb. **65**, 451 (1991)

[10.1160] E. A. Ferrán, P. Ferrara: In *Artificial Neural Networks*, ed. by T. Kohonen, K. Mäkisara, O. Simula, J. Kangas (North-Holland, Amsterdam, Netherlands 1991) p. II-1341

[10.1161] E. A. Ferrán, P. Ferrara: Computer Applications in the Biosciences **8**, 39 (1992)

[10.1162] E. A. Ferrán, B. Pflugfelder, P. Ferrara: In *Artificial Neural Networks, 2*, ed. by I. Aleksander, J. Taylor (North-Holland, Amsterdam, Netherlands 1992) p. II-1521

[10.1163] E. A. Ferrán, P. Ferrara: Physica A **185**, 395 (1992)

[10.1164] E. A. Ferrán, P. Ferrara, B. Pflugfelder: In *Proc. First Int. Conf. on Intelligent Systems for Molecular Biology*, ed. by L. Hunter, D. Searls, J. Shavlik (AAAI Press, Menlo Park, CA 1993) p. 127

[10.1165] E. A. Ferrán, B. Pflugfelder: Computer Applications in the Biosciences **9**, 671 (1993)

[10.1166] J. J. Merelo, M. A. Andrade, C. Urena, A. Prieto, F. Morán: In *Proc. IWANN'91, Int. Workshop on Artificial Neural Networks*, ed. by A. Prieto (Springer, Berlin, Germany 1991) p. 415

[10.1167] J. J. Merelo, M. A. Andrare, A. Prieto, F. Morán: In *Neuro-Nîmes '91. Fourth Int. Workshop on Neural Networks and Their Applications* (EC2 1991) p. 765

[10.1168] J. J. Merelo, M. A. Andrare, A. Prieto, F. Morán: Neurocomputing **6**, 1 (1994)

[10.1169] M. A. Andrare, P. Chacón, J. J. Merelo, F. Morán: Protein Engineering **6**, 383 (1993)

[10.1170] E. A. Ferrán, P. Ferrara: Int. J. Neural Networks **3**, 221 (1992)

[10.1171] M. Turner, J. Austin, N. M. Allinson, P. Thomson: In *Proc. ICANN'94, Int. Conf. on Artificial Neural Networks*, ed. by M. Marinaro, P. G. Morasso (Springer, London, UK 1994) p. II-1087

[10.1172] F. Menard, F. Fogelman-Soulié: In *Proc. INNC'90, Int. Neural Network Conf.* (Kluwer, Dordrecht, Netherlands 1990) p. 99

[10.1173] R. Goodacre, M. J. Neal, D. B. Kell, L. W. Greenham, W. C. Noble, R. G. Harvey: J. Appl. Bacteriology **76**, 124 (1994)

[10.1174] R. Goodacre: Microbiology Europe **2**, 16 (1994)

[10.1175] R. Goodacre, S. A. Howell, W. C. Noble, M. J. Neal: Zentralblatt für Microbiologie (1994)

[10.1176] M. Blanchet, S. Yoshizawa, N. Okudaira, S.-i. Amari: In *Proc. 7'th Symp. on Biological and Physiological Engineering* (Toyohashi University of Technology, Toyohashi, Japan 1992) p. 171

[10.1177] G. Dorffner, P. Rappelsberger, A. Flexer: In *Proc. ICANN'93, Int. Conf. on Artificial Neural Networks*, ed. by S. Gielen, B. Kappen (Springer, London, UK 1993) p. 882

[10.1178] P. Elo, J. Saarinen, A. Värri, H. Nieminen, K. Kaski: In *Artificial Neural Networks, 2*, ed. by I. Aleksander, J. Taylor (North-Holland, Amsterdam, Netherlands 1992) p. II-1147

[10.1179] P. Elo: Technical Report 1-92 (Tampere University of Technology, Electronics Laboratory, Tampere, Finland 1992)

[10.1180] S. Kaski, S.-L. Joutsiniemi: In *Proc. ICANN'93, of Int. Conf. on Artificial Neural Networks*, ed. by S. Gielen, B. Kappen (Springer, London, UK 1993) p. 974

[10.1181] P. E. Morton, D. M. Tumey, D. F. Ingle, C. W. Downey, J. H. Schnurer: In *Proc. IEEE Seventeenth Annual Northeast Bioengineering Conf.*, ed. by M. D. Fox, M. A. F. Epstein, R. B. Davis, T. M. Alward (IEEE Service Center, Piscataway, NJ 1991) p. 7

[10.1182] M. Peltoranta: PhD Thesis (Graz University of Technology, Graz, Austria 1992)

[10.1183] S. Roberts, L. Tarassenko: In *Proc. Second Int. Conf. on Artificial Neural Networks* (IEE, London, UK 1991) p. 210

[10.1184] S. Roberts, L. Tarassenko: IEE Proc. F [Radar and Signal Processing] **139**, 420 (1992)

[10.1185] S. Roberts, L. Tarassenko: In *IEE Colloquium on 'Neurological Signal Processing' (Digest No.069)* (IEE, London, UK 1992) p. 6/1

[10.1186] M. Pregenzer, G. Pfurtscheller, C. Andrew: In *Proc. ESANN'95, European Symp. on Artificial Neural Networks*, ed. by M. Verleysen (D Facto Conference Services, Brussels, Belgium 1995) p. 247

[10.1187] M. Süssner, M. Budil, T. Binder, G. Porental: In *Proc. EANN'95, Engineering Applications of Artificial Neural Networks* (Finnish Artificial Intelligence Society, Helsinki, Finland 1995) p. 461

[10.1188] D. Graupe, R. Liu: In *Proc. 32nd Midwest Symp. on Circuits and Systems* (IEEE Service Center, Piscataway, NJ 1990) p. II-740

[10.1189] C. N. Schizas, C. S. Pattichis, R. R. Livesay, I. S. Schofield, K. X. Lazarou, L. T. Middleton: In *Computer-Based Medical Systems*, Chap. 9.2, Unsupervised Learning in Computer Aided Macro Electromyography (IEEE Computer Soc. Press, Los Alamitos, CA 1991)

[10.1190] M. Bodruzzaman, S. Zein-Sabatto, O. Omitowoju, M. Malkani: In *Proc. WCNN'95, World Congress on Neural Networks*, (INNS, Lawrence Erlbaum, Hillsdale, NJ 1995) p. II-854

[10.1191] K. Portin, R. Salmelin, S. Kaski: In *Proc. XXVII Annual Conf. of the Finnish Physical Society, Turku, Finland*, ed. by T. Kuusela (Finnish Physical Society, Helsinki, Finland 1993) p. 15.2

[10.1192] S. Roberts, L. Tarassenko: Med. & Biol. Eng. & Comput. **30**, 509 (1992)

[10.1193] T. Conde: In *Proc. ICNN'94, Int. Conf. on Neural Networks* (IEEE Service Center, Piscataway, NJ 1994) p. 3552

[10.1194] M. Morabito, A. Macerata, A. Taddei, C. Marchesi: In *Proc. Computers in Cardiology* (IEEE Comput. Soc. Press, Los Alamitos, CA 1991) p. 181

[10.1195] Y. H. Hu, S. Palreddy, W. J. Tompkins: In *Proc. NNSP'95, IEEE Workshop on Neural Networks for Signal Processing* (IEEE Service Center, Piscataway, NJ 1995) p. 459

[10.1196] M. J. Rodríquez, F. d Pozo, M. T. Arredondo: In *Proc. WCNN'93, World Congress on Neural Networks* (Lawrence Erlbaum, Hillsdale, NJ 1993) p. II-469

[10.1197] K. Kallio, S. Haltsonen, E. Paajanen, T. Rosqvist, T. Katila, P. Karp, P. Malmberg, P. Piirilä, A. R. A. Sovijärvi: In *Artificial Neural Networks*, ed. by T. Kohonen, K. Mäkisara, O. Simula, J. Kangas (North-Holland, Amsterdam, Netherlands 1991) p. I-803

[10.1198] P. Morasso, A. Pareto, S. Pagliano, V. Sanguineti: In *Proc. ICANN'93, Int. Conf. on Artificial Neural Networks*, ed. by S. Gielen, B. Kappen (Springer, London, UK 1993) p. 806

[10.1199] T. Harris: In *Proc. IJCNN-93-Nagoya, Int. Joint Conf. on Neural Networks* (IEEE Service Center, Piscataway, NJ 1993) p. I-947

[10.1200] B. W. Jervis, M. R. Saatchi, A. Lacey, G. M. Papadourakis, M. Vourkas, T. Roberts, E. M. Allen, N. R. Hudson, S. Oke: In *IEE Colloquium on 'Intelligent Decision Support Systems and Medicine' (Digest No.143)* (IEE, London, UK 1992) p. 5/1

[10.1201] X. Liu, G. Cheng, J. Wu: In *Proc. ICNN'94, Int. Conf. on Neural Networks* (IEEE Service Center, Piscataway, NJ 1994) p. 649

[10.1202] S. Breton, J. P. Urban, H. Kihl: In *Proc. WCNN'95, World Congress on Neural Networks*, (Lawrence Erlbaum, Hillsdale, NJ 1995) p. II-406

[10.1203] M. Köhle, D. Merkl: In *Proc. ESANN'96, European Symp. on Artificial Neural Networks*, ed. by M. Verleysen (D Facto Conference Services, Brussels, Belgium 1996) p. 73

[10.1204] S. Lin, J. Si, A. B. Schwartz: In *Proc. ICANN'95, Int. Conf. on Artificial Neural Networks*, ed. by F. Fogelman-Soulié, P. Gallinari, (EC2, Nanterre, France 1995) p. I-133

[10.1205] W. W. v Osdol, T. G. Myers, K. D. Paull, K. W. Kohn, J. N. Weinstein: In *Proc. WCNN'95, World Congress on Neural Networks*, (Lawrence Erlbaum, Hillsdale, NJ 1995) p. II-762

[10.1206] F. Giuliano, P. Arrigo, F. Scalia, P. P. Cardo, G. Damiani: Comput. Applic. Biosci. **9**, 687 (1993)

[10.1207] J. N. Weinstein, T. G. Myers, Y. Kan, K. D. Paull, D. W. Zaharevitz, K. W. K. W. W. v Osdol: In *Proc. WCNN'95, World Congress on Neural Networks*, (INNS, Lawrence Erlbaum, Hillsdale, NJ 1995) p. II-750

[10.1208] R. H. Stevens, P. Wang, A. Lopo: In *Proc. WCNN'95, World Congress on Neural Networks*, (Lawrence Erlbaum, Hillsdale, NJ 1995) p. II-785

[10.1209] G. Pfurtscheller, D. Flotzinger, K. Matuschik: Biomedizinische Technik **37**, 122 (1992)

[10.1210] M. J. v. Gils, P. J. M. Cluitsman: In *Proc. ICANN'93, Int. Conf. on Artificial Neural Networks*, ed. by S. Gielen, B. Kappen (Springer, London, UK 1993) p. 1015

[10.1211] A. Glaría-Bengoechea, Y. Burnod: In *Artificial Neural Networks*, ed. by T. Kohonen, K. Mäkisara, O. Simula, J. Kangas (Elsevier, Amsterdam, Netherlands 1991) p. 501

[10.1212] G. Pfurtscheller, D. Flotzinger, W. Mohl, M. Peltoranta: Electroencephalography and Clinical Neurophysiology **82**, 313 (1992)

[10.1213] T. Pomierski, H. M. Gross, D. Wendt: In *Proc. ICANN-93, Int. Conf. on Artificial Neural Networks*, ed. by S. Gielen, B. Kappen (Springer, London, UK 1993) p. 142

[10.1214] E. Dedieu, E. Mazer: In *Toward a Practice of Autonomous Systems. Proc. First European Conf. on Artificial Life*, ed. by F. J. Varela, P. Bourgine (MIT Press, Cambridge, MA 1992) p. 88

[10.1215] G. Pfurtscheller, W. Klimesch: J. Clin. Neurophysiol. **9**, 120 (1992)

[10.1216] K. Obermayer, K. Schulten, G. G. Blasdel: In *Advances in Neural Information Processing Systems 4*, ed. by J. E. Moody, S. J. Hanson, R. P. Lippmann (Morgan Kaufmann, San Mateo, CA 1992) p. 83

[10.1217] K. Obermayer: In *Proc. Conf. on Prerational Intelligence – Phenomenology of Complexity Emerging in Systems of Agents Interagtion Using Simple Rules*, (Center for Interdisciplinary Research, University of Bielefeld, Bielefeld, Germany 1993) p. I-117

[10.1218] T. Kohonen: In *Proc. WCNN'94, World Congress on Neural Networks* (Lawrence Erlbaum, Hillsdale, NJ 1994) p. III-97

[10.1219] P. Morasso, V. Sanguineti: In *Proc. ICANN'94, Int. Conf. on Artificial Neural Networks*, ed. by M. Marinaro, P. G. Morasso (Springer, London, UK 1994) p. II-1247

[10.1220] H. Kita, Y. Nishikawa: In *Proc. WCNN'93, World Congress on Neural Networks* (Lawrence Erlbaum, Hillsdale, NJ 1993) p. II-413

[10.1221] H.-U. Bauer: In *Proc. ICANN'94, Int. Conf. on Artificial Neural Networks*, ed. by M. Marinaro, P. G. Morasso (Springer, London, UK 1994) p. I-42

[10.1222] H.-J. Boehme, U.-D. Braumann, H.-M. Gross: In *Proc. ICANN'94, Int. Conf. on Artificial Neural Networks*, ed. by M. Marinaro, P. G. Morasso (Springer, London, UK 1994) p. II-1189

[10.1223] T. Grönfors: In *Proc. Conf. on Artificial Intelligence Res. in Finland*, ed. by C. Carlsson, T. Järvi, T. Reponen, Number 12 in Conf. Proc. of Finnish Artificial Intelligence Society (Finnish Artificial Intelligence Society, Helsinki, Finland 1994) p. 44

[10.1224] T. Martinetz, H. Ritter, K. Schulten: In *Connectionism in Perspective*, ed. by R. Pfeifer, Z. Schreter, F. Fogelman-Soulié, L. Steels (North-Holland, Amsterdam, Netherlands 1989) p. 403

[10.1225] K. Obermayer, H. Ritter, K. Schulten: In *Proc. IJCNN-90-WASH-DC, Int. Joint Conf. of Neural Networks* (IEEE Service Center, Piscataway, NJ 1990) p. 423

[10.1226] K. Obermayer, H. J. Ritter, K. J. Schulten: Proc. Natl Acad. of Sci., USA **87**, 8345 (1990)

[10.1227] K. Obermayer, H. Ritter, K. Schulten: IEICE Trans. Fund. Electr. Comm. Comp. Sci. **E75-A**, 537 (1992)

[10.1228] K. Obermayer, G. G. Blasdel, K. Schulten: In *Artificial Neural Networks*, ed. by T. Kohonen, K. Mäkisara, O. Simula, J. Kangas (Elsevier, Amsterdam, Netherlands 1991) p. 505

[10.1229] K. Obermayer, H. Ritter, K. Schulten: In *Advances in Neural Information Processing Systems 3*, ed. by R. P. Lippmann, J. E. Moody, D. S. Touretzky (Morgan Kaufmann, San Mateo, CA 1991) p. 11

[10.1230] K. Obermayer, G. G. Blasdel, K. Schulten: Physical Review A [Statistical Physics, Plasmas, Fluids, and Related Interdisciplinary Topics] **45**, 7568 (1992)

[10.1231] K. Obermayer: Annales du Groupe CARNAC **5**, 91 (1992)

[10.1232] K. Obermayer: *Adaptive neuronale Netze und ihre Anwendung als Modelle der Entwicklung kortikaler Karten* (Infix Verlag, Sankt Augustin, Germany 1993)

[10.1233] G. G. Sutton III, J. A. Reggia, S. L. Armentrout, C. L. D'Autrechy: Neural Computation **6**, 1 (1994)

[10.1234] N. V. Swindale: Current Biology **2**, 429 (1992)

[10.1235] J. B. Saxon: M.Sc. Thesis (Texas A&M University, Computer Science Department, College Station, TX 1991)

[10.1236] E. B. Werkowitz: M.Sc. Thesis (Air Force Inst. of Tech., School of Engineering, Wright-Patterson AFB, OH, USA 1991)

[10.1237] S. Garavaglia: In *Proc. WCNN'93, World Congress on Neural Networks* (Lawrence Erlbaum, Hillsdale, NJ 1993) p. I-362

[10.1238] A. Ultsch, H. Siemon: In *Proc. INNC'90, Int. Neural Network Conf.* (Kluwer, Dordrecht, Netherlands 1990) p. 305

[10.1239] A. Ultsch: In *Information and Classification* ed.by O Opitz, B. Lausen, R. Klar (Springer, London UK 1993) p. 307

[10.1240] A. Varfis, C. Versino: In *Artificial Neural Networks, 2*, ed. by I. Aleksander, J. Taylor (North-Holland, Amsterdam, Netherlands 1992) p. II-1583

[10.1241] X. Zhang, Y. Li: In *Proc. IJCNN-93-Nagoya, Int. Joint Conf. on Neural Networks* (IEEE Service Center, Piscataway, NJ 1993) p. III-2448

[10.1242] B. Back, G. Oosterom, K. Sere, M. v. Wezel: In *Proc. Conf. on Artificial Intelligence Res. in Finland*, ed. by C. Carlsson, T. Järvi, T. Reponen, Number 12 in Conf. Proc. of Finnish Artificial Intelligence Society (Finnish Artificial Intelligence Society, Helsinki, Finland 1994) p. 140

[10.1243] D. L. Binks, N. M. Allinson: In *Artificial Neural Networks*, ed. by T. Kohonen, K. Mäkisara, O. Simula, J. Kangas (North-Holland, Amsterdam, Netherlands 1991) p. II-1709

[10.1244] F. Blayo, P. Demartines: Bull. des Schweizerischen Elektrotechnischen Vereins & des Verbandes Schweizerischer Elektrizitätswerke **83**, 23 (1992)

[10.1245] B. Martín-del-Brío, C. Serrano-Cinca: Neural Computing & Application **1**, 193 (1993)

[10.1246] L. Vercauteren, R. A. Vingerhoeds, L. Boullart: In *Parallel Processing in Neural Systems and Computers*, ed. by R. Eckmiller, G. Hartmann, G. Hauske (North-Holland, Amsterdam, Netherlands 1990) p. 503

[10.1247] C. L. Wilson: In *Proc. ICNN'94, Int. Conf. on Neural Networks* (IEEE Service Center, Piscataway, NJ 1994) p. 3651

[10.1248] A. Varfis, C. Versino: Neural Network World **2**, 813 (1992)

[10.1249] C. Serrano, B. Martín, J. L. Gallizo: In *Proc. 16th Annual Congress of the European Accounting Associatian* (1993)

[10.1250] K. Marttinen: In *Proc. of the Symp. on Neural Networks in Finland, Åbo Akademi, Turku, January 21.*, ed. by A. Bulsari, B. Saxén (Finnish Artificial Intelligence Society, Helsinki, Finland 1993) p. 75

[10.1251] E. Carlson: In *Artificial Neural Networks*, ed. by T. Kohonen, K. Mäkisara, O. Simula, J. Kangas (North-Holland, Amsterdam, Netherlands 1991) p. II-1309

[10.1252] E. A. Riskin, L. E. Atlas, S.-R. Lay: In *Proc. Workshop on Neural Networks for Signal Processing*, ed. by B. H. Juang, S. Y. Kung, C. A. Kamm (IEEE Service Center, Piscataway, NJ 1991) p. 543

[10.1253] C.-Y. Shen, Y.-H. Pao: In *Proc. WCNN'95, World Congress on Neural Networks*, (Lawrence Erlbaum, Hillsdale, NJ 1995) p. I-142

[10.1254] A. Ultsch: In *Information and Classification*, ed. by O. Opitz, B. Lausen, R. Klar (Springer, London, UK 1993) p. 301

[10.1255] A. Ultsch, D. Korus: In *Proc. ICNN'95, IEEE Int. Conf. on Neural Networks*, (IEEE Service Center, Piscataway, NJ 1995) p. IV-1828

[10.1256] P. Demartines: PhD Thesis (Grenoble University, Grenoble, France 1995)

[10.1257] A. Guérin-Dugué, C. Aviles-Cruz, P. M. Palagi: In *Proc. ESANN'96, European Symp. on Artificial Neural Networks*, ed. by M. Verleysen (D Facto Conference Services, Brussels, Belgium 1996) p. 229

[10.1258] N. Mozayyani, V. Alanou, J. F. Dreyfus, G. Vaucher: In *Proc. ICANN'95, Int. Conf. on Artificial Neural Networks*, ed. by F. Fogelman-Soulié, P. Gallinari, (EC2, Nanterre, France 1995) p. II-75

[10.1259] M. Budinich: Neural Computation **7**, 1188 (1995)

[10.1260] E. Cervera, A. P. del Pobil: In *Proc. CAEPIA'95, VI Conference of the Spanish Association for Artificial Intelligence* (Spain 1995) p. 129

[10.1261] J. P. Bigus: In *Proc. ICNN'94, Int. Conf. on Neural Networks* (IEEE Service Center, Piscataway, NJ 1994) p. 2442

[10.1262] K. M. Marks: In *Proc. 1st Interface Prolog User Day* (Interface Computer GmbH, Munich, Germany 1987)

[10.1263] D. Unlu, U. Halici: In *Proc. IASTED Int. Symp. Artificial Intelligence Application and Neural Networks – AINN'90*, ed. by M. H. Hamza (IASTED, ACTA Press, Anaheim, CA 1990) p. 152

[10.1264] S. Heine, I. Neumann: In *28th Universities Power Engineering Conf. 1993* (Staffordshire University, Stafford, UK 1993)

[10.1265] J. B. Arseneau, T. Spracklen: In *Proc. ICANN'94, Int. Conf. on Artificial Neural Networks*, ed. by M. Marinaro, P. G. Morasso (Springer, London, UK 1994) p. II-1384

[10.1266] J. B. Arseneau, T. Spracklen: In *Proc. WCNN'94, World Congress on Neural Networks* (Lawrence Erlbaum, Hillsdale, NJ 1994) p. I-467

[10.1267] A. Ultsch, G. Halmans: In *Proc. IJCNN'91, Int. Joint Conf. on Neural Networks* (IEEE Service Center, Piscataway, NJ 1991)

[10.1268] D. Merkl, A. M. Tjoa, G. Kappel: In *Proc. 2nd Int. Conf. of Achieving Quality in Software, Venice, Italy* (1993) p. 169

[10.1269] D. Merkl: In *Proc. IJCNN-93-Nagoya, Int. Joint Conf. on Neural Networks* (IEEE Service Center, Piscataway, NJ 1993) p. III-2468

[10.1270] D. Merkl, A. M. Tjoa, G. Kappel: *Retrieval of Reusable Software Based on Semantic Similarity: An Artificial Neural Network Approach*. Technical Report (Institut für Angewandte Informatik und Informationssysteme, Universität Wien, Vienna, Austria 1993)

[10.1271] D. Merkl, A. M. Tjoa, G. Kappel: In *Proc. 5th Australian Conf. on Neural Networks*, ed. by A. C. Tsoi, T. Downs (Univ. Queensland, St Lucia, Australia 1994) p. 13

[10.1272] D. Merkl, A. M. Tjoa, G. Kappel: In *Proc. ICNN'94, Int. Conf. on Neural Networks* (IEEE Service Center, Piscataway, NJ 1994) p. 3905

[10.1273] A. Dekker, P. Farrow: *Artificial Intelligence and Creativity*, Creativity, Chaos and Artificial Intelligence. (Kluwer, Dordrecht, The Netherlands 1994)

[10.1274] P. Morasso, J. Kennedy, E. Antonj, S. di Marco, M. Dordoni: In *Proc. INNC'90, Int. Neural Network Conf.* (Kluwer, Dordrecht, Netherlands 1990) p. 141

[10.1275] M. B. Waldron, S. Kim: In *Proc. ICNN'94, Int. Conf. on Neural Networks* (IEEE Service Center, Piscataway, NJ 1994) p. 2885

[10.1276] P. Wittenburg, U. H. Frauenfelder: In *Twente Workshop on Language Technology 3: Connectionism and Natural Language Processing*, ed. by M. F. J. Drossaers, A. Nijholt (Department of Computer Science, University of Twente, Enschede, Netherlands 1992) p. 5

[10.1277] S. Finch, N. Chater: In *Artificial Neural Networks, 2*, ed. by I. Aleksander, J. Taylor (North-Holland, Amsterdam, Netherlands 1992) p. II-1365

[10.1278] T. Hendtlass: In *Proc. 5th Australian Conf. on Neural Networks*, ed. by A. C. Tsoi, T. Downs (University of Queensland, St Lucia, Australia 1994) p. 169

[10.1279] H. Ritter, T. Kohonen: *Self-Organizing Semantic Maps* (Helsinki Univ. of Technology, Lab. of Computer and Information Science, Espoo, Finland 1989)

[10.1280] J. C. Scholtes: In *Worknotes of the AAAI Spring Symp. Series on Machine Learning of Natural Language and Ontology, Palo Alto, CA, March 26-29* (American Association for Artificial Intelligence 1991)

[10.1281] J. C. Scholtes: In *Proc. IJCNN'91, Int. Conf. on Neural Networks* (IEEE Service Center, Piscataway, NJ 1991) p. I-107

[10.1282] P. G. Schyns: In *Connectionist Models: Proc. of the 1990 Summer School* (Morgan-Kaufmann, San Mateo, CA 1990) p. 228

[10.1283] P. G. Schyns: Cognitive Science **15**, 461 (1991)

[10.1284] T. Honkela, V. Pulkki, T. Kohonen: In *Proc. ICANN'95, Int. Conf. on Artificial Neural Networks*, ed. by F. Fogelman-Soulié, P. Gallinari, (EC2, Nanterre, France 1995) p. II-3

[10.1285] R. Paradis, E. Dietrich: In *Proc. ICNN'94, Int. Conf. on Neural Networks* (IEEE Service Center, Piscataway, NJ 1994) p. 2339

[10.1286] R. Paradis, E. Dietrich: In *Proc. WCNN'94, World Congress on Neural Networks* (Lawrence Erlbaum, Hillsdale, NJ 1994) p. II-775

[10.1287] S. W. K. Chan, J. Franklin: In *Proc. ICNN'95, IEEE Int. Conf. on Neural Networks*, (IEEE Service Center, Piscataway, NJ 1995) p. VI-2965

[10.1288] J. C. Bezdek, N. R. Pal: IEEE Transactions on Neural Networks **6**, 1029 (1995)

[10.1289] J. Scholtes: In *Proc. 3rd Twente Workshop on Language Technology* (University of Twente, Twente, Netherlands 1992)

[10.1290] J. C. Scholtes: In *Proc. 2nd SNN, Nijmegen, The Netherlands, April 14-15* (1992) p. 86

[10.1291] J. C. Scholtes: In *Proc. First SHOE Workshop* (University of Tilburg, Tilburg, Netherlands 1992) p. 279

[10.1292] J. C. Scholtes, S. Bloembergen: In *Proc. IJCNN-92-Baltimore, Int. Joint Conf. on Neural Networks* (IEEE Service Center, Piscataway, NJ 1992) p. II-69

[10.1293] J. C. Scholtes, S. Bloembergen: In *Proc. IJCNN-92-Beijing, Int. Joint Conf. on Neural Networks* (IEEE Service Center, Piscataway, NJ 1992)

[10.1294] J. C. Scholtes: In *Artificial Neural Networks, 2*, ed. by I. Aleksander, J. Taylor (North-Holland, Amsterdam, Netherlands 1992) p. II-1347

[10.1295] T. Honkela, A. M. Vepsäläinen: In *Artificial Neural Networks*, ed. by T. Kohonen, K. Mäkisara, O. Simula, J. Kangas (North-Holland, Amsterdam, Netherlands 1991) p. I-897

[10.1296] T. Honkela: In *Proc. ICANN'93, Int. Conf. on Artificial Neural Networks*, ed. by S. Gielen, B. Kappen (Springer, London, UK 1993) p. 408

[10.1297] W. Pedrycz, H. C. Card: In *IEEE Int. Conf. on Fuzzy Systems* (IEEE Service Center, Piscataway, NJ 1992) p. 371

[10.1298] J. C. Scholtes: In *Proc. Informatiewetenschap 1991, Nijmegen* (STINFON, Nijmegen, Netherlands 1991) p. 203

[10.1299] J. C. Scholtes: In *Proc. 2nd Australian Conf. on Neural Nets* (University of Sydney, Sydney, Australia 1991) p. 38

[10.1300] J. C. Scholtes: In *Proc. SNN Symposium* (STINFON, Nijmegen, Netherlands 1991) p. 64

[10.1301] J. C. Scholtes: In *Proc. CUNY 1991 Conf. on Sentence Processing, Rochester, NY, May 12-14* (1991) p. 10

[10.1302] J. C. Scholtes: In *The Annual Conf. on Cybernetics: Its Evolution and Its Praxis, Amherst, MA, July 17-21* (1991)

[10.1303] J. C. Scholtes: In *Worknotes of the Bellcore Workshop on High Performance Information Filtering* (Bellcore, Chester, NJ 1991)

[10.1304] J. C. Scholtes: In *Artificial Neural Networks*, ed. by T. Kohonen, K. Mäkisara, O. Simula, J. Kangas (North-Holland, Amsterdam, Netherlands 1991) p. II-1751

[10.1305] J. C. Scholtes: Technical Report (Department of Computational Linguistics, University of Amsterdam, Amsterdam, Netherlands 1991)

[10.1306] J. C. Scholtes: In *Proc. SPIE Conf. on Applications of Artificial Neural Networks III, Orlando, Florida, April 20-24* (SPIE, Bellingham, WA 1992)

[10.1307] J. C. Scholtes: In *Proc. 3rd Australian Conf. on Neural Nets, Canberra, Australia, February 3-5* (1992)

[10.1308] J. C. Scholtes: In *Proc. Symp. on Document Analysis and Information Retrieval, Las Vegas, NV, March 16-18* (UNLV Publ. 1992) p. 151

[10.1309] J. C. Scholtes: In *Proc. First SHOE Workshop, Tilburg, Netherlands, February 27-28* (1992) p. 267

[10.1310] J. M. Campanario: Scientometrics **33**, 23 (1995)

[10.1311] G. Cheng, X. Liu, J. X. Wu: In *Proc. WCNN'94, World Congress on Neural Networks* (Lawrence Erlbaum, Hillsdale, NJ 1994) p. IV-430

[10.1312] K. G. Coleman, S. Watenpool: AI Expert **7**, 36 (1992)

[10.1313] R. Kohlus, M. Bottlinger: In *Proc. ICANN'93, Int. Conf. on Artificial Neural Networks*, ed. by S. Gielen, B. Kappen (Springer, London, UK 1993) p. 1022

[10.1314] P. G. Schyns: In *Proc. IJCNN-90-WASH-DC, Int. Joint Conf. on Neural Networks* (Lawrence Erlbaum, Hillsdale, NJ 1990) p. I-236

[10.1315] H. Tirri: New Generation Computing **10**, 55 (1991)

[10.1316] A. Ultsch, G. Halmans, R. Mantyk: In *Proc. Twenty-Fourth Annual Hawaii Int. Conf. on System Sciences*, ed. by V. Milutinovic, B. D. Shriver (IEEE Service Center, Piscataway, NJ 1991) p. I-507

[10.1317] A. Ultsch: In *Artificial Neural Networks, 2*, ed. by I. Aleksander, J. Taylor (North-Holland, Amsterdam, Netherlands 1992) p. I-735

[10.1318] A. Ultsch, R. Hannuschka, U. H. M. Mandischer, V. Weber: In *Artificial Neural Networks*, ed. by T. Kohonen, K. Mäkisara, O. Simula, J. Kangas (North-Holland, Amsterdam, Netherlands 1991) p. I-585

[10.1319] X. Lin, D. Soergel, G. Marchionini: In *Proc. 14th. Ann. Int. ACM/SIGIR Conf. on R & D In Information Retrieval* (1991) p. 262

[10.1320] J. C. Scholtes: In *Proc. IJCNN'91, Int. Joint Conf. on Neural Networks* (IEEE Service Center, Piscataway, NJ 1991) p. 18

[10.1321] J. C. Scholtes: ITLI Prepublication Series for Computational Linguistics CL-91-02 (University of Amsterdam, Amsterdam, Netherlands 1991)

[10.1322] B. Fritzke, C. Nasahl: In *Artificial Neural Networks*, ed. by T. Kohonen, K. Mäkisara, O. Simula, J. Kangas (North-Holland, Amsterdam, Netherlands 1991) p. 1375

[10.1323] J. C. Scholtes: In *Proc. IEEE Symp. on Neural Networks, Delft, Netherlands, June 21st* (IEEE Service Center, Piscataway, NJ 1990) p. 69

[10.1324] J. C. Scholtes. Computational Linguistics Project, (CERVED S.p.A., Italy, 1990)

[10.1325] A. J. Maren: IEEE Control Systems Magazine **11**, 34 (1991)

[10.1326] E. Wilson, G. Anspach: In *Applications of Neural Networks, Proc. of SPIE Conf. No. 1965, Orlando, Florida* (SPIE, Bellingham, WA 1993)

[10.1327] A. B. Baruah, L. E. Atlas, A. D. C. Holden: In *Proc. IJCNN'91, Int. Joint Conf. on Neural Networks* (IEEE Service Center, Piscataway, NJ 1991) p. I-596

[10.1328] J. Buhmann, H. Kühnel: In *Proc. IJCNN'92, Int. Conf. on Neural Networks* (IEEE Service Center, Piscataway, NJ 1992) p. IV-796

[10.1329] W. Snyder, D. Nissman, D. Van den Bout, G. Bilbro: In *Advances in Neural Information Processing Systems 3*, ed. by R. P. Lippmann, J. E. Moody, D. S. Touretzky (Morgan Kaufmann, San Mateo, CA 1991) p. 984

[10.1330] C.-D. Wann, S. C. A. Thomopoulos: In *Proc. WCNN'93, World Congress on Neural Networks* (Lawrence Erlbaum, Hillsdale, NJ 1993) p. II-545

[10.1331] C. Ambroise, G. Govaert: In *Proc. ICANN'95, Int. Conf. on Artificial Neural Networks*, ed. by F. Fogelman-Soulié, P. Gallinari, (EC2, Nanterre, France 1995) p. I-425

[10.1332] S. Schünemann, B. Michaelis: In *Proc. ESANN'96, European Symp. on Artificial Neural Networks*, ed. by M. Verleysen (D Facto Conference Services, Brussels, Belgium 1996) p. 79

[10.1333] G. Tambouratzis: Pattern Recognition Letters **15**, 1019 (1994)

[10.1334] V. Venkatasubramanian, R. Rengaswamy. *Neural Networks for Chemical Engineers, Computer-Aided Chemical Engineering, 6*, Clustering and statistical techniques in neural networks. (Elsevier, Amsterdam 1995) p. 659

[10.1335] H. Lari-Najafi, V. Cherkassky: In *Proc. NIPS'93, Neural Information Processing Systems* (1993)

[10.1336] M. Alvarez, J.-M. Auger, A. Varfis: In *Proc. ICANN'95, Int. Conf. on Artificial Neural Networks*, ed. by F. Fogelman-Soulié, P. Gallinari, (EC2, Nanterre, France 1995) p. II-21

[10.1337] T. Heskes, B. Kappen: In *Proc. ICANN'95, Int. Conf. on Artificial Neural Networks*, ed. by F. Fogelman-Soulié, P. Gallinari, (EC2, Nanterre, France 1995) p. I-81

[10.1338] S. Zhang, R. Ganesan, Y. Sun: In *Proc. WCNN'95, World Congress on Neural Networks*, (INNS, Lawrence Erlbaum, Hillsdale, NJ 1995) p. I-747

[10.1339] P. Hannah, R. Stonier, S. Smith: In *Proc. 5th Australian Conf. on Neural Networks*, ed. by A. C. Tsoi, T. Downs (University of Queensland, St Lucia, Australia 1994) p. 165

[10.1340] J. Walter, H. Ritter, K. Schulten: In *Proc. IJCNN-90-San Diego, Int. Joint Conf. on Neural Networks* (IEEE Service Center, Piscataway, NJ 1990) Vol. 1, p. 589

[10.1341] F. Mulier, V. Cherkassky: Neural Computation **7**, 1165 (1995)

[10.1342] F. M. Mulier, V. S. Cherkassky: Neural Networks **8**, 717 (1995)

[10.1343] M. Kurimo: Licentiate's Thesis (Helsinki University of Technology, Espoo, Finland 1994)

[10.1344] A. Hämäläinen: PhD Thesis (Jyväskylä University, Jyväskylä, Finland 1995)

[10.1345] L. Holmström, A. Hottinen, A. Hämäläinen: In *Proc. EANN'95, Engineering Applications of Artificial Neural Networks* (Finnish Artificial Intelligence Society, Helsinki, Finland 1995) p. 445

[10.1346] D. Hamad, S. Delsert: In *Proc. EANN'95, Engineering Applications of Artificial Neural Networks* (Finnish Artificial Intelligence Society, Helsinki, Finland 1995) p. 457

[10.1347] M. A. Kraaijveld, J. Mao, A. K. Jain: IEEE Trans. on Neural Networks **6**, 548 (1995)

[10.1348] J. Joutsensalo, A. Miettinen, M. Zeindl: In *Proc. ICANN'95, Int. Conf. on Artificial Neural Networks*, ed. by F. Fogelman-Soulié, P. Gallinari, (EC2, Nanterre, France 1995) p. II-395

[10.1349] J. Joutsensalo, A. Miettinen: In *Proc. ICNN'95, IEEE Int. Conf. on Neural Networks*, (IEEE Service Center, Piscataway, NJ 1995) p. I-111

[10.1350] K. Kopecz: In *Proc. ICANN'95, Int. Conf. on Artificial Neural Networks*, ed. by F. Fogelman-Soulié, P. Gallinari, (EC2, Nanterre, France 1995) p. I-431

[10.1351] J. C. Principe, L. Wang: In *Proc. NNSP'95, IEEE Workshop on Neural Networks for Signal Processing* (IEEE Service Center, Piscataway, NJ 1995) p. 11

[10.1352] C. M. Privitera, R. Plamondon: In *Proc. ICNN'95, IEEE Int. Conf. on Neural Networks*, (IEEE Service Center, Piscataway, NJ 1995) p. IV-1999

[10.1353] J. Lampinen, S. Smolander: In *Proc. ICANN'95, Int. Conf. on Artificial Neural Networks*, ed. by F. Fogelman-Soulié, P. Gallinari, (EC2, Nanterre, France 1995) p. II-315

[10.1354] M. Alvarez, A. Varfis: In *Proc. ESANN'94, European Symp. on Artificial Neural Networks*, ed. by M. Verleysen (D Facto Conference Services, Brussels, Belgium 1994) p. 245

[10.1355] R. M. V. França, B. G. A. Neto: In *Proc. EANN'95, Engineering Applications of Artificial Neural Networks* (Finnish Artificial Intelligence Society, Helsinki, Finland 1995) p. 481

[10.1356] P. Knagenhjelm: In *Proc. ICSPAT-92, Int. Conf. on Signal Processing Applications and Technology* (1992) p. 948

[10.1357] B. J. Oommen, I. K. Altinel, N. Aras: In *Proc. ICNN'95, IEEE Int. Conf. on Neural Networks*, (IEEE Service Center, Piscataway, NJ 1995) p. VI-3062

[10.1358] V. R. de Sa: In *Proc. NIPS'93, Neural Information Processing Systems*, ed. by J. D. Cowan, G. Tesauro, J. Alspector (Morgan Kaufmann Publishers, San Francisco, CA 1993) p. 112

[10.1359] V. R. de Sa: PhD Thesis (University of Rochester, Department of Computer Science, Rochester, New York, NY 1994)

[10.1360] K. Smith: In *Proc. ICNN'95, IEEE Int. Conf. on Neural Networks*, (IEEE Service Center, Piscataway, NJ 1995) p. IV-1876

[10.1361] G. Tambouratzis, D. Tambouratzis: Network: Computation in Neural Systems **5**, 599 (1994)

[10.1362] S. A. Khaparde, H. Gandhi: In *Proc. ICNN'93, Int. Conf. on Neural Networks* (IEEE Service Center, Piscataway, NJ 1993) p. II-967

[10.1363] J. Meister: J. Acoust. Soc. of America **93**, 1488 (1993)

[10.1364] J. Rubner, K. Schulten, P. Tavan: In *Proc. Int. Conf. on Parallel Processing in Neural Systems and Computers (ICNC), Düsseldorf* (Elsevier, Amsterdam, Netherlands 1990) p. 365

[10.1365] J. Rubner, K. J. Schulten: Biol. Cyb. **62**, 193 (1990)

[10.1366] S.-Y. Lu: In *Proc. IJCNN-90-San Diego, Int. Joint Conf. on Neural Networks* (IEEE Service Center, Piscataway, NJ 1990) p. III-471

[10.1367] V. Pang, M. Palaniswami: In *IEEE TENCON'90: 1990 IEEE Region 10 Conf. on Computer and Communication Systems* (IEEE Service Center, Piscataway, NJ 1990) p. II-562

[10.1368] I. Bellido, E. Fiesler: In *Proc. ICANN'93, Int. Conf. on Artificial Neural Networks*, ed. by S. Gielen, B. Kappen (Springer, London, UK 1993) p. 772

[10.1369] T. M. English, L. C. Boggess: In *Proc. ICASSP'92, Int. Conf. on Acoustics, Speech, and Signal Processing* (IEEE Service Center, Piscataway, NJ 1992) p. III-357

[10.1370] S. N. Kavuri, V. Venkatasubramanian: In *Proc. IJCNN'92, Int. Joint Conf. on Neural Networks* (IEEE Service Center, Piscataway, NJ 1992) p. I-775

[10.1371] K. Matsuoka, M. Kawamoto: In *Proc. WCNN'93, World Congress on Neural Networks* (Lawrence Erlbaum, Hillsdale, NJ 1993) p. II-501

[10.1372] S.-Y. Oh, J.-M. Song: Trans. of the Korean Inst. of Electrical Engineers **39**, 985 (1990)

[10.1373] S.-Y. Oh, I.-S. Yi: In *IJCNN-91-Seattle: Int. Joint Conf. on Neural Networks* (IEEE Service Center, Piscataway, NJ 1991) p. II-1000

[10.1374] G. L. Tarr: M.Sc. Thesis (Air Force Inst. of Tech., Wright-Patterson AFB, OH 1988)

[10.1375] Z. Bing, E. Grant: In *Proc. IEEE Int. Symp. on Intelligent Control* (IEEE Service Center, Piscataway, NJ 1991) p. 180

[10.1376] A. Di Stefano, O. Mirabella, G. D. Cataldo, G. Palumbo: In *Proc. IS-CAS'91, Int. Symp. on Circuits and Systems* (IEEE Service Center, Piscataway, NJ 1991) p. III-1601

[10.1377] M. Cottrell, P. Letrémy, E. Roy: In *New Trends in Neural Computation, IWANN'93* (Springer, Berlin, Germany 1993)

[10.1378] M. Cottrell, P. Letrémy, E. Roy: Technical Report 19 (Université Paris 1, Paris, France 1993)

[10.1379] D. DeMers, K. Kreutz-Delgado: In *Proc. WCNN'94 World Congress on Neural Networks* (Lawrence Erlbaum, Hillsdale, NJ 1994) p. II-54

[10.1380] A. L. Perrone, G. Basti: In *Proc. 3rd Int. Conf. on Fuzzy Logic, Neural Nets and Soft Computing* (Fuzzy Logic Systems Institute, Iizuka, Japan 1994) p. 501

[10.1381] I. Grabec: In *Artificial Neural Networks*, ed. by T. Kohonen, K. Mäkisara, O. Simula, J. Kangas (North-Holland, Amsterdam, Netherlands 1991) p. I-151

[10.1382] M. D. Alder, R. Togneri, Y. Attikiouzel: IEE Proc. I [Communications, Speech and Vision] **138**, 207 (1991)

[10.1383] O. Sarzeaud, Y. Stephan, C. Touzet: In *Neuro-Nîmes '90. Third Int. Workshop. Neural Networks and Their Applications* (EC2, Nanterre, France 1990) p. 81

[10.1384] O. Sarzeaud, Y. Stephan, C. Touzet: In *Artificial Neural Networks*, ed. by T. Kohonen, K. Mäkisara, O. Simula, J. Kangas (North-Holland, Amsterdam, Netherlands 1991) p. II-1313

[10.1385] W. K. Tsai, Z.-P. Lo, H.-M. Lee, T. Liau, R. Chien, R. Yang, A. Parlos: In *Proc. IJCNN-91, Int. Joint Conf. on Neural Networks* (IEEE Service Center, Piscataway, NJ 1991) p. II-1003

[10.1386] M. Gera: In *Artificial Neural Networks, 2*, ed. by I. Aleksander, J. Taylor (North-Holland, Amsterdam, Netherlands 1992) p. II-1357

[10.1387] S. Herbin: In *Proc. ICANN'95, Int. Conf. on Artificial Neural Networks*, ed. by F. Fogelman-Soulié, P. Gallinari, (EC2, Nanterre, France 1995) p. II-57

[10.1388] F. Murtagh: International Statistical Review **64**, 275 (1994)

[10.1389] F. Murtagh: Pattern Recognition Letters **16**, 399 (1995)

[10.1390] F. Murtagh, M. Hernández-Pajares: Statistics in Transition – Journal of the Polish Statistical Association **2**, 151 (1995)

[10.1391] F. Murtagh, M. Hernández-Pajares: Journal of Classification **12** (1995)

[10.1392] S. Ibbou, M. Cottrell: In *Proc. ESANN'95, European Symp. on Artificial Neural Networks*, ed. by M. Verleysen (D Facto Conference Services, Brussels, Belgium 1995) p. 27

[10.1393] A. Furukawa, N. Ishii: In *Proc. ICNN'95, IEEE Int. Conf. on Neural Networks*, (IEEE Service Center, Piscataway, NJ 1995) p. III-1316

[10.1394] K. Fujimura, H. Tokutaka, S. Kishida, K. Nishimori, N. Ishihara, K. Yamane, M. Ishihara: In *Proc. IJCNN-93-Nagoya, Int. Joint Conf. on Neural Networks* (IEEE Service Center, Piscataway, NJ 1993) p. III-2472

[10.1395] K. Fujimura, H. Tokutaka, S. Kishida: Technical Report NC93-146 (The Inst. of Electronics, Information and Communication Engineers, Tottori University, Koyama, Japan 1994)

[10.1396] K. Fujimura, H. Tokutaka, S. Kishida, K. Nishimori, N. Ishihara, K. Yamane, M. Ishihara: Technical Report NC92-141 (The Inst. of Electronics, Information and Communication Engineers, Tottori University, Koyama, Japan 1994)

[10.1397] K. Fujimura, H. Tokutaka, Y. Ohshima, S. Kishida: Technical Report NC93-147 (The Inst. of Electronics, Information and Communication Engineers, Tottori University, Koyama, Japan 1994)

[10.1398] M. K. Lutey: M.Sc. Thesis (Air Force Inst. of Tech., Wright-Patterson AFB, OH 1988)

[10.1399] H. E. Ghaziri: In *Artificial Neural Networks*, ed. by T. Kohonen, K. Mäkisara, O. Simula, J. Kangas (North-Holland, Amsterdam, Netherlands 1991) p. I-829

[10.1400] E. H. L. Aarts, H. P. Stehouwer: In *Proc. ICANN'93, Int. Conf. on Artificial Neural Networks*, ed. by S. Gielen, B. Kappen (Springer, London, UK 1993) p. 950

[10.1401] B. Angèniol, G. D. L. C. Vaubois, J. Y. L. Texier: Neural Networks **1**, 289 (1988)

[10.1402] M. Budinich: In *Proc. ICANN'94, Int. Conf. on Artificial Neural Networks*, ed. by M. Marinaro, P. G. Morasso (Springer, London, UK 1994) p. I-358

[10.1403] D. Burr: In *Proc. ICNN'88, Int. Conf. on Neural Networks* (IEEE Service Center, Piscataway, NJ 1988) p. I-69

[10.1404] F. Favata, R. Walker: Biol. Cyb. **64**, 463 (1991)

[10.1405] J.-C. Fort: Biol. Cyb. **59**, 33 (1988)

[10.1406] M. Geraci, F. Sorbello, G. Vassallo: In *Fourth Italian Workshop. Parallel Architectures and Neural Networks*, ed. by E. R. Caianiello (Univ. Salerno; Inst. Italiano di Studi Filosofici, World Scientific, Singapore 1991) p. 344

[10.1407] G. Hueter: In *Proc. ICNN'88, Int. Conf. on Neural Networks* (IEEE Service Center, Piscataway, NJ 1988) p. I-85

[10.1408] S.-C. Lee, J.-M. Wu, C.-Y. Liou: In *Proc. ICANN'93, Int. Conf. on Artificial Neural Networks*, ed. by S. Gielen, B. Kappen (Springer, London, UK 1993) p. 842

[10.1409] H. Tamura, T. Teraoka, I. Hatono, K. Yamagata: Trans. Inst. of Systems, Control and Information Engineers **4**, 57 (1991)

[10.1410] S. Amin: Neural Computing & Applications **2**, 129 (1994)

[10.1411] M. Budinich: Neural Computation **8**, 416 (1996)

[10.1412] C. S.-T. Choy, W.-C. Siu: In *Proc. ICNN'95, IEEE Int. Conf. on Neural Networks*, (IEEE Service Center, Piscataway, NJ 1995) p. V-2632

[10.1413] K. Fujimura, H. Tokutaka, Y. Ohshima, S.-I. Tanaka, S. Kishida: Trans. IEE of Japan **116-C**, 350 (1996)

[10.1414] S.-I. Tanaka, K. Fujimura, H. Tokutaka, S. Kishida: Technical Report NC95-70 (The Inst. of Electronics, Information and Communication Engineers, Tottori University, Koyama, Japan 1995)

[10.1415] H. Tokutaka, A. Tanaka, K. Fujimura, T. Koukami, S. Kishida, H. Hase: Technical Report NC94-79 (The Inst. of Electronics, Information and Communication Engineers, Tottori University, Koyama, Japan 1995)

[10.1416] M. Goldstein: In *Proc. INNC'90, Int. Neural Network Conf.* (Kluwer, Dordrecht, Netherlands 1990) p. I-258

[10.1417] C.-Y. Hsu, M.-H. Tsai, W.-M. Chen: In *Proc. Int. Symp. on Circuits and Systems* (IEEE Service Center, Piscataway, NJ 1991) p. II-1589

[10.1418] P. S. Khedkar, H. R. Berenji: In *Proc. WCNN'93, World Congress on Neural Networks* (Lawrence Erlbaum, Hillsdale, NJ 1993) p. II-18

[10.1419] H. R. Berenji: In *Proc. Int. Conf. on Fuzzy Systems* (IEEE Service Center, Piscataway, NJ 1993) p. 1395

[10.1420] C. V. Buhusi: In *Proc. IJCNN-93-Nagoya, Int. Joint Conf. on Neural Networks* (IEEE Service Center, Piscataway, NJ 1993) p. I-786

[10.1421] C. Isik, F. Zia: In *Proc. WCNN'93, World Congress on Neural Networks* (Lawrence Erlbaum, Hillsdale, NJ 1993) p. II-56

[10.1422] A. O. Esogbue, J. A. Murrell: In *Proc. Int. Conf. on Fuzzy Systems* (IEEE Service Center, Piscataway, NJ 1993) p. 178

[10.1423] B. Michaelis, O. Schnelting, U. Seiffert, R. Mecke: In *Proc. WCNN'95, World Congress on Neural Networks*, (INNS, Lawrence Erlbaum, Hillsdale, NJ 1995) p. III-103

[10.1424] J. H. L. Kong, G. P. M. D. Martin: In *Proc. ICNN'95, IEEE Int. Conf. on Neural Networks*, (IEEE Service Center, Piscataway, NJ 1995) p. III-1397

[10.1425] C. Andrew, M. Kubat, G. Pfurtscheller: In *Proc. ESANN'95, European Symp. on Artificial Neural Networks*, ed. by M. Verleysen (D Facto Conference Services, Brussels, Belgium 1995) p. 291

[10.1426] A. Baraldi, F. Parmiggiani: In *Proc. ICNN'95, IEEE Int. Conf. on Neural Networks*, (IEEE Service Center, Piscataway, NJ 1995) p. V-2444

[10.1427] R. Dogaru, A. T. Murgan, C. Cumaniciu: In *Proc. ESANN'96, European Symp. on Artificial Neural Networks*, ed. by M. Verleysen (D facto conference services, Bruges, Belgium 1996) p. 309

[10.1428] S.-J. Huang, C.-C. Hung: In *Proc. ICNN'95, IEEE Int. Conf. on Neural Networks*, (IEEE Service Center, Piscataway, NJ 1995) p. II-708

[10.1429] R. M. Kil, Y. I. Oh: In *Proc. WCNN'95, World Congress on Neural Networks*, (Lawrence Erlbaum, Hillsdale, NJ 1995) p. I-778

[10.1430] A. R. M. Ramos, D. A. C. Barone: In *Proc. WCNN'95, World Congress on Neural Networks*, (Lawrence Erlbaum, Hillsdale, NJ 1995) p. I-770

[10.1431] T. Ojala, V. T. Ruoppila, P. Vuorimaa: In *Proc. WCNN'95, World Congress on Neural Networks*, (Lawrence Erlbaum, Hillsdale, NJ 1995) p. II-713

[10.1432] C. N. Manikopoulos, J. Li: In *Proc. IJCNN'89, Int. Joint Conf. on Neural Networks* (IEEE Service Center, Piscataway, NJ 1989) p. II-573

[10.1433] C. Manikopoulos, G. Antoniou, S. Metzelopoulou: In *Proc. IJCNN'90, Int. Joint Conf. on Neural Networks* (IEEE Service Center, Piscataway, NJ 1990) p. I-481

[10.1434] C. N. Manikopoulos, J. Li, G. Antoniou: J. New Generation Computer Systems **4**, 99 (1991)

[10.1435] C. N. Manikopoulos, G. E. Antoniou: J. Electrical and Electronics Engineering,Australia **12**, 233 (1992)

[10.1436] M. Hernandez-Pajares, E. Monte: In *Proc. IWANN'91, Int. Workshop on Artificial Neural Networks*, ed. by A. Prieto (Springer, Berlin, Germany 1991) p. 422

[10.1437] A. Langinmaa, A. Visa: Tekniikan näköalat (TEKES, Helsinki, Finland 1990) p. 10

[10.1438] M. Stinely, P. Klinkhachorn, R. S. Nutter, R. Kothari: In *Proc. WCNN'93, World Congress on Neural Networks* (Lawrence Erlbaum, Hillsdale, NJ 1993) p. I-597

[10.1439] X. Magnisalis, E. Auge, M. G. Strintzis: In *Parallel and Distributed Computing in Engineering Systems. Proc. IMACS/IFAC Int. Symp.*, ed. by S. Tzafestas, P. Borne, L. Grandinetti (North-Holland, Amsterdam, Netherlands 1992) p. 383

[10.1440] L. D. Giovanni, M. Fedeli, S. Montesi: In *Artificial Neural Networks*, ed. by T. Kohonen, K. Mäkisara, O. Simula, J. Kangas (North-Holland, Amsterdam, Netherlands 1991) p. II-1803

[10.1441] Y. Kojima, H. Yamamoto, T. Kohda, S. Sakaue, S. Maruno, Y. Shimeki, K. Kawakami, M. Mizutani: In *Proc. IJCNN-93-Nagoya, Int. Joint Conf.*

on Neural Networks (IEEE Service Center, Piscataway, NJ 1993) p. III-2161

[10.1442] F. Togawa, T. Ueda, T. Aramaki, A. Tanaka: In *Proc. IJCNN'93, Int. Joint Conf. on Neural Networks* (IEEE Service Center, Piscataway, NJ 1991) p. II-1490

[10.1443] R. Togneri, M. D. Alder, Y. Attikiouzel: In *Proc. AI'90, 4th Australian Joint Conf. on Artificial Intelligence*, ed. by C. P. Tsang (World Scientific, Singapore 1990) p. 274

[10.1444] H. K. Kim, H. S. Lee: In *EUROSPEECH-91. 2nd European Conf. on Speech Communication and Technology* (Assoc. Belge Acoust.; Assoc. Italiana di Acustica; CEC; et al, Istituto Int. Comunicazioni, Genova, Italy 1991) p. III-1265

[10.1445] E. McDermott, S. Katagiri: In *Proc. ICASSP'89, Int. Conf. on Acoustics, Speech and Signal Processing* (IEEE Service Center, Piscataway, NJ 1989) p. I-81

[10.1446] E. McDermott: In *Proc. Acoust. Soc. of Japan* (March, 1990) p. 151

[10.1447] E. McDermott, S. Katagiri: IEEE Trans. on Signal Processing **39**, 1398 (1991)

[10.1448] J. T. Laaksonen: In *Proc. EUROSPEECH-91, 2nd European Conf. on Speech Communication and Technology* (Assoc. Belge Acoust.; Assoc. Italiana di Acustica; CEC; et al, Istituto Int. Comunicazioni, Genova, Italy 1991) p. I-97

[10.1449] R. Alpaydin, U. Ünlüakin, F. Gürgen, E. Alpaydin: In *Proc. IJCNN'93-Nagoya, Int. Joint Conf. on Neural Networks* (IEEE Service Center, Piscataway, NJ 1993) p. I-239

[10.1450] S. Nakagawa, Y. Ono, K. Hur: In *Proc. IJCNN-93-Nagoya, Int. Joint Conf. on Neural Networks* (IEEE Service Center, Piscataway, NJ 1993) p. III-2223

[10.1451] K. S. Nathan, H. F. Silverman.: In *Proc. ICASSP'91, Int. Conf. on Acoustics, Speech and Signal Processing* (IEEE Service Center, Piscataway, NJ 1991) p. I-445

[10.1452] Y. Bennani, N. Chaourar, P. Gallinari, A. Mellouk: In *Proc. Neuro-Nîmes '90, Third Int. Workshop. Neural Networks and Their Applications* (EC2, Nanterre, France 1990) p. 455

[10.1453] Y. Bennani, N. Chaourar, P. Gallinari, A. Mellouk: In *Proc. ICASSP'91, Int. Conf. on Acoustics, Speech and Signal Processing* (IEEE Service Center, Piscataway, NJ 1991) p. I-97

[10.1454] C. Zhu, L. Li, C. Guan, Z. He: In *Proc. WCNN'93, World Congress on Neural Networks* (Lawrence Erlbaum, Hillsdale, NJ 1993) p. IV-177

[10.1455] K. Kondo, H. Kamata, Y. Ishida: In *Proc. ICNN'94, Int. Conf. on Neural Networks* (IEEE Service Center, Piscataway, NJ 1994) p. 4448

[10.1456] L. D. Giovanni, R. Lanuti, S. Montesi: In *Proc. Fourth Italian Workshop. Parallel Architectures and Neural Networks*, ed. by E. R. Caianiello (World Scientific, Singapore 1991) p. 238

[10.1457] A. Duchon, S. Katagiri: J. Acoust. Soc. of Japan **14**, 37 (1993)

[10.1458] S. Olafsson: BT Technology J. **10**, 48 (1992)

[10.1459] T. Komori, S. Katagiri: J.Acoust.Soc.Japan **13**, 341 (1992)

[10.1460] Y. Cheng, et al.: In *Proc. ICASSP'92, Int. Conf. on Acoustics, Speech and Signal Processing* (IEEE Service Center, Piscataway, NJ 1992) p. I-593

[10.1461] J.-F. Wang, C.-H. Wu, C.-C. Haung, J.-Y. Lee: In *Proc. ICASSP'91, Int. Conf. on Acoustics, Speech and Signal Processing* (IEEE Service Center, Piscataway, NJ 1991) p. I-69

[10.1462] C.-H. Wu, J.-F. Wang, C.-C. Huang, J.-Y. Lee: Int. J. Pattern Recognition and Artificial Intelligence **5**, 693 (1991)

[10.1463] G. Yu, et al.: In *Proc. ICASSP'90, Int. Conf. on Acoustics, Speech and Signal Processing* (IEEE Service Center, Piscataway, NJ 1990) p. I-685

[10.1464] M. Kurimo: In *Proc. Int. Symp. on Speech, Image Processing and Neural Networks* (IEEE Hong Kong Chapter of Signal Processing, Hong Kong 1994) p. II-718

[10.1465] M. W. Koo, C. K. Un: Electronics Letters **26**, 1731 (1990)

[10.1466] F. Schiel: In *Proc. EUROSPEECH-93, 3rd European Conf. on Speech, Communication and Technology* (ESCA, Berlin, Germany 1993) p. III-2271

[10.1467] Y. Bennani, F. Fogelman-Soulié, P. Gallinari: In *Proc. ICASSP'90, Int. Conf. on Acoustics, Speech and Signal Processing* (IEEE Service Center, Piscataway, NJ 1990) p. I-265

[10.1468] Y. Bennani, F. Fogelman-Soulié, P. Gallinari: In *Proc. INNC'90, Int. Neural Network Conf.* (Kluwer, Dordrecht, Netherlands 1990) p. II-1087

[10.1469] G. C. Cawley, P. D. Noakes: In *Proc. IJCNN-93-Nagoya, Int. Joint Conf. on Neural Networks* (IEEE Service Center, Piscataway, NJ 1993) p. III-2227

[10.1470] H. Fujita, M. Yamamoto, S. Kobayashi, X. Youheng: In *Proc. IJCNN-93-Nagoya, Int. Joint Conf. on Neural Networks* (IEEE Service Center, Piscataway, NJ 1993) p. I-951

[10.1471] D. Flotzinger, J. Kalcher, G. Pfurtscheller: Biomed. Tech. (Berlin) **37**, 303 (1992)

[10.1472] D. Flotzinger, J. Kalcher, G. Pfurtscheller: In *Proc. WCNN'93, World Congress on Neural Networks* (Lawrence Erlbaum, Hillsdale, NJ 1993) p. I-224

[10.1473] D. Flotzinger: In *Proc. ICANN'93, Int. Conf. on Artificial Neural Networks*, ed. by S. Gielen, B. Kappen (Springer, London, UK 1993) p. 1019

[10.1474] S. C. Ahalt, T. Jung, A. K. Krishnamurthy: In *Proc. IEEE Int. Conf. on Systems Engineering* (IEEE Service Center, Piscataway, NJ 1990) p. 609

[10.1475] M. M. Moya, M. W. Koch, R. J. Fogler, L. D. Hostetler: Technical Report 92-2104 (Sandia National Laboratories, Albuquerque, NM 1992)

[10.1476] P. Ajjimarangsee, T. L. Huntsberger: Proc. SPIE – The Int. Society for Optical Engineering **1003**, 153 (1989)

[10.1477] T. Yamaguchi, T. Takagi, M. Tanabe: Trans. Inst. of Electronics, Information and Communication Engineers **J74C-II**, 289 (1991)

[10.1478] Y. Sakuraba, T. Nakamoto, T. Moriizumi: In *Proc. 7'th Symp. on Biological and Physiological Engineering, Toyohashi, November 26-28, 1992* (Toyohashi University of Technology, Toyohashi, Japan 1992) p. 115

[10.1479] Y. Bartal, J. Lin, R. E. Uhrig: In *Proc. ICNN'94, Int. Conf. on Neural Networks* (IEEE Service Center, Piscataway, NJ 1994) p. 3744

[10.1480] T. Jukarainen, E. Kärpänoja, P. Vuorimaa: In *Proc. Conf. on Artificial Intelligence Res. in Finland*, ed. by C. Carlsson, T. Järvi, T. Reponen, Number 12 in Conf. Proc. of Finnish Artificial Intelligence Society (Finnish Artificial Intelligence Society, Helsinki, Finland 1994) p. 155

[10.1481] A. Kurz: In *Artificial Neural Networks, 2*, ed. by I. Aleksander, J. Taylor (North-Holland, Amsterdam, Netherlands 1992) p. I-587

[10.1482] V. S. Smolin: In *Artificial Neural Networks*, ed. by T. Kohonen, K. Mäkisara, O. Simula, J. Kangas (North-Holland, Amsterdam, Netherlands 1991) p. II-1337

[10.1483] T. Rögnvaldsson: Neural Computation **5**, 483 (1993)

[10.1484] M. Vogt: In *Proc. ICNN'93, Int. Conf. on Neural Networks* (IEEE Service Center, Piscataway, NJ 1993) p. III-1841

[10.1485] M. Tokunaga, K. Kohno, Y. Hashizume, K. Hamatani, M. Watanabe, K. Nakamura, Y. Ageishi: In *Proc. 2nd Int. Conf. on Fuzzy Logic and Neural Networks, Iizuka, Japan* (1992) p. 123

[10.1486] P. Koikkalainen, E. Oja: In *Proc. ICNN'88, Int. Conf. on Neural Networks* (IEEE Service Center, Piscataway, NJ 1988) p. 533

[10.1487] P. Koikkalainen, E. Oja: In *Proc. COGNITIVA'90* (North-Holland, Amsterdam, Netherlands 1990) p. II-769

[10.1488] P. Koikkalainen, E. Oja: In *Advances in Control Networks and Large Scale Parallel Distributed Processing Models*, ed. by M. Frazer (Ablex, Norwood, NJ 1991) p. 242

[10.1489] A. Cherubini, R. Odorico: Computer Phys. Communications **72**, 249 (1992)

[10.1490] T. Kohonen, J. Kangas, J. Laaksonen, K. Torkkola: In *Proc. IJCNN'92, Int. Joint Conf. on Neural Networks* (IEEE Service Center, Piscataway, NJ 1992) p. I-725

[10.1491] Y. Lirov: Neural Networks **5**, 711 (1992)

[10.1492] G. Whittington, C. T. Spracklen: In *IEE Colloquium on 'Neural Networks: Design Techniques and Tools' (Digest No.037)* (IEE, London, UK 1991) p. 6/1

[10.1493] T. Nordström: In *Proc. DSA-92, Fourth Swedish Workshop on Computer System Artchitecture* (1992)

[10.1494] T. Nordström: PhD Thesis (Lule University of Technology, Lule, Sweden 1995)

[10.1495] M. E. Azema-Barac: In *Proc. Sixth Int. Parallel Processing Symp.*, ed. by V. K. Prasanna, L. H. Canter (IEEE Computer Soc. Press, Los Alamitos, CA 1992) p. 527

[10.1496] C. Bottazzi: Informazione Elettronica **18**, 21 (1990)

[10.1497] V. Demian, J.-C. Mignot: In *Parallel Processing: CONPAR 92-VAPP V. Second Joint Int. Conf. on Vector and Parallel Processing*, ed. by L. Bouge, M. Cosnard, Y. Robert, D. Trystram (Springer, Berlin, Germany 1992) p. 775

[10.1498] V. Demian, J.-C. Mignot: In *Proc. IJCNN-93-Nagoya, Int. Joint Conf. on Neural Networks* (IEEE Service Center, Piscataway, NJ 1993) p. I-483

[10.1499] B. Dorizzi, J.-M. Auger: In *Proc. INNC'90, Int. Neural Network Conference* (Kluwer, Dordrecht, Netherlands 1990) p. II-681

[10.1500] D. Gassilloud, J. C. Grossetie (eds.). *Computing with Parallel Architectures: T.Node* (Kluwer, Dordrecht, Netherlands 1991)

[10.1501] R. Hodges, C.-H. Wu, C.-J. Wang: In *Proc. IJCNN-90-WASH-DC, Int. Joint Conf. on Neural Networks* (1990) p. II-141

[10.1502] G. Whittington, C. T. Spracklen: In *Proc. ICNN'94, Int. Conf. on Neural Networks* (IEEE Service Center, Piscataway, NJ 1994) p. 17

[10.1503] C.-H. Wu, R. E. Hodges, C. J. Wang: Parallel Computing **17**, 821 (1991)

[10.1504] K. Wyler: In *Proc. WCNN'93, World Congress on Neural Networks* (Lawrence Erlbaum, Hillsdale, NJ 1993) p. II-562

[10.1505] B. Martín-del-Brío: IEEE Trans. on Neural Networks **3**, 529 (1996)

[10.1506] D. Hammerstrom, N. Nguyen: In *Artificial Neural Networks*, ed. by T. Kohonen, K. Mäkisara, O. Simula, J. Kangas (North-Holland, Amsterdam, Netherlands 1991) p. I-715

[10.1507] A. König, X. Geng, M. Glesner: In *Proc. ICANN'93, Int. Conf. on Artificial Neural Networks*, ed. by S. Gielen, B. Kappen (Springer, London, UK 1993) p. 1046

[10.1508] M. Manohar, J. C. Tilton: In *DCC '92. Data Compression Conf.*, ed. by J. A. Storer, M. Cohn (IEEE Comput. Soc. Press, Los Alamitos, CA 1992) p. 181

[10.1509] J. M. Auger: In *Computing with Parallel Architectures: T.Node*, ed. by D. Gassilloud, J. C. Grossetie (Kluwer, Dordrecht, Netherlands 1991) p. 215

[10.1510] P. Koikkalainen, E. Oja: In *Proc. SteP-88, Finnish Artificial Intelligence Symp.* (Finnish Artificial Intelligence Society, Helsinki, Finland 1988) p. 621

[10.1511] K. Obermayer, H. Heller, H. Ritter, K. Schulten: In *NATUG 3: Transputer Res. and Applications 3*, ed. by A. S. Wagner (IOS Press, Amsterdam, Netherlands 1990) p. 95

[10.1512] K. Obermayer, H. Ritter, K. Schulten: Parallel Computing **14**, 381 (1990)

[10.1513] K. Obermayer, H. Ritter, K. Schulten: In *Parallel Processing in Neural Systems and Computers*, ed. by R. Eckmiller, G. Hartmann, G. Hauske (North-Holland, Amsterdam, Netherlands 1990) p. 71

[10.1514] B. A. Conway, M. Kabrisky, S. K. Rogers, G. B. Lamon: Proc. SPIE – The Int. Society for Optical Engineering **1294**, 269 (1990)

[10.1515] A. Singer: Parallel Computing **14**, 305 (1990)

[10.1516] T.-C. Lee, I. D. Scherson: In *Proc. Fourth Annual Parallel Processing Symp.* (IEEE Service Center, Piscataway, NJ 1990) p. I-365

[10.1517] T. Lee, A. M. Peterson: In *Proc. 13th Annual Int. Computer Software and Applications Conf.* (IEEE Comput. Soc. Press, Washington, DC 1989) p. 672

[10.1518] S. Churcher, D. J. Baxter, A. Hamilton, A. F. Murray, H. Reekie: In *Proc. 2nd Int. Conf. on Microelectronics for Neural Networks*, ed. by U. Ramacher, U. Ruckert, J. A. Nossek (Kyrill & Method Verlag, Munich, Germany 1991) p. 127

[10.1519] P. Heim, B. Hochet, E. Vittoz: Electronics Letters **27**, 275 (1991)

[10.1520] P. Heim, X. Arregvit, E. Vittoz: Bull. des Schweizerischen Elektrotechnischen Vereins & des Verbandes Schweizerischer Elektrizitaetswerke **83**, 44 (1992)

[10.1521] D. Macq, M. Verleysen, P. Jespers, J.-D. Legat: IEEE Trans. on Neural Networks **4**, 456 (1993)

[10.1522] J. R. Mann, S. Gilbert: In *Advances in Neural Information Processing Systems I*, ed. by D. S. Touretzky (Morgan Kaufmann, San Mateo, CA 1989) p. 739

[10.1523] B. J. Sheu, J. Choi, C. F. Chang: In *1992 IEEE Int. Solid-State Circuits Conf. Digest of Technical Papers. 39th ISSCC*, ed. by J. H. Wuorinen (IEEE Service Center, Piscataway, NJ 1992) p. 136

[10.1524] E. Vittoz, P. Heim, X. Arreguit, F. Krummenacher, E. Sorouchyari: In *Proc. Journées d'Électronique 1989, Artificical Neural Networks, Lausanne, Switzerland, October 10-12* (Presses Polytechniques Romandes, Lausanne, Switzerland 1989) p. 291

[10.1525] H. Wasaki, Y. Horio, S. Nakamura: Trans. Inst. of Electronics, Information and Communication Engineers A **J76-A**, 348 (1993)

[10.1526] Y. Kuramoti, A. Takimoto, H. Ogawa: In *Proc. IJCNN-93-Nagoya, Int. Joint Conf. on Neural Networks* (IEEE Service Center, Piscataway, NJ 1993) p. III-2023

[10.1527] T. Lu, F. T. S. Yu, D. A. Gregory: Proc. SPIE – The Int. Society for Optical Engineering **1296**, 378 (1990)

[10.1528] T. Lu, F. T. S. Yu, D. A. Gregory: Optical Engineering **29**, 1107 (1990)

[10.1529] F. T. S. Yu, T. Lu: In *Proc. IEEE TENCON'90, 1990 IEEE Region 10 Conf. Computer and Communication Systems* (IEEE Service Center, Piscataway, NJ 1990) p. I-59

[10.1530] D. Ruwisch, M. Bode, H.-G. Purwins: In *Proc. ICANN'94, Int. Conf. on Artificial Neural Networks*, ed. by M. Marinaro, P. G. Morasso (Springer, London, UK 1994) p. II-1335

[10.1531] Y. Owechko, B. H. Soffer: In *Proc. Optical Implementation of Information Processing, SPIE Vol 2565* (SPIE, Bellingham, WA 1995) p. 12

[10.1532] N. M. Allinson, M. J. Johnson, K. J. Moon: In *Advances in Neural Information Processing Systems I* (Morgan Kaufmann, San Mateo, CA 1989) p. 728

[10.1533] E. Ardizzone, A. Chella, F. Sorbello: In *Artificial Neural Networks*, ed. by T. Kohonen, K. Mäkisara, O. Simula, J. Kangas (North-Holland, Amsterdam, Netherlands 1991) p. I-721

[10.1534] S. M. Barber, J. G. Delgado-Frias, S. Vassiliadis, G. G. Pechanek: In *Proc. IJCNN'93-Nagoya, Int. Joint Conf. on Neural Networks* (IEEE Service Center, Piscataway, NJ 1993) p. II-1927

[10.1535] M. A. de Barros, M. Akil, R. Natowicz: In *Proc. IJCNN-93-Nagoya, Int. Joint Conf. on Neural Networks* (IEEE Service Center, Piscataway, NJ 1993) p. I-197

[10.1536] N. E. Cotter, K. Smith, M. Gaspar: In *Advanced Res. in VLSI. Proc. of the Fifth MIT Conf.*, ed. by J. Allen, F. T. Leighton (MIT Press, Cambridge, MA 1988) p. 1

[10.1537] F. Deffontaines, A. Ungering, V. Tryba, K. Goser: In *Proc. Neuro-Nîmes* (1992)

[10.1538] M. Duranton, N. Mauduit: In *First IEE Int. Conf. on Artificial Neural Networks* (IEE, London, UK 1989) p. 62

[10.1539] W.-C. Fang, B. J. Sheu, O. T.-C. Chen, J. Choi: IEEE Trans. Neural Networks **3**, 506 (1992)

[10.1540] M. Gioiello, G. Vassallo, A. Chella, F. Sorbello: In *Trends in Artificial Intelligence. 2nd Congress of the Italian Association for Artificial Intelligence, AI IA Proceedings*, ed. by E. Ardizzone, S. Gaglio, F. Sorbello (Springer, Berlin, Germany 1991) p. 385

[10.1541] M. Gioiello, G. Vassallo, A. Chella, F. Sorbello: In *Fourth Italian Workshop. Parallel Architectures and Neural Networks*, ed. by E. R. Caianiello (World Scientific, Singapore 1991) p. 191

[10.1542] M. Gioiello, G. Vassallo, F. Sorbello: In *Int. Conf. on Signal Processing Applications and Technology* (1992) p. 705

[10.1543] K. Goser: In *Tagungsband der ITG-Fachtagung "Digitale Speicher"* (ITG, Darmstadt, Germany 1988) p. 391

[10.1544] K. Goser: Z. Mikroelektronik **3**, 104 (1989)

[10.1545] K. Goser, I. Kreuzer, U. Rueckert, V. Tryba: Z. Mikroelektronik **me4**, 208 (1990)

[10.1546] K. Goser: In *Artificial Neural Networks*, ed. by T. Kohonen, K. Mäkisara, O. Simula, J. Kangas (North-Holland, Amsterdam, Nethderlands 1991) p. I-703

[10.1547] K. Goser, U. Ramacher: Informationstechnik (1992)

[10.1548] R. E. Hodges, C.-H. Wu, C.-J. Wang: In *Proc. ISCAS'90, Int. Symp. on Circuits and Systems* (IEEE Service Center, Piscataway, NJ 1990) p. I-743

[10.1549] T. Kohonen: Report A 15 (Helsinki Univ. of Technology, Lab. of Computer and Information Science, Espoo, Finland 1992)

[10.1550] P. Kotilainen, J. Saarinen, K. Kaski: In *Proc. ICANN'93, Int. Conf. on Artificial Neural Networks*, ed. by S. Gielen, B. Kappen (Springer, London, UK 1993) p. 1082

[10.1551] P. Kotilainen, J. Saarinen, K. Kaski: In *Proc. IJCNN-93-Nagoya, Int. Joint Conf. on Neural Networks* (IEEE Service Center, Piscataway, NJ 1993) p. II-1979

[10.1552] C. Lehmann: In *Proc. ICANN'93, Int. Conf. on Artificial Neural Networks*, ed. by S. Gielen, B. Kappen (Springer, London, UK 1993) p. 1082

[10.1553] M. Lindroos: Technical Report 10-92 (Tampere University of Technology, Electronics Laboratory, Tampere, Finland 1992)

[10.1554] N. Mauduit, M. Duranton, J. Gobert, J.-A. Sirat: In *Proc. IJCNN'91, Int. Joint Conf. on Neural Networks* (IEEE Service Center, Piscataway, NJ 1991) p. I-602

[10.1555] N. Mauduit, M. Duranton, J. Gobert, J.-A. Sirat: IEEE Trans. on Neural Networks **3**, 414 (1992)

[10.1556] H. Onodera, K. Takeshita, K. Tamaru: In *1990 IEEE Int. Symp. on Circuits and Systems* (IEEE Service Center, Piscataway, NJ 1990) p. II-1073

[10.1557] V. Peiris, B. Hochet, G. Corbaz, M. Declercq, S. Piguet: In *Proc. Journees d'Electronique 1989. Artificial Neural Networks* (Presses Polytechniques Romandes, Lausanne, Switzerland 1989) p. 313

[10.1558] V. Peiris, B. Hochet, S. Abdo, M. Declercq: In *Int. Symp. on Circuits and Systems* (IEEE Service Center, Piscataway, NJ 1991) p. III-1501

[10.1559] V. Peiris, B. Hochet, T. Creasy, M. Declercq: Bull. des Schweizerischen Elektrotechnischen Vereins & des Verbandes Schweizerischer Elektrizitaetswerke **83**, 41 (1992)

[10.1560] S. Rueping, U. Rueckert, K. Goser: In *Proc. IWANN-93, Int. Workshop on Artificial Neural Networks* (1993) p. 488

[10.1561] A. J. M. Russel, T. E. Schouten: In *Proc. ICANN'93, Int. Conf. on Artificial Neural Networks*, ed. by S. Gielen, B. Kappen (Springer, London, UK 1993) p. 456

[10.1562] J. Saarinen: PhD Thesis (Tampere University of Technology, Tampere, Finland 1991)

[10.1563] J. Saarinen, M. Lindroos, J. Tomberg, K. Kaski: In *Proc. Sixth Int. Parallel Processing Symp.*, ed. by V. K. Prasanna, L. H. Canter. (IEEE Comput. Soc. Press, Los Alamitos, CA 1991) p. 537

[10.1564] H. Speckmann, P. Thole, W. Rosentiel: In *Proc. IJCNN-93-Nagoya, Int. Joint Conf. on Neural Networks* (IEEE Service Center, Piscataway, NJ 1993) p. II-1983

[10.1565] H. Speckmann, P. Thole, M. Bogdan, W. Rosentiel: In *Proc. ICNN'94, Int. Conf. on Neural Networks* (IEEE Service Center, Piscataway, NJ 1994) p. 1959

[10.1566] H. Speckmann, P. Thole, M. Bogdan, W. Rosenstiel: In *Proc. WCNN'94, World Congress on Neural Networks* (Lawrence Erlbaum, Hillsdale, NJ 1994) p. II-612

[10.1567] J. Van der Spiegel, P. Mueller, D. Blackman, C. Donham, R. Etienne-Cummings, P. Aziz, A. Choudhury, L. Jones, J. Xin: Proc. SPIE – The Int. Society for Optical Engineering **1405**, 184 (1990)

[10.1568] J. Tomberg: PhD Thesis (Tampere University of Technology, Tampere, Finland 1992)

[10.1569] J. Tomberg, K. Kaski: In *Artificial Neural Networks, 2*, ed. by I. Aleksander, J. Taylor (North-Holland, Amsterdam, Netherlands 1992) p. II-1431

[10.1570] V. Tryba, H. Speckmann, K. Goser: In *Proc. 1st Int. Workshop on Microelectronics for Neural Networks* (1990) p. 177

[10.1571] M. A. Viredaz: In *Proc. of Euro-ARCH'93, Munich*, ed. by P. P. Spies (Springer, Berlin, Germany 1993) p. 99

[10.1572] J. Vuori, T. Kohonen: In *Proc. ICNN'95, IEEE Int. Conf. on Neural Networks*, (IEEE Service Center, Piscataway, NJ 1995) p. IV-2019

[10.1573] V. Peiris, B. Hochet, M. Declercq: In *Proc. ICNN'94, Int. Conf. on Neural Networks* (IEEE Service Center, Piscataway, NJ 1994) p. 2064

[10.1574] T. Ae, R. Aibara: IEICE Trans. Elecronics **E76-C**, 1034 (1993)

[10.1575] A. G. Andreou, K. A. Boahen: Neural Computation **1**, 489 (1989)

[10.1576] J. Choi, B. J. Sheu: IEEE J. Solid-State Circuits **28**, 579 (1993)

[10.1577] Y. He, U. Çilingiroğlu: IEEE Trans. Neural Networks **4**, 462 (1993)

[10.1578] B. Hochet, V. Peiris, G. Corbaz, M. Declercq: In *Proc. IEEE 1990 Custom Integrated Circuits Conf.* (IEEE Service Center, Piscataway, NJ 1990) p. 26.1/1

[10.1579] B. Hochet, V. Peiris, S. Abdo, M. J. Declerq: IEEE J. Solid-State Circuits **26**, 262 (1991)

[10.1580] P. Ienne, M. A. Viredaz: In *Proc. Int. Conf. on Application-Specific Array Processors (ASAP'93), Venice, Italy*, ed. by L. Dadda, B. Wah (IEEE Computer Society Press, Los Alamitos, CA 1993) p. 345

[10.1581] J. Mann, R. Lippmann, B. Berger, J. Raffel: In *Proc. Custom Integrated Circuits Conference* (IEEE Service Center, Piscataway, NJ 1988) p. 10.3/1

[10.1582] D. E. Van den Bout, T. K. Miller III: In *Proc. IJCNN'89, Int. Joint Conf. on Neural Networks* (IEEE Service Center, Piscataway, NJ 1989) p. II-205

[10.1583] D. E. Van den Bout, W. Snyder, T. K. Miller III: In *Advanced Neural Computers*, ed. by R. Eckmiller (North-Holland, Amsterdam, Netherlands 1990) p. 219

Chapter 11

[11.1] J. Anderson, J. Silverstein, S. Ritz, R. Jones: Psych. Rev. **84**, 413 (1977)

[11.2] K. Fukushima: Neural Networks **1**, 119 (1988)

[11.3] R. Penrose: Proc. Cambridge Philos. Soc. **51**, 406 (1955)

Index

Springer Series in Information Sciences

Editors: Thomas S. Huang   Teuvo Kohonen   Manfred R. Schroeder